Communication Electronics: RF Design with Practical Applications using Pathwave/ADS Software

RIVER PUBLISHERS SERIES IN ELECTRONIC MATERIALS, CIRCUITS AND DEVICES

The "River Publishers Series in Electronic Materials, Circuits and Devices" is a series of comprehensive academic and professional books which focus on theory and applications of advanced electronic materials, circuits and devices. This includes analog and digital integrated circuits, memory technologies, system-on-chip and processor design. Also theory and modeling of devices, performance and reliability of electron and ion integrated circuit devices and interconnects, insulators, metals, organic materials, micro-plasmas, semiconductors, quantum-effect structures, vacuum devices, and emerging materials. The series also includes books on electronic design automation and design methodology, as well as computer aided design tools.

Books published in the series include research monographs, edited volumes, handbooks and textbooks. The books provide professionals, researchers, educators, and advanced students in the field with an invaluable insight into the latest research and developments.

Topics covered in this series include:-

- Analog Integrated Circuits
- Data Converters
- Digital Integrated Circuits
- Electronic Design Automation
- Insulators
- Integrated Circuit Devices
- Interconnects
- Memory Design
- MEMS
- Nanoelectronics
- Organic Materials
- Power ICs
- Processor Architectures
- Quantum-effect Structures
- Semiconductors
- Sensors and Actuators
- System-on-Chip
- Vacuum Devices

For a list of other books in this series, visit www.riverpublishers.com

Communication Electronics: RF Design with Practical Applications using Pathwave/ADS Software

Stephen I Long

ECE Department, University of California, Santa Barbara, USA

River Publishers

Routledge
Taylor & Francis Group
NEW YORK AND LONDON

Published 2024 by River Publishers
River Publishers
Alsbjergvej 10, 9260 Gistrup, Denmark
www.riverpublishers.com

Distributed exclusively by Routledge
605 Third Avenue, New York, NY 10017, USA
4 Park Square, Milton Park, Abingdon, Oxon OX14 4RN

Communication Electronics: RF Design with Practical Applications using Pathwave/ADS Software / by Stephen I Long.

Routledge is an imprint of the Taylor & Francis Group, an informa business

ISBN 978-87-7022-856-5 (hardback)
ISBN 978-87-7004-054-9 (paperback)
ISBN 978-10-0381-079-7 (online)
ISBN 978-10-3262-977-3 (master ebook)

Contents

Preface

Communication Electronics: Practical Applications with ADS Software.

The topic of radio frequency electronics is both very old and at the same time very new. When products such as cellphones and computers are designed to be obsolete in a year or two, it would appear at first glance that learning older concepts of RF engineering would be superfluous. Beginning in the early 20th century with visionaries such as Armstrong, the applications today would have been seen as science fiction back then. Yet some of what he invented, as well as much discovered from the 1920s through 1940s by many brilliant engineers, is still in use today. It is my belief that in order to step up to the latest and greatest RF based circuits and systems, one must fully understand the fundamentals, many of which have been around for years.

The chapters in this book were developed from my lecture notes on a two-quarter class at the senior and graduate level at UCSB. The class had a significant lab component employing measurement techniques, board-level prototyping, and RFIC design. The emphasis of the material in this book is first to develop practical intuition into the art of RF circuit design and second to begin the learning curve for a widely used simulation tool, Pathwave ADS, from Keysight Technologies. Version 2022 Update 1 was used to generate the examples in this text. The intent is to build a strong foundation based upon experience with project-oriented assignments. There are a number of books that go into the detailed mathematical derivations of these topics. My intention was to not get bogged down in these but cut to the practical applications illustrated by examples, simulation tutorials, and homework problems. Learning through doing has proven to be an effective preparation for more advanced and complex applications. Thus, the book falls somewhere between a traditional textbook with all of its derivations and a practical handbook. The book will focus mainly on analog RF analysis and design. Little is said about DSP or digital communications, topics well beyond the scope of this work.

Also, due to space constraints and because many of the components used in the lab design assignments are no longer available, I have not included a lab manual. A few lab exercises are included in the text or as homework that make use of commercially available hardware or provide a sequence of steps that are applicable to any RF transistor. The goal is to begin to develop measurement expertise and to illustrate concepts described in the text through hardware applications. Homework solutions are available.

The UCSB course also included a significant review (many pages) of feedback theory for transistor circuits, using Bode plots to analyze phase margins, compensation and use of MATLAB. This was directed toward application to phase-locked loops. Due to space constraints, it has not been feasible to include these topics in the text.

Acknowledgements: I would like to gratefully acknowledge the contributions of Professor Mark Rodwell who also has taught these same courses. His lecture notes were also helpful to me for teaching this class and most likely have been partially included in some of the first few chapters. Our collaborations over many years have been one of the highlights of my time at UCSB. I am also very thankful for Andy Howard at Keysight Technologies who helped me climb the learning curve on ADS during a sabbatical leave in Santa Rosa. My friend Wes Hayward W7ZOI, now retired from Triquint Semiconductor, has been influential consultant in the content of the class and especially the labs. Laszlo Dobos (Maxim) helped identify the tapped inductor technique for improving phase noise. Joe Lai and Vikas Manan greatly benefitted from his mentoring. David Leeson (Stanford) provided a section on microwave links, antenna noise modeling, and satellite receivers. I also am much grateful for the contributions of other of my former graduate students who contributed much to the description of power amplifiers included in this book. To name a few: David Choi, Tony Long, Vamsi Paidi, David Schmelzer, Shouxuan Xie, and Jingshi Yao.

Additionally, my network of Amateur Radio friends and my lifetime interest in the hobby helped guide me into developing this class on RF design after transitioning from my former focus on high-speed GaAs digital circuits.

Finally, I am so thankful for my wife, Molly, my children, Chris and Betsy, and my late parents, Stan and Mae, for their love, encouragement and support for so many years (Col. 3:23).

Stephen Long, AC6T
Professor Emeritus, Electrical and Computer Engineering,
University of California Santa Barbara
March 2023

List of Figures

List of Tables

List of Abbreviations

ADS	Advanced Design System - Keysight Technologies
BJT	Bipolar Junction Transistor
CC	Common Collector
CD	Common Drain
CE	Common Emitter
CS	Common source
ESR	Effective series resistance
F	Noise figure (numerical)
FOM	Figure of merit of an oscillator
GA	Available Gain
Gamma opt	Input match for minimum noise
GP	Power Gain
GT	Transducer Gain
HB	Harmonic balance simulation
HD	Harmonic distortion
IF	Intermediate frequency
IIP3	Third-order input intercept
IMD	Intermodulation distortion
IMD3	Third-order intermodulation distortion
JFET	Junction FET
KCL	Kirchoff's current law
KVL	Kirchoff's voltage law
LNA	Low noise amplifier
MAG	Maximum available gain
MDS	Minimum detectable signal
MESFET	Metal Gate FET
MSG	Maximum stable gain
NF	Noise figure (dB)
OIP3	Third-order output intercept
P1dB	1 dB compression input power
PCB	Printed circuit board

POC	Power output capability
RFC	Radio frequency choke
RFIC	Radio frequency IC
SFDR	Spurious-free dynamic range
SINAD	Signal-to-noise-to distortion
SOI	Second-order intercept
SRF	Self-resonant frequency
THD	Total harmonic distortion
TOI	Third-order intercept
VCO	Voltage controlled oscillator
VSWR	Voltage Standing wave ratio

1

Nonideal Components

The design and implantation of hardware at radio frequencies requires extra care due to the non-idealities of passive and active components. This applies both to board-level and RFIC implementations. It is difficult or impossible to meet the design goals without accurate equivalent circuit models or one or two port parameter models. Measurements on a network analyzer are often needed to find the dependence of impedance with frequency. Then, an equivalent network might be constructed that emulates this measured behavior. A later chapter on S-parameters will describe how they might be used for the same purpose.

Chapter Goals:

1. Observe that familiar lumped element components when used at radio frequencies exhibit undesirable parasitics that significantly affect their impedance as a function of frequency.
2. Understand how equivalent circuit models may be used to adequately represent such components.
3. Become acquainted with the concept of Q, the quality factor, for components and for the components when used in resonant circuits.
4. Study the tutorial (Appendix 1.1) on the use of the Pathwave/Keysight ADS simulation tools included at the end of this chapter.
5. Study Appendix 1.2 describing one technique that has been effective in measuring non-ideal components.

1.1 Lumped Element Components

Many of the components that you have used in the past are considered "lumped elements". As you will see, at high frequencies these are less than ideal, and can have impedances that are different from what you might expect. For this reason, we will identify the sources of the non-ideal behavior for *R*, *L* and *C* and develop equivalent circuits to better represent them for circuit modeling. In the chapters that follow, we will also consider "distributed element" implementations that use transmission lines instead of *L* or *C*. But, first consider lumped elements (Figure 1.1).

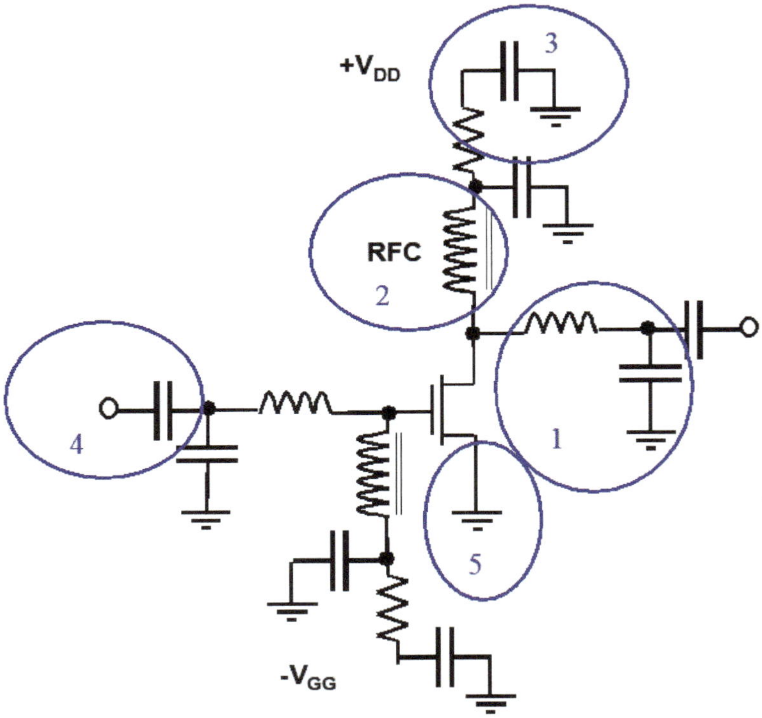

Figure 1.1 Representation of an amplifier circuit that illustrates five components that must be modeled with parasitics.

1. Narrow band matching networks. Design frequency is sensitive to component nonidealities.
2. Bias networks. RF choke. Must be broadband, presenting high *Z* to a device over a wide frequency range. Parasitic *C* causes resonances.
3. Bypass capacitors. Needed to keep the signal out of the power bus. Must present low *Z* over a wide frequency range. Parasitic *L* causes resonances.

4. DC block. Needed to keep the DC bias from measurement equipment or other stages. Must provide low Z at design frequency.
5. Ground. Parasitic series inductance can produce resonances, detune matching networks, and produce common mode feedback.

1.2 Component Equivalent Networks

1.2.1 Wire

In the past you have become accustomed to using wire to interconnect components when prototyping an analog or digital circuit. Most likely, you have considered the wire to be a zero-ohm ideal connection that can be ignored in the design of your circuit. This is definitely not a good assumption when building circuits that operate in the radio frequency or microwave spectrum (10 MHz to 26 GHz).

1.2.2 Resistance

The resistance of the wire can sometimes be important for the circuit as well. This is especially true if the wire is being used intentionally to fabricate an inductor for use in a series or parallel resonant circuit. The resistance increases loss, decreases unloaded Q (to be defined later) and will increase the bandwidth of the resonance.

The resistance of the wire *at low frequencies* is easily determined if the cross-sectional area is known. There are tables in handbooks that give cross-sectional area versus wire gauge (in AWG units). Or, of course, the diameter can be measured easily. The resistivity, ρ, must also be known for the particular material in use. Resistivity is given in units of ohm-meter or ohm-cm. Resistance is given by Equation (1.1), where l is the length of the wire and A the cross-sectional area.

$$R = \frac{\rho l}{A}. \tag{1.1}$$

For copper, $\rho = 1.7 \times 10^{-6}$ ohm-cm. For example, #24 copper wire has a cross sectional area of 2.047×10^{-3} cm^2. So, 10 cm of wire has a resistance of only 8.3 mΩ, probably negligible in most cases. But this is only partially true. In addition, there is a phenomenon called the *skin effect* that causes the resistance of a conductor to increase with frequency. The current is restricted to the surface of a conductor due to magnetic field effects [1]. Therefore, the effective cross-sectional area is reduced. On a wire with circular cross-section, the current flows through an annulus of skin depth δ. Or on a planar conductor, current would flow on the edges at this depth.

$$\delta = \sqrt{2\rho/\omega\mu}. \tag{1.2}$$

The factor μ is the magnetic permeability. Of free space, $\mu = 4\pi \times 10^{-7}$ Henry/meter. For copper at 1 GHz, the skin depth is about 2 μm. Then, for a circular conductor,

$$R = \frac{\rho l}{2\pi r\delta}. \tag{1.3}$$

This reduction in area would increase the resistance of our 10 cm wire to roughly 0.5 ohms. A substantial change.

1.2.3 Inductance of a wire

While resistance can be significant with thin on-chip conductors, the bigger issue for wires is the inductance. Because a current flows through the wire, a magnetic field is induced around the wire. With a high frequency AC current, the magnetic field causes an induced voltage in the wire that opposes the change in current flow (remember that you cannot change the current through an inductor instantaneously). This effect presents itself as a self-inductance, *L*, and can be modeled as shown in Figure 1.2.

Figure 1.2 Equivalent circuit model of a wire.

An estimate of the self-inductance can be obtained from this empirical formula [2]:

$$L = 0.002\,l\left[\ln\left(\frac{4l}{d} - 0.75\right)\right] \;\mu\text{H} \tag{1.4}$$

where l = length of wire in cm and d = diameter of the wire in cm. According to this formula, 1 inch (2.5 cm of wire) with a diameter of 1 mm will have 16 nH of inductance. This is 100 jΩ at 1 GHz, a very significant reactance.

Remember! 1 inch = 16 nH = 100 jΩ at 1 GHz

It is easy to estimate the reactance of an inductor. Note that:

$$\boxed{1\text{ nH} = 6\text{ j}\Omega\text{ at }1\text{ GHz}}$$

Likewise, 1 μH = 6 jΩ at 1 MHz, 60 jΩ at 10 MHz, and so on.

1.2.4 Resistors

You have also used leaded resistors extensively in the prototyping of transistor circuits. However, if you think about the leaded resistor in the context of the above discussion, you can see that the high frequency equivalent circuit of the resistor must also include some inductance. In addition, because the ends of the resistor are generally at different potentials, you might expect to see some parallel capacitance too. So, an equivalent circuit model of the resistor would look like those in Figure 1.3.

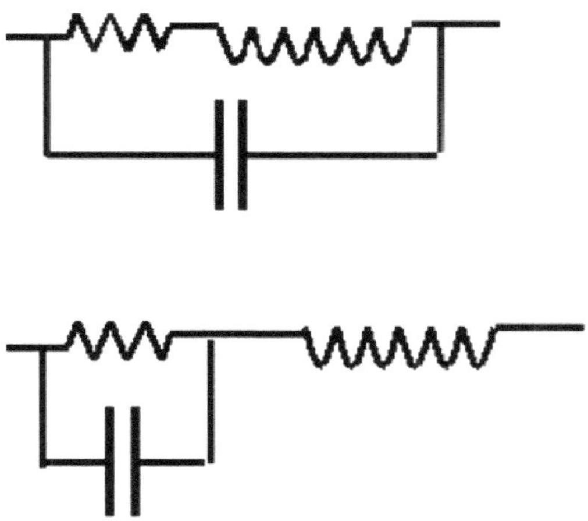

Figure 1.3 Equivalent circuit models of a chip and leaded resistor.

The frequency dependence of this circuit shows some surprising behavior at high frequencies, as shown in Figure 1.4. If you must use leaded resistors, keep the leads very short. Of course, for DC purposes, the leaded resistor is fine. A leaded resistor could have a series inductance of about 10 nH and a parallel capacitance of about 0.5 pF.

ADS can be used to simulate the magnitude of the impedance of a leaded resistor versus frequency. Assume about 0.5 pF capacitance and 10 nH

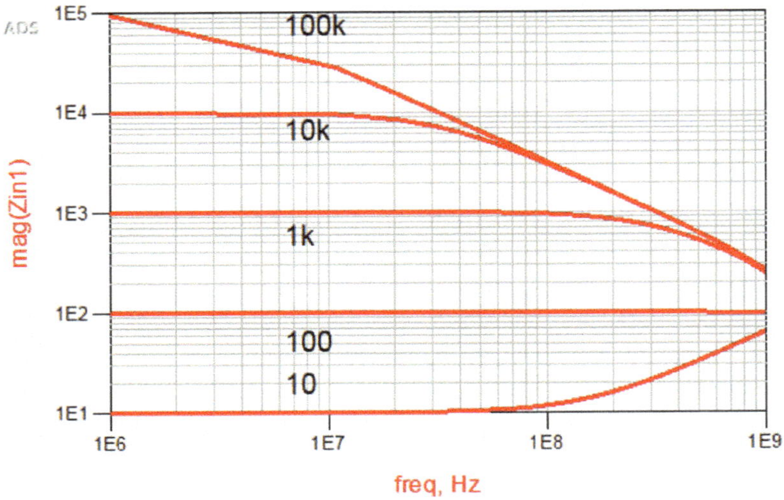

Figure 1.4 As shown, the effective magnitude of impedance is strongly frequency dependent. For higher values, capacitance dominates with Z decreasing with f. For small values, inductance takes over.

inductance. See the ADS Basics Tutorial section (Appendix 1.1) for details on how to set up a parameter sweep.

Because of the inductance problem, you will often use chip resistors (Figure 1.5) instead of leaded resistors if they are to be used at high frequencies. You will grow to appreciate these for their electrical performance and possibly grow to hate them because they are difficult to solder and they easily fracture.

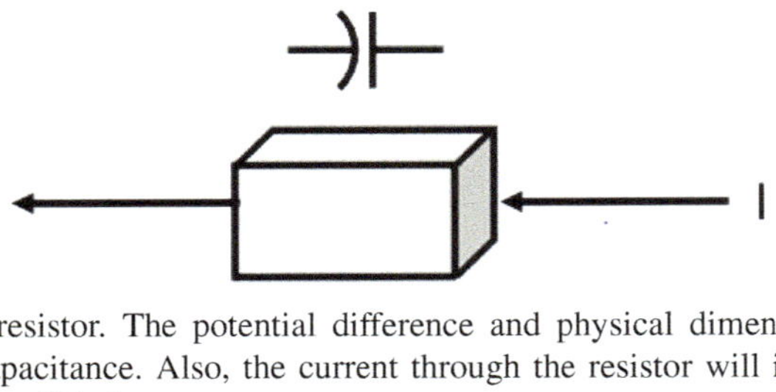

Figure 1.5 Chip resistor. The potential difference and physical dimensions result in an effective parallel capacitance. Also, the current through the resistor will induce a magnetic field, which leads to self-inductance.

The chip resistors come in various sizes. The smaller the size, the better the high frequency performance, but less power can be dissipated, and the soldering gets harder! The standard notation for the physical size of chip components is length-width in 10 mil increments. For example, chip resistors are available in 1206, 0805, 0603, 0402, 0201 dimensions where an 0402 would be 40×20 mils (1×0.5 mm).

Representative values for parasitic elements of a 1206 chip resistor are $L = 1.2$ nH and $C = 0.03$ pF.

1.2.5 Capacitors

Capacitors for use in high frequency circuits generally perform one of these functions:

Bypassing: needs to have low impedance over wide frequency range in order to provide a good AC ground.

Coupling: blocking of DC between stages.

Resonator: frequency control or filter applications.

Reactance: used to introduce a pole or zero at a certain complex frequency.

Capacitors are made from many different materials, but for high frequency applications, ceramic dielectrics are most frequently used. Some ceramics, however, have large variation of capacitance with temperature (or temperature coefficient, typically in ppm/°C), so care must be exercised in selecting dielectrics when resonators or filters are the application.

Capacitors have the same problem as resistors. When leaded, the self-inductance is often too high to be useful at radio frequencies. They also include some loss, sometimes referred to as ESR or effective series resistance. One possible equivalent circuit is shown in Figure 1.6.

Figure 1.6 Equivalent circuit model of a capacitor for high frequency simulation.

Since the equivalent circuit of the capacitor looks like a series *RLC* network, it will have a self-resonant frequency. Above this frequency, it actually behaves like an inductor. This would be a serious deficiency if you were trying to use the capacitor to build an *LC* filter!

$$\omega = \frac{1}{\sqrt{LC}} \tag{1.5}$$

For example, a 100 pF leaded cap might have a series inductance of 10 nH. That means that it will series resonate at 159 MHz and look like a small real resistance. This is referred to as its *self-resonant frequency or SRF.* If a larger *C* is needed or higher frequency operation is required, then chip capacitors are generally used. These have much lower inductance, typically of the order of 1 nH.

There are cases where the self-resonance should be avoided. For example, in digital circuits where there is a lot of transient behavior, resonances should be damped by adding a small series resistance.

Note that some familiar types of capacitors are not at all effective at high frequencies. They have large inductance. These should be avoided *except for DC supplies, audio frequency applications, and the occasional need for low frequency bypassing on RF amplifiers.* So, in general, for high frequency applications:

Avoid: Electrolytic and tantalum capacitors for RF applications

1.2.6 Bypass capacitor example

The equivalent circuit model in Figure 1.6 was simulated in ADS to plot magnitude and phase versus frequency (Figure 1.7). (Refer to Appendix 1.1 on ADS tutorial basics to see how this was set up.) Let's assume a typical series resistance of 1 ohm and inductance of 10 nH for a leaded ceramic capacitor. The capacitance is 10 nF. This might be a reasonable choice for a high frequency bypass capacitor whose function is to provide low impedance over a range of frequencies in order to prevent signal leakage into the power bus and to provide a good AC ground when needed.

We can see that the capacitor is self-resonant at 16 MHz and would provide a decent low impedance to ground from about 5 MHz to 50 MHz. For higher frequency use, a chip capacitor with much lower inductance would be a good choice. Again, self-resonance can be a source of ringing if there are transients that could excite them.

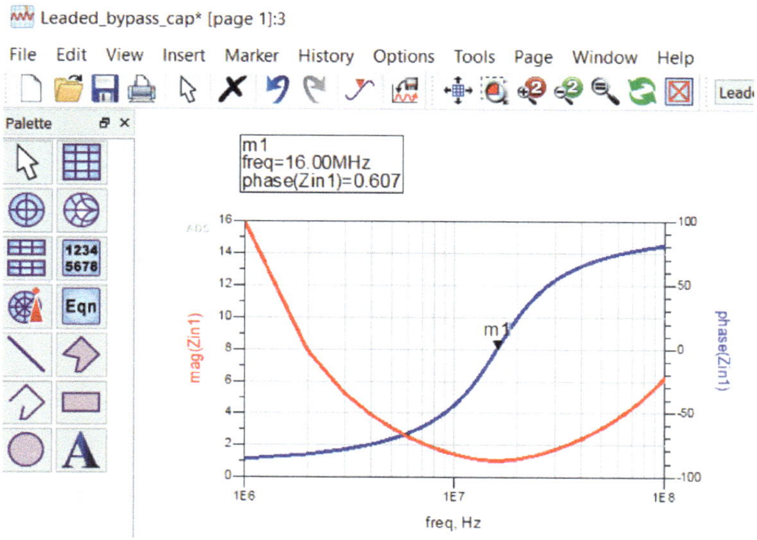

Figure 1.7 Magnitude and phase of the bypass capacitor versus frequency.

1.2.7 Inductor

Inductors in high frequency circuits generally perform one of the following roles:

RF choke: Block AC – often used for bias feed.

Resonator: frequency control or filter applications.

Reactance: used to introduce a pole or zero at a certain complex frequency.

As you by now will have anticipated, these are as non-ideal as the other components. An equivalent circuit is shown in Figure 1.8:

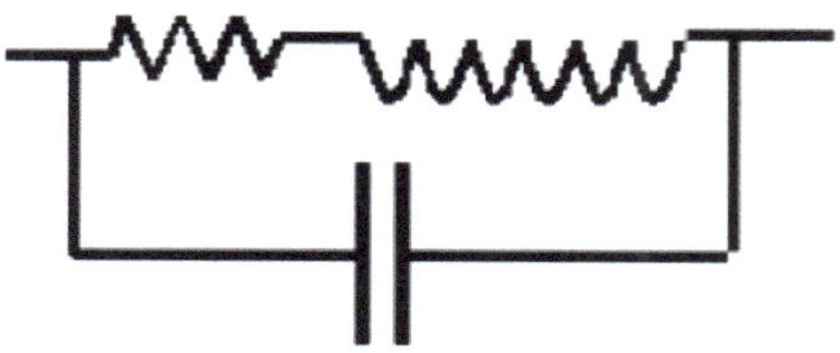

Figure 1.8 Equivalent circuit of an inductor at high frequencies.

The series resistance is due to wire IR loss, skin effect losses, radiation, magnetic core material losses (if any) and dielectric losses. Losses in inductors can be quite high. The capacitance in the circuit model is caused by the electric field between turns of the coils of wire since there will be a potential difference between turns. See Figure 1.9 for the frequency dependence of Zin1. The other port was connected to ground. In this example, $C = 5$ pF, $R =$ 1 ohm, and L is swept from 0.1 to 100 μH.

The inductor also is self-resonant and will exhibit capacitive reactance at high frequencies. This causes the apparent inductance to become quite frequency dependent as you approach the resonant frequency.

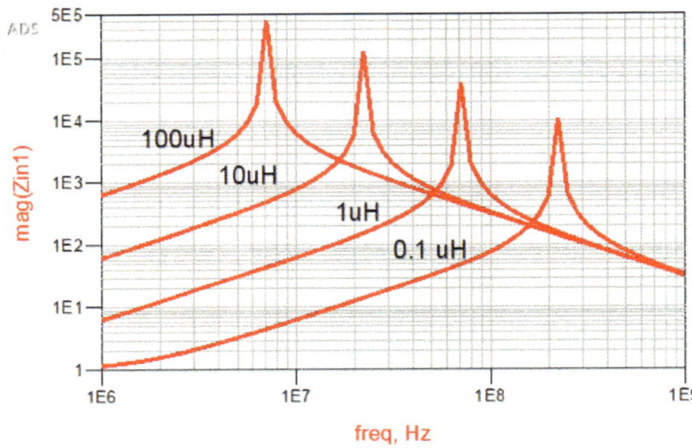

Figure 1.9 Parameter swept simulation of four values of L versus frequency. At lower frequencies, all inductors look inductive. Each has a clearly defined self-resonance. Above the self-resonant frequency they display the effect of the self-capacitance.

A filter example is shown in Figure 1.10.

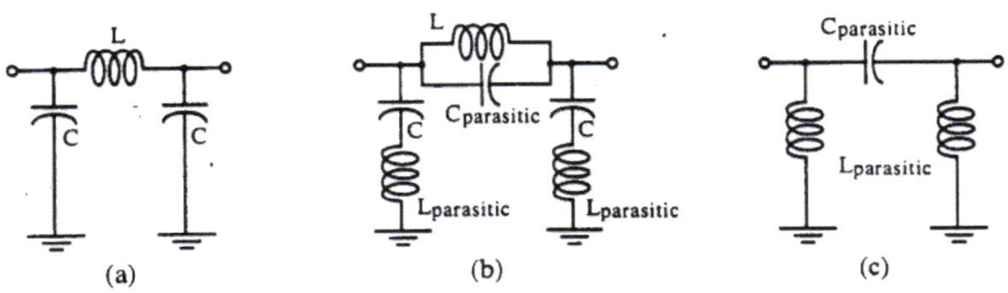

Figure 1.10 Effect of parasitic elements on a presumed lowpass filter structure: (a) as designed, (b) with parasitic elements added, (c) effective circuit above self-resonance of components becomes a high pass [3].

Inductors for use in the HF/VHF frequency range will often make use of magnetic core materials, powdered iron or ferrite, to reduce the length of wire needed to achieve the desired value of inductance. See Reference [4] for an extensive guide on selection of these materials. There are many sources for these in rod or toroidal form. They are also useful for interference suppression by providing common-mode chokes on coaxial or ribbon lines.

1.2.8 Quality factor, *Q*

Reactive components such as capacitors and inductors are often described with a figure of merit called Q. While it can be defined in many ways, its most fundamental description is:

$$Q = \omega \frac{\text{energy stored}}{\text{average power dissipated}}. \tag{1.6}$$

Thus, it is a measure of the ratio of stored versus lost energy per unit time. Note that this definition does not specify what type of system is required. Thus, it is quite general. Recall that an ideal reactive component (capacitor or inductor) stores energy

$$E = \frac{1}{2} CV_{pk}^2 \quad \text{or} \quad \frac{1}{2} LI_{pk}^2. \tag{1.7}$$

Since any real component also has loss due to the resistive component, the average power dissipated is

$$P_{avg} = \frac{1}{2} I_{pk}^2 R = \frac{V_{pk}^2}{2R}. \tag{1.8}$$

An example of a series resonant circuit is shown in Figure 1.11.

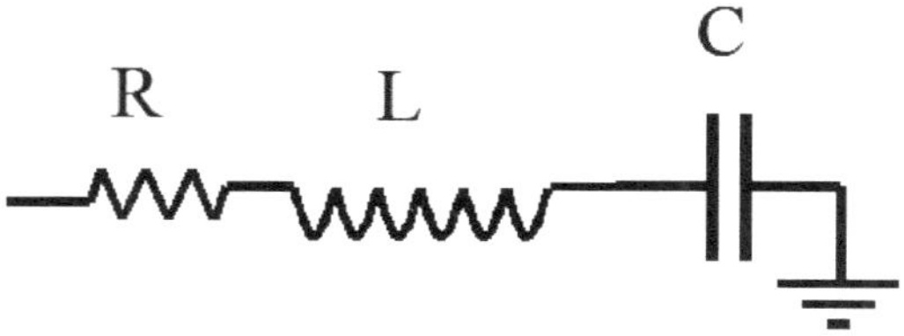

Figure 1.11 Series resonant LRC circuit.

At resonance, the reactances cancel out leaving just a peak voltage, V_{pk}, across the loss resistance, R. Thus, $I_{pk} = V_{pk}/R$ is the maximum current which passes through all elements. Then,

$$Q = \omega_O \frac{LI_{pk}^2/2}{I_{pk}^2 R/2} = \frac{\omega_O L}{R} = \frac{1}{\omega_O RC}. \tag{1.9}$$

In terms of the series equivalent network for a capacitor shown above, its Q is given by:

$$Q = \frac{1}{\omega RC} = \frac{X}{R} \tag{1.10}$$

where we pretend that the capacitor is resonated with an ideal inductor at frequency ω. X is the capacitive reactance, and R is the series resistance. Since this Q refers only to the capacitor itself, in isolation from the rest of the circuit, it is called *unloaded* Q. The higher the unloaded Q, the lower the loss. Notice that Q decreases with frequency.

The unloaded Q of an inductor in a series configuration is given by

$$Q = \frac{\omega L}{R} \tag{1.11}$$

where R is a series resistance as described above. Note that Q is proportional to frequency for an inductor. The Q of an inductor will depend upon the wire diameter, core material (air, powdered iron, ferrite) and whether or not it is in a shielded metal can.

It is easy to show that for a parallel resonant circuit, Q is given by:

$$Q = \frac{B}{G} \tag{1.12}$$

where B is the susceptance of the capacitor or inductor and G is the shunt conductance.

1.2.9 Loaded Q

When a resonant circuit is connected to the outside world, its total losses (let's call them R_P or G_P) are combined with the source and load resistances or conductances, R_S and R_L. An example is shown in Figure 1.12.

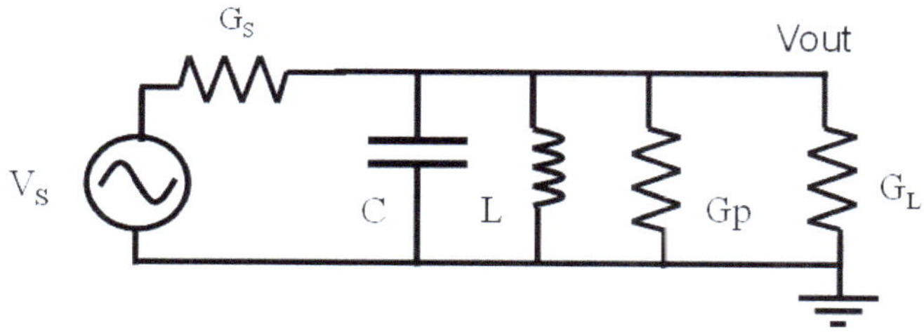

Figure 1.12 Parallel resonant circuit illustrating the concept of loaded Q.

Here is a parallel resonant circuit (C, L and G_P) connected to a load conductance G_L. The total Q of this circuit is called the *loaded* Q or Q_L and is given by

$$Q_L = \omega_o C\,(R_P\,||\,R_S\,||\,R_L) \tag{1.13}$$

or

$$Q_L = \frac{\omega_o C}{G_S + G_L + G_P} = \frac{\omega_o C}{G_{totcl}} = \frac{B}{G}. \tag{1.14}$$

The significance of this is that Q_L can be used to predict the bandwidth of a resonant circuit. We can see that higher Q_L leads to a narrower bandwidth:

$$BW = \frac{\omega_o}{Q_L}. \tag{1.15}$$

Figure 1.13 shows an example of a simple parallel resonant circuit. The unloaded Q is infinite, since no losses are included in the network. We see that there is no insertion loss in this case.

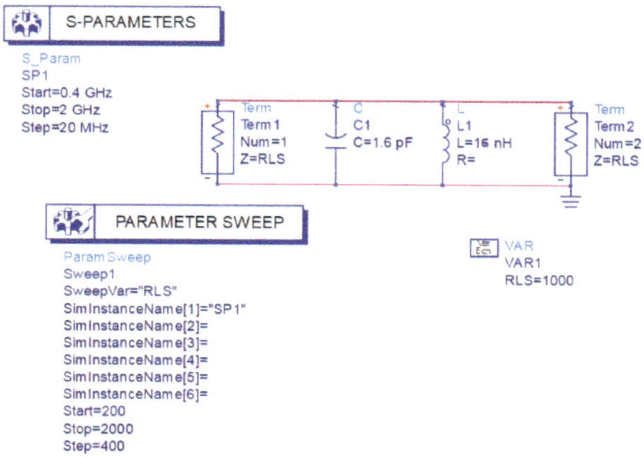

Figure 1.13 Loaded Q varies from 1 to 10 with the given parameter sweep.

Note that the S-parameters will be described in detail in Chapter 4. For now, think of S(2,1) as the transmission coefficient or insertion gain through the network where source and load resistances, $RLS = Z_0$. For a matched network, the insertion gain must be 1, therefore the factor of 2 is needed so that dB(S2,1) = 0.

$$dB((S2,1)) = 20\log(2V_{out}/V_{in}). \tag{1.16}$$

Figure 1.14 illustrates that at resonance we have 0 dB of insertion gain.

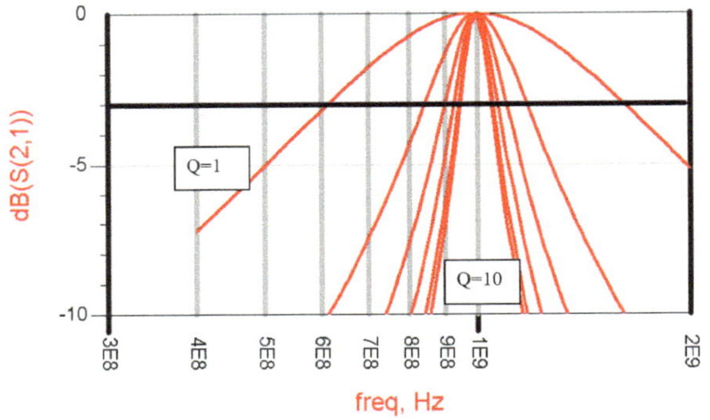

Figure 1.14 Frequency response. Note that the LC network is ideal, without any loss. The terminations load the network and result in the range of loaded Q shown.

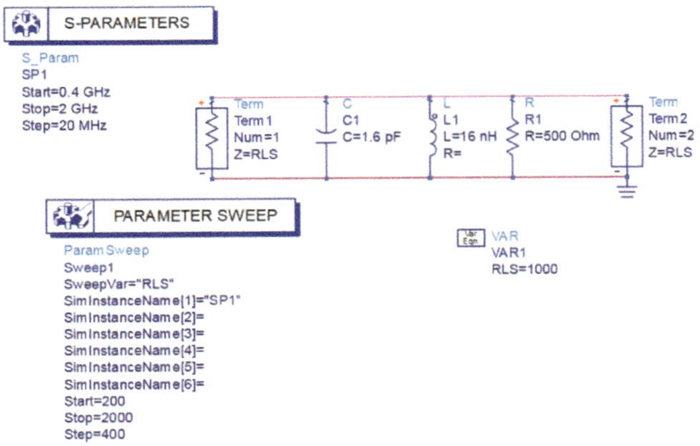

Figure 1.15 Now, the circuit is modified to include a 500 ohm resistor (R_1) in parallel with the *LC* network.

The resistance shown in Figure 1.15 represents the parallel equivalent loss due to both the L and the C. So, now we have a finite unloaded Q. We sweep the source and load term *RLS* from 200 to 2000 ohms using the ParamSweep function (Figure 1.16). Note that an initial value of *RLS* must be declared using the VarEqn icon in order for the parameter sweep to work.

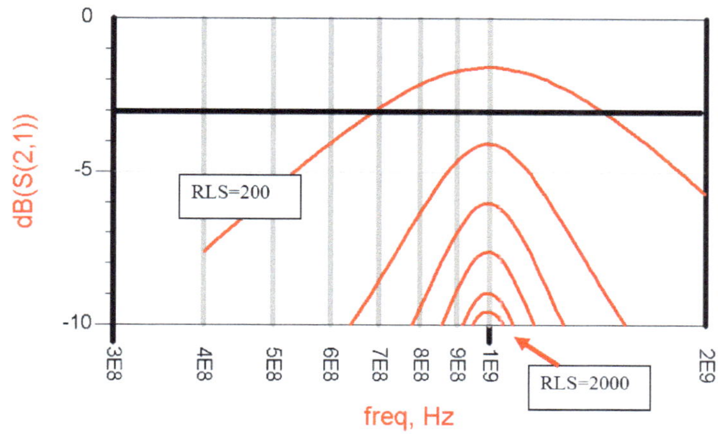

Figure 1.16 Note that the insertion loss increases as loaded Q, Q_L, approaches Q_U. Sweeping *RLS*, we see that, at resonance, the reactances cancel, and we are left with a resistive divider:

$$V_{out} = V_{in}[R_1/(2R_1 + R_{LS})]. \tag{1.17}$$

Also note that the 3 dB bandwidth $2\Delta\omega$ is inversely proportional to Q_L, as shown in Equation (1.18), where ω_0 is the resonant frequency.

$$2\Delta\omega = \frac{\omega_0}{Q_L} \tag{1.18}$$

Exercise 1.1

Design a series RLC network with a resonant frequency of 159 MHz and loss R = 2 ohms, Q_U = 10. Determine source and load resistances RLS such that dB(S(2,1)) as defined in Equation (1.13) is −3 dB. Calculate the loaded Q in this case. Simulate it on ADS to display dB(S(2,1)) versus frequency to verify your calculation.

1.3 Summary

1.3.1 Nonideal components

- What you see on the label is not always what you get at high frequencies.
- Models are needed at high frequencies for: wires, resistors, capacitors, inductors.

1.3.2 Quality factor (Q) of components and resonant circuits

- The definitions for unloaded Q and loaded Q were presented
- The effect of Q_L on insertion loss and bandwidth was described.
- A more comprehensive description of resonant circuits will be presented in Chapter 13.

References

[1] Lee T.H. (2004) *Planar Microwave Engineering*. Cambridge University Press, ch. 6.
[2] Bowick C. (2008) *RF Circuit Design*, 2nd edn. London: Newnes, ch.1.
[3] McWhorter M. et al. (1995) *EE344, High Frequency Laboratory*. Stanford University, ch. 1.
[4] Iron-Powder and Ferrite Coil Forms, available at www.amidoncorp.com

Homework

READ: Appendix 1.1, a short tutorial to using ADS.

1.

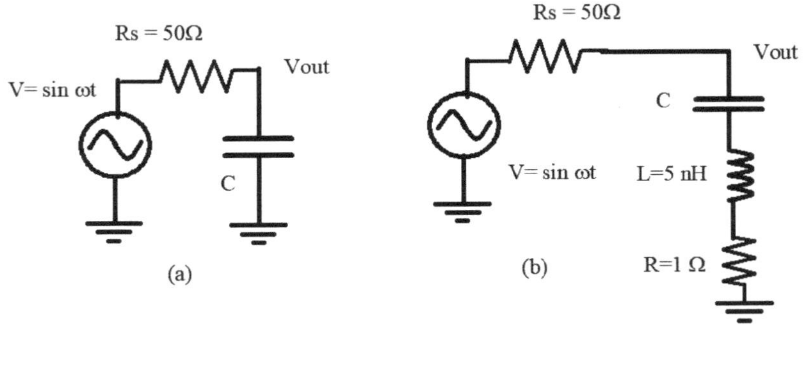

a. Design a simple low-pass RC filter as shown in Figure (a) with a cutoff frequency of 320 MHz. Plot the expected frequency response at V_{out}.
b. A leaded capacitor is used to implement this filter. The capacitor has an equivalent circuit model as shown in Figure (b). Sketch $|Z(jw)|$ and V_{out} versus frequency. Explain how and why the non-ideal capacitor affects the performance of the low pass filter. What are the minimum and maximum values of V_{out}?
c. Calculate the Q of the capacitor in (b) and the self-resonant frequency.
d. What would be the equivalent circuit model parameters associated with a capacitor that would be suitable for use in an *RC* lowpass with a 3 dB frequency of 320 MHz?
e. Use ADS (AC analysis mode) to simulate the response of Figures (a) and (b).

2. It is very convenient when analyzing circuits with nonideal components to transform between a series and parallel equivalent representation. There is a simple transformation

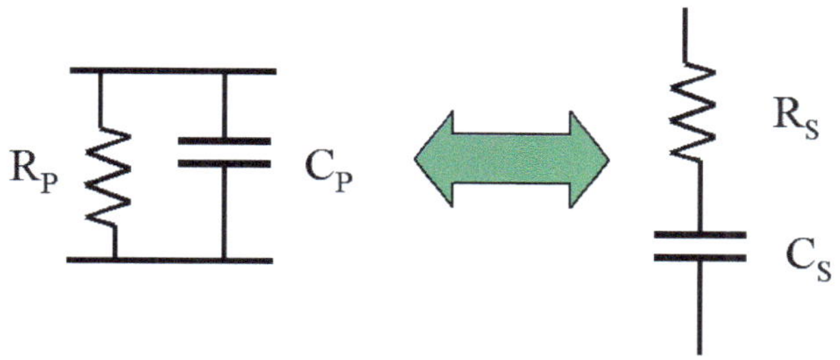

for this that is accurate *at a single frequency*. In many cases, this is sufficient.
Show that

$$R_P = R_S\left(Q^2+1\right)$$
$$X_P = X_S\left(\frac{Q^2+1}{Q^2}\right)$$

where X_P and X_S are the reactances of the parallel and series circuits respectively. Q is the same for both.
Note: This transformation will prove to be useful when designing matching circuits.

3. A leaded 1 nF capacitor was found to have the equivalent circuit as shown:

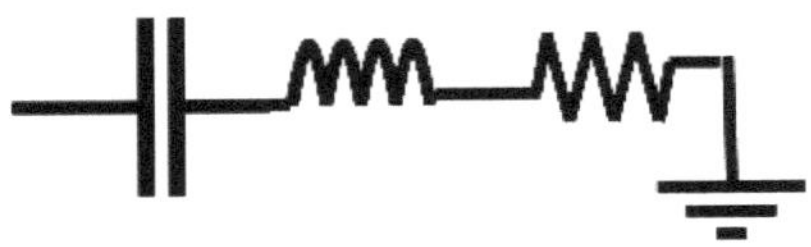

a. Determine the unloaded Q at 10 MHz and the self-resonant frequency.
b. Calculate and sketch the magnitude of the impedance of the capacitor versus frequency.
c. The capacitor is used as a bypass cap in the amplifier circuit shown. Plot $|V_{ac}/V_{out}|$ versus frequency. If V_{out} = 1 V ac, find the minimum and maximum values of V_{ac}.

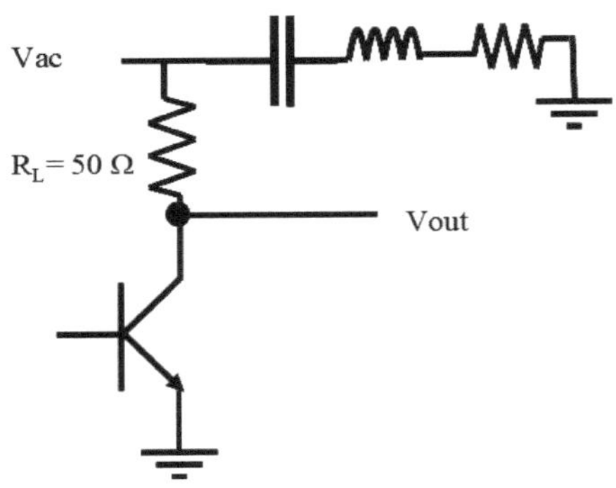

d. How would you modify the circuit so that the maximum V_{ac} at 10 MHz is less than: (i) 1 mV? (ii) 1 mV at 100 KHz?

Bypass capacitors. You have seen how wires act as inductors. About 16 nH per inch is typical. Connections to a power supply through wires of any significant length produce a high impedance at radio frequencies. Thus, any illusion that V_{CC} or V_{DD} is at AC ground on the circuit should be rejected. Even with short leads, a typical power supply does not present low output impedance at radio frequencies. Therefore, a local AC ground with low impedance is needed right at the circuit itself to keep the AC signal out of the power supply (V_{CC} or V_{DD}) connection on the circuit board or module or IC chip. This is the function that the bypass capacitor performs.

In order for the capacitor to be effective in attenuating AC signals on the DC voltage line, it must present low impedance at any frequency where the circuit under design exhibits gain. This is to avoid oscillation problems due to an unintended positive feedback locp through the power or ground connection. Thus, the capacitor must be selected so that both real and imaginary parts are small. The series resistance and series inductance are the main factors limiting the effectiveness. Choosing chip capacitors over leaded capacitors helps reduce inductance significantly. Low series resistance (ESR) is also important. Often, identical capacitors are connected in parallel to reduce both L and R while also increasing C.
Using a larger single capacitor is often less effective than using several smaller ones in parallel. This is especially true of electrolytic or tantalum capacitors which have large series inductances. These are fine for low frequency bypassing (below 100 KHz) but worse than useless at radio frequencies. If the bypass impedance must extend to such low frequencies, then a multi-stage bypass is needed that includes both ceramic (low inductance) and electrolytic/tantalum capacitors.

5. Consider a parallel resonant RLC network.

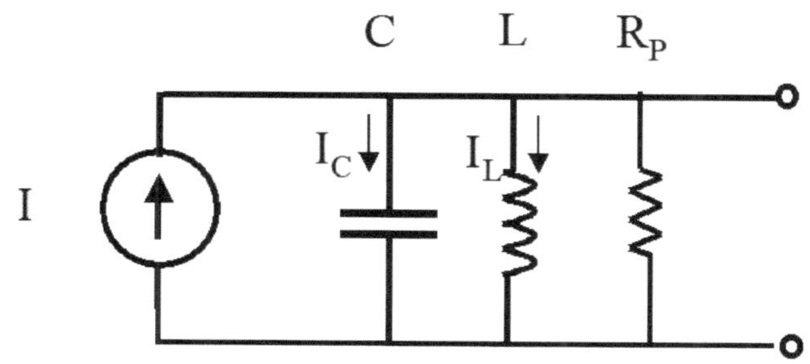

   a. Starting from the definition $Q = \omega$ energy stored/energy dissipated, derive an equation for the Q of the parallel network in terms of susceptance and conductance.
   b. Find the ratio $|I_C|/|I|$ or $|I_L|/|I|$ at the resonant frequency ω_o.
   c. For a series resonant circuit at the resonant frequency ω_o find $|V_C|/|V|$ or $|V_L|/|V|$ where $|V|$ is the magnitude of the voltage across the series network and $|V_C|$ is the voltage across the capacitor.

What does the outcome of parts (b) and (c) imply for components used in a high Q resonant network?

d. Assume a small frequency departure from resonance, $\Delta\omega$. Show that $2\Delta\omega = \omega_0/Q$ when $|V|$ (node voltage magnitude) is reduced by 3 dB.
e. Show that the phase slope at resonance is also related to Q and therefore the bandwidth.
f. Read the ADS tutorial. Simulate the parallel RLC circuit using the AC analysis mode and verify your result from parts (b), (d) and (e). Design the RLC for a 1 GHz resonant frequency and a Q of 10 with R_P = 1000 ohm.

Appendix 1.1: A Short Tutorial on using Pathwave/Keysight ADS

The Pathwave/Keysight Advanced Design System (ADS) is a powerful simulation tool, well suited for HF/microwave analog circuit analysis. The learning curve is a bit steep, so you will find several tutorials in various chapters that help you to get started using it. Here is an example that you can use to begin the S-parameter mode of analysis. Much detail on S-parameters can be found in Chapter 4.

Example A1: Here is a circuit with a 25 ohm source and a complex load. In this example, a matching network to match source to load is to be evaluated (Figure 1.17).

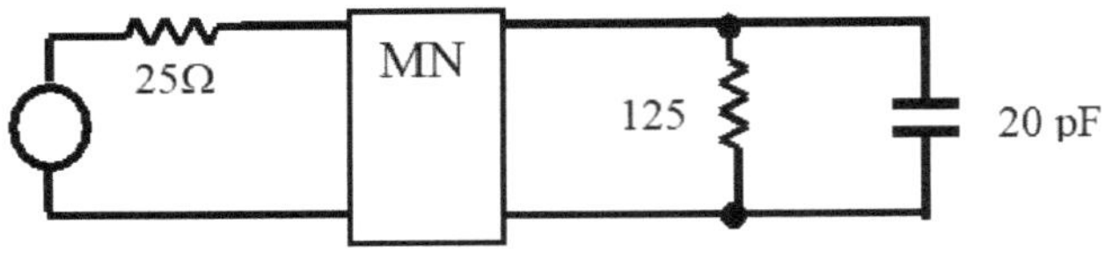

Figure 1.17 Match the 25 ohm source to the complex load by designing a matching network (MN).

The matching network can make use of either distributed or lumped elements. You will be learning how to design such networks soon. The procedure below illustrates how to simulate this circuit in ADS.

A1.1.1. Schematic window

1. Start up ADS: Pathwave Advanced Design System 2022 Update 1 (or current version). The main menu will open up.
2. Open up a new workspace. A simulation folder containing schematics, data files, and data displays is called a workspace. If the workspace has already been created, then open it up by double clicking on its name.

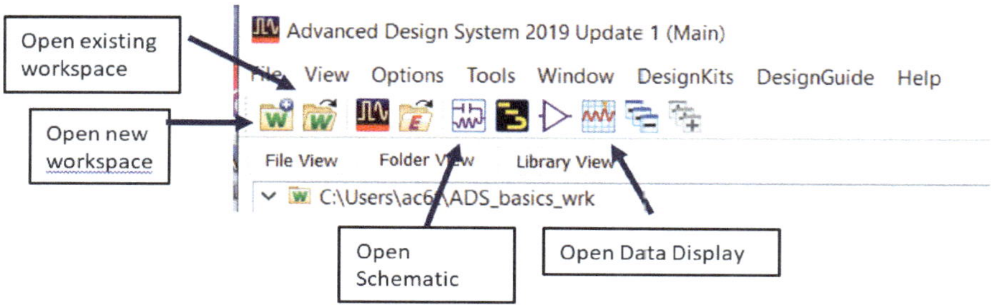

Figure 1.18 Start-up window in ADS.

If a schematic window doesn't open automatically, *open the window* with the Open Schematic button.

3. Construct the schematic diagram. We will use the S-Parameter simulation mode in ADS. This is a small-signal AC analysis. Rather than specifying a generator or load impedance, for the S-Parameter mode of simulation, terminations must be assigned to all inputs and outputs of the network. Terminations can be assigned any complex impedance, expressed as real + j* imaginary. While most of the time, 50 ohms would be used as provided by default, the input termination (Term) in this example is set to 25 ohms for Zo. The term icon is found in the Simulation-S_Param menu. This menu opens the palette of components shown in Figure 1.19.

   Most all of the circuit icons can be placed with the mouse cursor – just point and click. To complete the placement of a particular part, hit Escape

   Select components from the palette. Click to place them. Rotate if necessary. Connect the parts together using the wire (ctrl-w) and ground (G) icons.

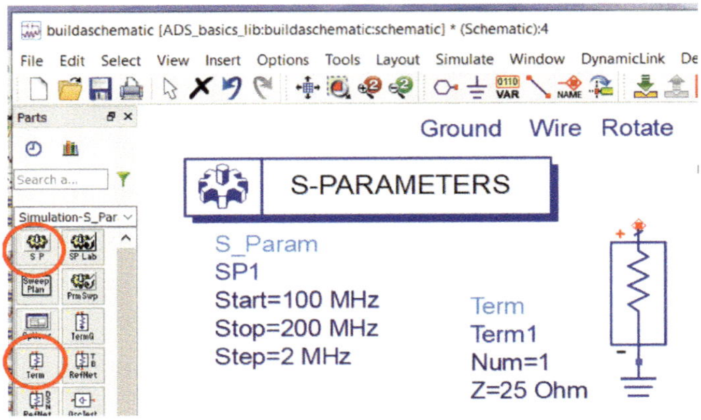

Figure 1.19 Schematic window showing S-Parameter controller and a termination, reset to 25 ohms.

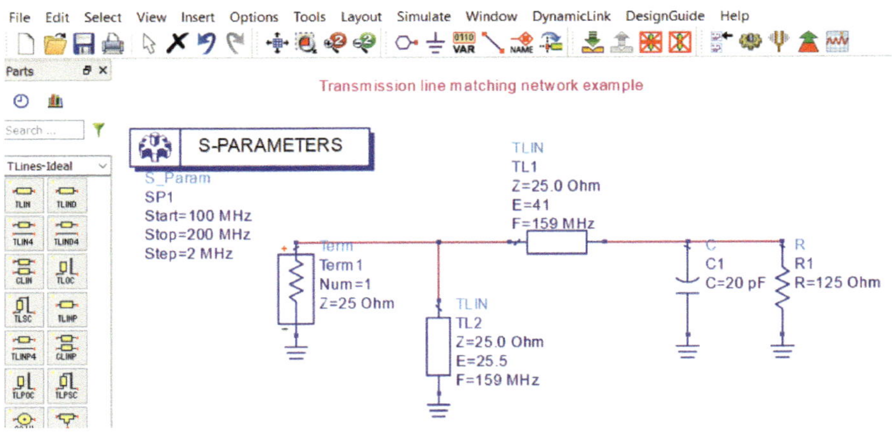

Figure 1.20 Edit the S PARAMETERS control icon. Set your Start Frequency, Stop Frequency, and Step Size. The transmission lines shown are TLIN (ideal trans lines) found in the *TLines-ideal* palette menu. The capacitor and resistor are in the *Lumped Components* palette menu.

4. Name and save the schematic file with the disk button or use Save As in the File menu.
5. Hit F7 or the gear icon to run the simulation. A data file will be saved with the same name as the schematic. The data display will open.

A1.1.2. Data display window

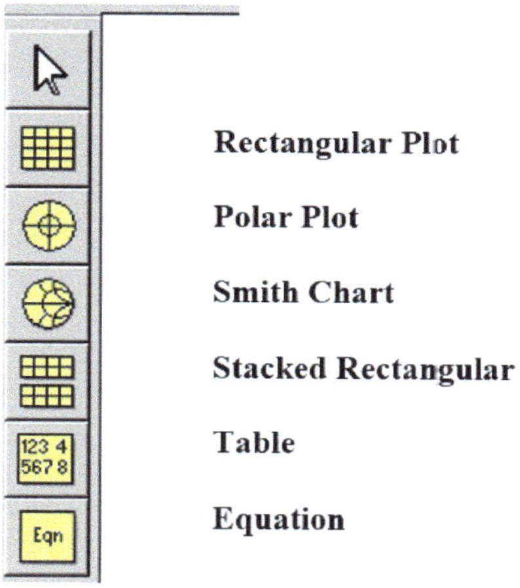

Figure 1.21 To select the how the output variable will be displayed, click on one of the icons.

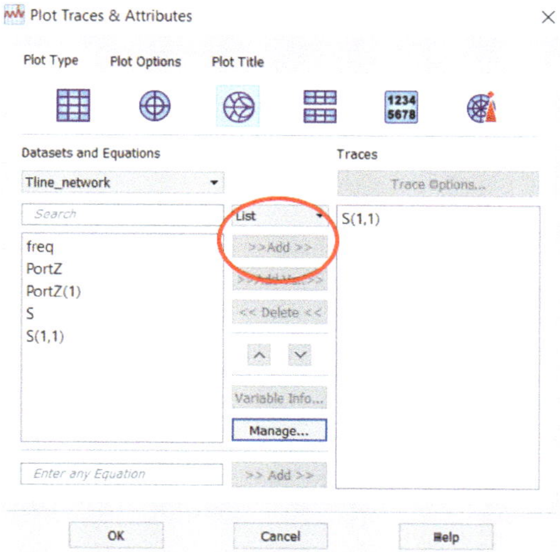

Figure 1.22 Select the Smith chart icon (a detailed description of the Smith Chart can be found in Chapter 5). Plot Traces & Attributes opens up with a menu that you can use to select the desired variable, in this case, S(1,1). Highlight the variable by clicking on it and click on Add.

A.1.1.3 Markers

Markers are used to read out frequency and the dependent variable off the plot as shown in Figure 1.23. In the case of S11, impedance or admittance can also be displayed (normalized to the Zo of the termination)

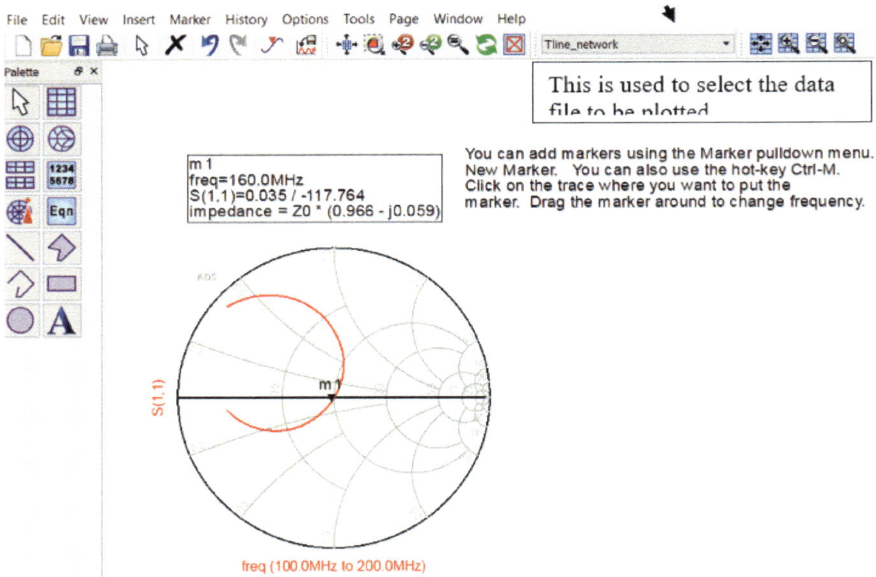

Figure 1.23 Data is displayed on the Smith Chart. Markers can be added.

Exercise 1.2

Here is an example of a circuit composed of lumped components. Construct the schematic, run the simulator, and plot S(1,1) on the Smith chart. Add a marker where magnitude of S(1,1) = 0.

Example A2: Simulate impedance or admittances: leaded 10nF bypass capacitor equivalent circuit model

The impedance or admittance can be easily plotted by placing a Zin or Yin equation icon on the schematic page. The button is found on the S-parameter palette. The example below shows calculation of *Z* for port 1.

You can then plot the impedance by clicking the rectangular plot type icon and placing a rectangular plot on the display panel. Select Zin1 and, because it is a complex number, you must choose real or imaginary part. In this example, choose magnitude (Figure 1.26).

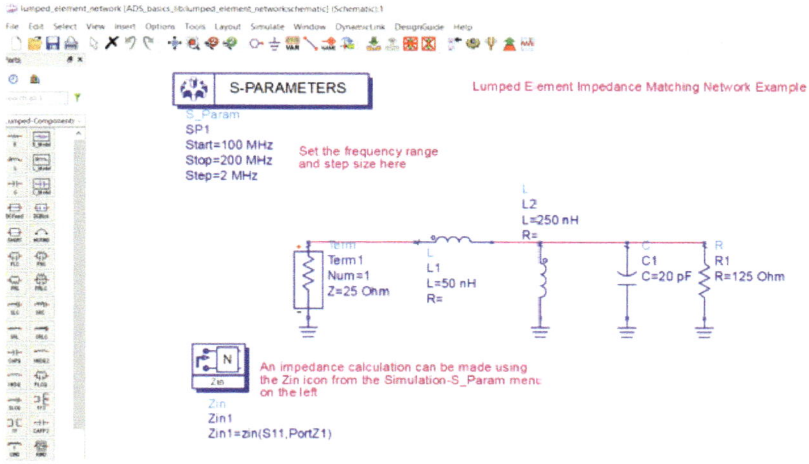

Figure 1.24 An example of a lumped element matching network.

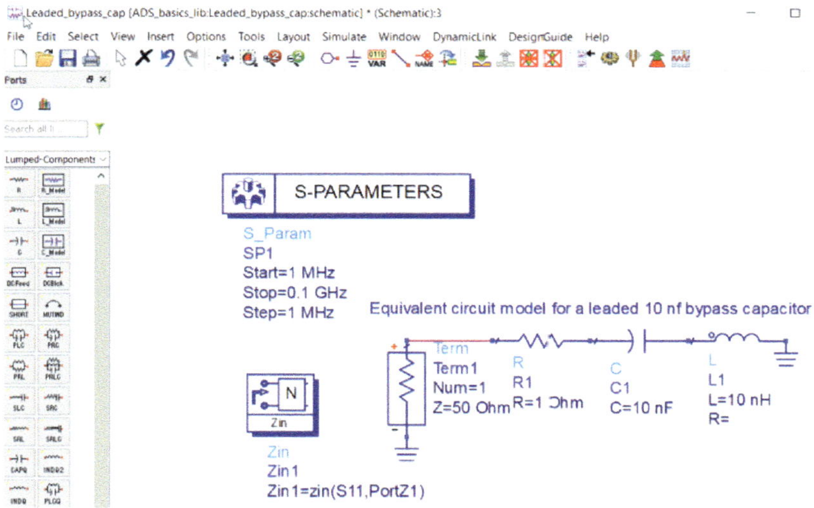

Figure 1.25 Schematic of a typical leaded capacitor.

Because of the wide frequency range, choosing the x-axis as a log plot would be useful.

Click on the Plot Options tab to change the scale of the axes or to plot versus log frequency.

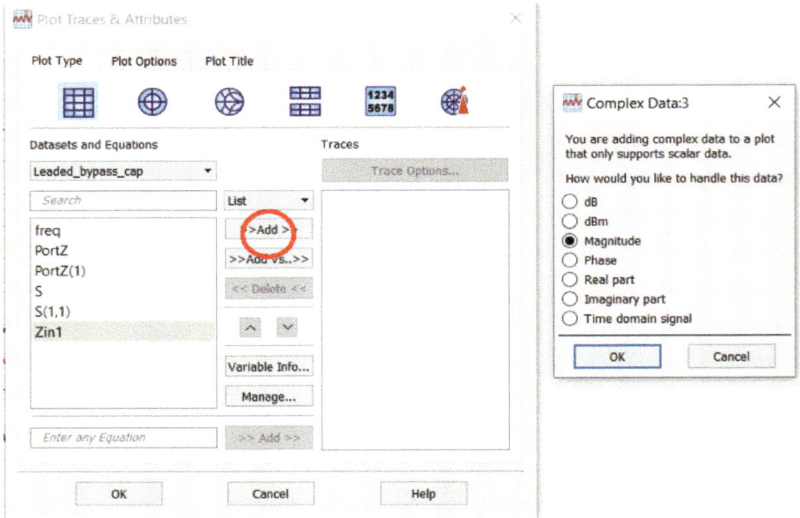

Figure 1.26 Rectangular plot selection.

We might also want to plot phase on the same rectangular plot. Add phase S(1,1) to the traces. Then, click on phase S(1,1), Trace Options, then choose Plot Axes. Select Right Y axis (Figure 1.27).

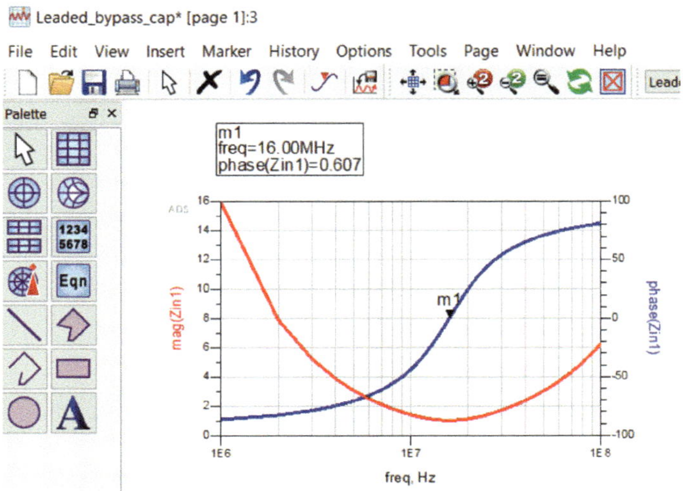

Figure 1.27 Magnitude and phase plot for the leaded capacitor.

By adding a marker, you can see that the capacitor is self-resonant at 16 MHz. It reaches its lowest magnitude of impedance at this frequency. At both lower and higher frequencies, it is less effective as a bypass capacitor. This illustrates the limitations of leaded components for RF/microwave applications.

A1.1.4. AC simulation mode

One can also obtain the same result using the AC simulation mode. An AC source must be selected from the *Sources-FreqDomain* palette. This is an ideal voltage source, so a source resistor must be added. A simulation controller can be found in the *Simulation-AC* palette.

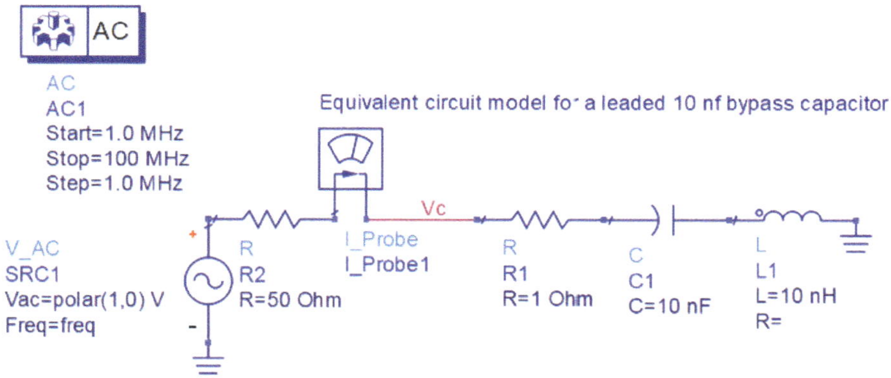

Figure 1.28 Setup to simulate the capacitor equivalent network using the AC mode.

In this case, a node label (Vc) must be added where voltage is to be measured. This can be done by right-clicking on the wire and selecting wire-pin label. Or, Alt-w can be used as a shortcut. As usual, the start, stop and step frequencies must be provided for the simulation controller, AC1. A current probe, I-probe, found in the probe components palette, is needed if the impedance is to be calculated.

In the data display, define an equation to calculate the impedance. Plot the magnitude on the left axis and the phase on the right axis by selecting a rectangular plot. Double click the plot, then add mag(Zcap) and phase(Zcap) from the Plot traces and attributes panel. To move the phase to the right axis, highlight phase(Zcap) then open the Plot Options tab to select right Y axis.

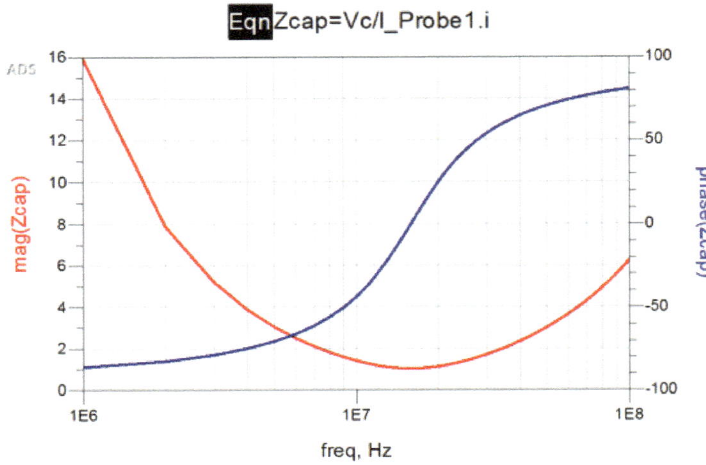

Figure 1.29 Plot of magnitude and phase of the impedance. Same result as found in Figure 1.27.

Next, a somewhat more complicated simulation with a swept component value.

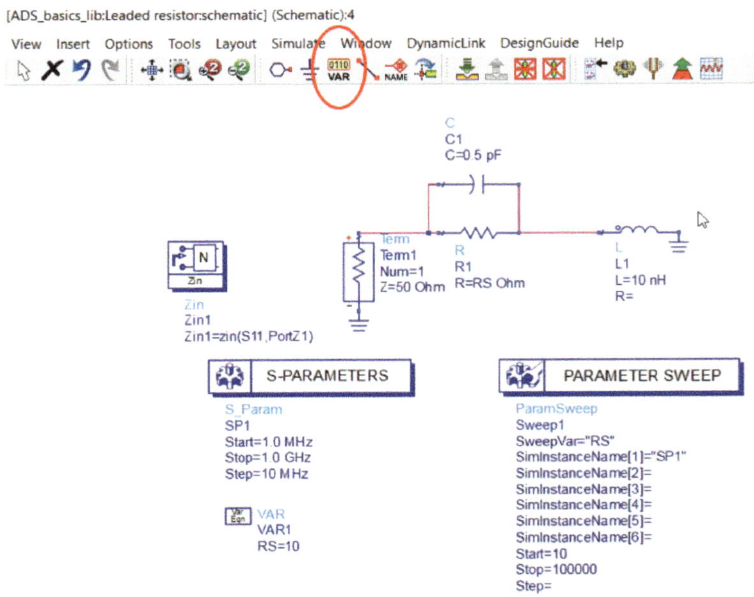

Figure 1.30 Frequency sweep of magnitude of impedance for a representative leaded resistor with parasitic C and L.

In this exercise, both frequency and resistor values are to be swept. Because there is a large range of resistor values, a logarithmic sweep is selected as shown in Figure 1.31. One point per decade was chosen in order to more clearly see the effect of frequency on each value of resistor. Note that, first, RS must be defined by the VAR icon (circled in red in the top palette).

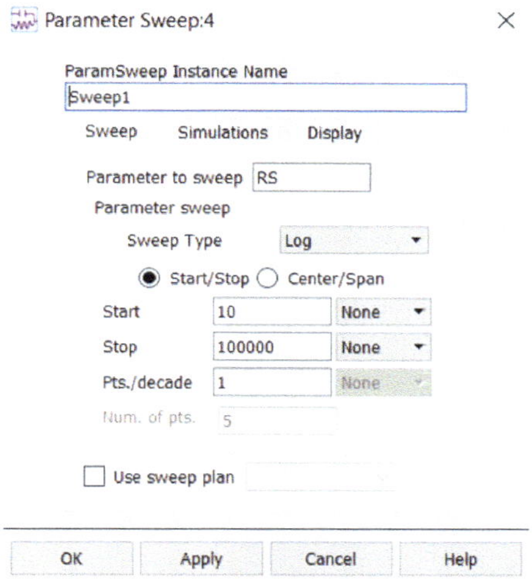

Figure 1.31 Setup for log parameter sweep.

Next, run the simulation and plot magnitude of Z(1,1) versus log frequency.

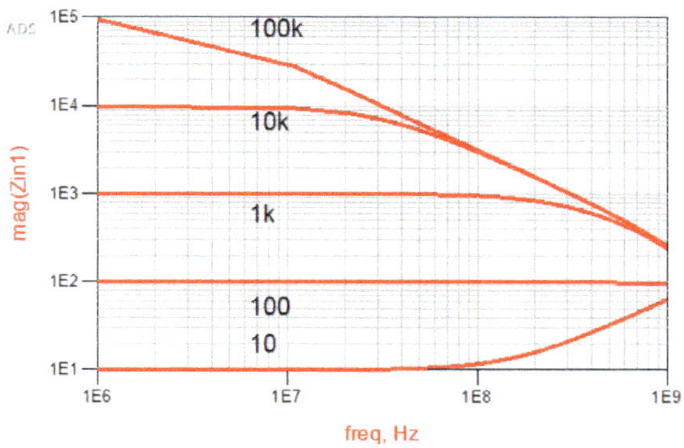

Figure 1.32. Magnitude of impedance versus log frequency.

Appendix 1.2. Measurement of Nonideal Components

After seeing how much the components such as resistors, capacitors, and inductors differ from what you expect, it is important to develop equivalent circuit models for each one. The reason for measuring components is to enable you to have realistic and accurate models that can be used in the design and simulation of your circuits. This greatly increases the likelihood that your project will work as expected.

How might you do that? It involves a vector network analyzer (VNA) and a suitable test fixture or board. First of all, the VNA must be calibrated using the standards provided in a suitable calibration kit.

The test board can consist of an SMA launcher block and a short length of circuit board (Figures 1.33 and 1.34). The center pin can be soldered to copper tape designed with a width for 50 ohm impedance (the width is approximately 0.185” for 50 ohms on Duroid with $\underline{\varepsilon r}$ = 2.3). This facilitates the evaluation of lumped components.

To measure components on the test board, you must extend the network analyzer reference plane (of port 1) to the board edge so that you will measure the S_{11} of the component without shifting the phase through the short length of 50 ohm transmission line. Note that this does not change the basic calibration of the analyzer, but just shifts the reference plane.[1]

To do this, attach the open circuited 50 ohm line extension block and dial in the extra delay until S_{11} looks as much like a dot as you can get at the right side of the plot. Setting the reference place with the open board helps compensate for the fringe capacitance across the open end of the board. Then, your component measurements will not include this additional capacitance.

Then, solder the component under test across the end of an identical open circuited board. Sweep the frequency over a wide range and fit the data to an equivalent circuit for the component. Chapters 3 and 4 explain the Smith chart and S-parameters.

[1]Port Extension is another mathematical adjustment that affects the data as though the measurement test ports were extended by a certain amount. It is commonly used to remove the delay of adapters or a fixture that is attached to the test port cables after a coaxial calibration is performed.

Figure 1.33 SMA launcher block. The center pin of the connector is elevated so that a Duroid or FR-4 one-sided board can slide underneath and a copper tape can be soldered to the pin.

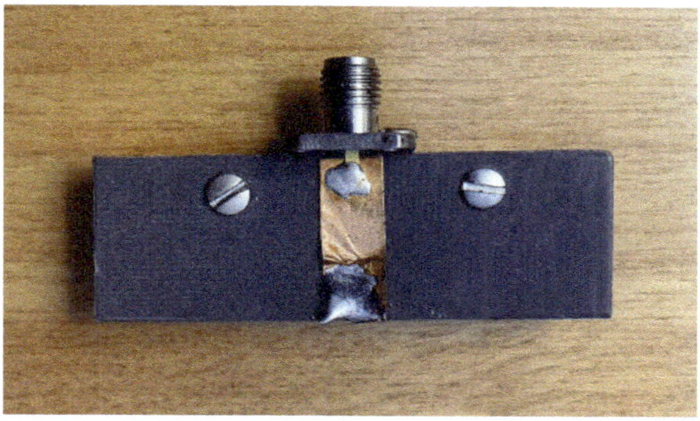

Figure 1.34 Test fixture board. A 50 ohm line is fabricated in copper tape. The component under test can be soldered to the end of the open-circuited line.

A1.2.1 Electrical delay versus port extension

Electrical delay causes the VNA's phase data to be mathematically shifted as though it were sent through a lossless delay line. It is useful to "unwind" a phase measurement or Smith chart measurement and look at the data's deviation from a linear phase response. Port extension is another mathematical adjustment that effects the data as though the measurement test ports were extended by a certain amount. It is commonly used to remove the delay of adapters or a fixture that is attached to the test port cables after a coaxial calibration is performed.

As an example, suppose that the port 1 extension is set to 10 ns, and the port 2 extension is 12 ns. This would cause the measured data to be adjusted with the following delays: S_{11}, 20 ns; S_{21}, 22 ns; S_{12}, 22 ns; and S_{22}, 24 ns. On the other hand, if the electrical delay is set to 15 ns, then the data for any measurement will be adjusted by the same amount, 15 ns.

For both techniques, the measurement data is always adjusted by the time delay value. However, the VNA often gives a display of the delay in units of distance, for convenience. If you know the relative velocity of the cable, adapter, or fixture you are using, you can enter the number and make your distance reading more accurate.

Exercise 1.3

1. Use the network analyzer to measure S_{11} between 5 MHz and 500 MHz (or higher) on a Smith Chart for the following components:
    a. 15 ohm chip resistor. Solder this resistor across the end of your test extension board (not the same one you use to set the port extension). These must be the same material and length as the short/open boards.
    b. 0.1 μF chip capacitor. Remove the resistor and replace with this capacitor.
    c. 0.1 μF leaded capacitor. Remove the chip capacitor and replace with the leaded cap. Keep the leads as short as possible.
    d. 10 μH leaded inductor. Keep the leads as short as possible.
2. How does each component's behavior differ from the ideal element? Determine, if possible, a self-resonant frequency for the component.
3. Propose an equivalent circuit model for each component that could represent the measured behavior. Refer to Appendix 1.1 and use ADS to plot S_{11} and the magnitude/ phase or real/imaginary parts of the impedances of your modeled components and compare your modeled results with the measurement data.

2

Transmission Lines – A Review and Explanation

Perspective:

1. We must quickly learn some foundational material on transmission lines. It is described in the books [1, 2] and in much of the literature in a highly mathematical way.

   DON'T GET LOST IN THE MATH!

   We want to use the Smith chart to cut through the boring math – but must understand first to know how to use the chart.
2. We are hoping to generate insight and interest – not pages of equations.

Goals:

1. Recognize various transmission line structures.
2. Define reflection and transmission coefficients and calculate propagation of voltages and currents on ideal transmission lines.
3. Learn to use S-parameters and the Smith chart to analyze circuits.
4. Learn to design impedance matching networks that employ both lumped and transmission line (distributed) elements.
5. Be able to model (Pathwave ADS) and measure (network analyzer) nonideal lumped components and transmission line networks at high frequencies.

This chapter will describe the function of transmission lines in the time domain (Figure 2.1). The chapter that follows introduces frequency domain behavior.

2.1 What is a Transmission Line?

2.1.1 Common features

- A pair of conductors

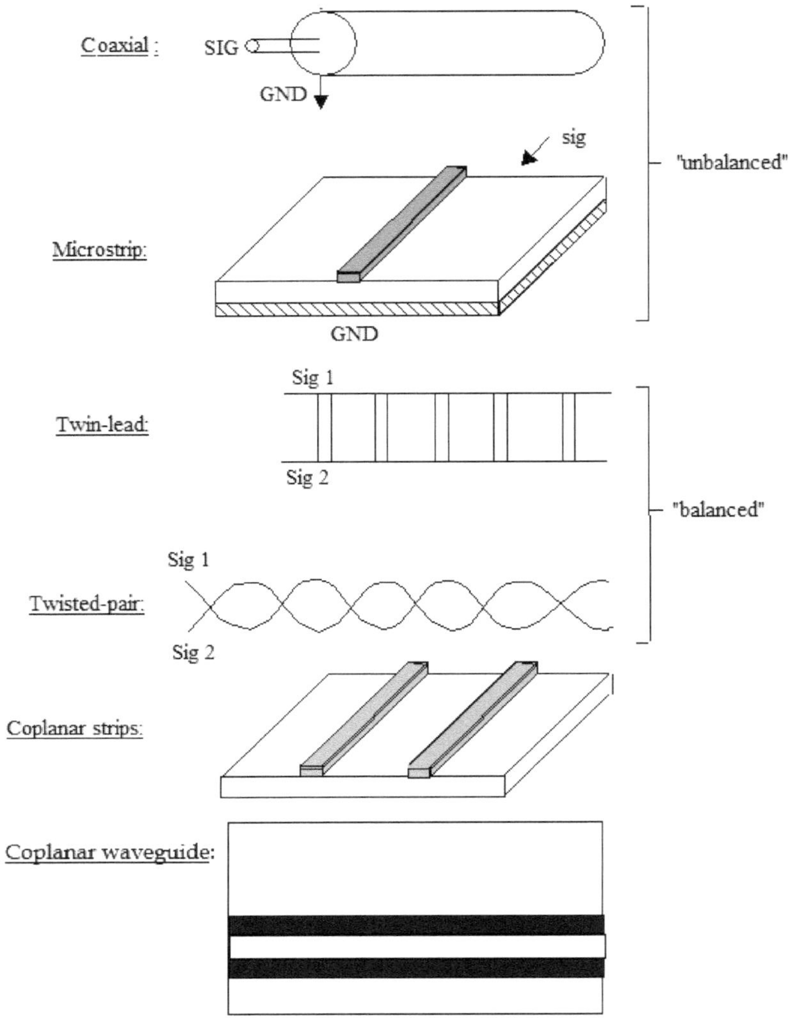

Figure 2.1 Typical transmission line structures.

- Geometry doesn't change with distance.

A guided wave will propagate on these lines.
An *unbalanced line* is characterized by:

1. A signal conductor and ground
2. Ground is at zero potential relative to distant objects.

True unbalanced line: coaxial line; nearly unbalanced: coplanar waveguide, microstrip.
Balanced Lines are characterized by:

1. Two symmetric conductors
2. Conductor potentials are symmetric with respect to distant objects.

Twin lead, twisted pair, coplanar strips.

2.2 Compare Microstrip to Stripline

A stripline is shown in Figure 2.2.

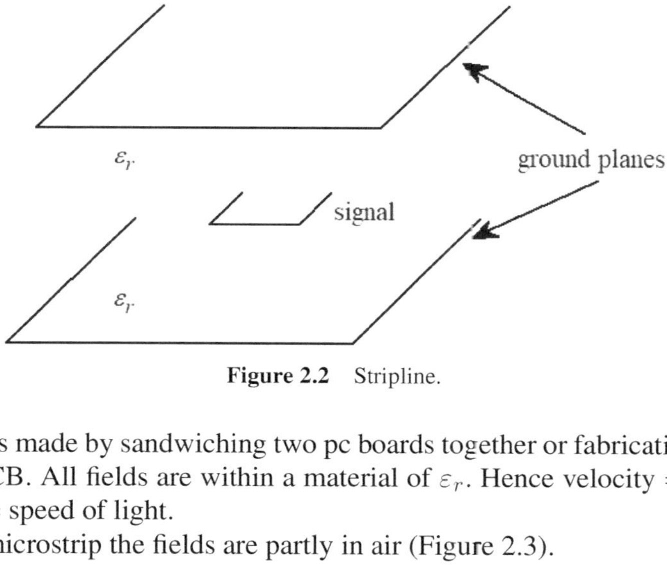

Figure 2.2 Stripline.

This is made by sandwiching two pc boards together or fabricating a multilayer PCB. All fields are within a material of ε_r. Hence velocity $= c/\sqrt{\varepsilon_r}$, where c = speed of light.

In a microstrip the fields are partly in air (Figure 2.3).

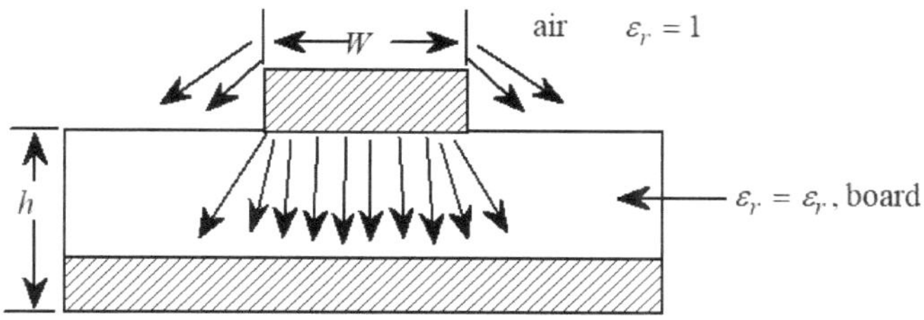

Figure 2.3 Microstrip.

So for wide lines, the fields are almost all in the board, while narrower lines will have proportionally more field energy in air.

Wide lines:

$$v \cong \frac{c}{\sqrt{\varepsilon_r}}. \tag{2.1}$$

For lines of noninfinite width (*w/H*) $v \neq \frac{c}{\sqrt{\varepsilon_r}}$. This leads to the idea of an *effective dielectric constant*

$$\varepsilon_{r,eff} = c^2/v^2, \tag{2.2}$$

so that

$$v = c/\sqrt{\varepsilon_{r,eff}}. \tag{2.3}$$

For microstrip, a typical formula given to approximately estimate the effective dielectric constant:

$$\varepsilon_{r,eff} \cong \frac{\varepsilon_r + 1}{2} - \frac{\varepsilon_r - 1}{2}\left(1 + 12\frac{h}{w}\right)^{-\frac{1}{2}} \tag{2.4}$$

where w = the line width and h = substrate thickness.

Advice: Ignore the above formula and use LINECALC (part of ADS). See the Appendix 2.1 at the end of this chapter.

2.3 Some Typical Material Parameters

Line parameters are for microstrip lines with $W = h$ (Table 2.1).

Table 2.1 Material parameters related to microstrip lines. Phase velocity is relative to *c*, the speed of light.

Dielectric	ε_r	$\varepsilon_{r,eff}$	v_{phase}	Z_0
Duroid	2.2	1.77	0.75*c*	94.3Ω
Epoxy-glass	4.8	3.4	0.54*c*	68Ω
Alumina (Al_2O_3)	9.8	6.5	0.39*c*	49Ω
GaAs	13	8.9	0.34*c*	43Ω
Quartz (SiO_2)	3.8	2.89	0.58*c*	73Ω

Duroid is available with a variety of dielectric constants. Sapphire (crystalline Al_2O_3) is sometimes used, but is anisotropic. SiO_2 is used both in its crystalline form (quartz) and in amorphous form (fused silica). Epoxy glass is often a default PCB material for less critical circuits. Alumina has good uniformity, but has a high effective dielectric constant which leads to narrower lines.

2.4 Voltage and Current on Transmission Lines

Transmission lines can be described from an EM viewpoint or equivalent circuit. We will use the circuit point of view since it will tie into later applications more easily (Figure 2.4).

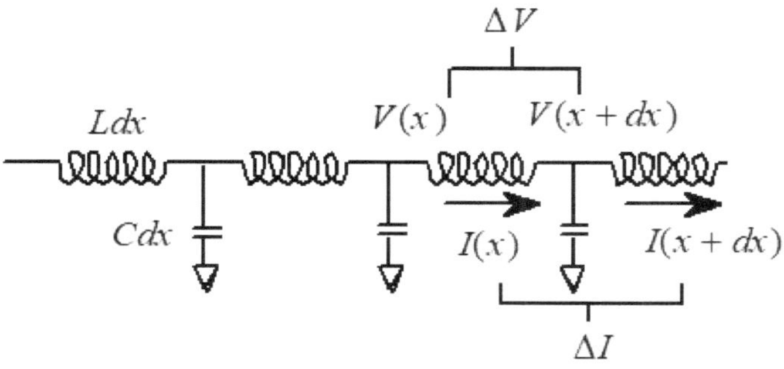

Figure 2.4 Lumped element model of a transmission line.

From Kirchoff's voltage law (KVL), the ideal lossless line can be represented by Equation (2.5)

$$-V(x+dx,t)+V(x,t)-Ldx\frac{\partial I}{\partial t} = 0 \quad (KVL) \tag{2.5}$$

C = capacitance/length
L = inductance/length

$$\left.\begin{aligned}\frac{\partial V}{\partial x} &= -L\frac{\partial I}{\partial t}\\ \frac{\partial I}{\partial x} &= -C\frac{\partial V}{\partial t}\end{aligned}\right\} \text{“telegrapher's equations”} \tag{2.6}$$

Using Kirchoff's current law, KCL:

$$I(x,t)-I(x+dx,t)-C\frac{\partial V}{\partial t} = 0 \quad (KCL). \tag{2.7}$$

Then, combining Equations (2.5)–(2.7), the wave equation (2.8) can be derived:

$$\frac{\partial^2 V}{\partial x^2} = LC\frac{\partial^2 V}{\partial t^2} \quad \text{wave equation.} \tag{2.8}$$

The wave equation has solutions of the form through superposition:

$$V(x,t) = V^{+}\left(t-{}^{x}\!/_{v}\right) + V^{-}\left(t+{}^{x}\!/_{v}\right) \tag{2.9}$$

$$I(x,t) = \left[V^{+}\left(t - {}^{x}/_{v}\right) - V^{-}\left(t + {}^{x}/_{v}\right)\right]/Z_O + I_{DC} \tag{2.10}$$

$$v = \frac{1}{\sqrt{LC}} = \text{velocity of propagation,} \tag{2.11}$$

$$Z_0 = \sqrt{L/C} = \text{characteristic impedance.} \tag{2.12}$$

The voltage at any point on the transmission line (Figure 2.5) is the superposition of the forward and reverse travelling waves:

$$V(x,t) = V^{+}(x,t) + V^{-}(x,t) \text{ voltage waves,} \tag{2.13}$$

where V^+ is the forward travelling wave and V^- the reverse wave.

$$I(x,t) = \frac{V^{+}(x,t) - V^{-}(x,t)}{Z_0} \quad \text{current.} \tag{2.14}$$

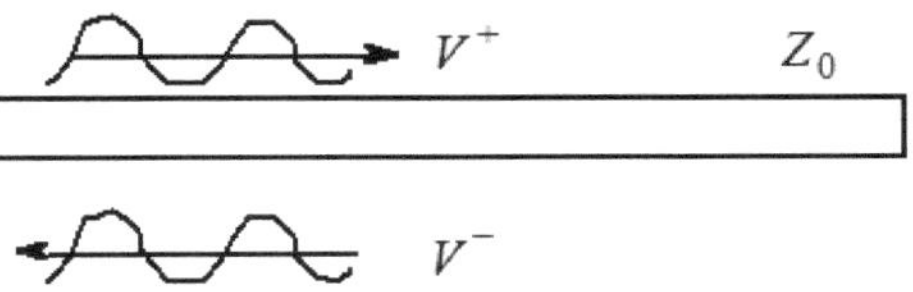

Figure 2.5 Forward and reverse voltage waves propagating on a transmission line.

2.5 Reflection Parameters

What happens if we connect an arbitrary resistive load to the line (Figure 2.6)?

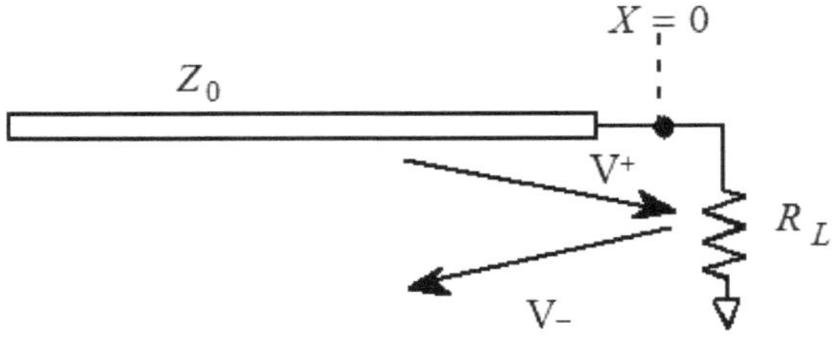

Figure 2.6 Termination R_L at $x = 0$.

The ratio of voltage/current = Z_0 on the line. At the load end at $x = 0$, it must equal R_L.

Thus,

$$\frac{V}{I} = R_L = \frac{V^+ + V^-}{V^+ - V^-} Z_0 \text{ at } x = 0. \tag{2.15}$$

If we define a *reflection coefficient* as:

$$\Gamma_L = \frac{V^-}{V^+} \text{ at } x = 0, \tag{2.16}$$

then,

$$R_L = \frac{(1 + \Gamma_L) Z_0}{1 - \Gamma_L}, \tag{2.17}$$

and

$$\Gamma_L = \frac{\left({}^{R_L}/_{Z_0}\right) - 1}{\left({}^{R_L}/_{Z_0}\right) + 1} = \frac{R_L - Z_0}{R_L + Z_0}. \tag{2.18}$$

We see that Γ varies between -1 and +1 for $R_L \geq 0$. If $R_L < 0$, then $|\Gamma| > 1$. This is a negative resistance and still can be described by Γ. The use of a reflection coefficient is critical to understanding the Smith chart.

2.6 Examples

1. **Open circuit**. $R_L = \infty,\ \Gamma_L = 1$

$$V^+ = V^-.$$

The reflected voltage wave has the same amplitude as the forward wave. Summing up the total voltage at $x = 0$,

$$V(x = 0, t) = V^+ + V^- = 2V^+. \tag{2.19}$$

Voltage is doubled at open circuit!
Therefore, the current must be equal to 0 because of the infinite load resistance.

$$I(x = 0, t) = \frac{V^+ - V^-}{Z_0}. \tag{2.20}$$

2. **Short circuit**. $R_L = 0,\ \Gamma_L = -1$

$$V^- = -V^+.$$

The reflected wave is the negative of the forward wave because the voltage at the short circuit must be zero. Therefore, the current at the short is doubled.

$$V\ (x = 0, t) = 0 \text{ short circuit.}$$

$$I(x = 0, t) = 2V^{+}/Z_0 \textit{ current is doubled.} \tag{2.21}$$

3. **Matched load.** $R_L = Z_0,\ \Gamma_L = 0$
   In a matched load there is no reflected voltage or current. The incident wave is absorbed into the resistor R_L.

2.7 Return Loss

Reflection coefficients can also be represented as a return loss:

$$\text{Return loss} = -20 \log |\Gamma|. \tag{2.22}$$

Notice that this does not provide information about phase. A matched load has a return loss = ∞ (no reflected power). A short or open has return loss = 0 dB; everything reflected.

2.8 What About the Source End?

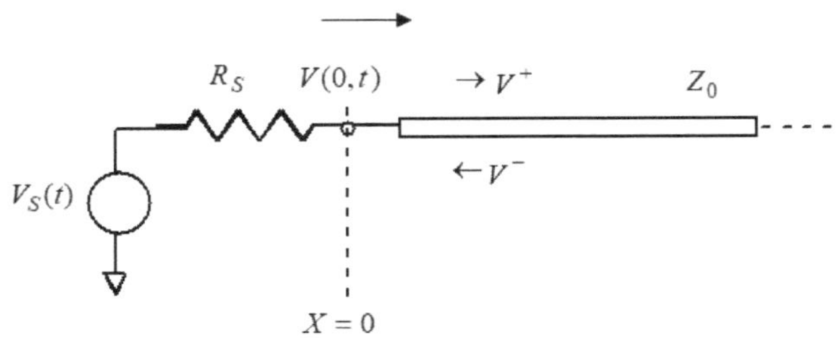

Figure 2.7 The source resistance R_s at the input end of the line affects the voltage and current launched on the line.

$$V(0) = V_s - I(0)R_s = V^{+} + V^{-} \tag{2.23}$$

$$I(0) = (V^{+} - V^{-})/Z_0. \tag{2.24}$$

To define a *source reflection coefficient*, Γ_S, use the Z_0 of the line on which the incident wave is traveling and the R_S of the generator.

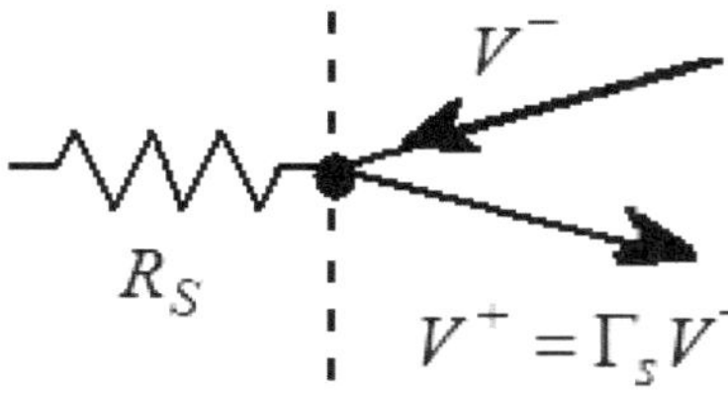

Figure 2.8 Reflected wave impinges on the generator.

$$V^+(0,t) = \Gamma_S V(0,t), \tag{2.25}$$

where

$$\Gamma_S = \frac{(R_S/Z_0) - 1}{(R_S/Z_0) + 1}. \tag{2.26}$$

Solve for V^+:

$$V^+(0,t) = \Gamma_S V^-(0,t) + V_S(t)T_S, \tag{2.27}$$

where

$$T_S = \text{source transmission coefficient} = \frac{Z_0}{Z_0 + R_S}(\text{voltage divider}). \tag{2.28}$$

Therefore, if a line is terminated with $R_L = Z_0, V^- = 0$ and so V^+ is the total voltage. But, if V^- is finite, it will affect the total voltage at that node and you must account for the source reflection coefficient. At some later time, a second reflected wave will return to $x = 0$, and the voltage at that node will change accordingly.

2.9 Transmission Parameters

Transmission through an interface or device

Transmission coefficient:

$$T = \frac{V_{transmitted}}{V_{incident}} = \tau \angle \phi. \tag{2.29}$$

Insertion loss:

$$IL\,(dB) \;=\; -\,20\,\log_{10}\;|T|\,. \tag{2.30}$$

Transmission gain:

$$\text{Gain}\,(dB) \;=\; 20\,\log_{10}\;|T|\,. \tag{2.31}$$

Also,

$$T = 1 + \Gamma. \tag{2.32}$$

2.10 Transmission Line Junction

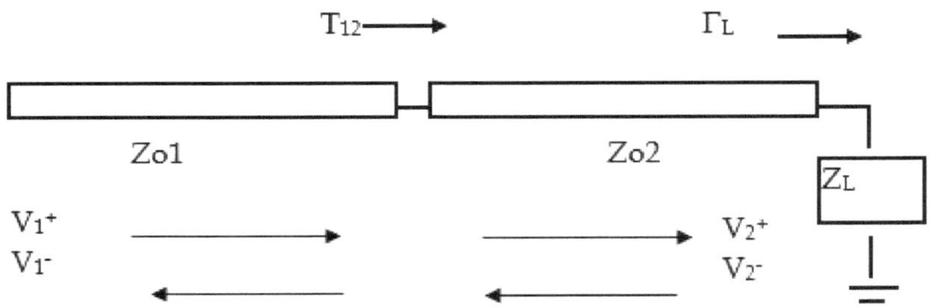

Figure 2.9 Determine the effect of a junction between two transmission lines.

Let's apply the transmission coefficient to a junction between two transmission lines with different characteristic impedances, Zo1 and Zo2.

$$V_2{}^{+} = T_{12}V_1{}^{+} + \Gamma_{21}V_2^{-}, \tag{2.33}$$

where T_{12} = transmission coefficient between Zo1 and Zo2

$$= 1 + \Gamma_{12} = 1 + [(Zo2/Zo1) - 1]/[(Zo2/Zo1) + 1] = 2Zo2/(Zo1 + Zo2) \tag{2.34}$$

and

$$\Gamma_{21} = -\Gamma_{12}, \tag{2.35}$$

$$V_2{}^{-} = V_2{}^{+}\Gamma_L. \tag{2.36}$$

As before, the reflected voltage at the end of line 2 depends on the termination R_L relative to Zo2.

ADS also can be used to simulate time-domain behavior of circuits including transmission lines. Appendix 2.2 presents a brief introduction to using this tool.

2.11 Transmission line laws

1. Source and load impedances should be equal to the characteristic impedance of the line if reflections are to be avoided.
2. Think about the voltages on transmission line conductors before connecting them.
3. Think about the currents on transmission line conductors before connecting them.

Example 2.1: T-junction in coplanar waveguide.
Circuit diagram:

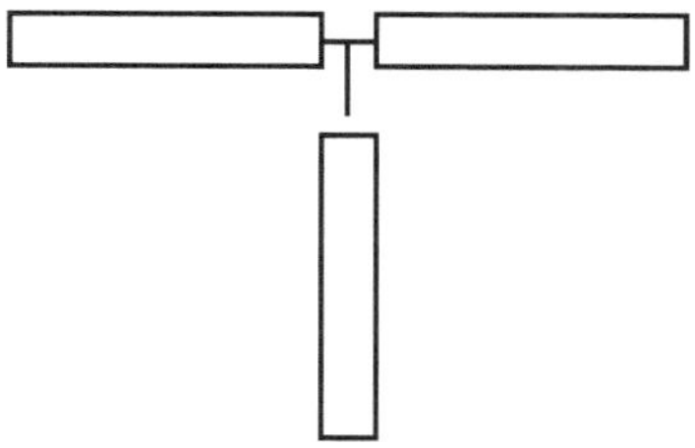

Figure 2.10 A T-junction.

Incorrect implementation:

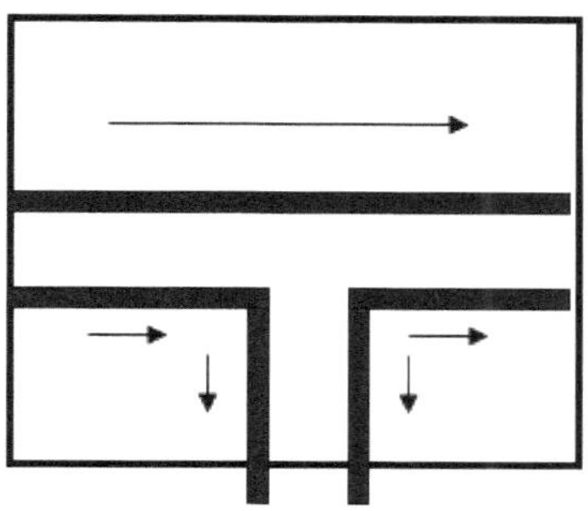

Figure 2.11 T-junction *interrupts* the current in the ground plane!

Correct implementation:

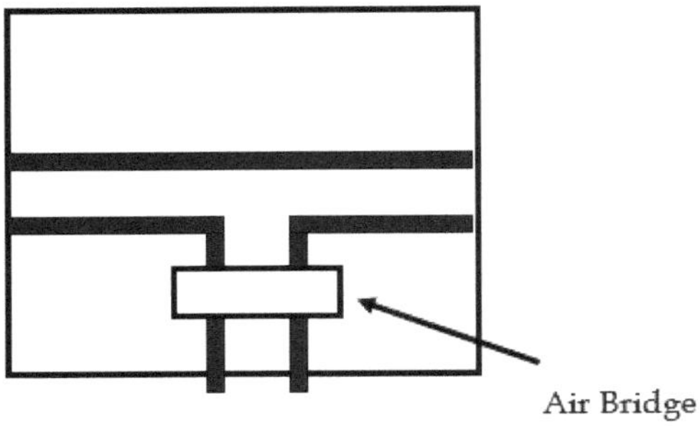

Figure 2.12 The air bridge connection between ground restores the current path.

Example 2.2: Junction between balanced and unbalanced lines.

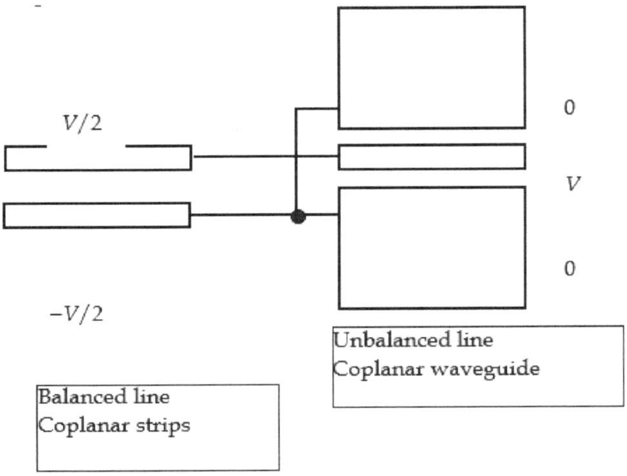

Figure 2.13 Transition between balanced and unbalanced lines. A "balun" is required.

This is a problem because the balanced line has potentials ($+V/2$ and $-V/2$) on both conductors while the coplanar waveguide has zero potential on the ground conductors.

A "balun" is needed to make this junction (more on this later).

2.12 Lumped-element Equivalent Circuit

Let's go back to the lumped-element equivalent circuit for a transmission line

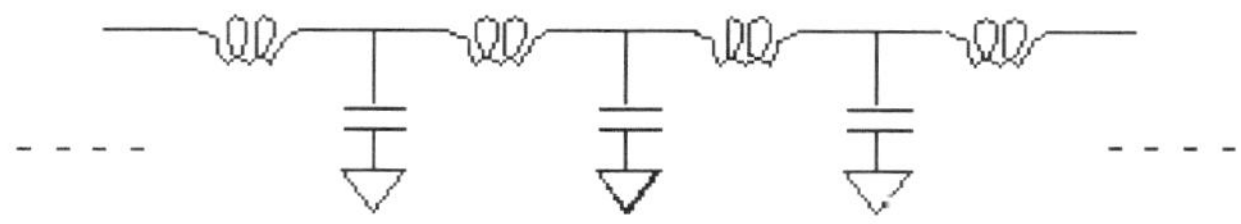

$$Z_0 = \sqrt{L/C} \qquad \text{characteristic impedance} \tag{2.37}$$

$$v = \frac{1}{\sqrt{LC}} \qquad \text{phase velocity.} \tag{2.38}$$

If the line is of length l:

$$L_T = \text{total inductance} = \frac{Z_0 l}{v} = Z_0 T \tag{2.39}$$

where

$$\frac{l}{v} = T \text{ is the propagation time delay on the line} \tag{2.40}$$

and

$$C_T = \frac{l}{Z_0 v} = \frac{T}{Z_0}. \tag{2.41}$$

So, a given line of length l can be modeled by a T or Π section or a series of these sections (Figure 2.14).

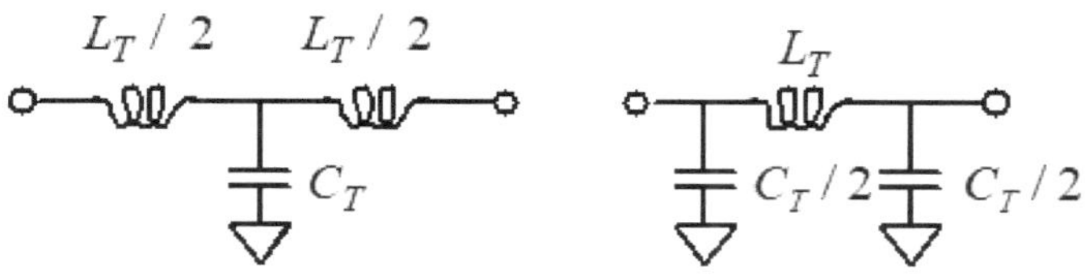

Figure 2.14 T or Π section equivalent networks.

Physical length is not relevant. Impedance and delay time describe electrical behavior.

If the line is short

$$l \ll \lambda$$

or $T \ll t_r, t_f$ (the reverse and forward propagation delay on the line) then the short line section can be used to represent a lumped element:

$$\text{high } Z_0 : L_T = Z_0 T \text{ big}; \quad C_T = \frac{T}{Z_0} \text{ small INDUCTOR,} \tag{2.42}$$

$$\text{low } Z_0 : L_T \text{ small;} \quad C_T \text{ large CAPACITOR.} \tag{2.43}$$

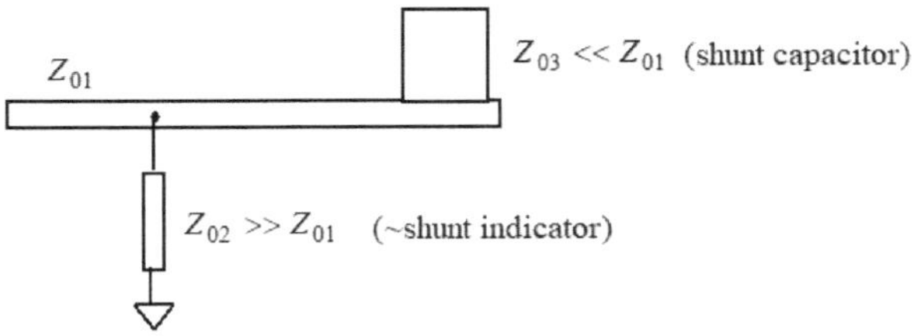

Figure 2.15 This is routinely used at microwave frequencies to synthesize inductors or capacitors that would be otherwise very hard to fabricate.

This property also presents a challenge when you are laying out an RF circuit on a PC board.

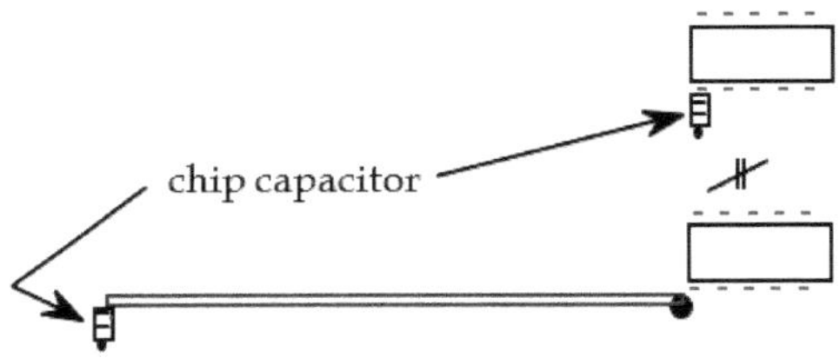

Figure 2.16 A lengthy conductor adds inductance. This will make the chip capacitor self-resonance drop to a much lower frequency and can make it ineffective as a bypass capacitor. Therefore the lumped capacitor should be located as close as possible to the point where it must provide a wideband AC ground.

2.13 Summary

1. A transmission line can be implemented in many different physical forms; whether it is balanced or unbalanced depends on its relationship to ground.
2. A transmission line possesses a characteristic impedance, Z_0, an effective dielectric coefficient, ε_{eff}, and a transmission delay depending upon its velocity of propagation and length.
3. Both forward and reverse travelling waves can exist by superposition. They depend on the source impedance, R_s, and the load termination, R_L.
4. The reflected wave can be eliminated when $R_L = Z_0$.

5. Transmission lines can be used to implement inductors and capacitors. This is a more effective method than lumped components when frequencies are high.

Acknowledgement

A grateful acknowledgement to Professor Mark Rodwell, whose lecture notes were very helpful to me in preparing this chapter.

References

[1] G. Gonzalez, Microwave Transistor Amplifiers, analysis and design. Chapter 1. Second Ed. Prentice-Hall, 1984.
[2] For a more mathematical and in-depth explanation of transmission line theory, see, D. Pozar, Microwave Engineering, chapter 2. Third Ed., Wiley, 2005.

Homework

READ: Chapters 1 and 2. ADS Tutorial in Chapter 1. LineCalc Appendix.

1. Plot the source voltage V1 and load voltage VL as a function of time for the network below.

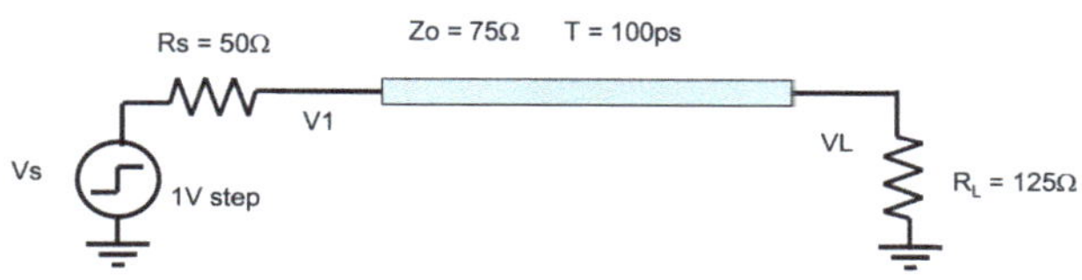

2. Trace the propagation of a 2 V step function from source to load for 3 round-trip reflections. Plot the source voltage and load voltage vs. time.

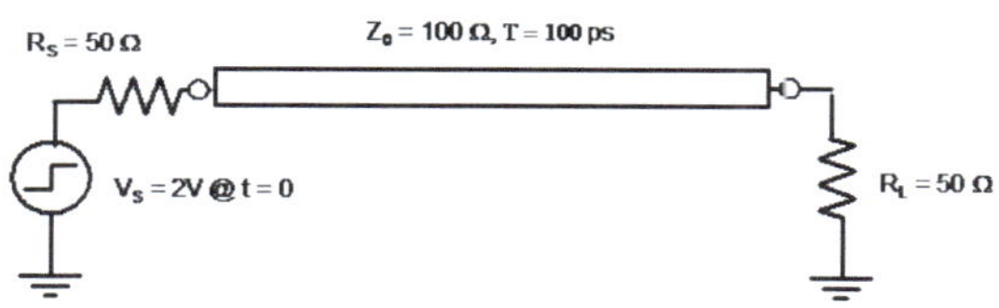

3. Read the brief appendix on LineCalc, a transmission line design and analysis tool which is part of ADS. Design a 50 ohm microstrip transmission line on a fiberglass board substrate. The board has the following characteristics:

   Thickness of board= 62 mils — Relative dielectric constant = 4.5

   Conductivity of copper = 5.8 e7 S/m — Thickness of copper = 0.15 mils

Determine the conductor width and length needed to produce a 1/8 wavelength line at 1 GHz.

Appendix 2.1. Linecalc Transmission Line Analysis Tool

Pathwave/Keysight ADS includes an analysis and design tool for transmission lines of many structures. To use this tool, open up a schematic page on ADS, and add a transmission line and line parameters. In the case of this example, we are designing a microstrip line on a fiberglass substrate (FR-4). In the schematic window, select the *TLines-Microstrip* pull-down menu. The line MLIN is a single conductor line. Also, add MSUB, the substrate parameters as shown below. Typical FR-4 parameters have been entered.

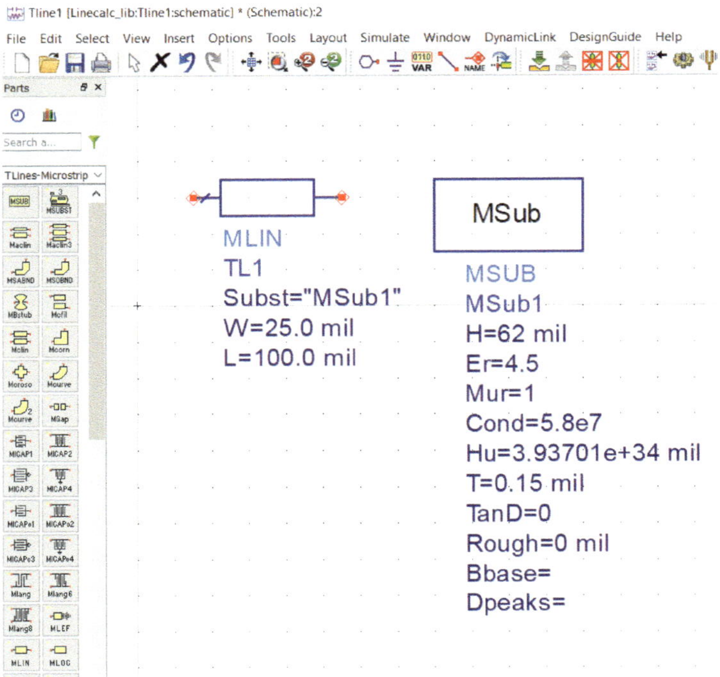

LineCalc can be enabled through the Tools Icon in the schematic window. Tools > LineCalc > Start LineCalc

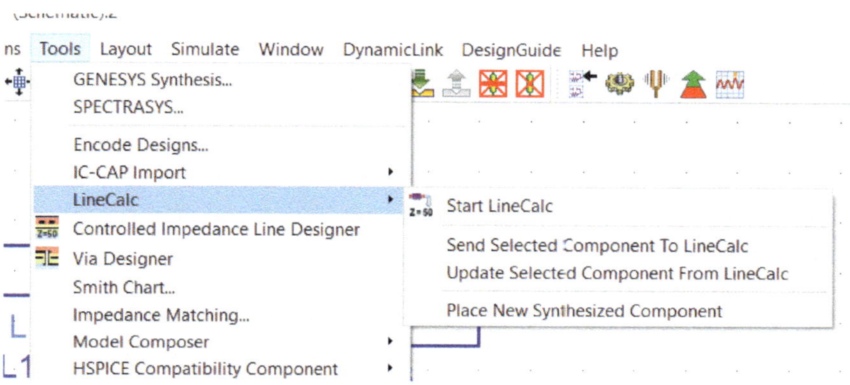

When LineCalc starts up, it imports the parameters from MSub.

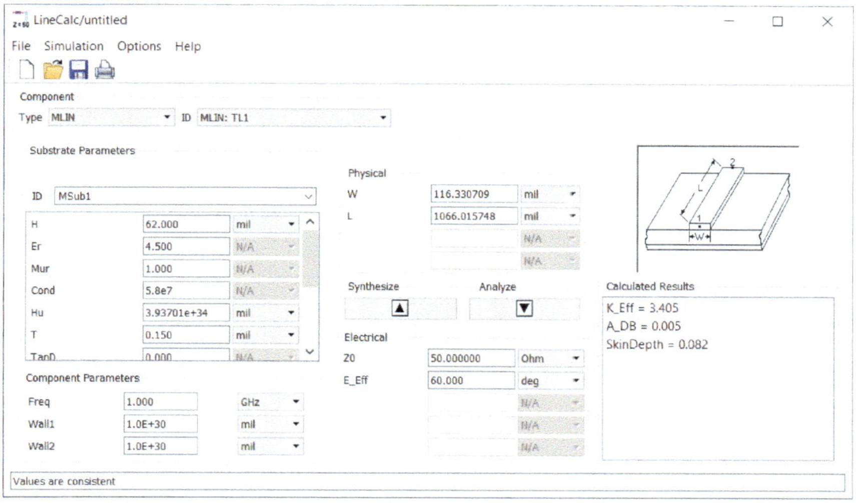

In this example, we start from the electrical specifications (50 ohms, 60 degree length). To determine the dimensions, W and L, click on Synthesize up arrow. Or, if we have specific W and L already, then Analyze will calculate the electrical specifications of that line dimension on the substrate chosen.

Appendix 2.2. Transient Analysis in ADS

Chapter 1 gave some examples of how to use S-parameter or AC simulation modes in ADS. These are inherently linear analysis modes and they do not provide any information on nonlinearities of circuits. Transient analysis is a time-domain simulation tool, reminiscent of SPICE. Driving a circuit with step, pulse or impulse functions can reveal information about settling time or even harmonic generation. However, in the case where there are harmonics and/or closely-spaced frequencies, it is very time and memory consuming since the minimum time step must be compatible with the highest frequency present while the simulation must be run for long enough to observe one full period of the lowest frequency present. Another more efficient method, harmonic balance, is generally used for nonlinear analysis of most radio frequency applications, especially when multiple frequencies are expected in the response. Harmonic balance (HB) is described in Chapter 8.

Because Chapter 2 deals with the time-domain response of transmission line circuits, a brief introduction to how to use ADS transient analysis will be presented. An example of multiple reflections from a poorly terminated line will be used for this purpose.

Figure 2.17 presents an ADS schematic of this poorly terminated line. The transient simulation setup is found in the simulation-transient menu palette. Our example uses the simplest setup. Other sweeps and options can be invoked for more detailed simulation requirements.

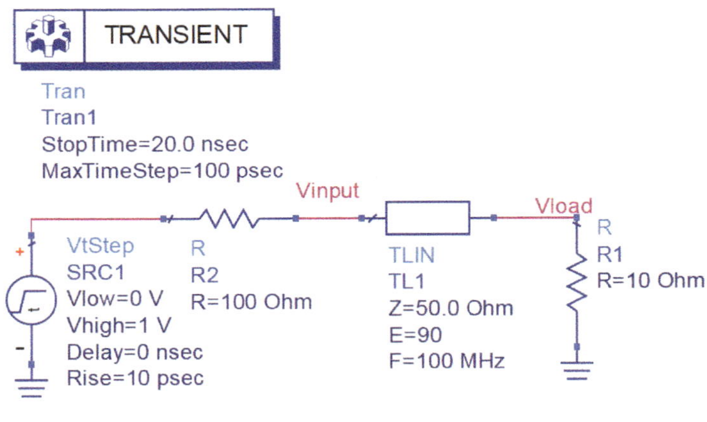

Figure 2.17 Schematic of a transient simulation in ADS.

Two parameters, StopTime and MaxTimeStep must be provided. StopTime is the length of time the simulation is allowed to run. MaxTimeStep selects the maximum allowed time step for the numerical integration. The simulator can select smaller time steps than this in order to achieve convergence when needed. In our example, a very short maximum time step, 100 ps, was required in order to accurately see the times of departure and arrival at the input and output of the line. Otherwise, the simulator automatically defaults to the maximum in order to reduce simulation time.

We can show (Chapter 4) that the delay time T on the line should be given by:

$$E = 360T f_{ref}. \tag{2A.1}$$

Because the ideal transmission line (Tlines-ideal palette) is specified with an electrical length, E, of 90° and a reference frequency of 100 MHz, the delay time T should be 2.5 ns.

Figure 2.17 also shows a voltage source, V_{tstep}, that can be selected from the sources-time domain palette. The step voltage range from V_{low} to V_{high}, delay time, and risetime are entered as parameters. In our example, a very short risetime was used to provide better resolution of the propagating signals on the transmission line.

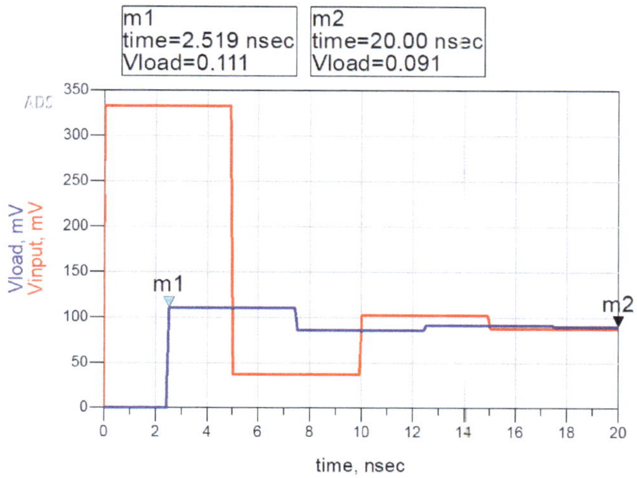

Figure 2.18 Output of the transient simulation.

Figure 2.18 shows the input voltage (red) and the load voltage (blue). The marker m1 shows that the expected transit delay of 2.5 ns was seen. The voltage on the input, 0.33 V is expected due to the voltage divider relationship between the 100 ohm source and the 50 ohm line. You can clearly see the back and forth reflections on the line. After 20 ns it has settled on a final steady state. The source and load reflection coefficients are given in Equations (2.18) and (2.26).

Exercise: Verify the various voltage steps at V_{input} and V_{load} shown in Figure 2A.2 using reflection coefficients.

3

Transmission Line Analysis in the Frequency Domain

Goals:

1. Define reflection and transmission coefficients.
2. Calculate propagation of voltages and currents on ideal transmission lines.
3. Define a reference plane.
4. Calculate the position dependence – phase constant, impedance and admittance.
5. Calculate the standing wave ratio.
6. Learn to use the Smith chart to transform impedances.
7. Learn to design and measure transmission lines in order to meet goals.

This chapter continues the discussion of transmission lines that began in Chapter 2. The representation and function of transmission lines in the frequency domain will be presented. Frequency domain analysis provides a clear description of how the phase of a sine wave is related to the length and propagation velocity of a line. The Smith chart will also be introduced as it is a widely utilized visual tool for design of circuits using either lumped elements or distributed elements.

Chapter 4 that follows introduces the important concept of S-parameters and how they are convenient for use with transmission lines.

3.1 Last Chapter: A Quick Review

Introduction to transmission lines:

1. Transmission lines are a linear system – superposition can be used.
2. Wave equation permits forward and reverse wave propagation on lines.

$$V(x,t) = V^+(t - x/v) + V^-(t + x/v)$$
$$I(x,t) = \frac{1}{Z_0}\left[V^+(t - x/v) + V^-(t + x/v)\right] + I_{DC} \tag{3.1}$$

Where phase velocity, v, and characteristic impedance Z_0 are:

$$v = \frac{1}{\sqrt{LC}}$$
$$Z_0 = \sqrt{L/C}$$
$$(L, C \text{ per unit length})$$

3. Reflections occur at discontinuities in impedance. A reflection coefficient Γ can be defined.

$$\Gamma = \frac{V^-}{V^+} = \frac{R_L/Z_0 - 1}{R_L/Z_0 + 1} \tag{3.2}$$

4. Transmission lines can be represented by an equivalent circuit (see Figure 3.1).

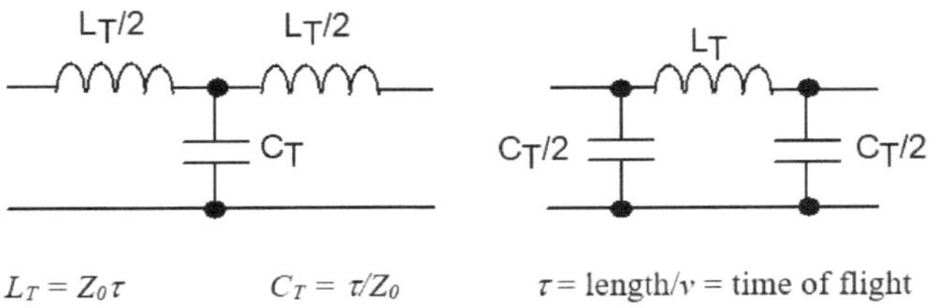

Figure 3.1 Lumped element equivalent circuit representing an ideal transmission line.

3.2 Transmission Lines in the Frequency Domain and the Smith Chart

Time domain analysis is intellectually clearer, the picture being forward and reverse waves propagating, reflecting, and re-reflecting. This analysis becomes intractable as soon as we introduce reactive impedances as multiple convolutions will be required for time domain reflection analysis.

So, we will analyze in the frequency domain instead. Frequency domain analysis of transmission lines is a classical approach to this problem (Figure 3.2).

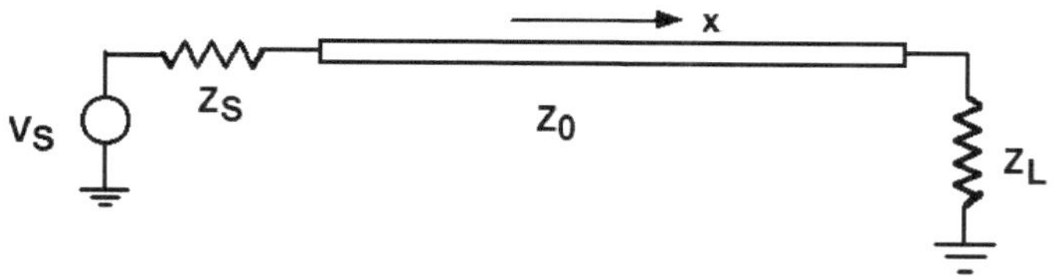

Figure 3.2 Transmission line with impedance Z_0 connects the source with Thevenin equivalent generator impedance Z_s to a load Z_L.

$$
\begin{aligned}
&V_s = \text{Re}\left\{V_o e^{j\omega t}\right\} \\
&V_o = |V_o|\, e^{j\phi_o} \\
&\text{so,} \\
&v_s(t) = |V_o| \cos\left(\omega t + \phi_o\right)
\end{aligned} \tag{3.3}
$$

On a transmission line at position x, waves travel in time as $x \pm vt$ where velocity v is the phase velocity. Equivalently, at a time t, waves vary with position x according to $t \pm x/v$.

Thus, we can represent $v_s(t)$ as:

$$
\begin{aligned}
&\cos\left(\omega t + \phi_o\right) \longrightarrow \cos\left[\omega(t \pm x/v) + \phi_o\right] \\
&= \cos\left[\omega t + \phi_o \pm x(\omega/v)\right] \\
&= \cos\left[\omega t + \phi_o \pm \beta x\right],
\end{aligned} \tag{3.4}
$$

where β is the *phase (or propagation) constant*

$$\beta = \omega/v = 2\pi/\lambda. \tag{3.5}$$

Here, λ is the wavelength. The sketches in Figure 3.3 illustrate the concept. Here, the function $\cos(\omega t - \beta x)$ is plotted.

1. Let's suppose that $t = t_0$ so that the wave appears frozen in time on the x axis. If distance $x_2 - x_1 = \lambda$ as shown, the corresponding phase change over this distance is 2π.

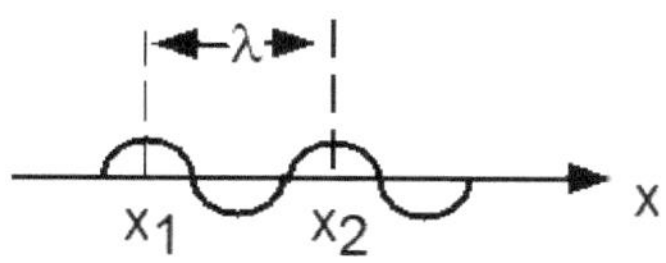

Figure 3.3A.

2. Now, if we plot this same cosine function as a function of βx, we see that the wavelength is equal to a phase $\beta x = 2\pi$ since $\lambda = 2\pi/\beta$.

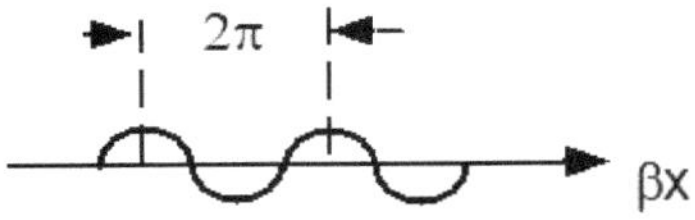

Figure 3.3B.

3. Also, we can set $x = 0$ and observe the wave function in the time domain for increasing time. In this case, one period requires a time interval $T = 2\pi/\omega = 1/f$ as shown.

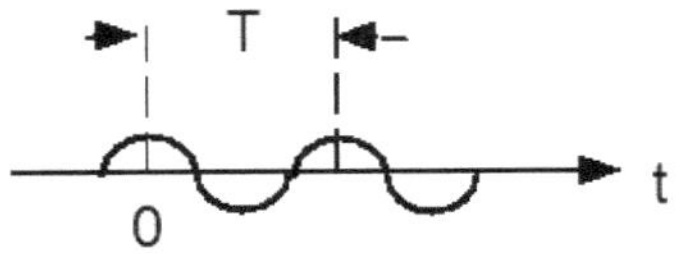

Figure 3.3C.

4. Finally, you might ask why the $\cos(\omega t - \beta x)$ wave is in the forward (to the right) travelling wave direction. To see why, track a point of constant phase with position on the x axis as time progresses from t_1 to t_2.

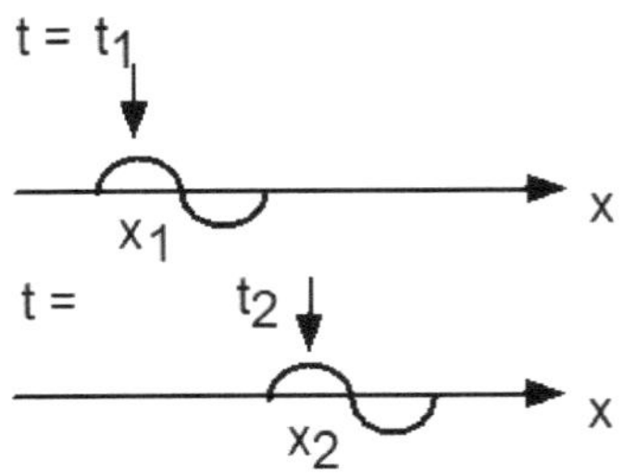

Figure 3.3D. Plot of wave propagation in the forward direction at times t_1 and t_2.

We can see that $t_2 > t_1$ and $x_2 > x_1$ for a forward direction of travel, and that the cosine function will have the same value at points of constant phase. Therefore,

$$\cos(\omega t_1 - \beta x_1) = \cos(\omega t_2 - \beta x_2) \tag{3.6}$$

Thus,

$$\omega t_1 - \omega t_2 = \beta x_1 - \beta x_2 < 0, \tag{3.7}$$

from the drawing. A forward wave must have a positive phase velocity.

$$\frac{\omega}{\beta} = v > 0. \tag{3.8}$$

From Equation (3.7),

$$v = \frac{x_2 - x_1}{t_2 - t_1} > 0. \tag{3.9}$$

Therefore, the wave is travelling in the forward direction for $\cos(\omega t - \beta x)$. Of course, these waves can also be described by complex exponentials:

$$\begin{array}{ccc} |V_o|\, e^{j\omega t} & e^{j\phi_o} & e^{\pm j\beta x} \\ \text{(sine wave)} & \text{(phase)} & \text{(position along line).} \end{array} \tag{3.10}$$

The $e^{j\omega t}$ time dependence is always taken as implicit and is frequently omitted when using phasor notation.

$$|V_O|\, e^{j\phi_o} e^{\pm j\beta x}. \tag{3.11}$$

3.3 Voltage and Current on Transmission Lines

Now, we can use this notation to describe the voltage and current on a transmission line at any location on the line.

$$V(x) = V^+(x) + V^-(x) = V^+(0)e^{-j\beta x} + V^-(0)e^{+j\beta x} \tag{3.12}$$

$$\begin{aligned} I(x) &= \frac{1}{Z_0}\left[V^+(x) - V^-(x)\right] \\ &= \frac{1}{Z_0}\left[V^+(0)e^{-j\beta x} - V^-(0)e^{+j\beta x}\right] \end{aligned} \tag{3.13}$$

3.4 Reflections in the Frequency Domain

In Figure 3.4 the forward wave is travelling in the positive x direction and reflects from the load Z_L. We set $x = 0$ at the load end of the transmission line as our reference plane. In frequency domain analysis, we assume that the wave amplitudes are steady state values.

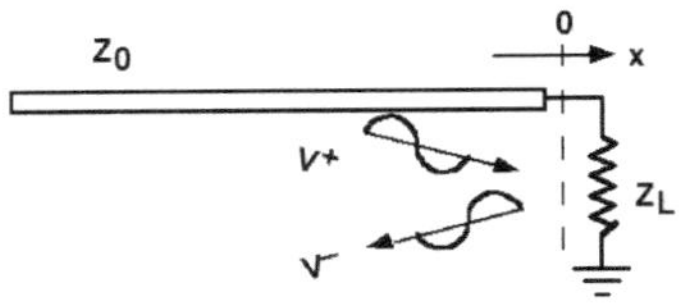

Figure 3.4 Reflection from a mismatched load impedance Z_L.

From the definition of reflection coefficient,

$$V^-(0) = V^+(0)\Gamma_L$$
$$\Gamma_L = \frac{z_L - 1}{z_L + 1}, \tag{3.14}$$

where z_L is the normalized load impedance Z_L/Z_0 . Γ_L is in general complex.

3.5 Movement of Reference Plane

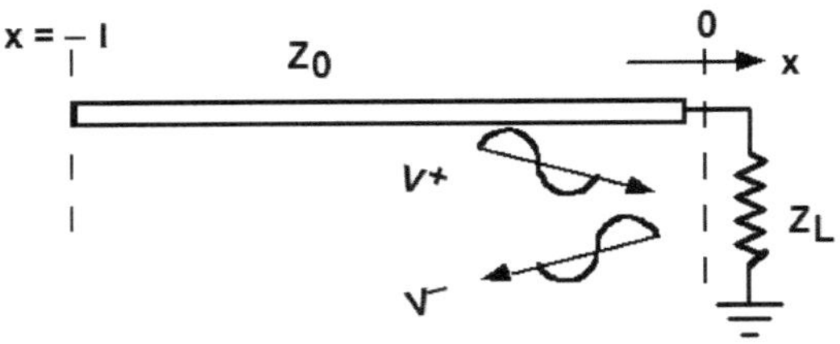

Figure 3.5 Voltage reflection at the load.

Now, we must determine voltage and current on the line as a function of position. This is often referred to as moving the reference plane. Here, we move from $x = 0$ at the load to $x = -l$ at the left end of Figure 3.5.

$$V(x) = V^+(x) + V^-(x) = V^+(x)[1 + \Gamma(x)] \tag{3.15}$$

where

$$\Gamma(x) = \frac{V^-(x)}{V^+(x)}$$

is the position-dependent reflection coefficient. Substituting for $V^-(x)$ and $V^+(x)$,

$$\Gamma(x) = \frac{V^-(0)e^{j\beta x}}{V^+(0)e^{-j\beta x}} = \Gamma(0)e^{2j\beta x} \tag{3.16}$$

and

$$V(x) = V^{+}(0)e^{-j\beta x}\left[1 + \Gamma(0)e^{2j\beta x}\right] \tag{3.17}$$

From this we can see that the reflection coefficient at position $-\ell$ from the load is given by

$$\Gamma(-l) = \Gamma(0)e^{-2j\beta l} \tag{3.18}$$

and the reflection coefficient goes through a phase shift of

minus $2\pi(2)(\ell/\lambda)$ radians
minus $2\beta\ell$ radians
minus $360(2)(\ell/\lambda)$ degrees.

Thus, the reflection coefficient changes phase with position.

3.6 Voltage Standing Wave Ratio (VSWR)

We have seen that, in general, there are two waves travelling in opposite directions on a transmission line. We also saw that the variation of voltage along the line at position x due to the sum of these two waves being given by:

$$V(x) = V^{+}(0)e^{-j\beta x}\left[1 + \Gamma(0)e^{2j\beta x}\right] \tag{3.19}$$

The magnitude is given by:

$$|V(x)| = \left|V^{+}(0)\right|\left|1 + \Gamma(0)e^{2j\beta x}\right| \tag{3.20}$$

So, we can easily determine the minimum and maximum voltage magnitude that will be found along the transmission line at some position x:

$$|V(x)|_{\max} = \left|V^{+}(0)\right|(1 + |\Gamma(0)|) \tag{3.21}$$

$$|V(x)|_{\min} = \left|V^{+}(0)\right|(1 - |\Gamma(0)|) \tag{3.22}$$

Taking the ratio of $\max$ to min gives us the VSWR:

$$VSWR = \frac{|V(x)_{\max}|}{|V(x)_{\min}|} = \frac{1 + \Gamma(0)|}{1 - |\Gamma(0)|} \tag{3.23}$$

An open or short-circuited line will give us an infinite VSWR because the minimum voltage on the line is zero; $|\Gamma(0)| = 1$ for both cases.

3.7 Impedance vs. Position

In the same way, the impedance at any point on the line can be found from the current and voltage or equivalently from the reflection coefficient:

$$Z(x) = \frac{V(x)}{I(x)} = \frac{[V^+(x) + V^-(x)]}{[(V^+(x) - V^-(x))/Z_0]} = Z_0 \frac{1 + \Gamma(x)}{1 - \Gamma(x)} \tag{3.24}$$

The normalized impedance at any point is easily found.

$$z(x) = Z(x)/Z_0 = \frac{1 + \Gamma(x)}{1 - \Gamma(x)} \tag{3.25}$$

So, impedance also depends on the position along the transmission line.

3.8 Introduction to the Smith Chart

While this is conceptually simple, there is a lot of math involved. This can become tedious. So, we could benefit from a graphical representation – this is called the *Smith chart* after Philip Smith who invented this convenient graph of transmission line parameters back in the 1930s [3].

The relationship for the normalized impedance, $z(x)$, is the key to the Smith chart. The Smith chart is just a polar plot of the reflection coefficient. Impedance is determined by

$$z = \frac{1 + \Gamma}{1 - \Gamma} \tag{3.26}$$

It is a one-to-one mapping between complex numbers Γ and z, and is in fact an analytic function and a conformal transformation. You can read about this in math books on complex analysis,

In the two-dimensional plane of Γ the Γ plane, a reflection coefficient $\Gamma$is represented by a point. Consider a plot of Γ in a unit circle (Figure 3.6).

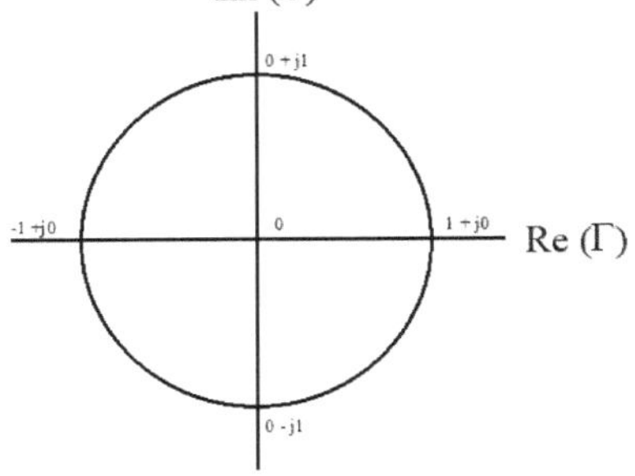

Figure 3.6 Plot of Γ on the complex plane.

As we move away from the load by a distance l on the transmission line, Γ rotates by an angle $\Delta\theta$ where:

$$\Delta\theta = -2\beta l(\text{ radians }) = -360^\circ \times 2(l/\lambda)(\text{degrees}). \tag{3.27}$$

Figure 3.7 illustrates that as we move away from the load by an angle $\Delta\theta$, Γ rotates in a clockwise direction. One whole rotation is required in the Γ plane for each half-wavelength movement on the line.

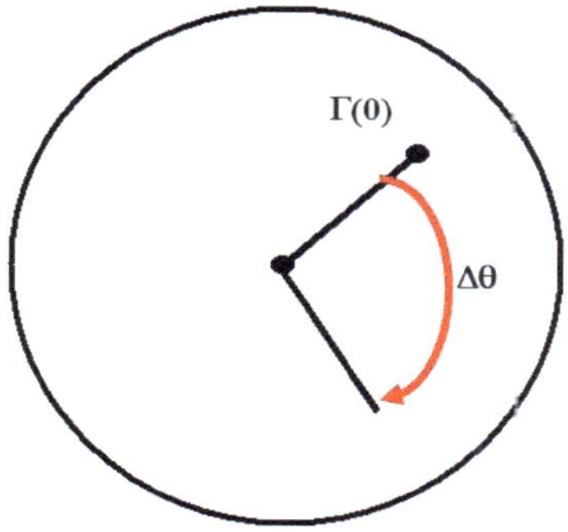

Figure 3.7 Phase shift $\Delta\theta$.

Note the following:

1. $\beta\ell$ is defined as "electrical length" in radians.
2. The degree scale on the edge of the Smith chart represents $\Delta\theta = 360 \times 2\beta\ell$, the angle of the reflection coefficient.

Lines of constant resistance can be plotted on the chart through the use of

$$z = \frac{1+\Gamma}{1-\Gamma} \tag{3.28}$$

These will take the form of circles whose centers are on a line across the center of the chart as shown in Figure 3.8 [1].

Note that the units of r are normalized to Z_0, in this case 50 ohms. So, the circle labeled $r = 0.5$ corresponds to the 25 ohm resistance circle in this case. However, Z_0 can be any convenient real value. If Z_0 were 100 ohms, $r = 0.5$ would represent 50 ohms.

Similarly, the reactances x can be represented by circles. These have their centers on the vertical axis at the right edge of the chart. These are also normalized to Z_0. Positive x corresponds to inductive reactance and is above the center line of the chart. Negative x represents capacitive reactance.

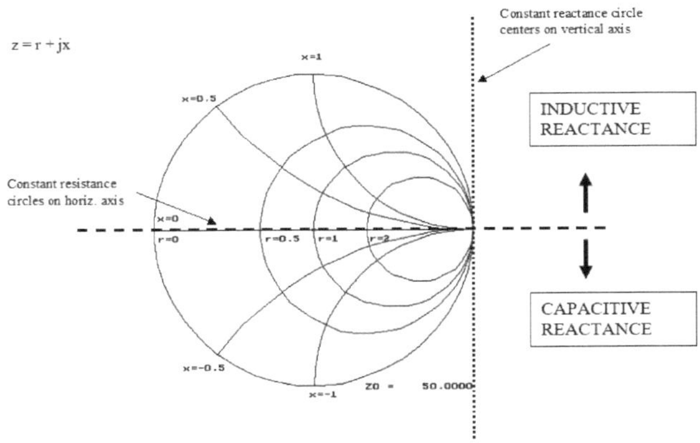

Figure 3.8 Normalized resistances r and reactances x are represented by circles [1] [4].

Given this mapping onto the Γ plane, we can associate any reflection coefficient (a point on the plane) with an impedance simply by reading the z coordinates of the point.

We can also associate the change of impedance with position with a rotation on the chart. Just rotate the Γ vector clockwise around the chart at the rate of one rotation for every half wavelength of movement on the line. Then read off the impedance directly from the chart.

Note that all points on this chart represent series equivalent impedances (Figure 3.9).

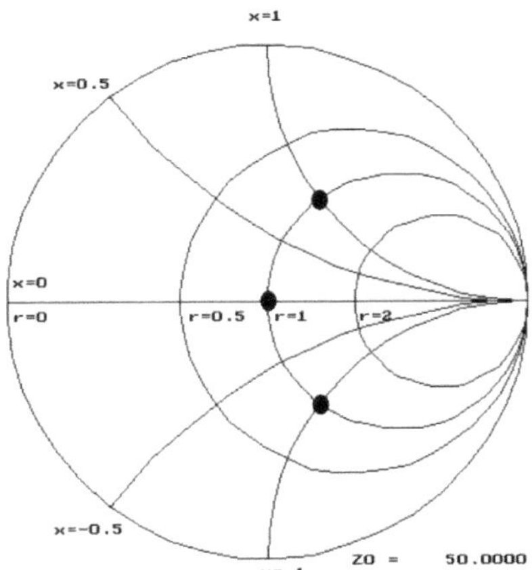

Figure 3.9 Example plots of normalized impedances on the Smith chart.

So, for example, $z = 1 + j1$ represents a series RL network. $z = 1 - j1$ is a series RC.

$z = 1$ is a point right in the center of the chart.

Normalization: Consider the point $z = 1 + j1$. If $Z_0 = 50$, then $Z = 50 + j50$ when the impedance is denormalized. If Z_0 were 1000 ohms, we would have $Z = 1000 + j1000$ for the same point. In this way, the chart can be used over an arbitrary range of impedance.

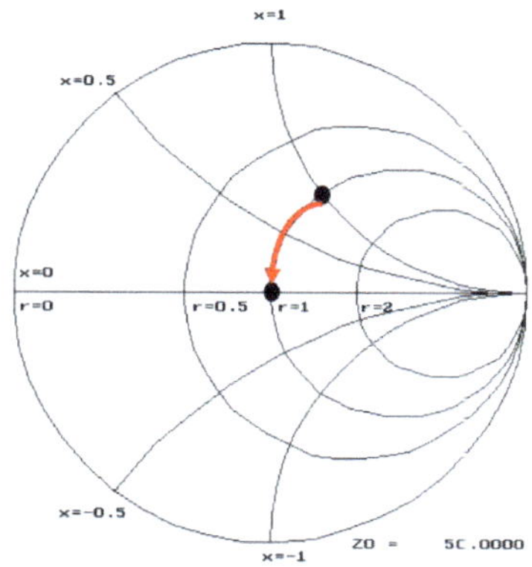

Figure 3.10 Capacitive reactance $-j1$ added to z.

On the chart in Figure 3.10, we see an impedance $1 + j1$ corresponding to the series RL network. If we add a capacitive reactance $-j1$ in series with this, the point will move along the constant resistance line $r = 1$ to the center. The reactance has been cancelled and the series circuit is self-resonant.

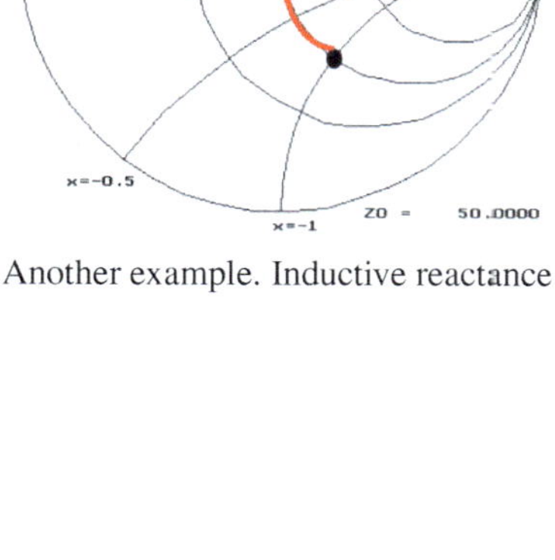

Figure 3.11 Another example. Inductive reactance $+j1$ added to z.

In the example given in Figure 3.11, we have started with a series RC network, $z = 1 - j1$. By adding inductive reactance $+j1$ in series, we can again move along the $r = 1$ circle up to the center of the chart.

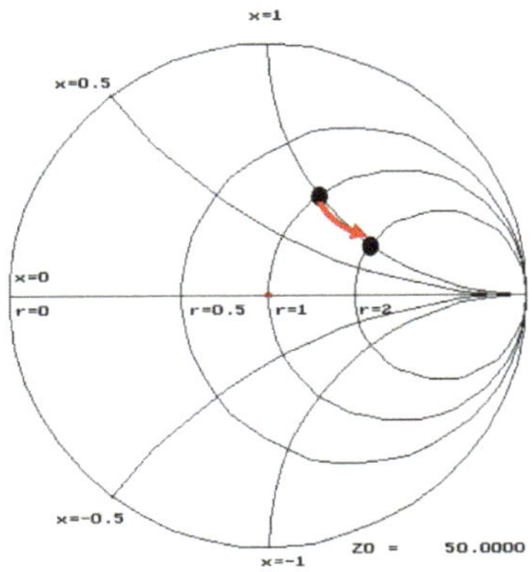

Figure 3.12 Added resistance r can also be plotted on the chart.

Series resistance. The chart can also represent the addition of resistance (Figure 3.12). This will correspond to movement along a circle of constant reactance. For example, if we begin with $z = 1 + j1$ again and add a normalized $r = 1$ to that, we arrive at $z = 2 + j1$.

3.9 Let's Look at Some Interesting Examples

3.9.1 Quarter wavelength transmission line

How will impedances be transformed by this line (Figure 3.13)?

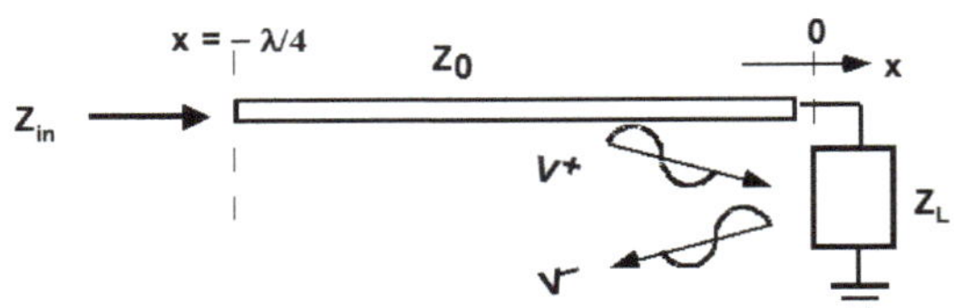

Figure 3.13 Quarter wavelength transmission line example.

$$\Gamma(x) = \Gamma(0)e^{2j\beta x}. \tag{3.29}$$

The reflection coefficient retains the same magnitude but changes phase with position. The variable x is defined as shown in Figure 3.13.

Let $x = -\lambda/4$. The propagation constant can be used to calculate the electrical length, βx :

$$\beta = \frac{2\pi}{\lambda} \text{ so } \quad \beta x = -\pi/2 \tag{3.30}$$

$$\Gamma(-\lambda/4) = \Gamma(0)e^{-j\pi}. \tag{3.31}$$

In other words, the angle of the reflection coefficient is rotated clockwise 180° for a quarter-wave line (negative angle = clockwise rotation).

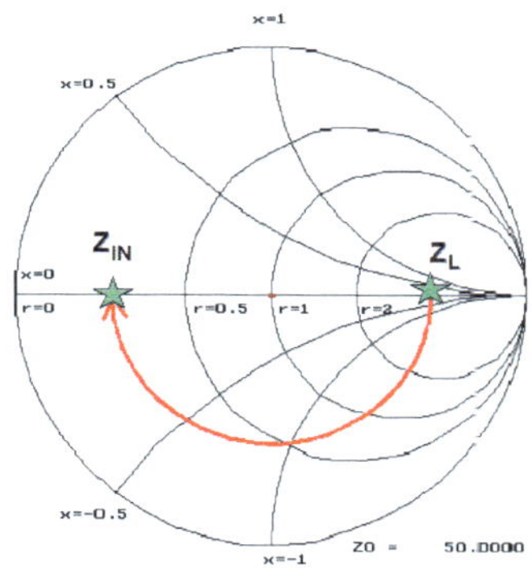

Figure 3.14 What does this do to the impedance?

$$\begin{aligned} Z(-\lambda/4) &= [1 + \Gamma(-\lambda/4)]/[1 - \Gamma(-\lambda/4)] \\ &= [1 + \Gamma(0)\angle 180]/[1 - \Gamma(0)\angle 180]. \end{aligned} \tag{3.32}$$

1. We rotate 180° on the Smith chart clockwise for the quarter wave line. This transforms a real Z_L into a real Z_{IN} in the example given in Figure 3.14.
   Example 3.1: $Z_L = Z_L/Z_0 = 200/50 = 4$. Rotating 180° takes us to $Z_{IN} = 0.25$. This is 12.5ohms when denormalized.
2. Or, using the above equation for normalized impedance,

$$\Gamma_L = \Gamma(0) = (z_L - 1)\,/\,(z_L + 1) = 3/5 \tag{3.33}$$

$$z_{IN} = (1 + 0.6\angle 180]/[1 - 0.6\angle 180] = 0.25.$$

3. You can also use the formula below, which is specific for the impedance transformation of a quarter-wave line. This obviously is good for either normalized or unnormalized impedances.

$$Z_{IN} = Z_0^2/Z_L \tag{3.34}$$

Clearly, the Smith chart is the easiest to use.

4. The Smith chart also can be used with complex loads (Figure 3.15).

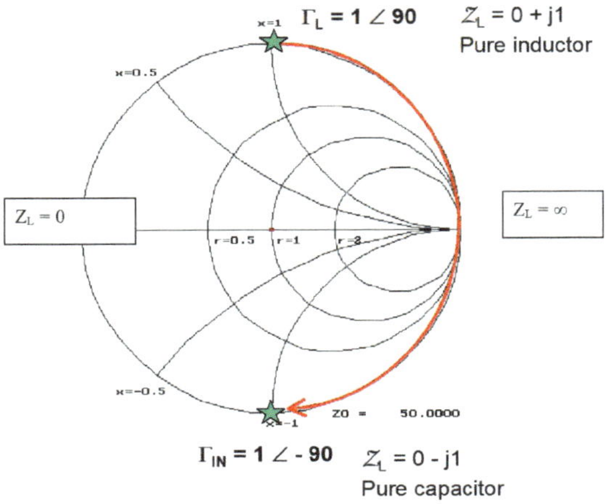

Figure 3.15 Quarter-wave line transforms an inductor into a capacitor.

One can also see how an eighth-wave line can be used to produce an inductive reactance $+j1$ when shorted at one end, and an open-circuited eighth-wave line produces a capacitive reactance $-j1$. And in Figure 3.15 we see that a quarter-wave line transforms an inductance into a capacitance.

So, an open circuited quarter wave line (Figure 3.16) produces a short circuit (when we ignore loss). Conversely, a shorted quarter wave line produces an open circuit.

Note that the electrical length of the line is a function of frequency, so these transformations will be exact only at the design frequency.

While the $\lambda/4$ line seems to do some magic as an impedance transformer, there are also applications in impedance matching networks, which we will be investigating soon for lines of other lengths.

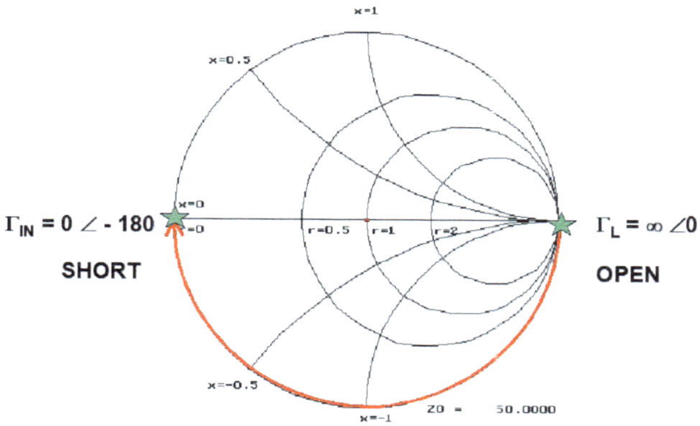

Figure 3.16 What about an open circuit at the load end?

On the Z_0 Smith chart, the angle of rotation through a transmission line of impedance Z_0 is always twice the electrical length: $-2\beta x$. Because the reference plane is defined at the load, and x is negative, the angle is positive (clockwise rotation) when moving away from the load toward the generator.

3.10 Admittance Chart

We can also use the Smith chart for admittances.

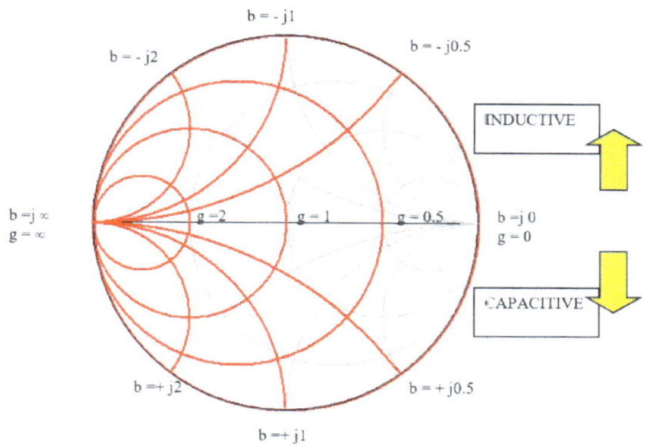

Figure 3.17 Admittance circles added to the Smith chart.

Impedance:

$$Z = R + jX. \tag{3.35}$$

Normalized impedance:

$$z = r + jx. \tag{3.36}$$

Admittance:

$$Y = 1/Z = G + jB. \tag{3.37}$$

Conductance + *j* susceptance
Normalized admittance:

$$y = g + jb, \tag{3.38}$$

where

$$y = \frac{Y}{Y_0}. \tag{3.39}$$

We can plot the values of both r and x, and g and b on the Γ plane by rotating the impedance coordinates by 180°.r and x then become constant conductance circles and constant susceptance circles. Since $Y = 1/Z$, taking $1/Z$ is equivalent to a 180° rotation in angle.

This is the impedance-admittance chart shown in Figure 3.17. The admittance coordinates are the bold red lines.

You can see from this that it is possible to use the chart to quickly translate between impedance and admittance. Every point on the chart can be interpreted in both ways.

The impedance Smith chart is convenient for evaluating the effect of adding components in series. The admittance chart is useful for components in parallel.

3.11 Examples

1. Use the *YZ* Smith chart to determine Z_{IN} of the circuit in Figure 3.18.

a. Normalize the reactances and admittances to Z_0 and Y_0. What would be a good choice for Z_0 in this case?
b. Starting from the resistor, add capacitive reactance –*j*1 to the impedance.
c. Using the admittance chart, add susceptance –*j*1.
d. Going back to impedance, add another –*j*1.

The result is Z_{IN} = 75 + *j*0 or normalized Z_{IN}/Z_0 = 1.

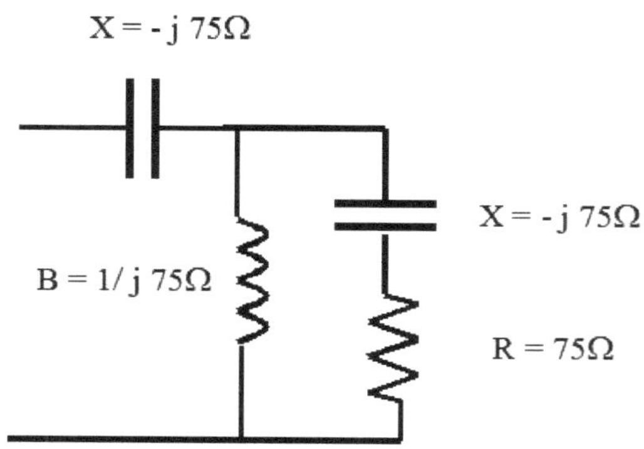

Figure 3.18 Use the Smith chart to determine Z_{IN}.

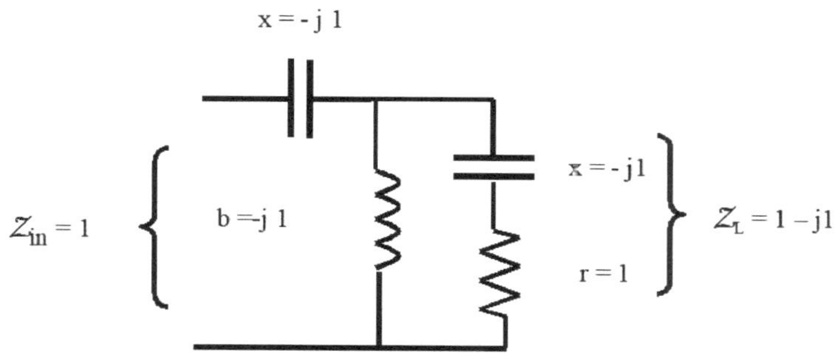

Figure 3.19 Example circuit with normalized impedances. Z_0 = 75 ohms.

3.12 An Important Observation

The path that we follow on the Smith chart will depend on the direction of the path through the circuit: load to source or source to load.

From our previous example, we proceeded from Z_L to Z_{IN} by starting with $Z_L = 1 - j1$, taking into account the shunt admittance of the inductor by moving up on a constant conductance circle to the unit resistance circle, then down to the center through the series capacitor.

What happens when we go the other way?

1. Start at the center (Z_{IN}). Go down in reactance $-j1$.
2. Go up in susceptance $-j1$
3. We are now at Z_L*.

This is just a graphical version of the statement: match the load with its complex conjugate impedance or admittance, see Figure 3.21.

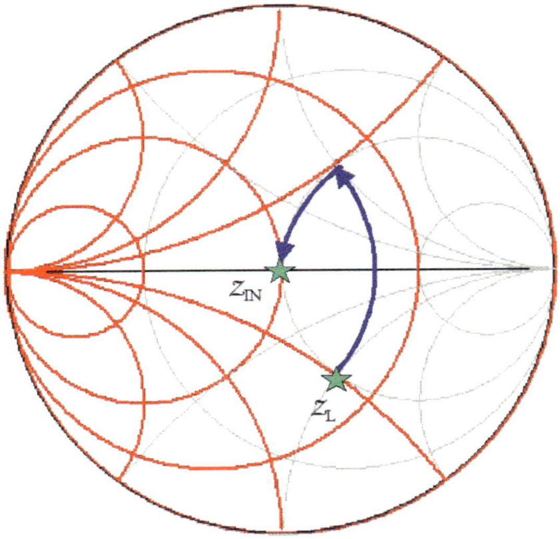

Figure 3.20 Use of the Smith chart to match load Z_L to Z_0.

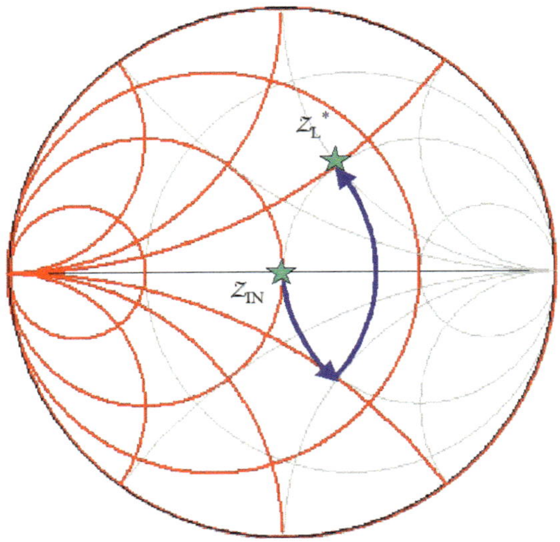

Figure 3.21 Traversing in the opposite direction produces the complex conjugate impedance. Starting from Z_{IN}, we reach Z_L*.

Beware: It is easy to get confused and design a network that will not match your load if you do not keep in mind this rule. For example, you could try to design a network that starts from $Z_{IN} = Z_0$ and stops at Z_L. This would not provide the desired match.

3.13 Measurement Exercises

3.13.1 Lab exercise

Here are some measurement exercises that complement the analysis and simulation methods described in this chapter. (You will need access to a vector network analyzer.)

Prototyping board. Circuits in this lab will be constructed on rectangular pieces of single sided printed circuit board, typically 2" wide. SMA connectors will provide measurement access using the launcher block shown in Chapter 1, Appendix 1.2. The board material will be 0.062" thick duroid with a dielectric constant of approximately 2.3. The bottom side, or ground plane, is covered with copper. Circuits are constructed on the top side by interconnecting components with strips of copper foil (Figure 3.22). This lab exercise explores the transmission line properties of these microstrip transmission lines.

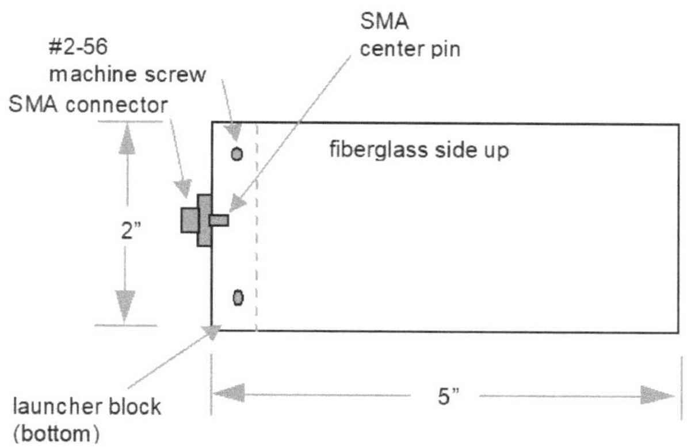

Figure 3.22 Prototyping board.

1. Use LINECALC (a transmission line simulation tool included with Pathwave/Keysight ADS. See the end of Chapter 2 for a description) to design a distributed transmission line stub that provides the following impedance at 500 MHz:

(i) Capacitive reactance of $-j60$ ohms.

Build the stub using copper tape on the 0.062" thick duroid board material. Use an Exacto knife and copper tape to construct a microstrip line with the appropriate length and width. Taper one end and solder to one of the SMA connectors. Examples of open and shorted stubs are shown in Figures 3.23 and 3.24.

Plot S_{11} on the Smith chart from 100 to 1000 MHz and use markers to display the impedance at 500 MHz and at the frequencies corresponding to $\lambda/8$ and $\lambda/4$.

2. Next, build a 50 ohm through line between connectors 1 and 2. Measure the magnitude and phase of S_{21} over the same frequency range.

Answer the following questions:

(i) Can you observe any loss on these lines?
(ii) Determine the phase velocity, effective dielectric constant, and Z_0 for each test structure from your measurements.
(iii) If there is disagreement between your design and measurement, propose an explanation. Perhaps simulation with an E/M tool such as Sonnet could lend some insight.

Figure 3.23 Example: microstrip open-circuited stub.

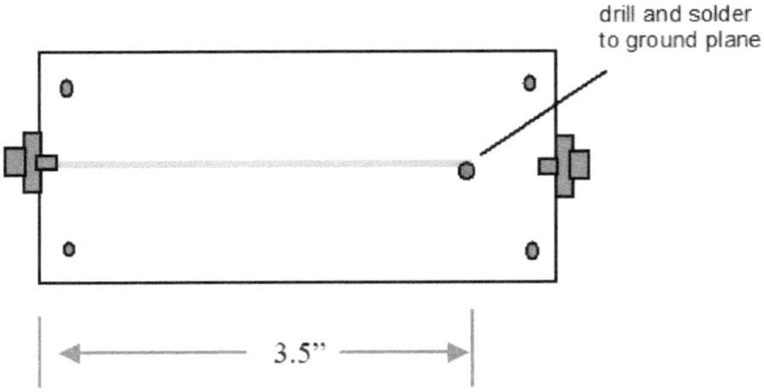

Figure 3.24 Example: microstrip shorted stub.

3.13.2 Quarter wave impedance transformer

Design, build and evaluate a quarter-wave series microstrip line impedance transformer to match a shunt 15 ohm chip resistor to 50 ohms at 450 MHz. Use the layout shown in Figure 3.24 to terminate the microstrip line. Measure and plot the magnitude and phase of S_{11} on rectangular plots with frequency swept from 50 to 500 MHz. Place markers at 450 MHz and at the actual quarter wave frequency. Also plot VSWR over the same frequency range. This can be done on ADS.

a. Over what frequency range are you able to keep the VSWR below 2.0?
b. Compare the simulation with measurement. If there is disagreement between your design and measurement, propose an explanation.

3.13.3 Two techniques for measuring Z_0 and v_p of a transmission line

1. **Open stub**. Use the layout shown in Figure 3.23 to create an open stub. Looking on the network analyzer at low frequencies, S_{11} should look like an open circuit, at the right real axis of the Smith chart, in this case at $\Gamma = 1\angle 0°$. Now, increase frequency until the phase changes by $90°$ ($180°$ clockwise rotation since the Smith chart angle corresponds to $2\beta l$). At this frequency, the line is one-quarter wavelength long.
The phase velocity can be determined from the physical length of the line. At this particular length l, and frequency $\omega, 2\beta l = \pi$. From the definition of $\beta = \omega / v_p$, the phase velocity can be obtained:

$$v_p = \frac{2\omega l}{\pi}$$

Z_0 can be obtained by measuring the reactive part of the impedance at half of the frequency above. At this $\lambda/8$ frequency, the impedance measured (neglecting losses imaginary part) should be $-jZ_0$ (open stub is recommended - lower unknown parasitics) because the impedance looking into an eighth-wave lossless open line is

$$jZ_{\text{in}} = -jZ_0 \cot(\pi/4).$$

2. **Through line.** When the frequency is set so that the phase of S_{21} is $180°$, the line is one-half wavelength long. From the physical length, you can

determine phase velocity. The magnitude of S_{21} gives the transmission loss.

At half that frequency, you have a quarter-wave line which will act as an impedance transformer, transforming the 50ohm termination into a measured impedance Z_{in}. The transmission line characteristic impedance, $Z_{line,}$ can be determined from the real part of S_{11} at that frequency. Given $Z_0 = 50$ ohms for the network analyzer ports, the impedance looking into the quarter wave line, Z_{in}, is

$$Z_{in} = Z_0 \frac{1 + \Gamma_{in}}{1 - \Gamma_{in}}$$

If the unknown line impedance Z_{line} is other than 50 ohms, $Z_{in} \neq 0$. Then,

$$Z_{line} = \sqrt{Z_0 Z_{in}}$$

3.14 Summary

1. We have shown how the length of the transmission line affects the reflection coefficient and impedance.
2. A propagation constant β was defined and a phasor-style analysis was used to calculate the effect of electrical length.
3. The Smith chart was explained as a mapping of impedance and admittance onto the complex plane. It allows analysis of a circuit at a single frequency without dealing with complicated math.

Acknowledgements

I am grateful to Professor Mark Rodwell, whose lecture notes were very helpful to me in preparing this chapter.

References

[1] Gonzalez, G. (1984) *Microwave Transistor Amplifiers, Analysis and Design*, 2nd edn. Prentice-Hall, ch. 2.

[2] For a more mathematical and in-depth explanation of transmission line theory, see, Pozar, D. (2005) *Microwave Engineering*, 3rd edn. Wiley, ch. 2.

[3] Smith, P.H. (1939) Transmission line calculator. *Electronics*. 12 Jan.

[4] Another description of the Smith chart can be found in: Hayward, W. (1994) *Introduction to Radio Frequency Design*. ARRL, ch. 4.

Homework

1. Use an analytical method to determine the electrical length of the transmission line in degrees.

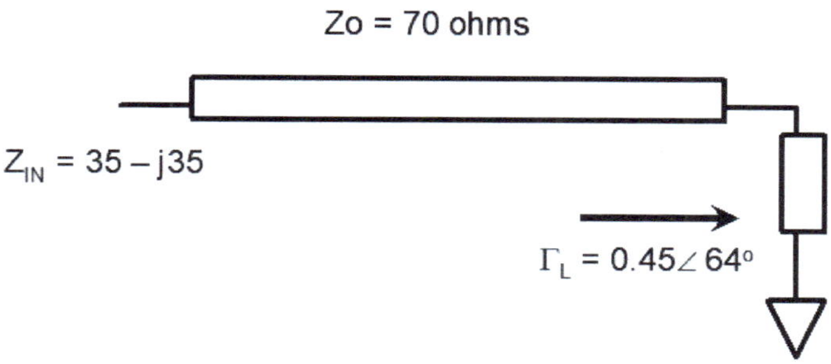

2. A component was measured over the frequency range of 10 MHz to 1 GHz. S11 (the reflection coefficient) is plotted on the figure below. Identify the component and determine values for the equivalent circuit representing that component.

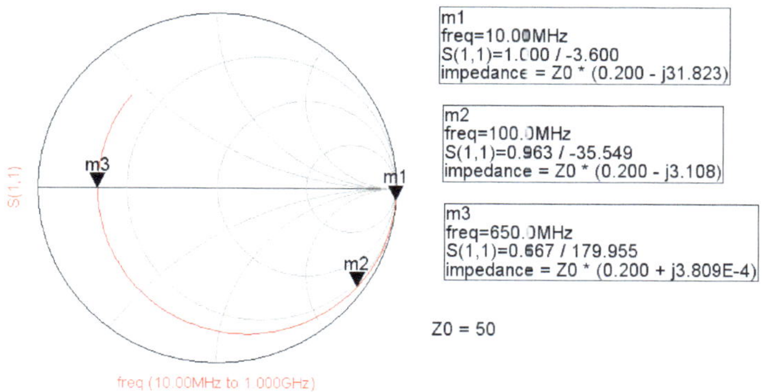

3. The plot below represents S11 (the input reflection coefficient) of an ideal 50 ohm transmission line. How is it connected at the load? Determine the electrical length in degrees. If the physical length is 5 cm, find the effective dielectric constant and the velocity of propagation.

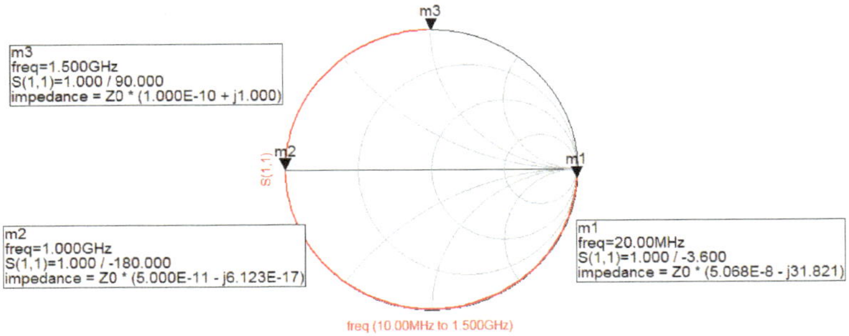

4. It is an interesting and useful fact that the load current I_L always has a magnitude equal to $V(-\lambda/4)/Z_0$ with a phase lag of 90° *independent of* Γ for a quarter wave transmission line. Prove it! (Hint: recall the relationship between position and voltages and currents on a transmission line.)

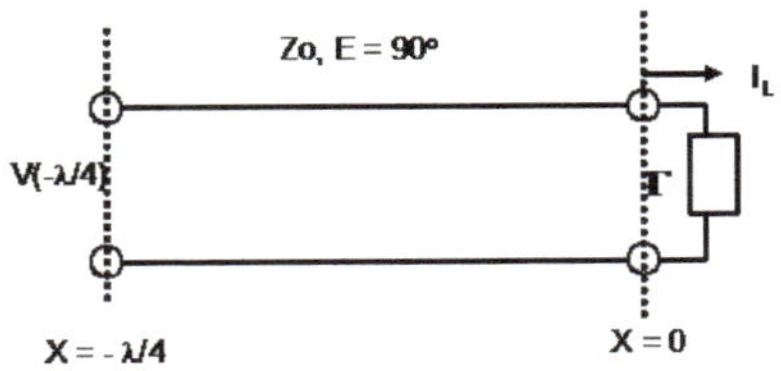

5. The impedance of a resistor was measured over frequency as shown below. Using the equivalent circuit model for the resistor, derive component values that make the impedance of the equivalent circuit match the data.

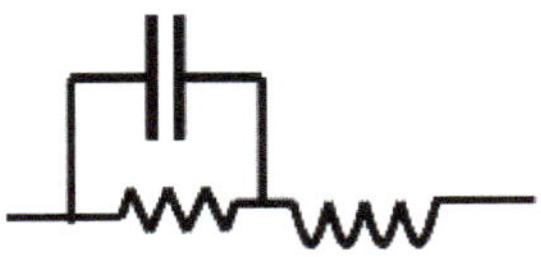

freq	Zin1
10.00 MHz	20.000 + j0.478
110.0 MHz	19.996 + j5.253
210.0 MHz	19.986 + j10.028
310.0 MHz	19.970 + j14.804
410.0 MHz	19.947 + j19.581
510.0 MHz	19.918 + j24.359
610.0 MHz	19.883 + j29.138
710.0 MHz	19.842 + j33.918
810.0 MHz	19.795 + j38.700
910.0 MHz	19.742 + j43.484

6.a. Calculate the input reflection coefficient and input impedance for this lumped element circuit at 1 GHz. Assume Z_0 = 50 ohms and ideal components.

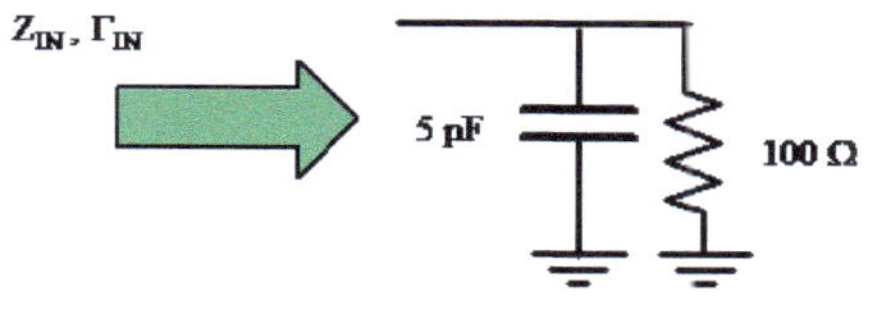

b. Use the parallel to series transformation below to derive a series equivalent network with the same impedance at 1 GHz.

$$R_P = R_S(Q^2 + 1)$$

$$X_P = X_S\left[\frac{Q^2 + 1}{Q^2}\right]$$

7.a. Calculate the input reflection coefficient and input impedance for this lumped element circuit at 1 GHz. Assume Z_0 = 25 ohms. Verify the result with the Smith chart and with ADS.

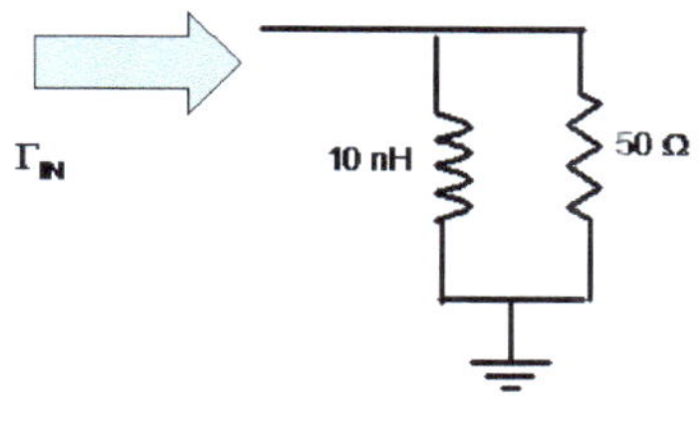

b. Add a 3/8 wavelength 25 ohm line ahead of the circuit and now find the input reflection coefficient and input impedance. Verify with the Smith chart and with ADS.

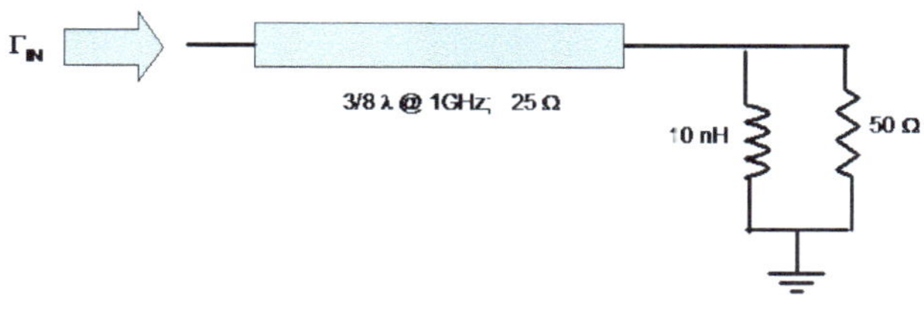

4

S-parameters

There are two commonly used ways of describing a circuit or a device:

A. Equivalent – Circuit Model

- Physically based
- Includes bias dependence (nonlinearity can be included)
- Includes frequency dependence
- Includes size dependence – *scalability*
- Ideal for IC design
- Weakness: Model necessarily simplified; some errors. Thus, weak for highly resonant designs.

B. Two – port Model

- Matrix of tabular data vs. frequency
- Need one matrix for each bias point and device size
- Clumsy – huge data sets required
- Traditional microwave method
- Exact but is a linear small-signal representation.

Goal:

1. Become proficient in the use of S-parameters for circuit analysis and measurement.

4.1 Two-port Descriptions

These are black box (mathematical) descriptions from the point of view of an input port (1) and an output port (2). The ratio of voltages and currents at each port are defined according to the termination on the opposite port (Figure 4.1).

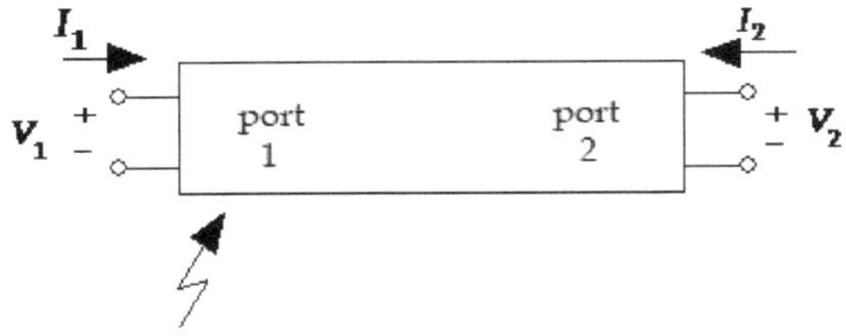

Figure 4.1 A two-port black box representation.

Inside might be a transistor, an FET, a transmission line, or peanut butter.

The terminal characteristics are V_1, V_2, I_1 and I_2 – there are *two* degrees of freedom.

4.1.1 Admittance parameters

$$\begin{bmatrix} I_1 \\ I_2 \end{bmatrix} = \begin{bmatrix} Y_{11} & Y_{12} \\ Y_{21} & Y_{22} \end{bmatrix} \begin{bmatrix} V_1 \\ V_2 \end{bmatrix} \tag{4.1}$$

Example 4.1: Simple equivalent circuit FET model (Figure 4.2). Note that some components are frequency dependent.

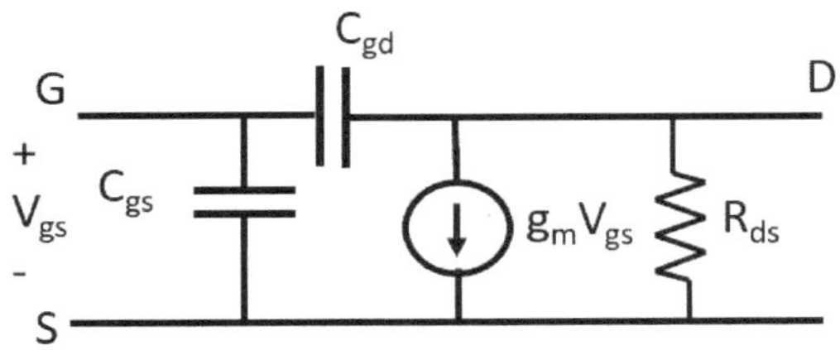

Figure 4.2 FET equivalent circuit diagram.

Y-parameters are derived by short circuiting the opposite port. For instance, Y_{11} requires shorting port 2. Y_{12} also. Y_{22} is derived by shorting port 1. This makes for an easy mathematical description but has its limitations, as we will see.

Therefore:

$$Y = \begin{bmatrix} j\omega C_{gs} + j\omega C_{gd} & -j\omega C_{gd} \\ g_m - j\omega C_{gd} & G_{ds} + j\omega C_{gd} \end{bmatrix} \tag{4.2}$$

Or, another way of describing the procedure is:

$$Y_{11} = \left.\frac{I_1}{V_1}\right|_{V_2=0} \qquad Y_{12} = \left.\frac{I_1}{V_2}\right|_{V_1=0} \tag{4.3}$$

and

$$Y_{21} = \left.\frac{I_2}{V_1}\right]_{V_2=0} \qquad Y_{22} = \left.\frac{I_2}{V_2}\right]_{V_1=0}$$

4.1.2 Impedance parameters

$$\begin{bmatrix} V_1 \\ V_2 \end{bmatrix} = \begin{bmatrix} Z_{11} & Z_{12} \\ Z_{21} & Z_{22} \end{bmatrix} \begin{bmatrix} I_1 \\ I_2 \end{bmatrix} \tag{4.4}$$

Example 4.2:

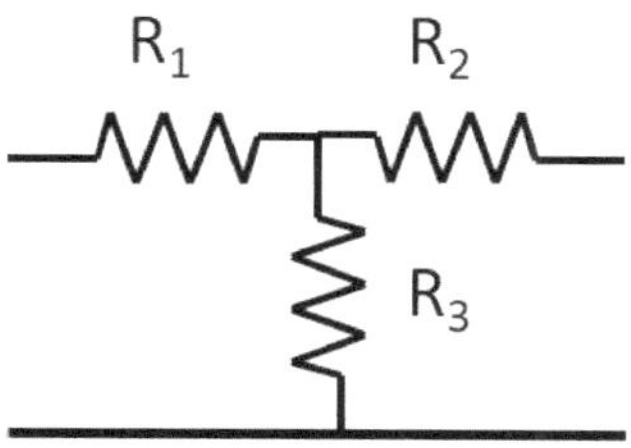

Figure 4.3 Simple resistive network. Z-parameters are easily derived, see Equation (4.5).

Z-parameters are derived by open circuiting the opposite port. For instance, to calculate Z_{11}, open circuiting port 2 is required.

$$Z_{11} = \left.\frac{V_1}{I_1}\right|_{I_2=0} \qquad Z_{21} = \left.\frac{V_2}{I_1}\right|_{I_2=0} \tag{4.5}$$

and

$$Z_{12} = \left.\frac{V_1}{I_2}\right|_{I_1=0} \qquad Z_{22} = \left.\frac{V_2}{I_2}\right]_{I_1=0} \tag{4.6}$$

From the resistive network in Figure 4.3, applying these definitions yields

$$Z = \begin{bmatrix} R_1 + R_3 & R_3 \\ R_3 & R_2 + R_3 \end{bmatrix} \tag{4.7}$$

But Y- and Z-parameters are not suitable for high frequency measurement. Why?

Problem:

How can you get a true open or short at the circuit terminals? Any real short is inductive. Any real open is capacitive.

To make matters worse, if you are trying to measure a high frequency active device, a short or open can make it oscillate!

Solution:

Terminate the ports in Z_0 instead. The advantages are many.

- Broadband
- Not very sensitive to parasitic L or C
- Kills reflections at each port
- Redefine parameters to use forward and reverse voltage waves
- Measurement can use directional couplers.

The common formulation with the terminated ports is referred to as the S-parameter.

4.2 S-parameters

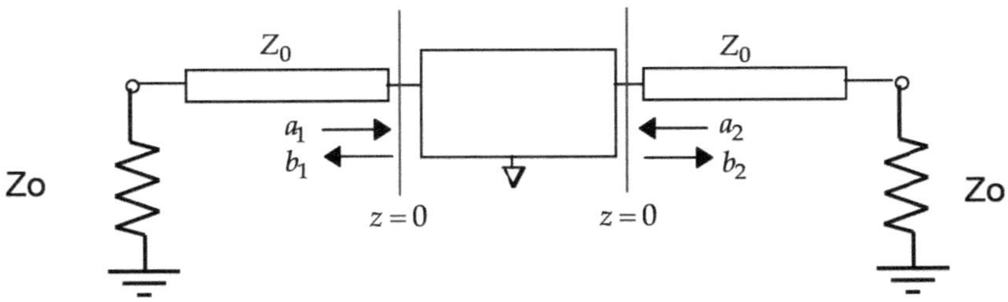

Figure 4.4 Two-port model describing S-parameters. Note that both ports are terminated in the characteristic impedance, Z_0. Transmission lines are added in the figure just to show the forward and reflected wave propagation.

Two-port model describing S-parameters is shown in Figure 4.4.

$$\begin{bmatrix} b_1 \\ b_2 \end{bmatrix} = \begin{bmatrix} S_{11} & S_{12} \\ S_{21} & S_{22} \end{bmatrix} \begin{bmatrix} a_1 \\ a_2 \end{bmatrix} \tag{4.8}$$

Note that Z_0 must be defined. We don't really need transmission lines. Also, recall that for amplitude quantities, power is defined as

$$P_{\text{avg}} = \frac{V^2}{2R} = \frac{I^2R}{2}. \tag{4.9}$$

Or if complex phasor quantities, VV^* or II^* must be used in place of V^2 or I^2.

Our objective now is to de-mystify S-parameters.

Recall from Chapter 3,

$$V(x) = V^+(x) + V^-(x) \quad \text{phasor quantities.}$$
$$I(x) = \frac{V^+(x)}{Z_0} - \frac{V^-(x)}{Z_0} \quad \text{amplitude, not rms values.} \tag{4.10}$$

We can normalize the amplitude of waves to Z_0 :

$$\begin{aligned} a(x) &= \frac{V^+(x)}{\sqrt{Z_0}} \quad \text{forward wave} \\ b(x) &= \frac{V^-(x)}{\sqrt{Z_0}} \quad \text{reverse wave} \end{aligned} \tag{4.11}$$

Why? So that $\frac{1}{2}a(x)a^*(x)$ is the power in the forward wave. And, likewise, $b(x)b^*(x)/2$ is the power in the reverse wave.

(Note: If $a = 1.414$ then the average power in the wave is 1 watt (or $a_{rms} = 1$).)

So, in terms of total voltage $V(x)$ and current $I(x)$,

$$\begin{aligned} v(x) &= \frac{V(x)}{\sqrt{Z_0}} = a(x) + b(x) \\ i(x) &= \sqrt{Z_0}I(x) = a(x) - b(x) \end{aligned} \tag{4.12}$$

or

$$\begin{aligned} a(x) &= \frac{1}{2}[v(x) + i(x)] = \frac{1}{2\sqrt{Z_0}}[V(x) + Z_0I(x)] \\ b(x) &= \frac{1}{2}[v(x) - i(x)] = \frac{1}{2\sqrt{Z_0}}[V(x) - Z_0I(x)] \end{aligned} \tag{4.13}$$

4.2.1 Reflection coefficient

So, how is Γ defined in terms of the S-parameters? At port 1,

$$\Gamma_1 = \frac{b_1}{a_1} \tag{4.14}$$

However,

$$b_1 = S_{11}a_1 + S_{12}a_2 \tag{4.15}$$

We need to eliminate a_2. How?

If $Z_L = Z_0$

$$\Gamma_L = 0 = \frac{a_2}{b_2}, \tag{4.16}$$

therefore $a_2 = 0$ if port 2 is terminated in Z_0. Then, we see that the input reflection coefficient is none other than S_{11}.

$$\Gamma_1 = \left.\frac{b_1}{a_1}\right|_{a_2=0} = S_{11} \tag{4.17}$$

Same with at port 2 with S_{22} :

$$S_{22} = \left.\frac{b_2}{a_2}\right|_{a_1=0} = \Gamma_2 \tag{4.18}$$

4.2.2 Transmission coefficient

S_{21} is the forward transmission through the network; S_{12} the reverse.

$$b_2 = S_{21}a_1 + S_{22}a_2 \tag{4.19}$$

So, the forward transmission S_{21} can be found by setting $a_2 = 0$ (terminate output in Z_0)

$$S_{21} = \left.\frac{b_2}{a_1}\right|_{a_2=0} \tag{4.20}$$

Reverse transmission, similarly, is found by setting $a_1 = 0$ (terminate input in Z_0)

$$b_1 = S_{11}a_1 + S_{21}a_2 \tag{4.21}$$

$$S_{12} = \left.\frac{b_1}{a_2}\right|_{a_1=0} \tag{4.22}$$

4.3 Some Comments on Power Measurement

Power can vary over a very large range, therefore it is often specified on a logarithmic scale. There must always be a point of reference on the scale; the power measurements are usually with reference to $1\ \mathrm{mW}$.

The unit is called dBm meaning dB relative to 1 mW of power. Thus,

$$\begin{aligned} 0\ \text{dBm} &= 1\ \text{mW} \\ 10\ \text{dBm} &= 10\ \text{mW} \\ -10\ \text{dBm} &= 0.1\ \text{mW} \\ &\text{etc.} \end{aligned}$$

To convert mW to dBm :

$$\text{dBm} = 10\log_{10}(P). \tag{4.23}$$

To convert dBm to mW :

$$P = 10^{\text{dBm}/10}. \tag{4.24}$$

One could also define a dBW referenced to 1 watt instead.

We must be careful to distinguish between dB and dBm. What is the difference between dB and dBm?

dB is a power ratio - used to describe a gain or loss for example.

$$\begin{aligned} G &= 10\log_{10}\left(P_{\text{out}}/P_{\text{in}}\right)\text{dB} \\ \text{Return loss} &= -20\log_{10}|\Gamma|\text{dB} \end{aligned}$$

However, dB says nothing about the absolute power level. *Don't confuse their usage!*

4.4 Define Available Power and Actual Load Power

P_{AVS} = max power output from a source with impedance Z_s that can be absorbed into a load.

If we assign

$$Z_S = Z_0, Z_L = Z_S^* = Z_0 \quad \text{(in this case)}$$

Then maximum power transfer will occur when we have a conjugate match; half of the source voltage V_{gen} appears across the load due to the voltage divider (Figure 4.5).

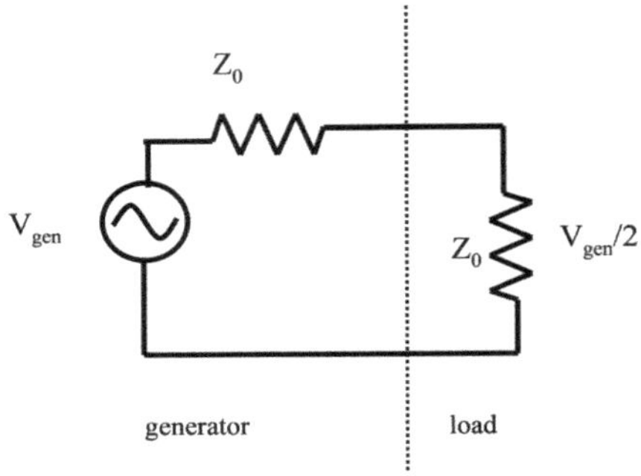

Figure 4.5 Maximum power transfer requires terminating the generator in the source impedance.

In this instance,

$$P_{load} = P_{AVS} = \frac{1}{8}\frac{V_{gen}^2}{Z_0} \tag{4.25}$$

Or, in terms of *a* and *b* (Figure 4.6):

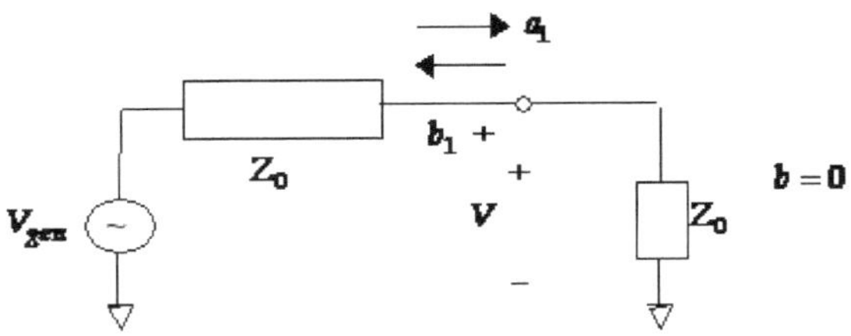

Figure 4.6 Same concept with forward and reflected wave definitions *a* and *b*.

$$a_1 = \frac{V^+}{\sqrt{Z_0}} \tag{4.26}$$

and $b_1 = 0$.

$$V^+ = V_{gen}\left(\frac{Z_0}{Z_0 + Z_0}\right) = \frac{V_{gen}}{2} \text{ and } V^- = 0 \tag{4.27}$$

so,

$$\boxed{P_{load} = P_{AVS} = \frac{1}{2}a_1 a_1^* = \frac{V_{gen}^2}{8Z_0}} \tag{4.28}$$

We see that the available power is independent of the actual load impedance. Even if the load is not matched, available power remains constant by definition. Actual power in the load is reduced however.

In the laboratory, a signal generator's output power is calibrated and displayed as available power.

4.4.1 Actual load power

The actual power received in the load depends upon the load Z_L. When there is a reflected wave, that power is not absorbed in the load. This is why impedance matching usually is employed (Chapter 5).

$$\mathrm{P_{Load}} = \frac{1}{2}|a_1|^2 - \frac{1}{2}|b_1|^2 = \frac{1}{2}\mathrm{Re}\left[I_1 V_1^*\right] \tag{4.29}$$

or

$$\boxed{P_{Load} = P_{AVS}\left(1 - |S_{11}|^2\right)} \tag{4.30}$$

4.4.2 Reflected power

By definition, reflected power is calculated from b $b_1 = a_1 S_{11}$ and $b_2 = a_2 S_{22}$:

$$P_R = \frac{1}{2}|b_1|^2 = \frac{1}{2}|a_1|^2 |S_{11}|^2 = P_{AVS}|S_{11}|^2 \tag{4.31}$$

$$\begin{aligned} |S_{11}|^2 &= \frac{\text{Power reflected from input}}{\text{Power incident on input}} = \frac{|b_1|^2}{|a_1|^2} \\ |S_{22}|^2 &= \frac{\text{Power reflected from network output}}{\text{Power incident on output}} = \frac{|b_2|^2}{|a_2|^2} \end{aligned} \tag{4.32}$$

Note that $|a|^2$ or $|b|^2$ is actually aa^* and bb^* in most cases.
Similarly (Figure 4.7),

$$\begin{aligned} \frac{1}{2}|a_2|^2 &= \text{Power incident on output} \\ &= \text{Reflected power from load} \end{aligned} \tag{4.33}$$

$$\frac{1}{2}|b_1|^2 = \text{Power reflected from input port} \tag{4.34}$$

$$\frac{1}{2}|b_2|^2 = \text{Power incident on load from the network} \tag{4.35}$$

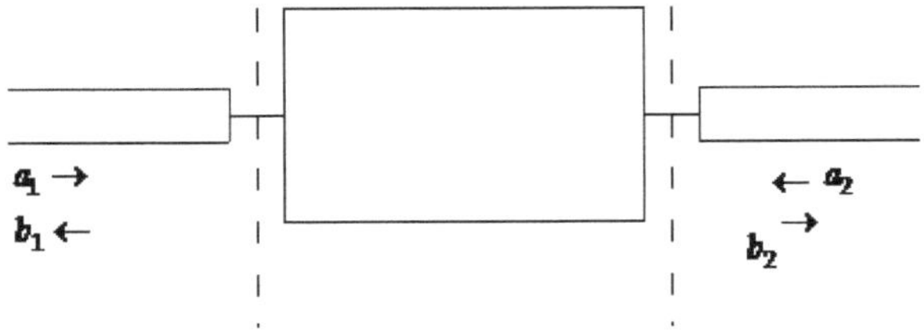

Figure 4.7 Incident and reflected waves at ports 1 and 2.

4.4.3 Transducer gain

Also, by definition,

$$\text{transducer gain} = \frac{P_{load}}{P_{avs}} = G_T \tag{4.36}$$

even if

1. Load isn't matched to network, and
2. Input of network not matched to generator.

Here,

$$P_{Load} = |b_2|^2 \left(1 - |\Gamma_L|^2\right) \tag{4.37}$$

S_{21} is defined in terms of transducer gain for the special case of where $Z_L = Z_0$:

$$|S_{21}|^2 = \left.\frac{|b_2|^2}{|a_1|^2}\right|_{a_2=0} \tag{4.38}$$

$$\frac{1}{2}|b_2|^2 = \text{ power incident on load (and is absorbed since } \Gamma_L = 0\text{)}$$

$$\frac{1}{2}|a_1|^2 = \text{ source available power.}$$

Therefore,

$$|S_{21}|^2 = \text{ transducer gain with source and load } Z_0.$$

Similarly,

$$|S_{12}|^2 = \text{ reverse transducer power gain.}$$

4.5 Reference Planes

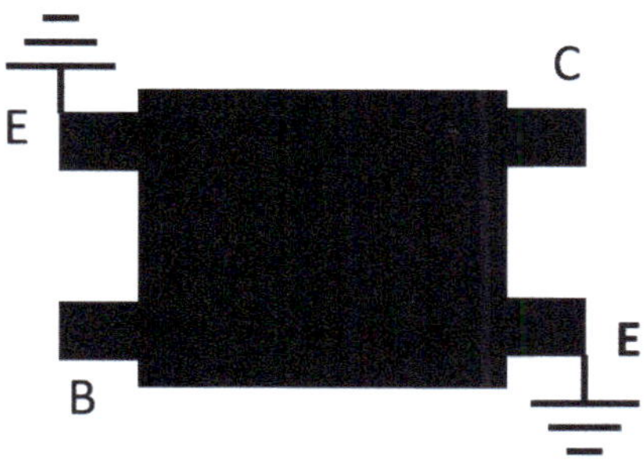

Figure 4.8 Drawing of a surface mount microwave transistor.

Define $x = 0$ at each port. Reference planes will shift with connections to the board on which it is mounted (Figure 4.8).

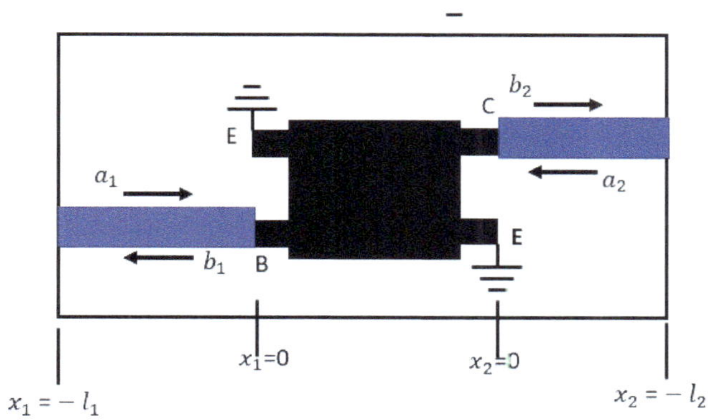

Figure 4.9 As was discussed in Chapter 3, shifting the reference planes changes the reflection coefficient and thus S-parameters.

Here the transmission lines on the board in Figure 4.9 introduce a phase shift θ_1 and θ_2.

$$\begin{aligned}\theta_1 &= \beta x_1 = -\frac{2\pi\ell_1}{\lambda}\\ \theta_2 &= \beta x_2 = -\frac{2\pi\ell_2}{\lambda}\end{aligned} \tag{4.39}$$

The S-parameters are modified by the phase shift. Thus, impedances are also affected.

$$S' = \begin{pmatrix} S_{11}e^{2\theta_1} & S_{12}e^{j(\theta_1+\theta_2)} \\ S_{21}e^{j(\theta_1+\theta_2)} & S_{22}e^{j2\theta_2} \end{pmatrix} \tag{4.40}$$

The reflection parameters are shifted in phase by twice the electrical length because the incident wave travels twice over this length upon reflection. The transmission parameters have the sum of the electrical lengths, since the transmitted wave must pass through both lengths.

4.5.1 Electrical length

There are measures of both physical length and electrical length of a transmission line. Physical length is obvious, but electrical length can be defined in more than one way. Physical length ℓ and electrical length E are related by the phase velocity v, wavelength λ_{ref} and the frequency f_{ref}.

The microwave literature might say a line is 43° long at say 5 GHz. What does this mean?

Recall the definition of phase velocity from Chapter 2:

$$v = \frac{1}{\sqrt{LC}} \tag{4.41}$$

$$\left.\begin{aligned} &\text{Electrical length} = \text{E} = \frac{\ell}{\lambda_{ref}} \cdot 360^\circ \\ &\text{Recall } f \cdot \lambda = v \text{ so } f_{ref}\lambda_{ref} = v \\ &\rightarrow E = \frac{\ell}{v/f_{ref}} \cdot 360^\circ = \frac{\ell}{v} \cdot f_{ref} \cdot 360^\circ \end{aligned}\right\} \tag{4.42)-(4.44}$$

Or if f_{ref} is changed to 500 MHz, then E = 4.3° for the same physical length and phase velocity.

Alternatively, the delay on the line can be described by a transit time T.

$$\boxed{E = T \cdot f_{ref} \cdot 360^\circ} \tag{4.45}$$

A line which is $1\,\text{ns}$ long has an electrical length $E = 360^\circ$ at $f_{ref} = 1\text{GHz}$ and an electrical length $E = 36^\circ$ at $f_{ref} = 100\text{MHz}$.

You should be conversant with both terminologies!

4.5.2 Converting from electrical length to physical length

Recall from Equation (4.43),

$$f\lambda_{ref} = v \tag{4.46}$$

thus:

$$\text{physical length} = \frac{E(\text{deg})\lambda_{ref}}{360} \tag{4.47}$$

in terms of wavelength λ_{ref}.

4.5.3 How to calculate S-parameters quickly

Recall:

$$S_{11} = \left.\frac{b_1}{a_1}\right|_{a_2=0} \tag{4.48}$$

$$b_1 = S_{11}a_1 + S_{12}a_2 \tag{4.49}$$

we must kill a_2 in order to measure or calculate S_{11} (Figure 4.10).

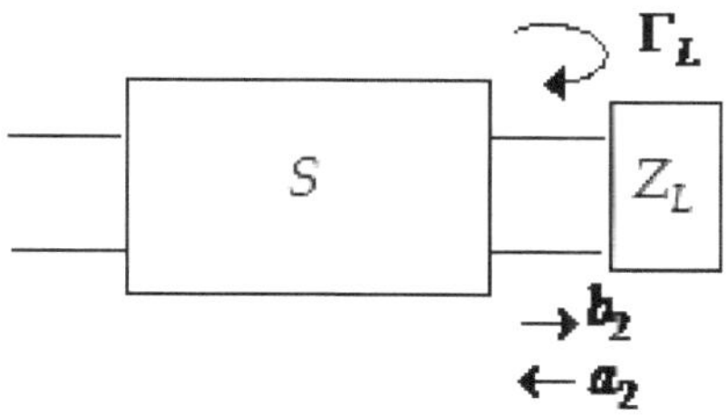

Figure 4.10 You must eliminate a_2.

If $Z_L = Z_0$, then Γ_L is zero and $a_2 = 0$. By definition,

$$S_{11} = \left.\frac{b_1}{a_1}\right|_{z_L=z_0} \tag{4.50}$$

so if we say that $Z_{in}|_{Z_L=Z_0}$ is the input impedance with $Z_0 = Z_L$ then

$$\boxed{S_{11} = \frac{Z_{in}|_{Z_L=Z_0} - Z_0}{Z_{in}|_{Z_L=Z_0} + Z_0} = \Gamma_{in}} \tag{4.51}$$

or

$$\boxed{Z_{in}|_{Z_L=Z_0} = \frac{1+S_{11}}{1-S_{11}}} \tag{4.52}$$

The same comment clearly applies for S_{22}. The Smith chart is often used to plot S_{11}, S_{22}.

Example 4.3: Find S_{11}:

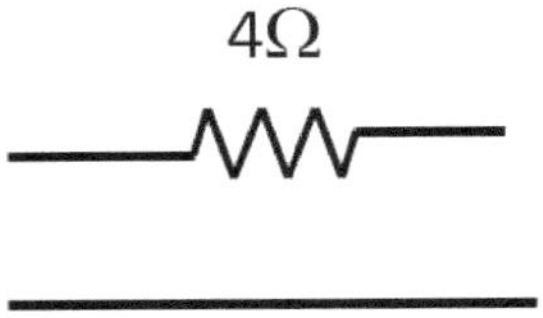

Figure 4.11 Example of a simple resistive two-port network.

Given Z_0 = 50 ohms, what is S_{11}?

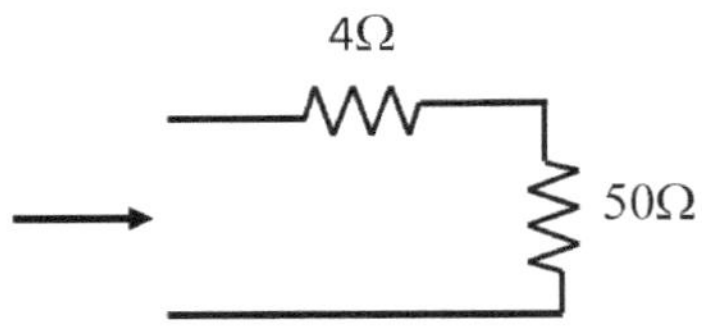

Figure 4.12 Terminate port 2 in Z_0.

$$Z_{in}|_{Z_L=Z_0} = 54\Omega$$

$$S_{11} = \frac{54-50}{54+50} = \frac{4}{104}$$

Similar arguments give $S_{22} = \frac{4}{104}$.

Find S_{21}:

$$S_{21} = \frac{b_2}{a_1}|_{a_2=0} \tag{4.53}$$

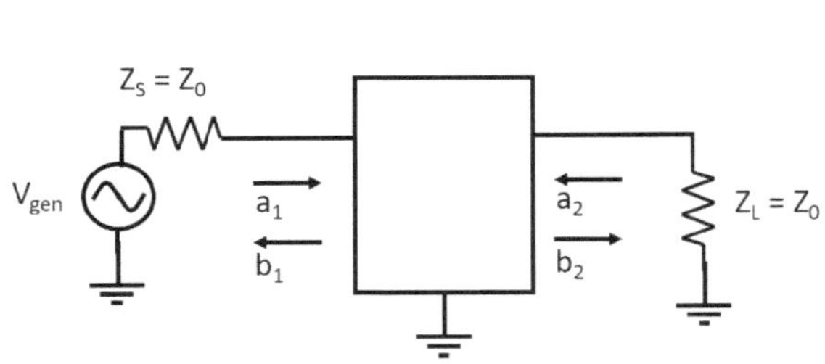

Figure 4.13 Block diagram of a network terminated at both ports in Z_0.

What is a_1 in this case?
We know that:

$$a_1 = \frac{V_1^+}{\sqrt{Z_o}} \text{ and } V_1^+ = \frac{V_{gen}}{2} \tag{4.54}$$

So,

$$a_1 = \frac{V_{gen}}{2\sqrt{Z_o}} \tag{4.55}$$

Consider the load:

$$b_2 = \frac{V_{out}}{\sqrt{Z_0}} \tag{4.56}$$

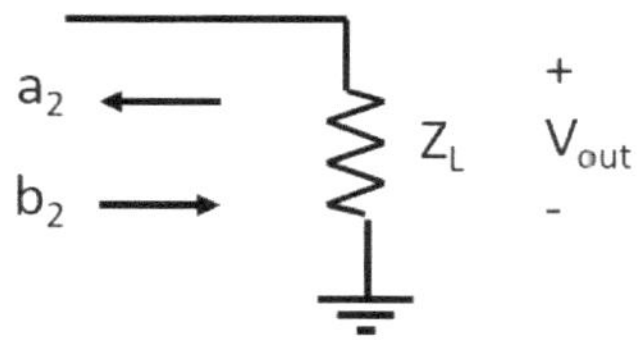

$$a_2 = \Gamma_L b_2 \tag{4.57}$$

But, $\Gamma_L = 0$ because $Z_L = Z_0$, so $a_2 = 0$.

$$\begin{aligned} V_{out} &= V^+ + V^- = \sqrt{Z_0}a_2 + \sqrt{Z_0}b_2 \\ &= \sqrt{Z_0}b_2 \end{aligned} \tag{4.58}$$

Now, calculate V_{out}/V_{gen}

$$V_{out} = \sqrt{Z_0}\, b_2 = \sqrt{Z_0}\,(S_{21}a_1 + S_{22}a_2) \tag{4.59}$$

$$V_{out} = \sqrt{Z_0}S_{21}a_1 \tag{4.60}$$

Substitute for a_1

$$a_1 = \frac{V_{gen}}{2\sqrt{Z_0}} \tag{4.61}$$

Then,

$$\frac{V_{out}}{V_{gen}} = \frac{\sqrt{Z_0}S_{21}}{2\sqrt{Z_0}} = \frac{S_{21}}{2} \tag{4.62}$$

Or,

$$\boxed{S_{21} = \frac{2V_{out}}{V_{gen}}} \tag{4.63}$$

when $Z_L = Z_S = Z_0$.

Why the factor of 2 ?

Referring again to Figure 4.5, we see that the generator voltage is split between the source and load in the matched case. Here, we see that $V_{out}/V_{gen} = 1/2$, but the transducer gain must be equal to 1. $(P_{load}/P_{avs}) \cdot |S_{21}|^2$ is the transducer gain in this situation. If we insert an amplifier into the network, the signal will be increased by an amount S_{21} (Figure 4.14).

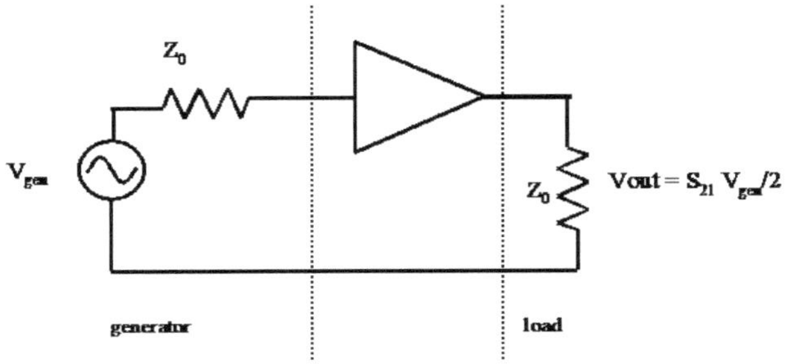

Figure 4.14 Amplifier with transducer gain S_{21} inserted.

So, $|S_{21}|^2$ is the *forward insertion gain* or *forward transducer gain* in a system of impedance Z_0.

Example 4.4: Find S_{21}:

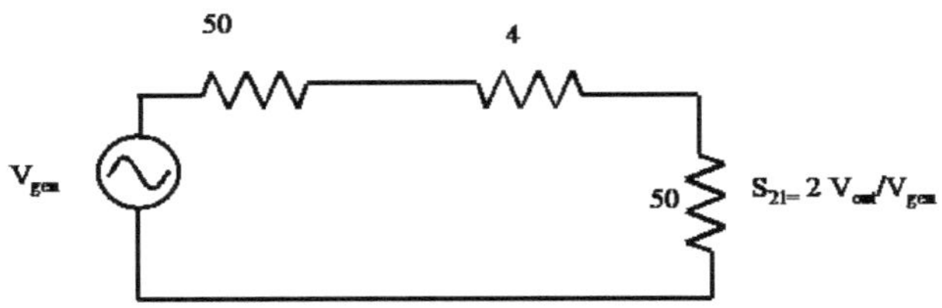

Figure 4.15 Ports 1 and 2 must both see the Z_0 termination to find transducer gain.

$$V_{out}/V_{gen} = 50/104 = 0.48,$$

therefore

$$S_{21} = 0.96.$$

Or, we could let $V_{gen} = 2$. Then, $S_{21} = V_{out}$.

4.5.4 What happens when there is a reference plane extension?

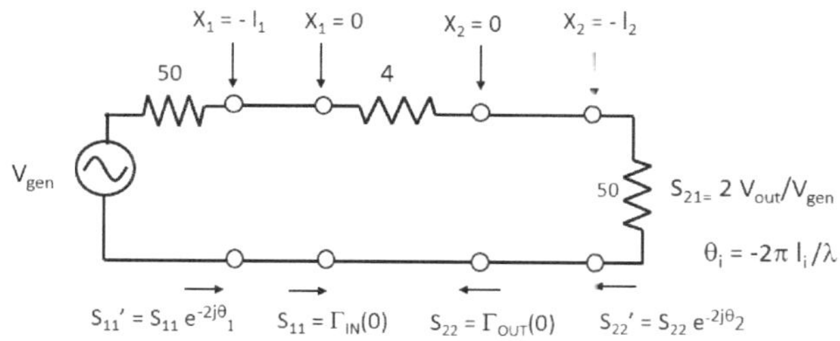

Figure 4.16 There is a θ_1 phase shift on S_{11} and θ_2 on S_{22}.

$$\theta_1 = -\beta\ell_1 = -\frac{2\pi}{\lambda}\ell_1 \quad \theta_2 = -\beta\ell_2 = -\frac{2\pi}{\lambda}\ell_2 \tag{4.64}$$

$$S'_{21} = S_{21}e^{j(\theta_1+\theta_2)} = S_{21}e^{-2\pi j(\ell_1+\ell_2)/\lambda} \tag{4.65}$$

Example 4.5: Find the four S-parameters of the circuit in Figure 4.17.

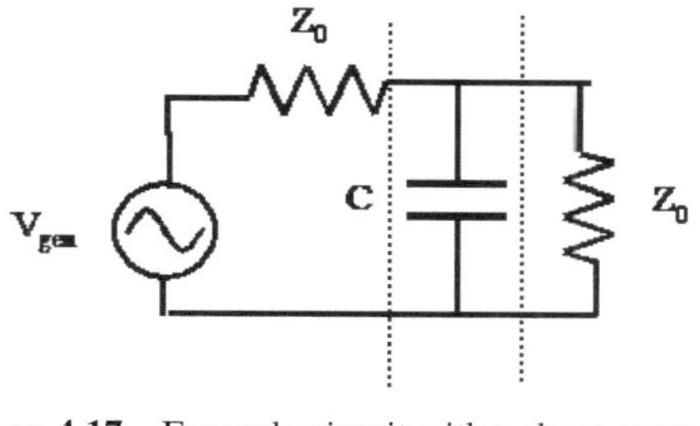

Figure 4.17 Example circuit with a shunt capacitor.

S_{11}: Find Z_{in} (with $Z_L = Z_0$), then calculate input reflection coefficient. C and Z_0 are in parallel so add admittances.

$$Z_{IN} \mid Z_{L=Z_0} = 1/\left(sC + 1/Z_0\right)$$
$$S_{11} = \frac{Z_{IN} - Z_0}{Z_{IN} + Z_0} = \frac{Z_{IN}/Z_C - 1}{Z_{IN}/Z_C + 1} \tag{4.66}$$

Let $s = j\omega$, then

$$S_{11} = \frac{-j\omega C Z_0/2}{1 + j\omega C Z_0/2}$$

S_{22} will be the same due to symmetry. Note that we calculated Z_{in} with port 2 terminated in Z_0. This is part of the definition of S_{11} that is essential.

Now find S_{21}:

First use the Thevenin-Norton transformation and then add admittances:

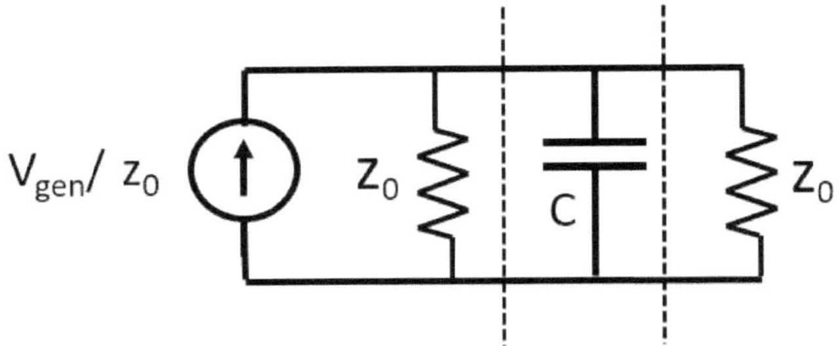

Figure 4.18 Parallel transformation of the circuit in Figure 4.17.

$$\begin{aligned} V_{out} &= \frac{V_{gen}}{Z_0}\frac{1}{2/Z_0 + sC} = \text{I/Y} \\ S_{21} &= \frac{2V_{out}}{V_{gen}} = \frac{1}{1 + j\omega C Z_0/2} = S_{12} \end{aligned} \tag{4.67}$$

S_{12} will be the same by symmetry.

4.6 Summary

1. S-parameters are the most effective way to characterize two-port networks. Terminating the ports in Z_0 kills reflections and enables convenient measurement.
2. Available power, P_{avs}, represents the maximum power that a source can deliver to a matched load. It is independent of the load impedance Z_L.
3. The actual power received in the load depends upon the load impedance Z_L. When there is a reflected wave, that reflected power is not absorbed in the load.
4. Transducer gain, G_T, is defined as the ratio of the actual power at the load, P_L to P_{avs}.
5. $|S_{21}|$ = G_T in the special case where source and load impedances are equal to Z_0.
6. Be familiar with the difference between dB and dBm.

Acknowledgement

I am grateful to Professor Mark Rodwell, whose lecture notes were very helpful to me in preparing this chapter.

References

[1] Gonzalez, G. (1984) *Microwave Transistor Amplifiers, Analysis and Design*, 2nd edn. Prentice-Hall, Sections 1.4–1.6.
[2] Pozar, D. (2005) *Microwave Engineering*, 3rd edn. Wiley, Section 4.3.

Homework

1.a. Calculate the four S-parameters for the network below at 2 GHz. Z_0 = 50 ohms.

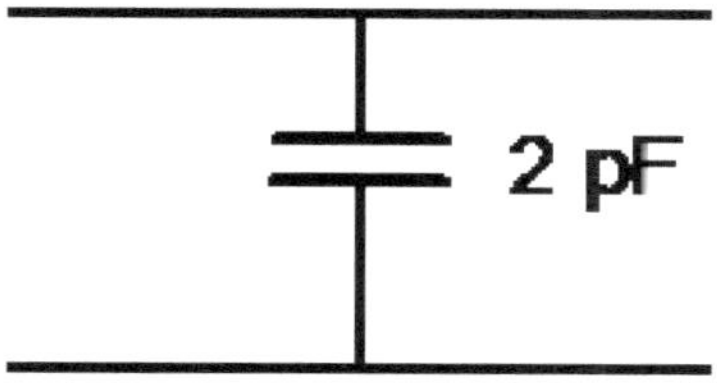

b. Calculate the forward transducer gain in dB.
c. Calculate the power delivered to a 50 ohm load at the output when driven by a 2 V source.
d. Calculate the reflected power from the input under the same conditions.

2. Calculate the four S-parameters for the network below.

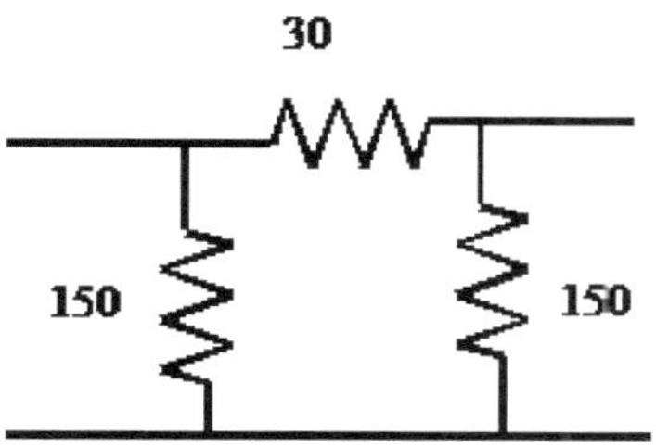

3.a. Find Γ_0 at 1 GHz.

b. Find the four S-parameters for this network at 1 GHz.

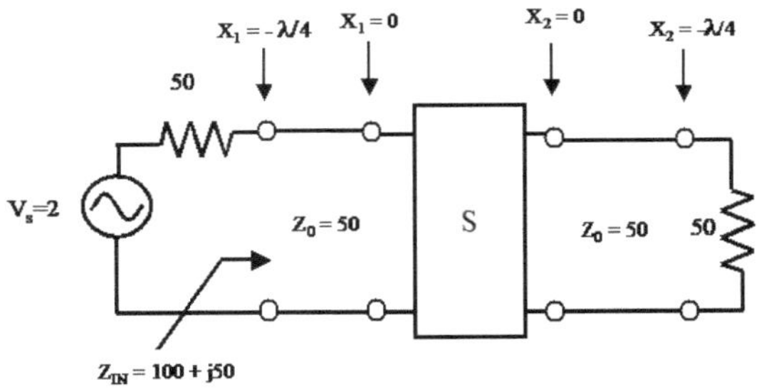

4. $S_{12} = 0; S_{21} = 3\angle 60°; S_{22} = 0.5\angle - 30°$.
a. Find S_{11} at $x_1 = 0$.
b. Find a_1 and b_1 at $x_1 = -\lambda/4$.
c. Find the actual power delivered to the load.
d. Find the actual power absorbed by the input of the network.

5. The two port below has the following S-parameters:
$S_{11} = 0\angle 0 \quad S_{12} = 0.1\angle 0 \quad S_{21} = 3\angle 0 \quad S_{22} = 0.6\angle 180 \quad Z_0 = 50$ ohms.
a. Determine the power delivered to the load, P_L, and the power reflected from the input of the network, $P_{reflected}$
b. Calculate the VSWR seen by the generator:

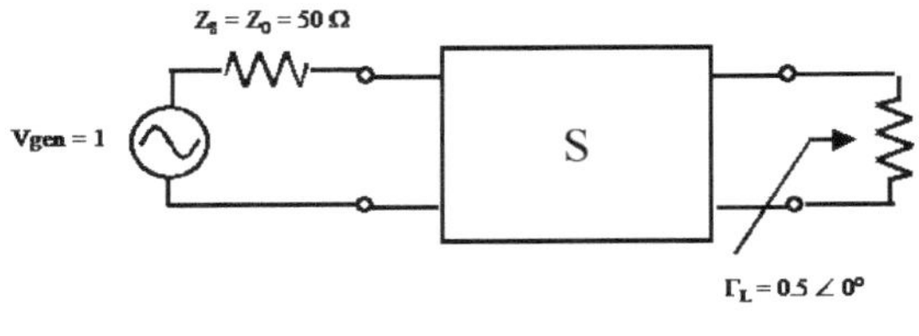

6.a. Calculate the input reflection coefficient and input impedance for this lumped element circuit at 1 GHz. Assume Z_0 = 50 ohms.

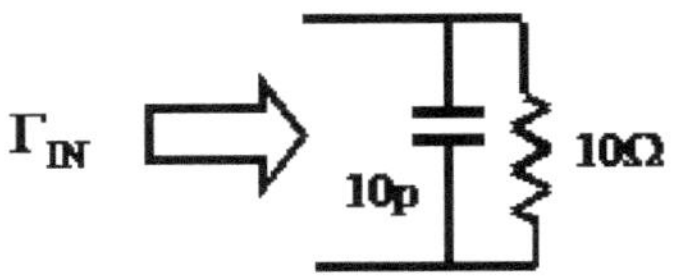

b. Add a one-eighth wavelength line ahead of the circuit and now find the input reflection coefficient and input impedance.

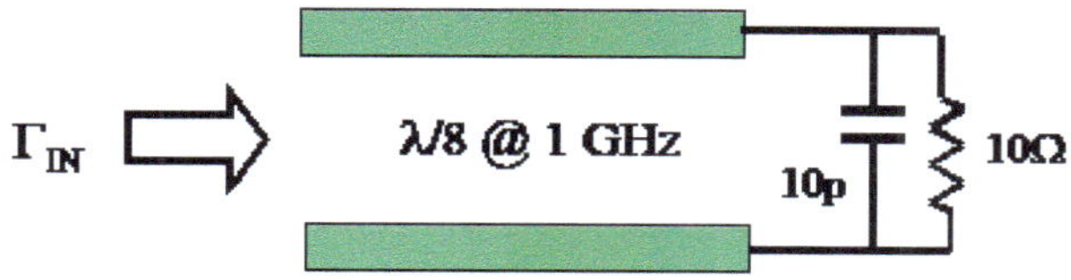

c. Use ADS to verify that your calculations are correct for parts (a) and (b). Using the S-parameter mode of ADS, plot the impedance between 10 MHz and 1 GHz on stacked rectangular plots of magnitude and phase.

7. Calculate the four S-parameters of this network at 1 GHz. Assume Z_0 = 50 ohms.

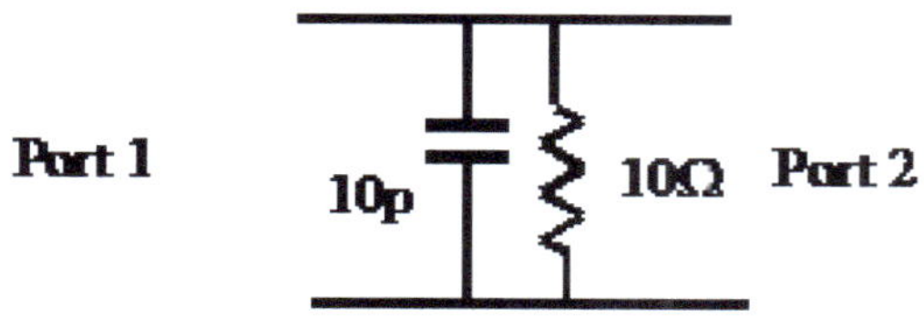

5

Matching Network Design

Goals:

1. Understand how impedance matching affects the gain and available power of a radio frequency circuit.
2. Understand that, at lower frequencies (HF, VHF and some UHF), lumped element matching networks can be constructed using chip components with low parasitics. However, these components must be modeled to include the parasitic L, R and C.
3. Know that, at higher frequencies, distributed matching networks consist of short transmission line sections in series and shunt configurations can be used.

And again, don't get lost in the math....

5.1 Impedance Matching

Why do we impedance match?

- Power transfer is reduced when we have a mismatch.

Example 5.1: Suppose we have a 1 V source with a 100 ohm source resistance, R_S. The available power is the largest power that can be extracted from the source, and this is ouly possible when source and load are matched: $R_L = R_S$.

$$P_{avs} = \frac{V_{gen}^2}{8R_S} = 1.25 \text{ mW}. \tag{5.1}$$

If we were to attach a 1000Ω load,

$$P_{Load} = \frac{1}{2} \text{Re} \{V_L I_L{}^*\} \tag{5.2}$$

where $V_L = V_{gen}(1000/1100)$, $I_L = V_{gen}/1100$, and $P_{Load} = 0.41$ mW

Alternatively, we could calculate the reflection coefficient.

$$\Gamma_L = \frac{R_L/Z_0 - 1}{R_L/Z_0 + 1} = 0.818 \tag{5.3}$$

$$P_L = P_{avs}\left(1 - |\Gamma_L|^2\right) = P_{avs}(0.33) = 0.41 \text{ mW}. \tag{5.4}$$

So, if the source and load impedances are not matched, we can lose lots of power.

In this example, we have delivered only 33% of the available power to the load.

Therefore, if we want to deliver the available power into a load with a non-zero reflection coefficient, a matching network is necessary.

5.2 Lumped element "L" Matching Networks

There are eight possibilities for single frequency (narrow-band) lumped element matching networks. These are commonly referred to as L networks due to their shape. These include series–shunt and shunt–series forms as seen below in Figures 5.1 and 5.2. X_s and X_p can be either an L or a C [1].

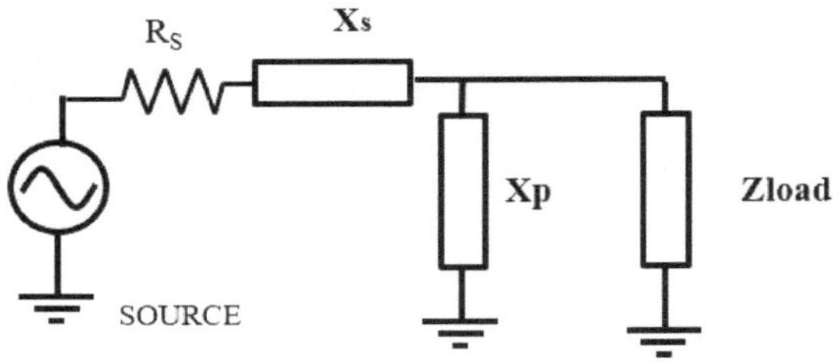

Figure 5.1 Series–shunt L network.

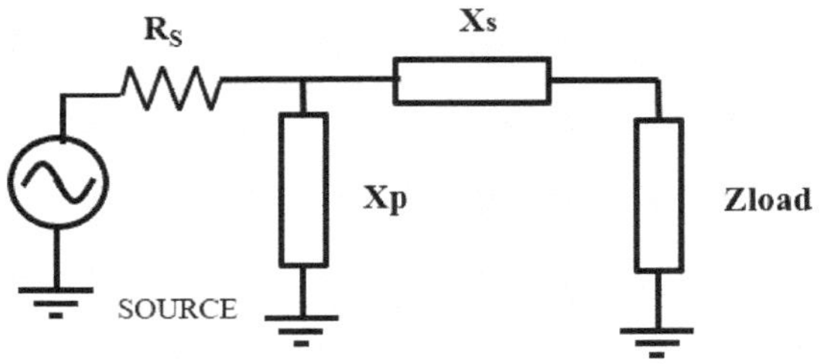

Figure 5.2 Shunt–series L network.

These networks are used to cancel the reactive component of the load and transform the real part so that the full available power is delivered into the real part of the load impedance. This can be accomplished by:

1. Absorbing or resonating the imaginary part of Z_S and Z_L.
2. Transforming the real part as needed to obtain maximum power transfer.
3. In other words, construct a network that presents a conjugate match Z_L^* to the load.

5.2.1 How to proceed

Recall the series-parallel transformations that you derived in the homework in Chapter 1:

$$\boxed{\begin{aligned} R_P &= R_S\left(Q^2+1\right) \\ X_P &= X_S\left(\frac{Q^2+1}{Q^2}\right) \end{aligned}} \tag{5.5}$$

Remember that these relationships between the series circuit and parallel circuit elements (Figure 5.3) are valid only at one frequency. And Q is the unloaded Q as defined in Chapter 1.

$$Q = \frac{1}{\omega RC} = \frac{X}{R} \tag{5.6}$$

or

$$Q = \frac{B}{G}. \tag{5.7}$$

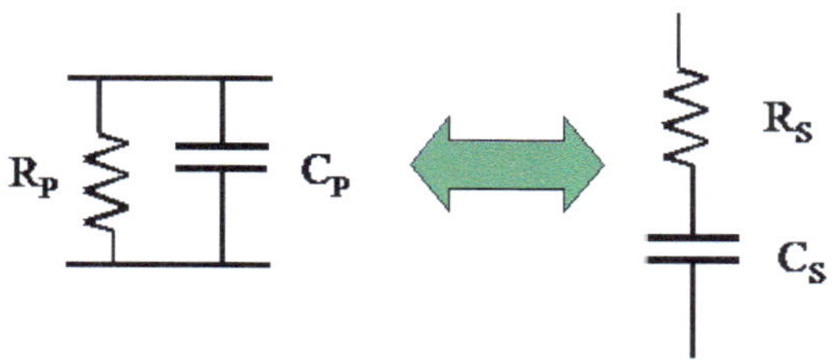

Figure 5.3 Parallel and series RC networks.

$$\text{Here, of course,} \quad X_P = \frac{1}{\omega C_P} \text{ and } X_S = \frac{1}{\omega C_S}. \tag{5.8}$$

5.2.2 Design a matching network

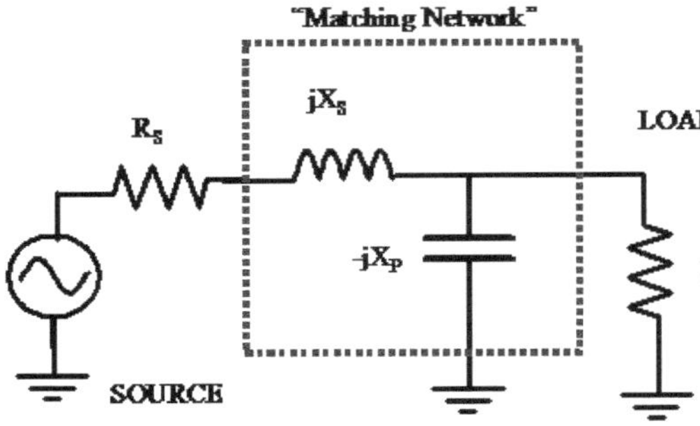

Figure 5.4 We want to match R_P to R_S and cancel reactances with a conjugate match. We could use this lowpass L network form for consideration.

For the configuration of a L network shown in Figure 5.4, R_P must be greater than R_S.

1. We know R_S and R_P (given). Use the first series-parallel transforming equation to determine Q such that R_P will be transformed into R_S.

$$\text{We can know } Q \text{ because: } Q^2 + 1 = \frac{R_P}{R_S} \text{ or } Q = \sqrt{\frac{R_P}{R_S} - 1} \tag{5.9}$$

2. Now, using the definition of unloaded Q for the series and parallel branches, compute

$$X_S = QR_S \tag{5.10}$$

$$X_P = R_P/Q. \tag{5.11}$$

3. Then determine their values:

$$L = X_S/\omega \quad C = 1/\omega X_P \tag{5.12}$$

Note that these reactive elements must be of opposite types.

Now, to show that it works, convert the parallel $Rp|| - jX_P$ into its series equivalent (Figure 5.5). We started by determining the Q based on the relationship between R_S and R_P, so we know that $R_{S1} = R_{S2}$.

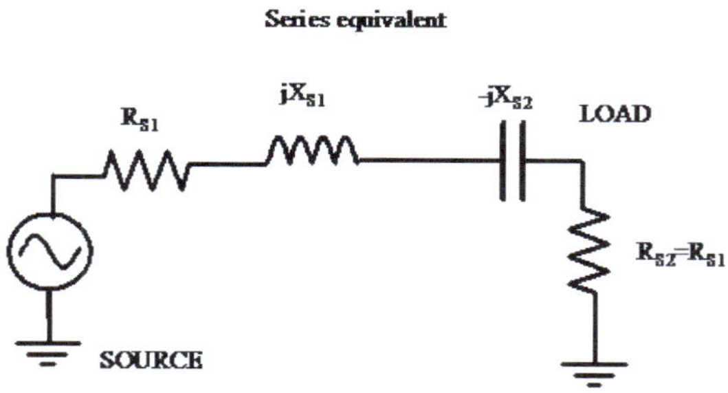

Figure 5.5 Series equivalent transformation.

Then,

$$X_{S2} = X_P\left(\frac{Q^2}{Q^2+1}\right) = \frac{QR_P}{Q^2+1} = QR_{S1} = X_{S1} \tag{5.13}$$

So, we see that $-X_{S2} = X_{S1}$, and we have cancelled the reactance as well as transforming the real part (Figure 5.6).

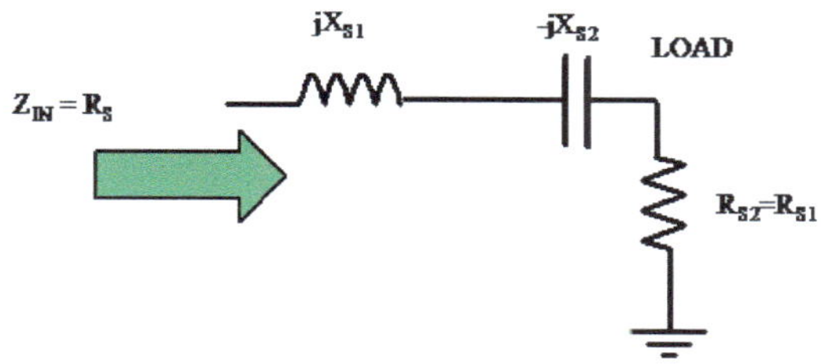

Figure 5.6 The input impedance is simply R_S.

The same process applies with a high pass form (Figure 5.7). The same X_S and X_P but different C and L values are required.

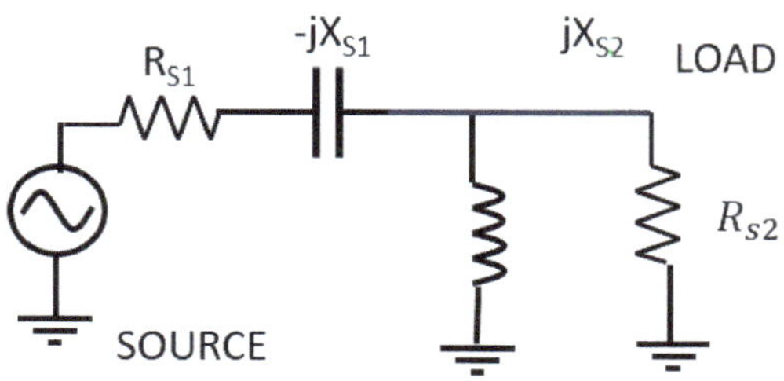

Figure 5.7 High pass configuration of an L network.

Example 5.2: Let's complete our matching network design. Refer to Figure 5.4. This will be the low pass form.

Suppose $f = 1590$ MHz, $\omega = 1 \times 10^{10}$rad/sec, $R_P = 500\ \Omega$, and $R_S = 50\ \Omega$,

$$Q = \sqrt{\frac{500}{50} - 1} = 3$$

$$X_S = 3R_S = 150\ \Omega \text{ and } X_P = R_P/Q = 500/3 = 167\ \Omega$$

Then evaluate at $\omega : C = 0.6$ pF; $L = 15$ nH,

Of course, we can also do this quite nicely on the Smith chart (Figure 5.8). Refer to the description of the Smith chart in Chapter 3. This tutorial will guide you to see how this remarkable tool can give insight and cut through much of the math.

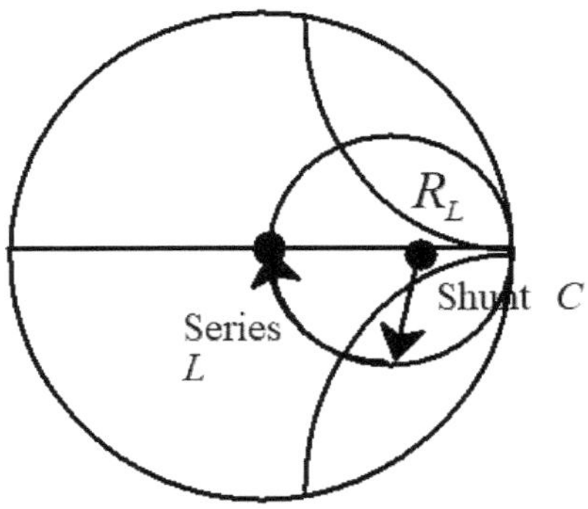

Figure 5.8 A Smith chart can be used to design the network.

Exercise: Plot this on the Smith chart. Normalize to 50 Ω Then $r_p = 10$ on real axis.

You can see that a constant conductance circle will intersect the unit resistance circle $r = 1$. Add shunt capacitive susceptance $+j0.3$ to reach this point of intersection.

So: $b_p = 0.3$. Denormalize:

$$B_P = 0.3/50 = 0.006 = \omega C$$
$$X_P = 1/B_P = 167\ \Omega$$
$$C = 0.6 \text{ pF}.$$

Next the series branch. We find that the normalized impedance at this point of intersection is $1 - j3.0$. Therefore, by adding a normalized inductive reactance of $+j3.0$ we will ohtain $z_s = 1$, the desired match. Then, move on the constant resistance circle from $1 - j3$ to center (inductive reactance).

Denormalize: $X_S = 3.0 \times 50 = 150\ \Omega = \omega L$

$$L = 15\ \text{nH}$$

Also note that Q can be read off the Smith chart (Figure 5.9):
$Q = x/r = 3.0/1.0 = b/g = 0.3/0.1 = 3$.

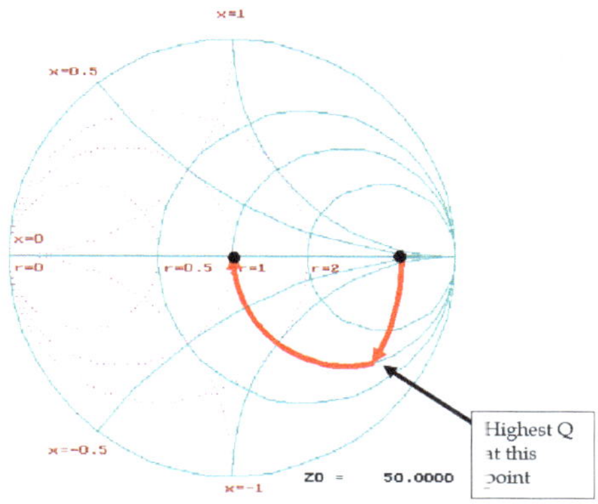

Figure 5.9 Largest ratio of x/r or b/g gives Q.

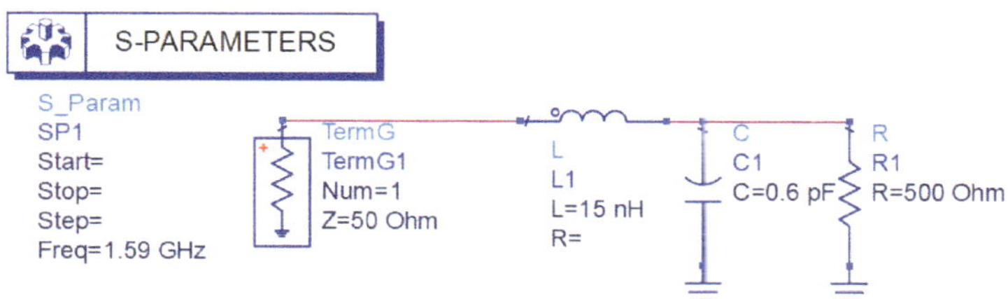

Figure 5.10 Schematic of the L network example.

Figure 5.10 illustrates how the circuit could be constructed in ADS for simulation using the S-Parameter controller. The simulation result is shown in the table in Figure 5.11.

|S(1,1)| is very close to the desired value of zero.

freq	S(1,1)
1.590 GHz	0.003 / -71.398

Figure 5.11 Simulation result of the L network design of Example 5.2.

Exercise: Now implement a high pass version of this same circuit. Use the Smith chart. Simulate on ADS.

5.2.3 Why choose one form (high pass vs. low pass) over the other?

1. *Absorb* load reactance into matching network (Figure 5.12).

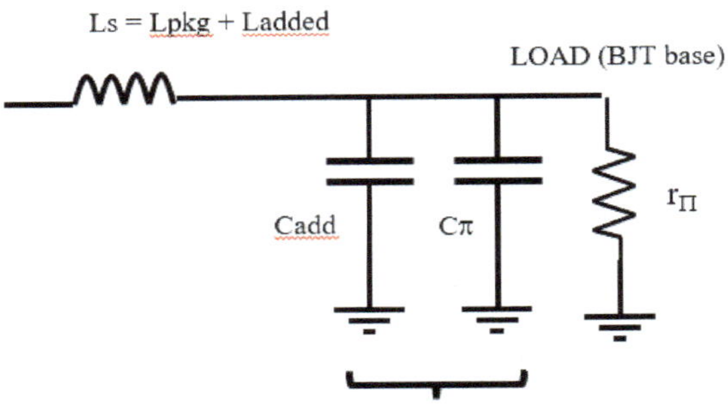

$$Cp = Cadd + C\pi$$

Figure 5.12 Input admittance of the transistor can be folded into the matching network.

2. OR: Resonate load reactance, as seen in Figure 5.13: (necessary if $C_\pi > C_P$).

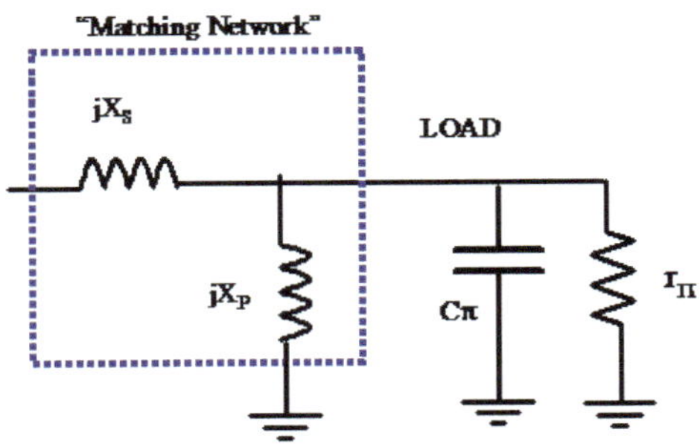

Figure 5.13 Cancel shunt reactance.

3. Harmonic suppression. The lowpass form will provide 40 dB/ decade harmonic suppression.
4. We can use the Smith chart and get the answer directly, or,
5. We can calculate the "Q" of the network. X_S, X_P can be determined from R_S and R_P.

Example 5.3: Suppose $C_\pi = 1\text{pF}, r_\pi = 500\ \Omega$. This could be the base of a bipolar transistor.

We know from the example above that $jX_P = -j167\ \Omega$ is required to achieve the match in a low pass L network.

Convert to susceptance: $B_P = 1/X_P = +0.006\ \text{S}$. This is the total susceptance required in the parallel branch.

But, we have already from C_π

$$B_P = \omega \times 10^{-12} = +0.01\ S.$$

This is more than we need. So, we must subtract $B_L = -0.004\ \text{S}$ by putting an inductor in parallel as shown in Figure 5.13.

$$L = 1/\omega B_L = 25\ \text{nH}.$$

Then add the required series *Xs* to bring to 50 ohms.

EXERCISE: Check the result on a Smith chart.

Also, note that there are other solutions possible.

IMPORTANT: Note that the lumped element components used to fabricate these networks also include parasitic *R*, *L*, or *C*. These must be included in the model used for simulation or hand analysis. Refer to Chapter 1 for information.

5.3 Matching with Distributed Elements

There are cases where transmission line elements are more effective than lumped elements in the design of matching networks:

- at higher frequencies
- when parasitics of lumped elements cannot be controlled
- when very small capacitors or inductors are required.

Suppose we have designed a lumped impedance matching network (Figure 5.14).

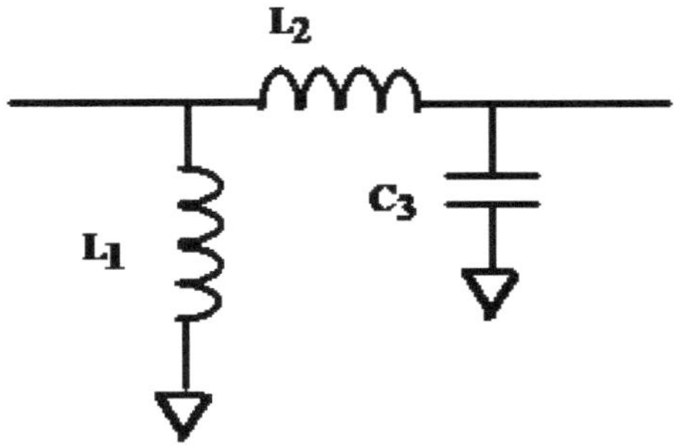

Figure 5.14 A lumped impedance matching network. Let's implement this using transmission lines.

This example has shunt and series inductors and a shunt capacitor. Think for a moment as to why no series capacitor has been chosen.

We may not have L and C available to us in realizable values when the frequency is very high and the values needed are quite small. In this case, the network can be implemented using transmission lines of characteristic impedances over the range Z_{min} to Z_{max}. This range would be dictated by the dielectric constant of the substrate material and the minimum and maximum line widths that would be practical. Figure 5.15 shows that a short transmission line can be represented either by a T or a Pi equivalent network.

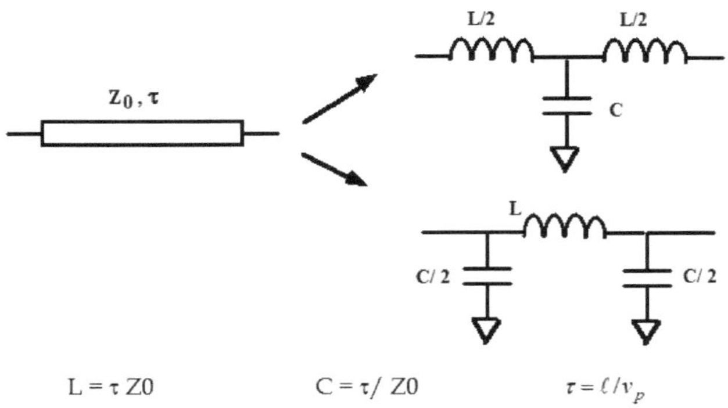

Figure 5.15 Basis for distributed matching using transmission line segments: the equivalent circuit model of a *short* transmission line. Refer to Chapters 2 and 3 for the properties of transmission lines.

5.3.1 Shunt inductor

Let's approximate a *shunt inductor* with a transmission line section (Figure 5.16) where v_p is the velocity of propagation.

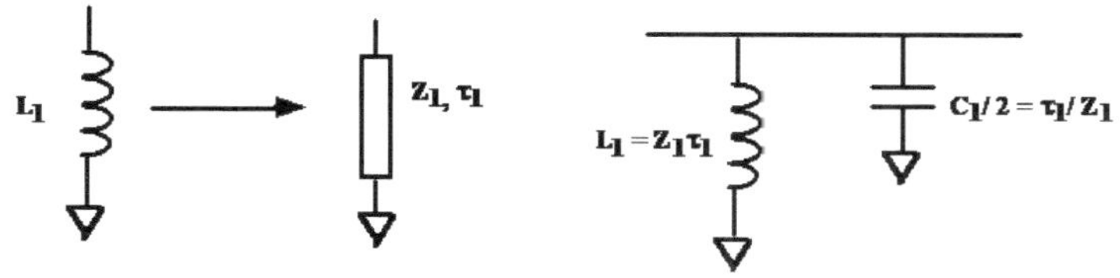

Figure 5.16 Equivalent circuit of the shunt transmission line.

So, we obtain the inductor L_1 we desire, together with a $C_1/2$ which we do not want. C_1 does vary as $1/Z_1$ and L_1 as Z_1, so using a high impedance line greatly helps to reduce C_1 relative to L_1. To make a good inductor, we need to keep C_1 small.

5.3.2 Series inductor

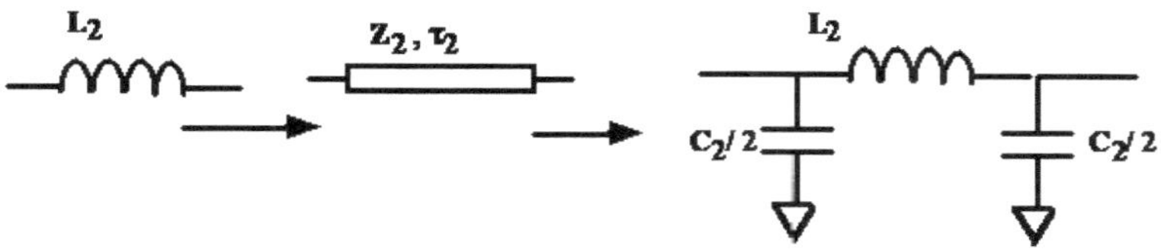

Figure 5.17 Equivalent circuit of a series transmission line.

Again, $L_2 = Z_2\tau_2 \quad C_2 = \tau_2/Z_2$ So, Z_2 shculd be high.

5.3.3 Shunt capacitor

A shunt capacitor is shown in Figure 5.18.

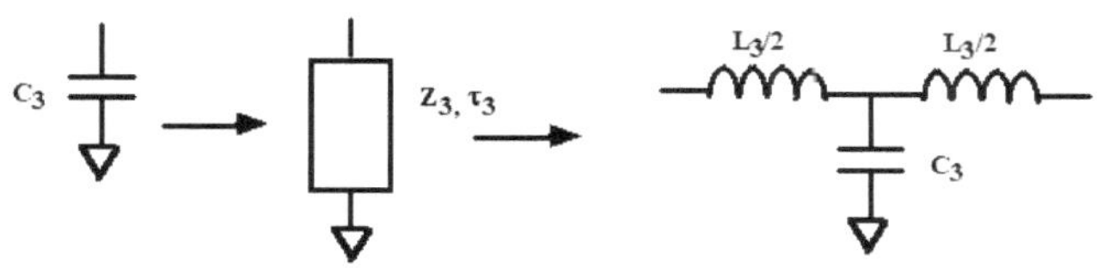

$C_3 = \tau_3/Z_3$ $L_3 = \tau_3 Z_3$ So, Z_3 should be kept low to minimize L_3.

Figure 5.18 Representation of a shunt capacitor with a transmission line.

We started with the circuit shown in Figure 5.19.

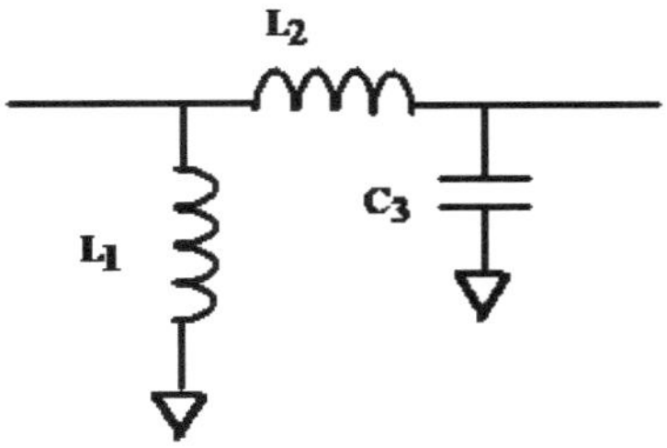

Figure 5.19 Lumped impedance network to be approximated with distributed elements.

And approximated it with transmission lines (Figure 5.20).

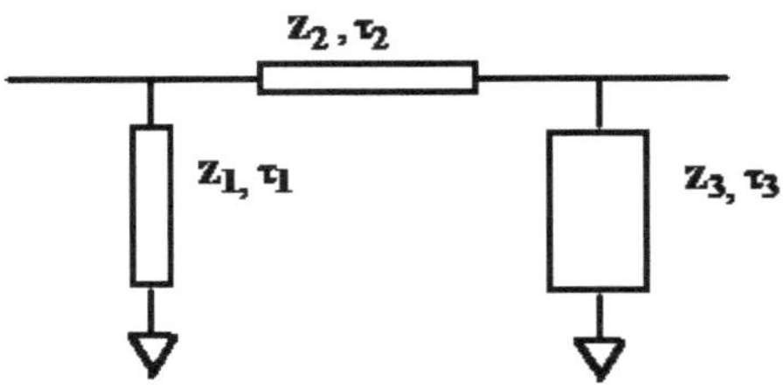

Figure 5.20 Transmission line implementation of Figure 5.12.

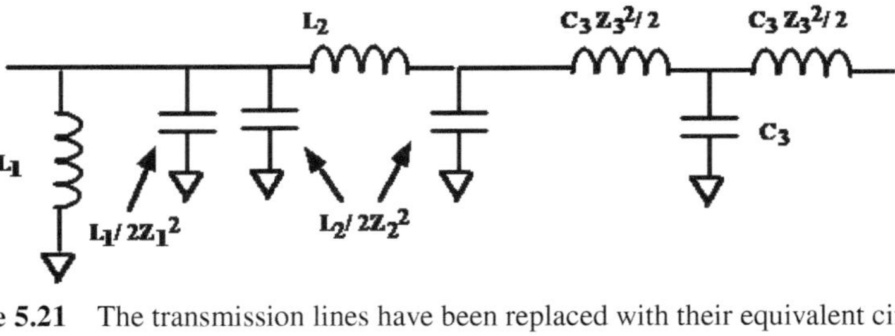

Figure 5.21 The transmission lines have been replaced with their equivalent circuits.

Which has an equivalent circuit approximately like that in Figure 5.21.

If Z_1 and Z_2 are sufficiently high and Z_3 sufficiently low, this will approximate the desired network.

5.4 Design of Transmission Line Matching Networks

As shown in the Section 5.3, T-line sections can substitute for lumped matching elements in L networks. Instead of starting from the lumped equivalent circuit to represent the transmission line, it is often easier to utilize the Smith chart to determine the required line lengths and impedances. Starting from the known load impedance Z_L,

we have two choices for shunt elements. Typically 90° or less:

$$\text{open or shorted} \quad |\Gamma| = 1 \angle = 0 \text{ or } -180°.$$

Then, a length l can be selected that can compensate for reactance or susceptance.

Series elements will present a constant $|\Gamma|$ when normalized to the Z_0 of the Smith chart. The trajectory then is a circle arcund the center of the chart.

Before we do an example, let's go back to a few equations that provide additional insight into the process of selecting line impedances Z_0 and lengths l.

Here is the equation that relates the input impedance seen by a length of transmission line of impedance Z_0, terminated in Z_L, of electrical length βl (Figure 5.22).

$$\boxed{Z_{IN} = Z_0 \frac{Z_L + jZ_0 \tan \beta \ell}{Z_0 + jZ_L \tan \beta \ell}} \tag{5.14}$$

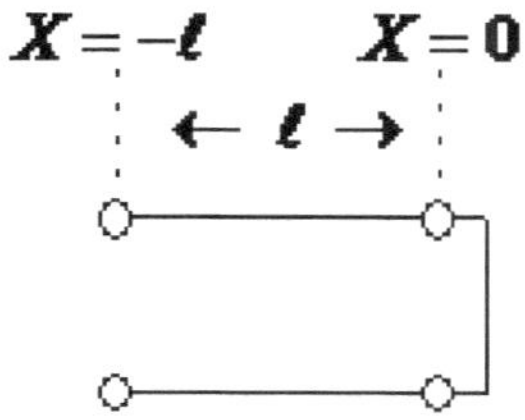

Figure 5.22 A shorted stub has a load impedance: Z_L = 0.

Therefore from Equation (5.14),

$$Z_{IN} = jZ_0 \tan \beta \ell \text{ for shorted stub} \tag{5.15}$$

where

$$\beta = \frac{2\pi}{\lambda}. \text{ So, if } \ell = \frac{\lambda}{8}, \beta \ell = \frac{\pi}{4}, \tan \beta \ell = 1 \tag{5.16}$$

and we get an inductor with $jX_L = jZ_0$. Of course, other lengths can provide an impedance between ∞ for a quarter wave stub (useful for bias network insertion) and for zero length (not very useful)

Figure 5.23 Open stub: $Z_L = \infty$.

An open-circuited line or "stub" has Z_L = infinity (Figure 5.23) therefore, the input impedance can be represented in equation 5.17.

And again from Equation (5.14),

$$Z_{IN} = -jZ_0 \cot \beta\ell. \tag{5.17}$$

Therefore, if

$$\ell = \frac{\lambda}{8},$$

we get a shunt capacitor with

$$jX_c = jZ_0. \tag{5.18}$$

An open stub can provide a short for a quarter wave length and provide other impedances in between.

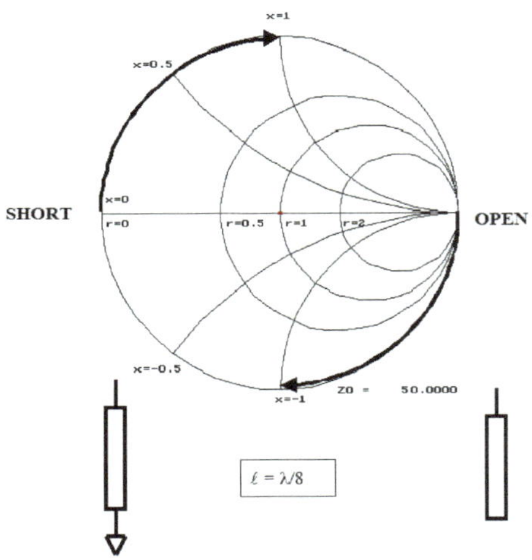

Figure 5.24 The Smith chart illustrates the impedance transformations for eighth-wave open and shorted stubs.

These equations can be represented graphically by the Smith chart. Figure 5.24 shows the case where an open or shorted line has a length of an eighth wavelength. If the lines are of lengths other than eighth-wave, the angle $2\beta l$ or the wavelength scale on the perimeter of the chart can be used to find the desired Z_{IN} for the given length.

5.4.1 Series transmission lines

While the derivation was described in Chapter 3 , let's look at again at how a load impedance can be transformed by a series line. Here, it may be easier to grasp if we think in terms of reflection coefficient.

$$\begin{aligned} \boldsymbol{V}^{-}(0) &= V^{+}(0)\Gamma_L \\ \boldsymbol{\Gamma}_L &= \frac{z_L - 1}{z_L + 1}, \end{aligned} \tag{5.19}$$

where Z_L is the normalized load impedance $Z_L/Z_0.\Gamma_L$ is in general complex. From Chapter 3,

$$\Gamma(x) = \Gamma(0)e^{2j\beta x}. \tag{5.20}$$

The reflection coefficient retains the same magnitude but changes phase with position (Figure 5.25).

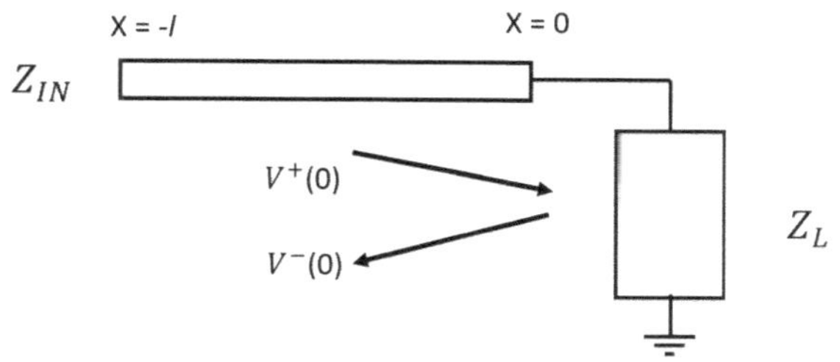

Figure 5.25 The variable x is defined as shown.

Figure 5.26 illustrates the effect of a quarter-wave line length. This special case is often referred to as a quarter-wave transformer.

$$\Gamma(-\lambda/4) = \Gamma(0)e^{-j\pi}. \tag{5.21}$$

In other words, the angle of the reflection coefficient is rotated clockwise 180° for a quarter-wave line (negative angle = clockwise rotation).

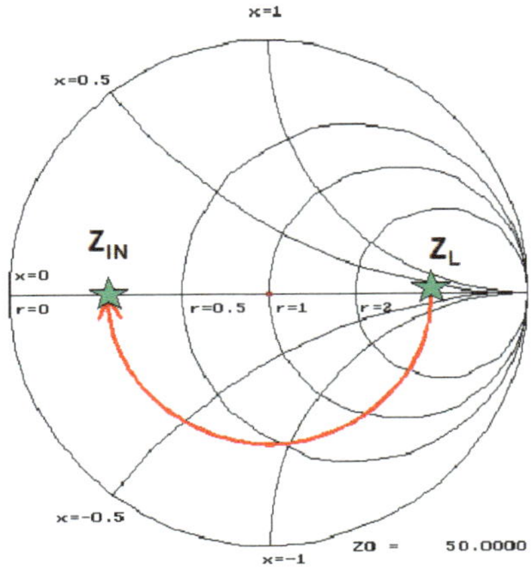

Figure 5.26 The phase of the reflection coefficient changes in a clockwise circle as we shift the reference plane in the negative direction.

However, we are not limited to this specific length. In the general case, the reflection coefficient goes through a phase shift of:

minus $\mathbf{2\pi(2)(\ell/\lambda)}$ radians

minus $\mathbf{2\beta\ell}$ radians

minus $\mathbf{360(2)(\ell/\lambda)}$ degrees.

5.5 Transmission Line L Network Design Examples

Our goal as usual is to match the load to the source. Let's start with a normalized load impedance $z_L = 1.8 + j1.9$ at a design frequency of 1 GHz. We want to match this load to a 50 Ω source impedance. There are many possible solutions to this design.

1. The first example, shown in Figure 5.27, uses a combination of series and shunt transmission lines, all of characteristic impedance $Z_0 = 50$.

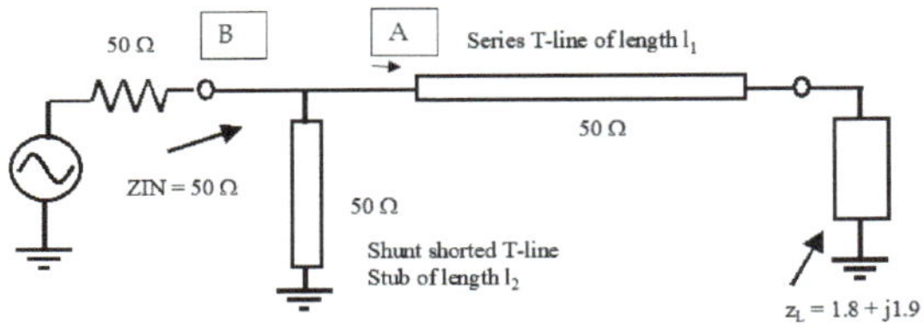

Figure 5.27 Matching network to be determined.

First step (Figure 5.28): Determine length of the series T-line ℓ_1 necessary to transform the load impedance so that it intersects the unit conductance circle. Using a 50Ω YZ Smith chart, draw a circle with radius $|\Gamma_L|$ around the center of the chart. Moving clockwise from the load impedance (negative angle $2\beta\ell_1$ since $\Gamma(x) = \Gamma(0)e^{2j\beta x}$ and $x = -\ell_1$), we arrive at point A on the unit conductance circle. The length in wavelengths can be determined from the outside wavelength scale around the perimeter of the Smith chart. If $\beta = 2\pi/\lambda_{ref}$, then the wavelength scale represents ℓ (in units of wavelength λ_{ref}). We can later determine the physical length of the line from the frequency and phase velocity.
Draw a straight line from the center of the chart through Z_L. This intersects the wavelength scale at 0.204λ. Add a series line until the unit conductance circle is reached at point A. Next, draw another straight line through point A. This intersects the scale at 0.427λ. So, the electrical length of the required series line in wavelengths is $0.427 - 0.204 = 0.223\lambda$. Converting to electrical length in degrees: $0.223 \times 360 = 80.2°$.

$$E = 360\ell/\lambda_{ref.} \tag{5.22}$$

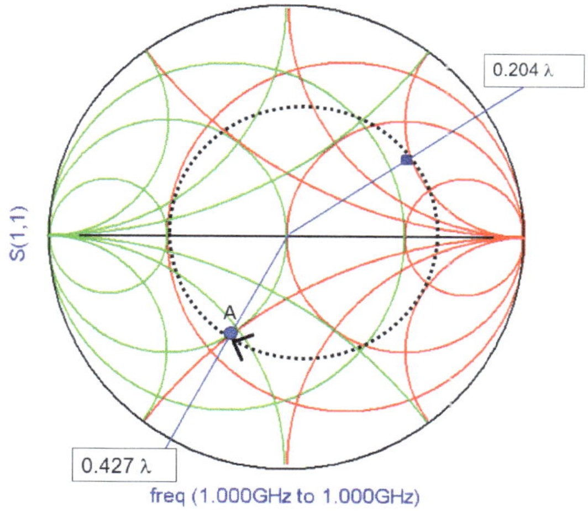

Figure 5.28 Illustration of first step through the series transmission line of length 0.223 λ_{ref}.

The **second step** (Figure 5.29) is to apply shunt susceptance from the shorted stub. According to the chart, we now have a normalized admittance $y_A = 1.0 + j1.53$. Thus, we must add $b = -1.53$ to cancel the susceptance. We will then arrive at point B, 50 Ω.

To determine the required line length, start from the short circuit and find the line of constant susceptance corresponding to $b = -1.53$. The difference in wavelength $0.092 \times 360 = 33.1°$ gives us the required length in degrees.

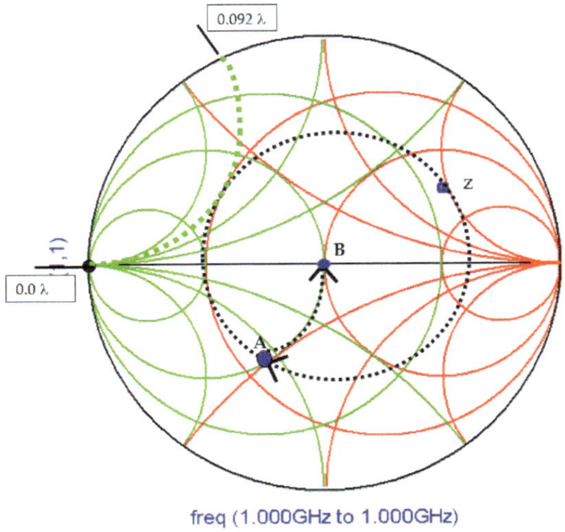

Figure 5.29 Second step. Add shunt shorted stub at the end of the series line. $b = -1.53$.

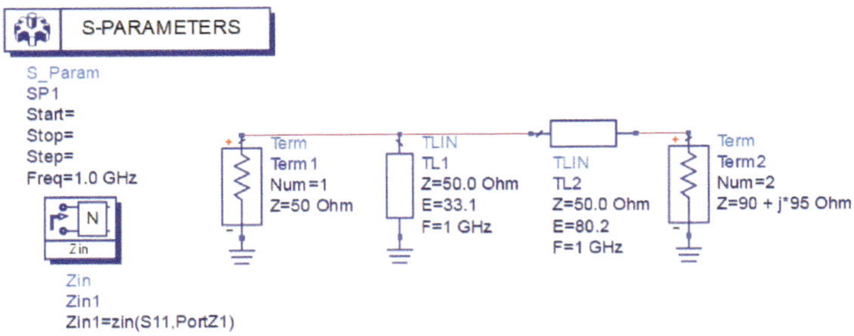

Figure 5.30 Schematic of ADS simulation of the above example.

The S-parameter simulation in ADS shown in Figure 5.30 used ideal transmission lines (TLIN). Input impedance can easily be calculated using the Zin function as seen in Figure 5.31. Of course, more accurate T-line models could have been substituted using LINECALC.

freq	Zin1	S(1,1)
1.000 GHz	50.190 / 0.019	0.002 / 4.872

Figure 5.31 Simulation output Zin1 and S(1,1).

We see that this design was very effective in matching the load impedance.

5.6 Summary

1. Impedances must be matched if the full available power from the source can be delivered to the load (we will see in subsequent chapters that this is not always the correct approach if other parameters such as noise figure or stability are critical).
2. Both lumped element and transmission line distributed networks can be employed to provide the needed impedance transformation.
3. In many cases, use of the Smith chart can provide a quick and intuitive solution. The design can always be verified using ADS.

References

[1] Gonzalez, G. (1984) Microwave Transistor Amplifiers, 2nd edn. Prentice-Hall, Section 2.4
[2] Bowick, C. (2008) RF Circuit Design, 2nd edn. Elsevier/Newnes, Ch. 4.

Homework

1a. Determine Z_{IN} and Q of the three-element T matching network below. $Z_0 = 50\ \Omega$ (assuming a frequency of 1 GHz).

b. Design a PI network matching circuit that provides the same normalized input impedance Z_{IN} and same Q. You can leave the element values for the new network as normalized susceptances and reactances. (See Appendix 5.2 for three element matching networks).

c. Verify your result on the Smith chart and on ADS.

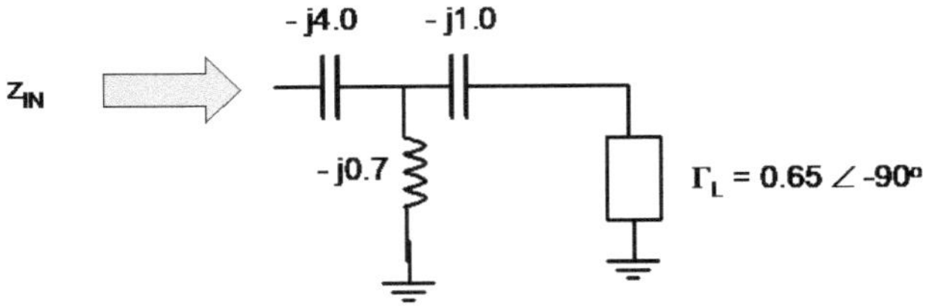

2a. Design the distributed matching network shown below to match Z_S to Z_L for maximum power transfer. You need to determine the length l_1 of series 50 Ω line (in wavelength or degrees) and choose the characteristic impedance Z_{02} of the $\lambda/8$ shunt stub. You must also decide whether to use an open or shorted shunt stub. Use the ADS simulator to verify the design.

b. What is the maximum Q for this design?

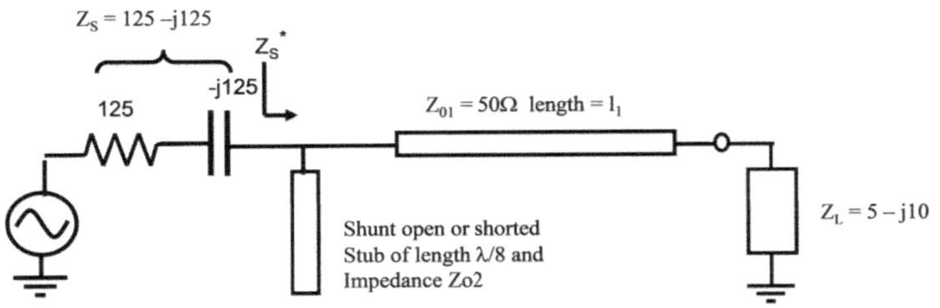

3a. The pi network below has been designed to operate at a frequency $\omega = 1 \times 10^9$ rad/sec. Find the input impedance Z_{IN} at this frequency.

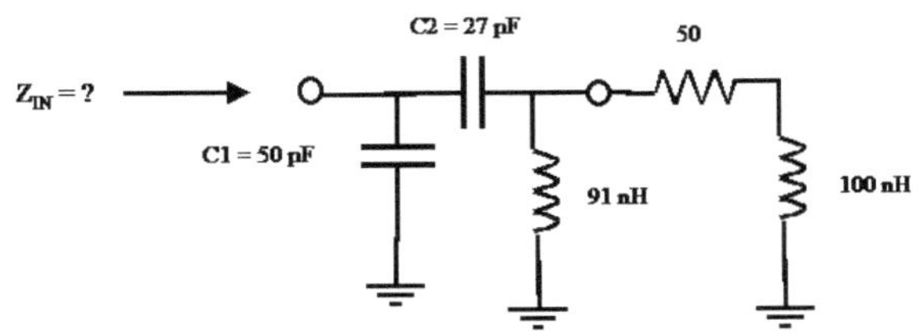

b. What is Q of this network?

4. Design a matching network using 50 Ω transmission lines that will provide for maximum power transfer between source with reflection coefficient Γ_S and load impedance, Z_L.

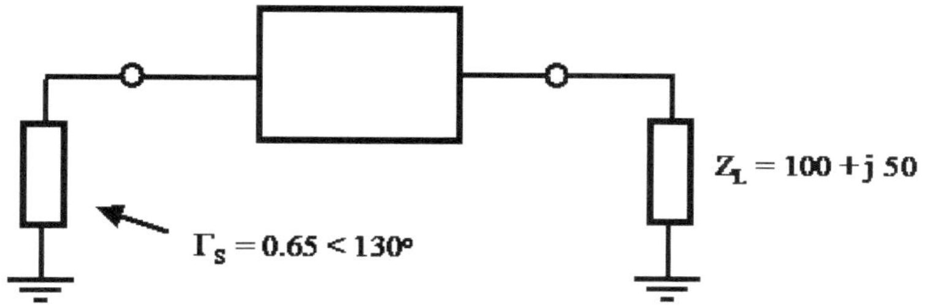

5.

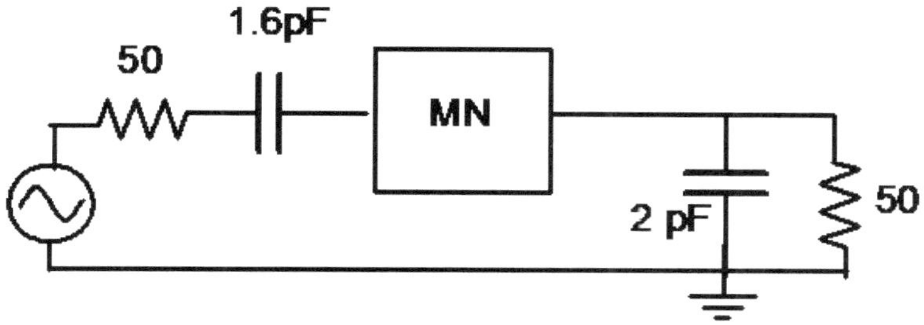

a. Design a distributed matching network (MN) that will allow the full available source power to be delivered to the load at a frequency of 2 GHz. Specify the electrical lengths of the components in wavelengths. Show your design on the Smith chart.
b. What is the Q of this matching network?
c. Now design an equivalent MN at 2 GHz using ideal lumped element components to perform the same function. Again, show your design on the Smith chart.
d. Verify each design using ADS.

Appendix 5.1. Discrete Distributed Matching Networks[1]

It is helpful to think of transmission lines in both their equivalent circuit form and in a distributed form.

As seen in Figure 5.32, when we merge all of these sections together we obtain the ordinary transmission line in Figure 5.34.

---

[1] Professor Mark Rodwell's notes on lumped-distributed networks for absorbing input capacitance of FETs into an artificial transmission line were used. This technique is used for the travelling-wave wideband amplifier design.

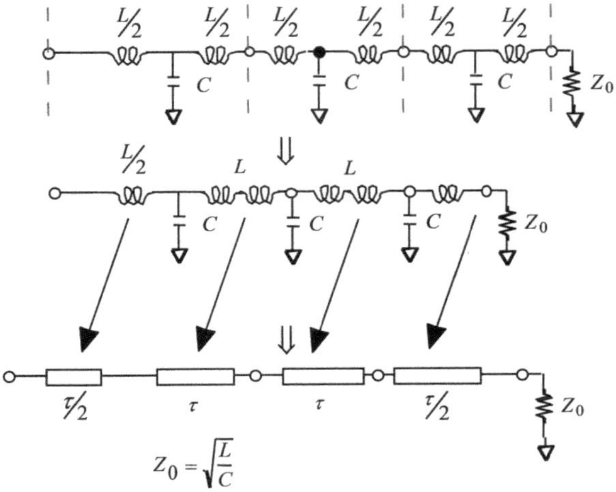

Figure 5.32 Correspondence between lumped element and transmission line representations.

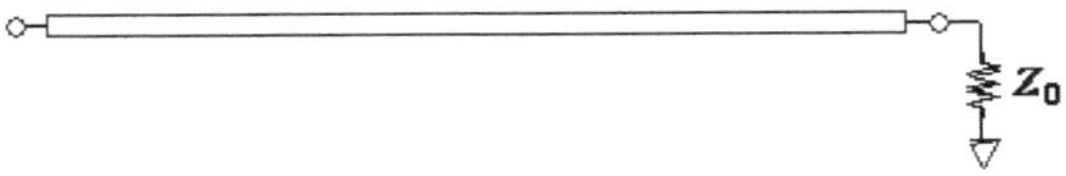

Figure 5.33 Equivalent transmission line from Figure 5.32.

What would happen new if we add extra capacitance to the line?

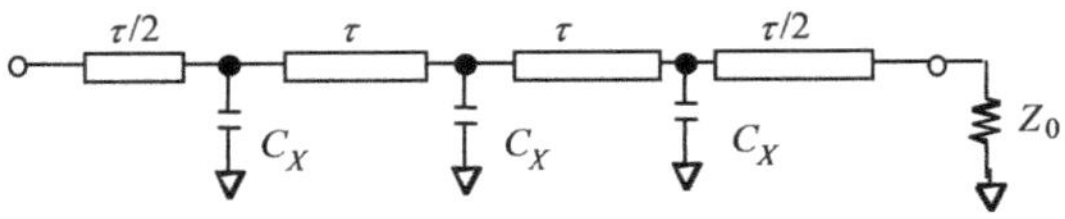

We have changed Z_0 of the composite line:

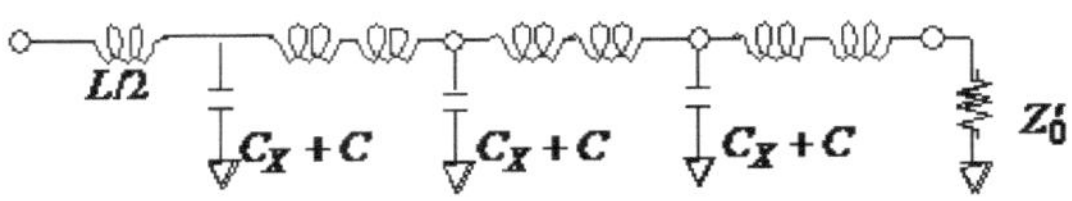

$$Z_0' \cong \sqrt{\frac{L}{C+C_x}} \quad \tau \cong \sqrt{L\,(C+C_x)} \text{ (per section)} \tag{5.23}$$

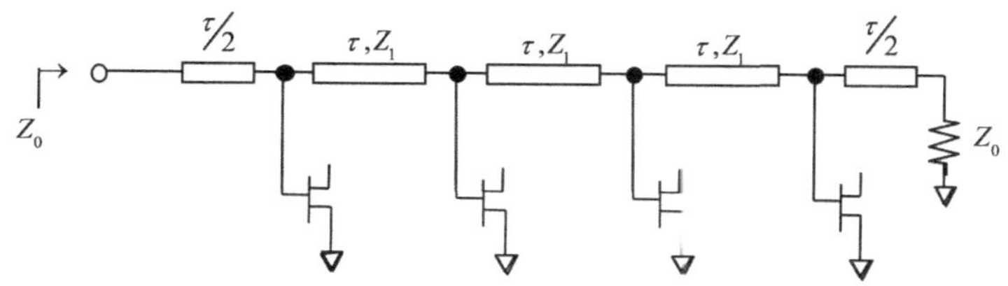

Figure 5.34 Add capacitive loading at periodic intervals on the line segments.

It can be seen from Equation (5.23) that the extra capacitance will have the effect of reducing the line characteristic impedance.

We also have now a frequency limitation on the transmission line: the Bragg cutoff frequency:

$$\omega_C = \frac{2}{\sqrt{L(C + C_X)}}. \tag{5.24}$$

This equation is limited to $\omega \ll \omega_C$.

This occurs when you construct an artificial line with discrete L and C.

$$L = \tau Z_0 \qquad\qquad C = \tau / Z_0 \tag{5.25}$$

Shorter line sections (small T) lead to higher ω_C.

A5.1.2. Why do we care?

Nice trick for broadband designs.

1. Distributed or traveling wave amplifier

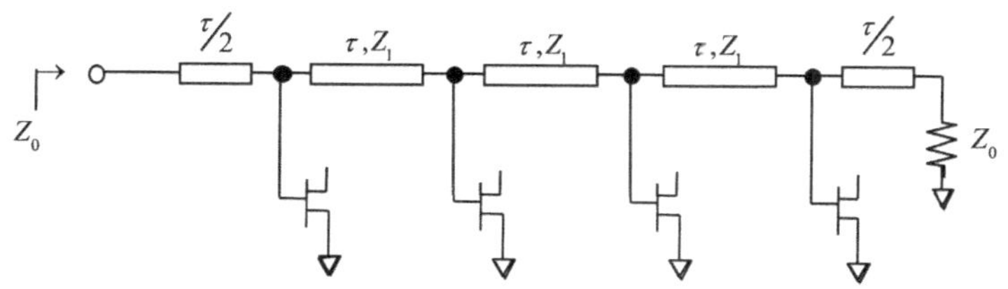

Figure 5.35 Cgs of FETs is absorbed into transmission line

$$L = Z_1 \tau$$

2. Wideband input match

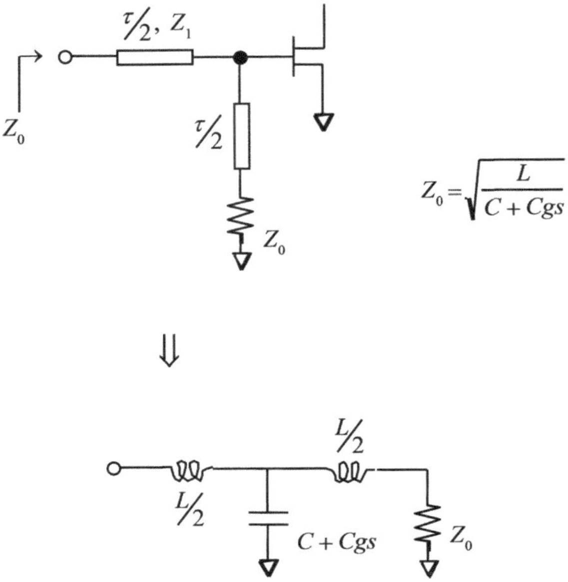

Figure 5.36 The FET Cgs is inserted into a transmission line. It changes the impedance Z_0. Equation (5.23).

Appendix 5.2: Three-element Matching Networks

Why three elements instead of two?

Q of an L network is determined by the resistance ratio $Q = \sqrt{\frac{R_2}{R_1} - 1}$.

There is no freedom to change Q. If *higher* Q is desired, then a three-element network is needed.

Why would we want higher Q?

For narrow bandwidth applications. We will get better suppression of out-of-band frequencies.

It also provides more opportunity for parasitic absorption in active circuits.

A5.2.1._PI network

X_2 must be opposite to X_1, X_3. In other words, If X_1 and X_3 are capacitors, then X_2 must be an inductor. This would produce a low-pass type of match. If X_1 and X_3 are inductors, then X_2 is a capacitor. This will produce a high pass.

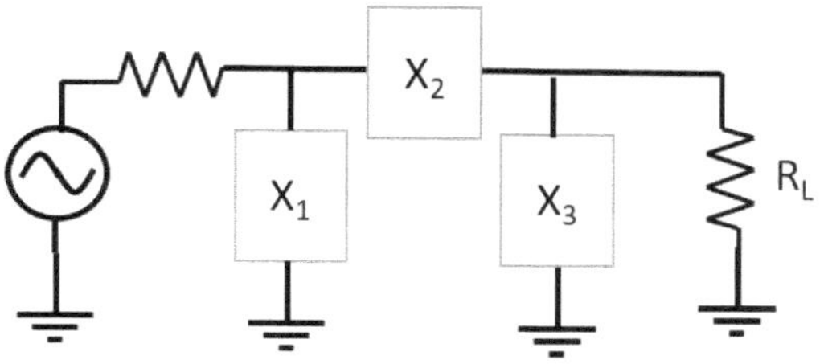

Figure 5.37 Pi network.

The Pi network (Figure 5.37) can be considered as two back-to-back L networks. Determine a virtual resistance at the middle of the L networks and match R_s and R_L to this common fake resistance. Note that Q will be different for the two L networks. The desired Q should be determined by the largest R to R_s or R_L ratio.

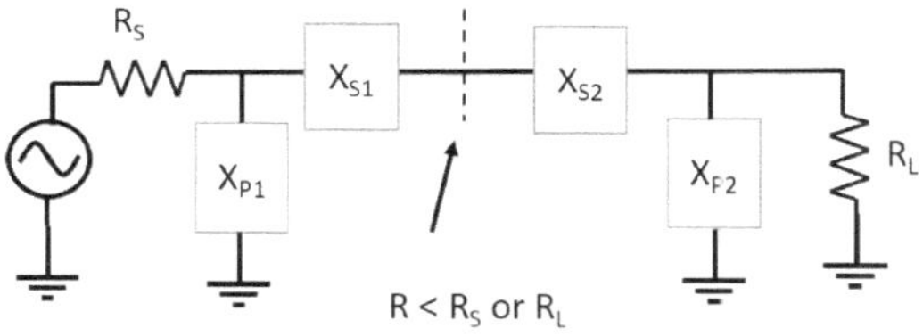

Figure 5.38 Use the desired Q to determine a virtual resistance R.

For π-network in Figure 5.37, use $Q \cong \sqrt{\frac{R_H}{R}} - 1$ where R_H is the higher of the two resistances R_L or R_S.

$$R \cong \frac{R_H}{Q^2 + 1}. \tag{5.26}$$

Design both sides to match to this R at the center of the two networks [2].

Example 5.4: Let's redesign the Example 5.2 L network for a low pass pi network.

Suppose $f = 1590$ MHz, $\omega = 1 \times 10^{10}$ rad/sec, $R_L = 500\ \Omega, R_S = 50\ \Omega$.

The Q would be 3 for the L network. Now, find the virtual resistance for a new $Q = 5$.

Use $R_L = R_H.R = 500/(25 + 1) = 19.2\Omega$

Then, $X_{P2} = R_L/Q = 100, C_2 = 1/\omega X_{P2} = 1\text{pF}, X_{S2} = QR = 96, L_2 = X_{S2}/\omega = 9.6$ nH.

Next, determine the Q of the left-hand network. This will be lower than 5.

$$\sqrt{\frac{R_S}{R}} - 1 = 1.26. \tag{5.27}$$

Then, $X_{P_1} = R_S/Q_1 = 39.7, C = 2.5$ pF$, X_{S1} = RQ_1 = 24.2, L_1 = 2.4$ nH.

Now, combine $L_1 + L_2 = 12$ nH.

Figure 5.39 shows the final design. Figure 5.40 shows that it worked!

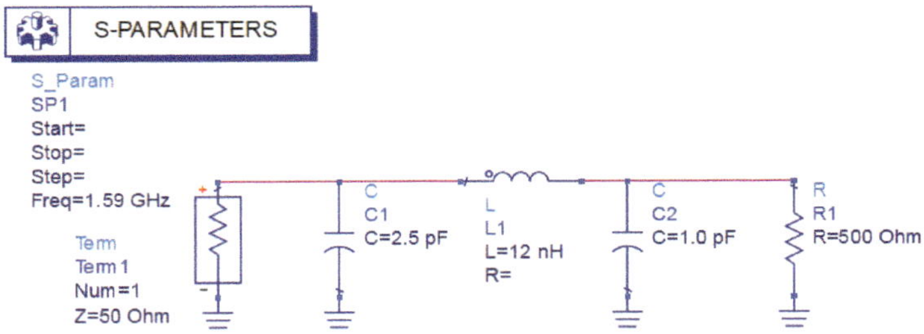

Figure 5.39 ADS schematic of the Pi Network.

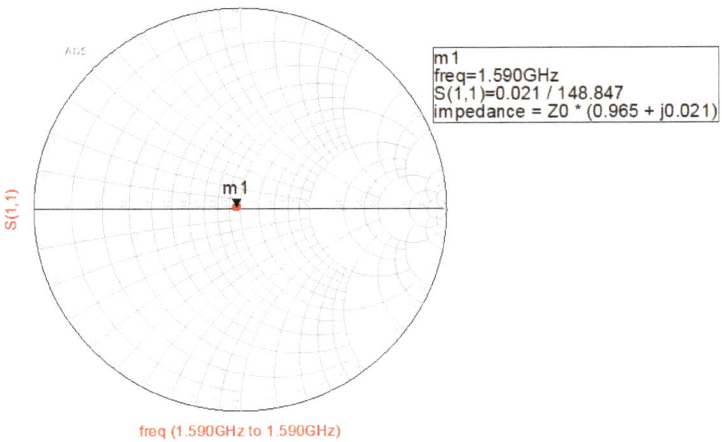

Figure 5.40 Marker m1 shows that S(1,1) is nearly a perfect match.

A5.2.2. T network

X_2 must be opposite to X_1, X_3 (Figure 5.41).

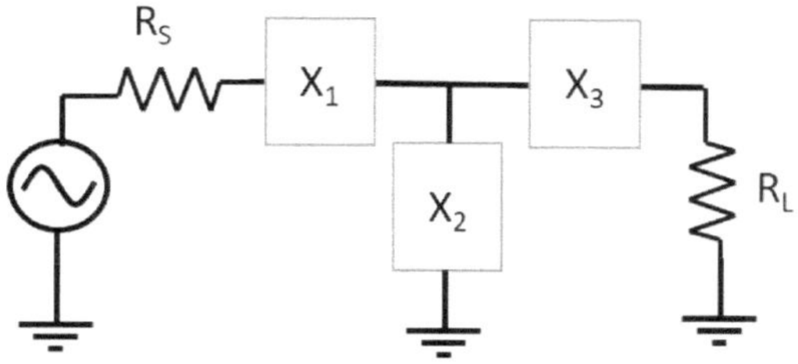

Figure 5.41 T-network.

Consider again as two back-to-back L networks (Figure 5.42).

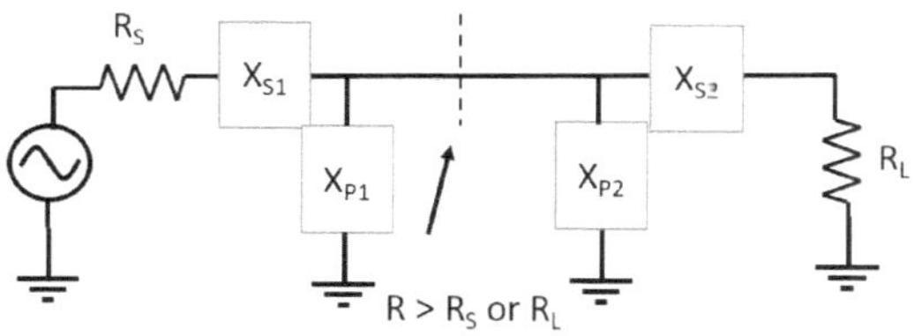

Figure 5.42 Virtual resistance *R'* at center of the two L sections.

R' $> R_S$ or R_L in this topology.

$$Q \cong \sqrt{\frac{R}{R_{\text{small}}} - 1}$$
$$R \cong \left(Q^2 + 1\right) R_{\text{small}} \tag{5.28}$$

$$R_{\text{small}} = \text{least of the two resistances } R_P \text{ or } R_S. \tag{5.29}$$

Can you still use the design equations when the source and load is complex?

Yes. Just absorb the series or parallel reactance/susceptance into the design.

1. Convert series Z_L to parallel equivalent.

Example 5.5:

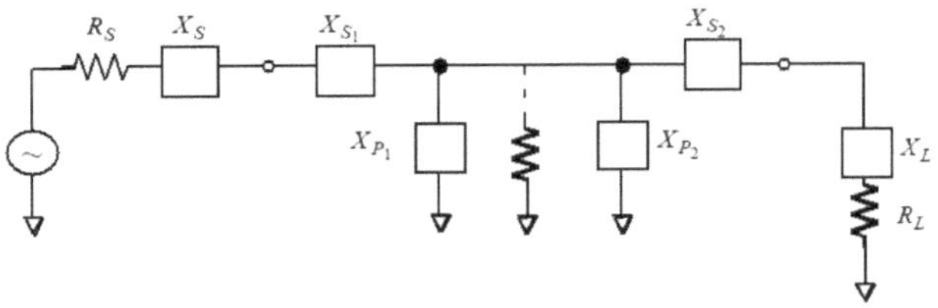

Figure 5.43 T network with included source and load reactances.

(a) Set the desired Q.

$$R = \left(Q^2 + 1\right) R_{\text{small}} \tag{5.30}$$

Then if $R_S < R_L$

$$\begin{aligned} X_{S_1\text{total}} &= X_S + X_{S_1} = QR_S \\ X_{P_1} &= \frac{R}{Q} \end{aligned} \tag{5.31}$$

(b) Find the Q for the other half L network.

$$\begin{aligned} Q' &= \sqrt{\frac{R}{R_L} - 1} \\ X_{S_{2\text{total}}} &= X_{S_2} + X_L = Q'R_L \\ X_{P_2} &= \frac{R}{Q'} \end{aligned} \tag{5.32}$$

(c) Combine X_{P1} and X_{P2} to absorb the load reactance X_L.

6

Small Signal Amplifier Design

In other courses you will have learned to design amplifiers using small signal models for devices. This works reasonably well at lower frequencies, but at high frequencies often the device small signal model is not accurate enough. Then, measured S-parameters can be used to accurately design the amplifier. These are only useful for small signal analysis. Large signal design requires a device model with DC biasing and includes nonlinear equivalent circuit components.

The S-parameter design technique employs relationships between input and output powers, and forward and reflected powers, which look scary at first but can easily be derived using the signal flow graph method and Mason's gain rules [1] .

Our topics will include:

1. Power gain definitions
2. Stability of amplifiers
3. Source and load stability circles
4. Transducer gain circles
5. Bilateral design
6. Wideband stabilization
7. Limitations of Rollett's [3, 4, 5] stability factor.

Goal: Learn to design stable narrowband amplifiers using S-parameters

6.1 Transducer Power Gain

Recall the definition of the S-parameters in Chapter 4 (Figure 6.1):

$$\begin{aligned} b_1 &= S_{11}a_1 + S_{12}a_2 \\ b_2 &= S_{21}a_1 + S_{22}a_2 \end{aligned} \tag{6.1}$$

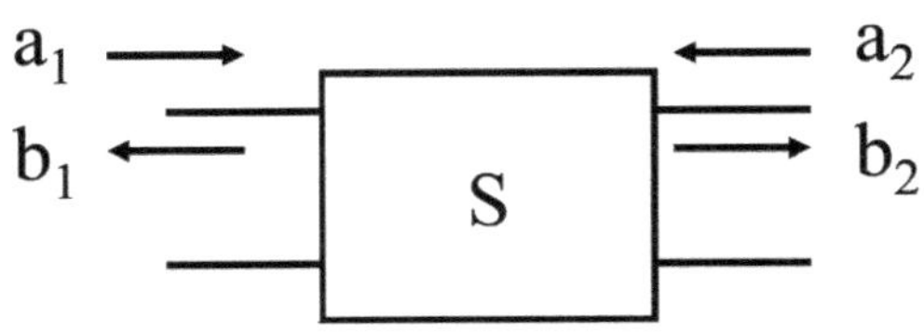

Figure 6.1 S-parameter definitions applied to a transistor. Recall that S-parameters are small-signal definitions.

Consider the forward transmission and calculate the transducer power gain:

$$S_{21} = \frac{2V_{out}}{V_{gen}} = \left.\frac{b_2}{a_1}\right|_{a_2=0}. \tag{6.2}$$

In general, for an arbitrary R_S and R_L,

$$P_{AVS} = \frac{V_{gen}^2}{8R_S} \qquad P_L = \frac{V_{out}^2}{2R_L}. \tag{6.3}$$

The definition of transducer power gain, power delivered to the load over available source power is:

$$G_T = \frac{P_L}{P_{AVS}}. \tag{6.4}$$

So, for the special case where $R_S = R_L = Z_0$,

$$|S_{21}|^2 = \frac{4V_{out}^2}{V_{gen}^2} = \frac{V_{out}^2}{2Z_O}\frac{8Z_O}{V_{gen}^2} = G_T. \tag{6.5}$$

However, life is generally not that straightforward because $|S_{21}|^2$ is often much less than the optimum gain that you could obtain from a given transistor. You must add matching networks to transform Z_0 to a more suitable Γ_S and Γ_L (Figure 6.2).

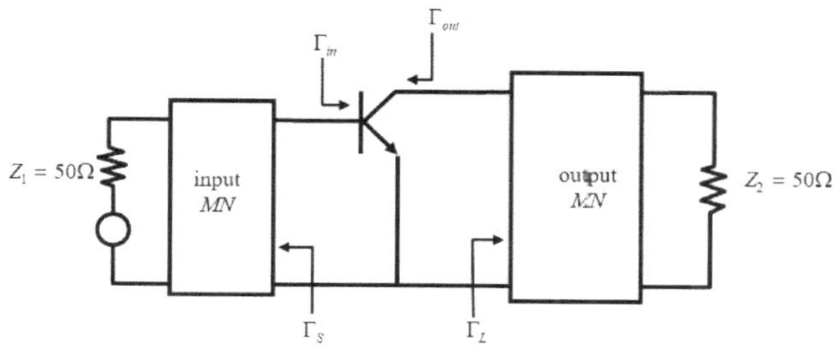

Figure 6.2 Amplifier block diagram showing reflection coefficients at each port.

How do we calculate gain from S-parameters?

Evaluate the appropriate gain equation. *Transducer gain* is defined as power delivered to the load divided by the available power from the source.

$$G_T = \frac{P_L}{P_{AVS}} \tag{6.6}$$

$$G_T = \frac{1 - |\Gamma_S|^2}{|1 - S_{11}\Gamma_S|^2} |S_{21}|^2 \frac{1 - |\Gamma_L|^2}{|1 - \Gamma_{OUT}\Gamma_L|^2} \tag{6.7}$$

where:

$$\Gamma_{OUT} = S_{22} + \frac{S_{12}S_{21}\Gamma_S}{1 - S_{11}\Gamma_S} \tag{6.8}$$

So, if you are given the S-parameters and Γ_S, Γ_L then you can calculate the gain. But this may not be valid unless the amplifier is known to be stable.

Note also that Γ_{out} depends on Γ_S unless $S_{12} = 0$!

Other gain definitions can also be used for specific purposes

$$\text{Operating power gain } = G_P = \frac{\text{Power delivered to load}}{\text{Power input to network}} \tag{6.9}$$

$$G_P = \frac{1}{1 - |\Gamma_{IN}|^2} |S_{21}|^2 \frac{1 - |\Gamma_L|^2}{|1 - S_{22}\Gamma_L|^2}. \tag{6.10}$$

- This can be useful because it eliminates the dependence of gain on Γ_s – helpful when the device is *bilateral* – and passes signal both ways. It assumes that the input is matched.

$$\text{Available power gain } = G_A = \frac{\text{Power available from network}}{\text{Power available from source}} \quad (6.11)$$

$$G_A = \frac{1-|\Gamma_S|^2}{|1-S_{11}\Gamma_S|}|S_{21}|^2\frac{1}{1-|\Gamma_{OUT}|^2}. \quad (6.12)$$

- Used in noise calculations, it eliminates dependence of gain on Γ_L – helpful for low noise amplifier design. It assumes that the output is matched.

Derivations can be found in Section 2.6 and equations 3.2.1–3.2.4 of [2] .

Calculating power gains from S-parameters is a mechanical process – or you can use CAD tools such as Pathwave/Keysight ADS.

6.2 Stability of Amplifiers

Suppose the amplifier specifications are presented in a design sense: given a device, design input and output matching networks for a particular value of G_T.

Here, we find many possible solutions. To determine the best solution, we need to first consider the stability of the amplifier – we must guarantee that the amplifier does not oscillate under the expected source and load impedances. This is often easier said than done. The traditional methodology using Rollett's stability factor is based on S-parameters, but is incomplete [3, 5]. It is useful, however, because it provides insight into how to select source and load impedances that are likely to lead to stability. Further tests are often necessary to verify the design, especially for fully integrated RFIC designs which do not provide access to modifying the design once they are fabricated. Also, there may be more than one stage to the amplifier. In that case, access to internal nodes from the external amplifier S-parameters is not possible. Reference will be made later to some tools that improve upon stability verification. But first, look at:

- Stability factor
- Stability circles
- Gain circles.

Then, once a stable matching condition region in the Γ_S and Γ_L planes is identified, gain circles can be plotted to assist in selecting a Γ_S and Γ_L that is least susceptible to circuit variances.

Goal:

- Robust design
- Stable and repeatable

6.2.1 Basis for stability of amplifiers

First let's review the concept of negative resistance (Figure 6.3).

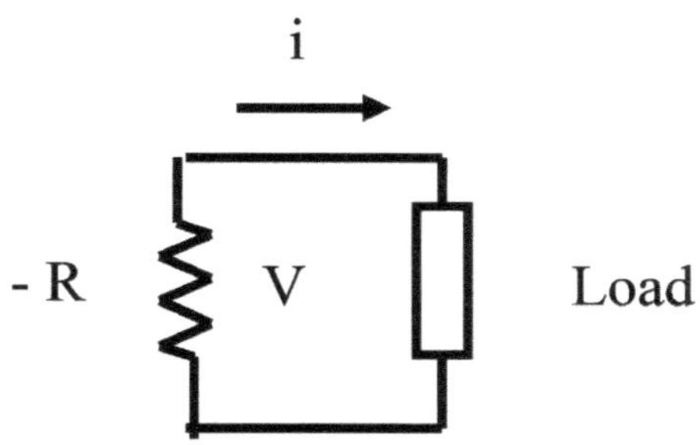

Figure 6.3 Note that the negative resistor has the opposite to the passive sign convention.

The negative resistor delivers power into the load rather than dissipating power as a positive resistor does.

Why does this lead to instability? Given the definition of negative resistance in Figure 6.3, the series network R_L, L, and C in Figure 6.4 will receive current from $-R$.

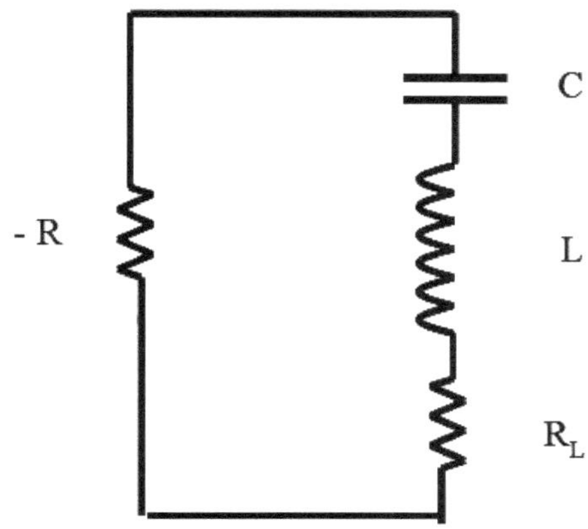

Figure 6.4 RLC circuit where $-R$ is equal to or greater than R_L.

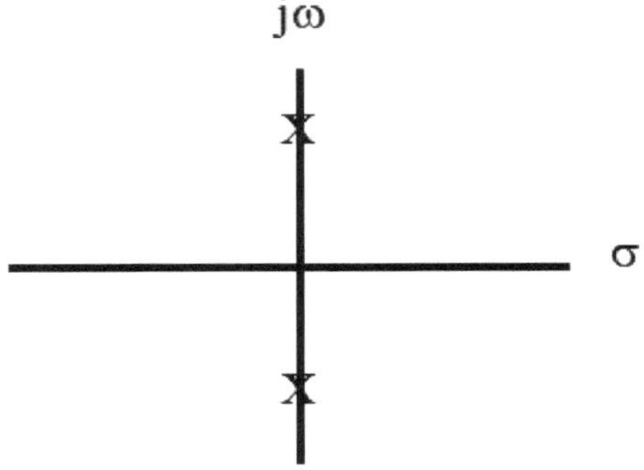

Figure 6.5 S-plane. If $|-R| = R_L$, there is no net resistance in the loop. This locates the complex poles right on the $j\omega$ axis.

The transient response (from Laplace transform) then has the form $Ae^{j\omega t}$, a sustained sinusoidal oscillation. Or if negative *R* exceeds positive, right-hand plane poles may exist. Poles in the right-hand plane are even worse and will produce an unstable system, but usually at certain frequencies, often outside the desired center frequency of the narrowband design. Loop gain methods using Bode plots and Nyquist diagrams can also reveal instability [4] .

6.2.2 Stability

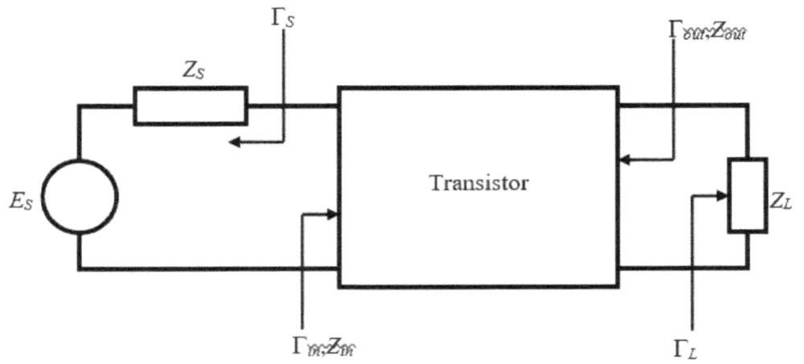

Figure 6.6 Oscillation is possible if either input or output port has a negative resistance.

Referring to Figure 6.6, if a net negative real part exists, that is if Re$\{Z_S + Z_{in}\}$ or Re$\{Z_{out} + Z_L\}$<0, then transient response will grow and oscillation will occur.

Unconditional stability:

$$\left.\begin{array}{l}|\Gamma_S| < 1\\ |\Gamma_L| < 1\end{array}\right\} \text{for any passive source and load,}$$

$$|\Gamma_{in}| = \left|S_{11} + \frac{S_{12}S_{21}\Gamma_L}{1 - S_{22}\Gamma_L}\right| < 1 \tag{6.13}$$

and

$$|\Gamma_{out}| = \left|S_{22} + \frac{S_{12}S_{21}\Gamma_S}{1 - S_{11}\Gamma_S}\right| < 1 \tag{6.14}$$

This unconditional stability is rather unusual for most microwave- or millimeter-wave devices of interest which will have lots of gain at lower frequencies well below f_T. So, we must establish a method to determine regions in the Γ_S and Γ_L plane that are stable. We can then either avoid the unstable regions or modify the transistor with resistive loading to make it unconditionally stable.

Conditional stability: This is the usual case. Also known as potentially unstable.

$\text{Re}\{Z_{in}\}$ and $\text{Re}\{Z_{out}\} > 0$ *for some* $|\Gamma_s| \leq 1$ and $|\Gamma_L| \leq 1$ at *some specific* frequency.

Stability factor: The first step in designing the amplifier with the S-parameter method is to determine whether the transistor is likely to be *unconditionally stable* or *potentially unstable*. This can be easily estimated using the *k* stability factor (Rollett [3]) and delta, the determinant of *S* [2, 3]. The amplifier will in most cases be unconditionally stable if:

$k > 1$ and $\text{mag}(\Delta) < 1$

$$k = \frac{1 + |\overbrace{S_{11}S_{22} - S_{12}S_{21}}^{\Delta}|^2 - |S_{11}|^2 - |S_{22}|^2}{2\,|S_{12}|\,S_{21}\,|} > 1 \tag{6.15}$$

$$\text{and } |\Delta| = |S_{11}S_{22} - S_{12}S_{21}| = \det S < 1. \tag{6.16}$$

1. If a transistor is potentially unstable, typically: $|\Delta| < 1$ and 0<k<1
2. Negative *k* values can occur, but result in most of the Smith chart producing instability.

   So, life can be much easier when you choose a device that will be unconditionally stable. But, this may lead to designs that don't push the edge of performance.

3. Also, you must check for stability *at all frequencies* for which the device has $k < 1$. This is valid only when the S-parameter set for the device has been measured over the required range of frequencies.
4. ADS can be used to plot these factors over a range of frequencies, as seen in Figures 6.7 and 6.8.

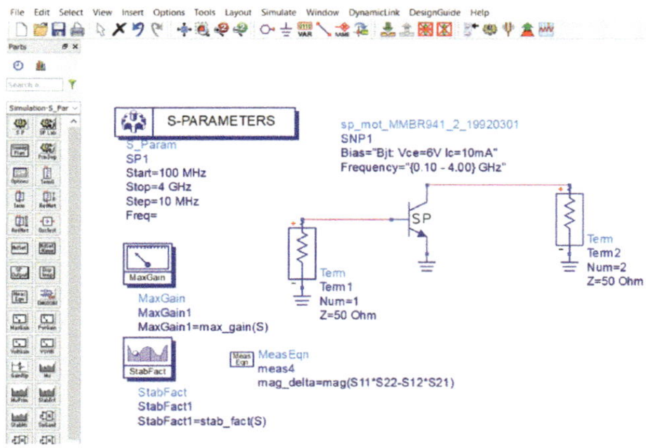

Figure 6.7 Wide frequency sweep of a device (100 MHz to 4 GHz) in order to determine gain and whether the device is potentially unstable and at what frequencies.

These factors are easily calculated using Measurement Equations in the ADS schematic panel. *K* is pre-programmed in as StabFact which can be selected from the S-parameter palette on the left. Delta can be programmed yourself using a blank MeasEqn.

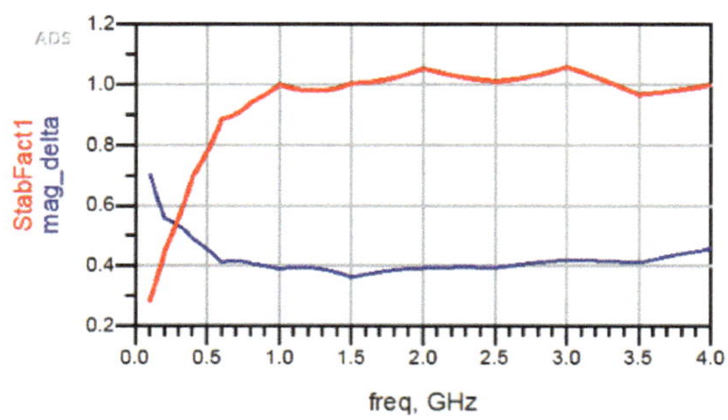

Figure 6.8 Plot of *K*-factor (StabFact1) and delta.

K-factor (StabFact1) and delta are plotted in a frequency sweep from 100 MHz to 4 GHz. You can see that the device is potentially unstable below 1.5 GHz where $k < 1$. The use of the k-factor to assess stability is convenient and is a necessary condition for stability; however it has been shown to be incomplete. Gray et al. [4] describe additional criteria that may affect the prediction of stability solely from the S-parameter methods. It is possible to build a two-port network that would have $|S_{11}| < 1$ and $|S_{22}| < 1$ with purely passive source and load terminations while still having poles in the right-hand plane. This reference also provides an analysis on how to assess this potential problem. ADS now also has a tool, WSProbe, that will be briefly referenced later.

6.2.3 More gain definitions [6]

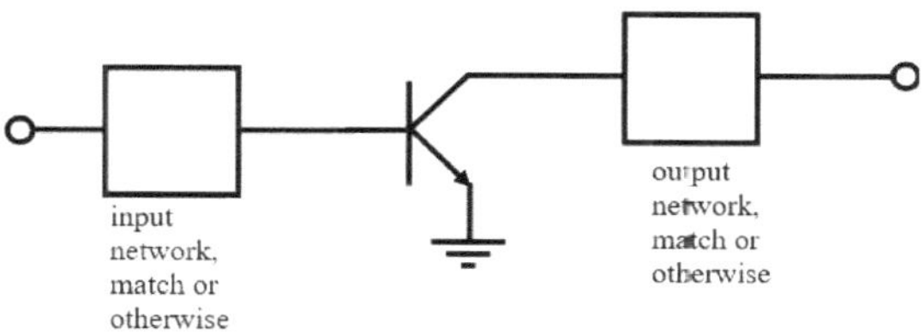

Figure 6.9 Picture of amplifier circuit of *any* type, but no feedback.

The amplifier block diagram in Figure 6.9 does not include external feedback, however, S(1,2) of the transistor is typically finite and the internal feedback of the device is included. The overall power gain must be $\leq G_{max}$. This is called the *maximum available gain* or MAG.

Power gain is equal to G_{max} *if* input and output are conjugately matched using lossless matching networks. However, MAG can only be obtained if the device is sufficiently stable under matched conditions.

Why not lossy networks? Figure 6.10 shows the insertion of a series resistor to the input of a transistor. In some cases, this can lead to conditional stability. Some gain (and noise figure) will be sacrificed.

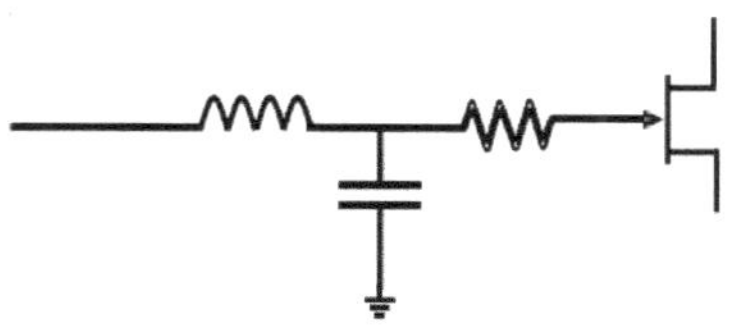

Figure 6.10 Add loss to input (maybe for stabilization).

$G < G_{max}$ because P_{in} device $= P_{gen} - P_{resistor} < P_{generator}$.

So, amplifiers fail to attain G_{max} because:

1. They fail to match on both input and output.
2. They use lossy elements (resistors) to attain a match.
3. They are potentially unstable. In that case, G_{max} is not possible due to oscillation. Then, G_{MSG}, the *maximum stable gain* (MSG), is the upper useful limit for gain.

For a potentially unstable transistor

Define *maximum stable gain* (MSG) as,

$$G_{MSG} = \frac{|S_{21}|}{|S_{12}|} \tag{6.17}$$

which is the $G_{T,\max}$ when $k = 1$.

This is used to describe the maximum gain which could possibly be obtained from the device. There is no Z_S or Z_L that can make the amplifier oscillate. Stabilization with resistive loading may still be required.

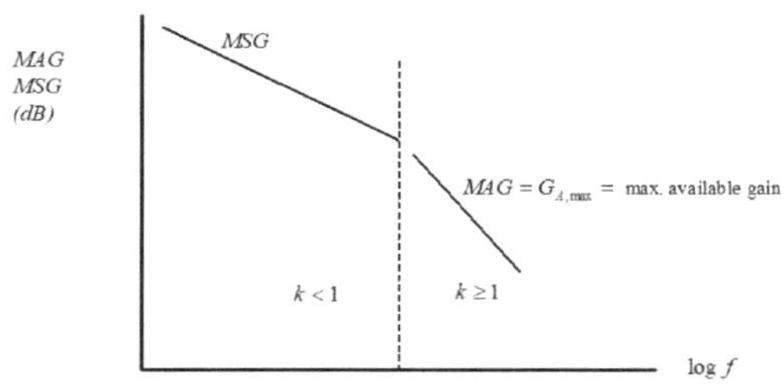

Figure 6.11 MAG and MSG plotted for a generic matched amplifier.

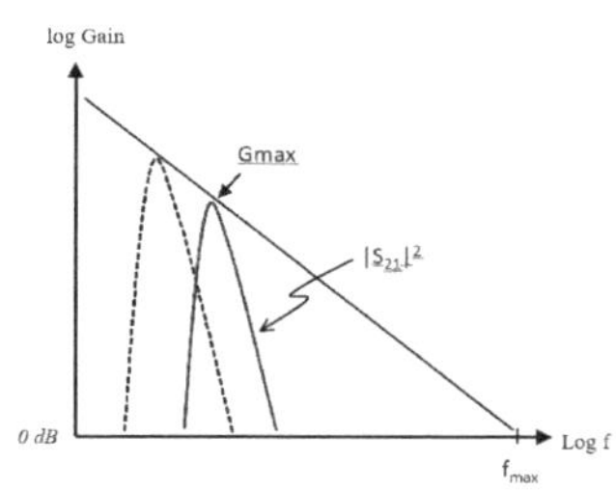

Figure 6.12 The G_{max} of a device is the upper limit of gain for a tuned amplifier.

Figure 6.11 is a plot of gain vs. frequency. It illustrates that when $k > 1$ we obtain the maximum available gain. G_{max}, typically MAG, is desirable for a tuned system (radio, receiver, etc.). Note that G_{max} is a function of frequency. f_{max} is the intersection with 0 dB gain – sometimes called the maximum frequency of oscillation. In Figure 6.12 we see that a tuned amplifier (with frequency dependent matching networks) can achieve G_{max} at only one frequency.

As can be seen in Figure 6.13, the maximum available gain (MAG) and maximum stable gain (MSG) can also be calculated using the MaxGain function and S(2,1), S(1,2). These are widely used to predict the limit of power gain for a given active device.

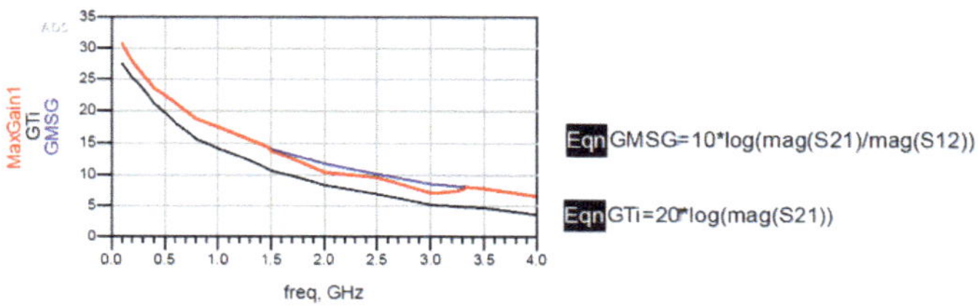

Figure 6.13 Swept frequency plot of MAG, MSG, and insertion transducer gain GTi.

Note that the user-defined equation capability of the display panel can be used to also calculate and plot MSG and Gti, the transducer gain when both Γ_S and Γ_L = 0. In the regions where $k < 1$, the Max_Gain function plots MSG. When unconditionally stable, it plots MAG.

In the case where $k > 1$, then MAG < MSG. The maximum transducer gain possible in this case is given by

$$G_{T\,\max} = \frac{|S_{21}|}{|S_{12}|}\left(k - \sqrt{k^2 - 1}\right) = MSG\left(k - \sqrt{k^2 - 1}\right) \tag{6.18}$$

How do we obtain the maximum unilateral transducer gain, G_{TU} or MAG?

$$G_{TU} = \underbrace{\frac{1-|\Gamma_S|^2}{|1-S_{11}\Gamma_S|^2}}_{G_S} \; |S_{21}|^2 \; \underbrace{\frac{1-|\Gamma_L|^2}{|1-S_{22}\Gamma_L|^2}}_{G_L} = G_S \cdot G_0 \cdot G_L \tag{6.19}$$

G_S: Fraction of gain/loss due to input network

G_L: fraction of gain/loss due to output network

If unilateral, G_T = G_{TU} = unilateral transducer power gain.

We could then apply simultaneous input/output conjugate match, $\Gamma_S = S_{11}^*$ and $\Gamma_L = S_{22}^*$.

Then, Equation (6.19) reduces to:

$$G_{TU,\max} = \frac{1}{1-|S_{11}|^2}|S_{21}|^2\frac{1}{1-|S_{22}|^2} \quad \text{if } S_{12}=0 \tag{6.20}$$

The unity gain frequency for G_{max} or MAG or $G_{TU,max}$ is called f_{max}. This represents the upper limit – the highest frequency that the device could ever have a power gain of 1.

Note, however, that for nearly all devices of interest, $|S_{12}| > 0$. And, if $k < 1$, you must consider stability circles and MSG applies. Conjugate match will result in instability. Nevertheless, f_{max} is often used as a figure of merit for an active device.

6.2.4 Stability circles

Now that we know the frequency boundary between unconditional and conditional stability, stability circles can be calculated to provide guidance with the design of input and output matching at our design frequency [2] .

First, we will consider the Γ_L or the "load plane" on the Smith chart. Define load stability circles which locate the boundary (values of Γ_L) between $|\Gamma_{in}| < 1$ and $|\Gamma_{in}| > 1$.
(stable) (unstable)

$$\text{To do this, set } |\Gamma_{\text{in}}| = \left|S_{11} + \frac{S_{12}S_{21}\Gamma_L}{1-S_{22}\Gamma_L}\right| = 1 \text{ and solve for } \Gamma_L. \tag{6.21}$$

The solution lies on a circle. Clearly, the circles are frequency dependent as are the S-parameters:

radius:

$$r_L = \left|\frac{S_{12}S_{21}}{|S_{22}|^2 - |\Delta|^2}\right| \tag{6.22}$$

center:

$$c_L = \left|\frac{(S_{22}-\Delta S_{11}^*)^*}{|S_{22}|^2 - |\Delta|^2}\right| \tag{6.23}$$

where:

$$\Delta = S_{11}S_{22} - S_{12}S_{21} \tag{6.24}$$

These circles form the boundary between stable and unstable operation. Plot on Γ_L the Smith chart because $|\Gamma_{IN}|$ depends on Γ_L .

In a similar way, we can set $|\Gamma_{out}| = 1$ and solve for Γ_S. The boundary circles which can be plotted on the Γ_S or "source plane" are defined by:

$$r_S = \left| \frac{S_{12}S_{21}}{|S_{11}|^2 - |\Delta|^2} \right| \tag{6.25}$$

$$c_S = \left| \frac{(S_{11} - \Delta S_{22}^*)^*}{|S_{11}|^2 - |\Delta|^2} \right| . \tag{6.26}$$

Plot the source stability circles on the Γ_S plane because $|\Gamma_{OUT}|$ depends on Γ_S.

1. If the circle intersects the chart, there is a region of instability.
2. If no intersection, device or amplifier is likely to be unconditionally stable.

Fortunately, we can use ADS to plot these for us, as described in Figures 6.14 and 6.15.

These are plotted at a specific frequency, 500 MHz in this example. If the circles intersect the chart as seen in Figure 6.15, there are regions of

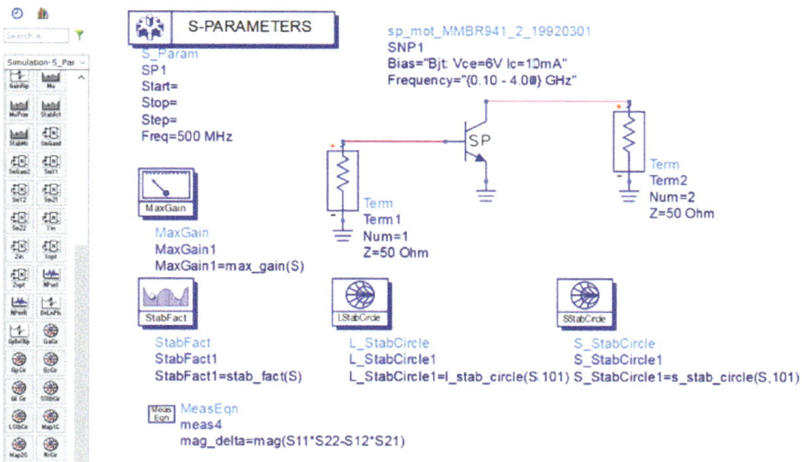

Figure 6.14 ADS schematic for a single frequency (500 MHz) analysis. The functions L_StabCircle and S_StabCircle, shown as Smith chart icons, are pre-programmed and can be selected from the Simulation-S_Par menu.

freq	MaxGain1	mag_delta	StabFact1
500.0 MHz	22.442	0.455	0.779

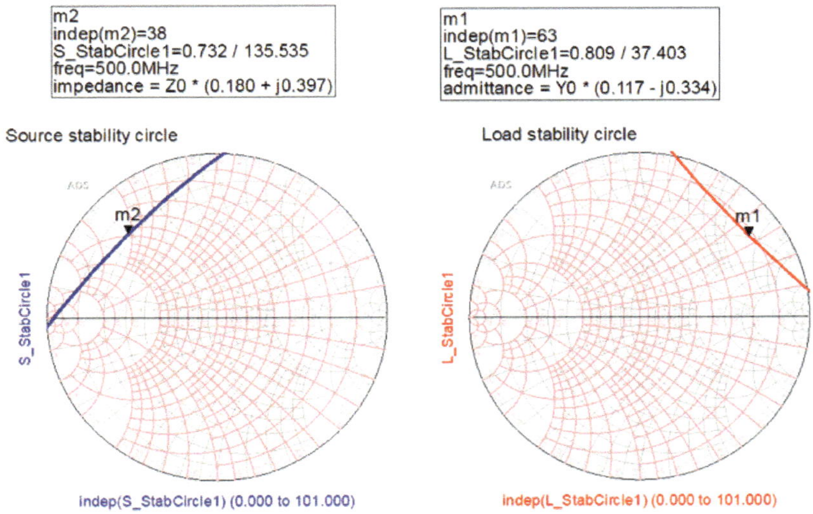

Figure 6.15 Source and load stability circles plotted on the Smith chart.

instability that must be avoided when designing the matching networks. Or, passive resistance must be added to eliminate any negative resistance.

Now, we must determine whether the inside or the outside of the circle is stable. Consider the load plane. Note that the load stability circle represents Equation (6.27) set equal to 1.

Let $\Gamma_L = 0$ (center of chart)
then $|\Gamma_{in}| = |S_{11}|$ (by definition)

$$|\Gamma_{\text{in}}| = \left| S_{11} + \frac{S_{12}S_{21}\Gamma_L}{1 - S_{22}\Gamma_L} \right| = 1 \tag{6.27}$$

if $|S_{11}|<1$, then this point represents a stable operating condition. So, the region inside the chart excluding the circle is stable, as shown in Figure 6.16.

Stability circles for all possible values of Γ_S must also be considered. Instability can be induced at either port. The source stability circle represents Equation (6.28) when set equal to 1 . With $\Gamma_S = 0$, center of chart, and if $|S_{22}| < 1$, the stable region is inside of the chart as was the case above.

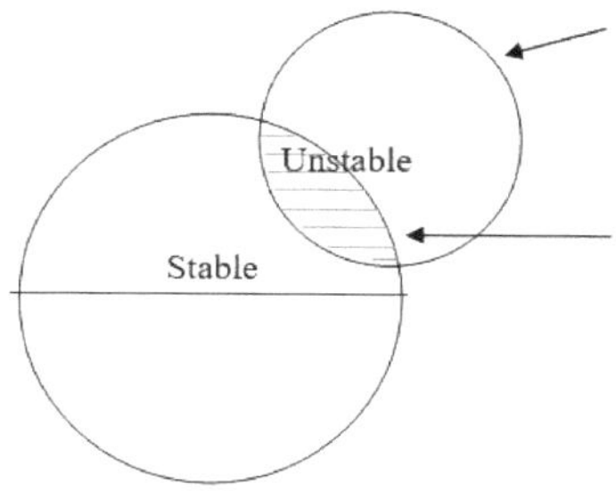

Figure 6.16 Unstable region of the Γ_L plane is defined. When $|S_{11}| < 1$, the inside of the chart corresponds to the stable region. If , then we have the opposite case.

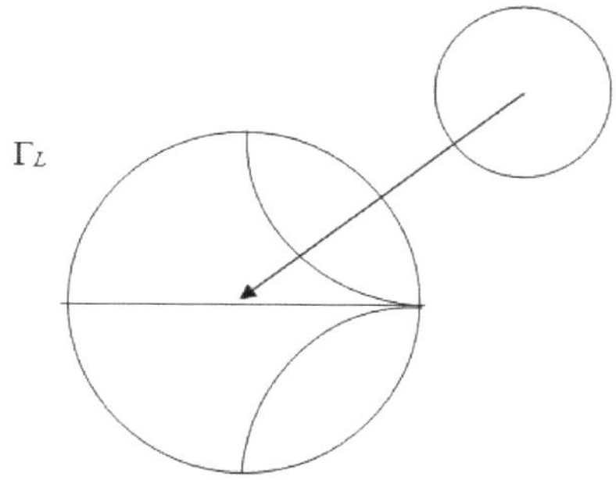

Figure 6.17 Case where $|C_L| - r_L > 1$.

$$|\Gamma_{out}| = \left| S_{22} + \frac{S_{12}S_{21}\Gamma_S}{1 - S_{11}\Gamma_S} \right| = 1. \tag{6.28}$$

The device is unconditionally stable if the load stability circle does not intersect the load plane. Similar for the source plane and source stability circle. Figure 6.17 shows a case where a stability circle could be entirely outside the $|\Gamma| = 1$ boundary.

There are several ways to stabilize the device over a range of frequencies.

1. Resistive loading on either input or output of the transistor.
2. Design matching network to avoid the unstable regions.
3. Design the bias network to kill gain at low frequencies.

Let's first experiment with option 1. We want $\text{Re}\,(Z_S + Z_{IN}) > 0$ and $\text{Re}\,(Z_L + Z_{OUT}) > 0$. Also, ADS tells us for this device, $|S_{11}| < 1$ at 500 MHz. Therefore, the center of the chart is stable.

How do we proceed with a design?

Consider Figure 6.15 again. Markers have been placed on the load and the source stability circle at 500 MHz. The load circle has an admittance real part of 0.117 normalized to 0.02 S. This is the point closest to the center of the chart. If we load the output of the transistor with a shunt admittance of $0.117 \times 0.02 = 0.00234$ S or a resistance of 427 ohms, which should push the stability circle outside of the Smith chart. Let's choose 400 ohms just so there is some margin for device variation.

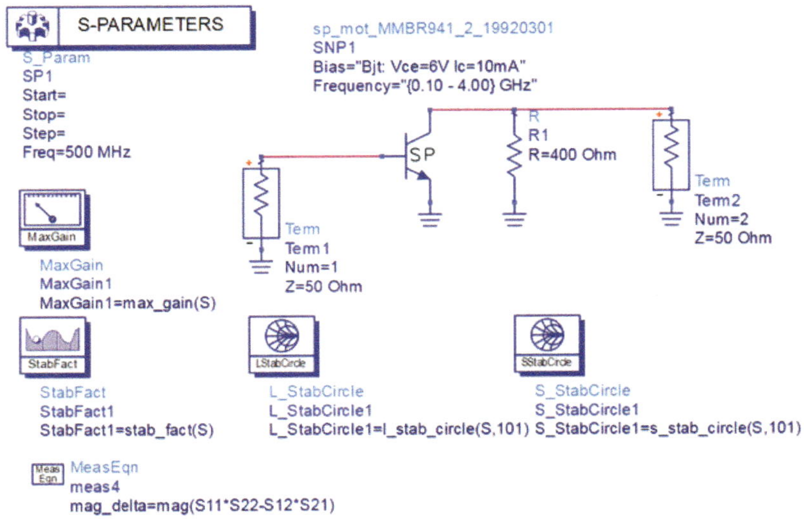

Figure 6.18 Schematic showing the added stabilizing resistor to the S-parameter simulation.

In Figure 6.18 we see how ADS is used to simulate stability circles, *k*-factor, delta. A stabilizing resistor of 400 ohms was added to the output side of the device.

In Figure 6.19 we can see that the stability circles are pushed outside the Smith chart. Therefore, conditionally stable.

But, what about stability below 500 MHz?

In Figure 6.20 we can see that we still have problems below 500 MHz even after resistive loading of the output.

More aggressive resistive loading might address this. But there are disadvantages (to be discussed later):

- Gain reduction
- Increased noise
- Reduced frequency response.

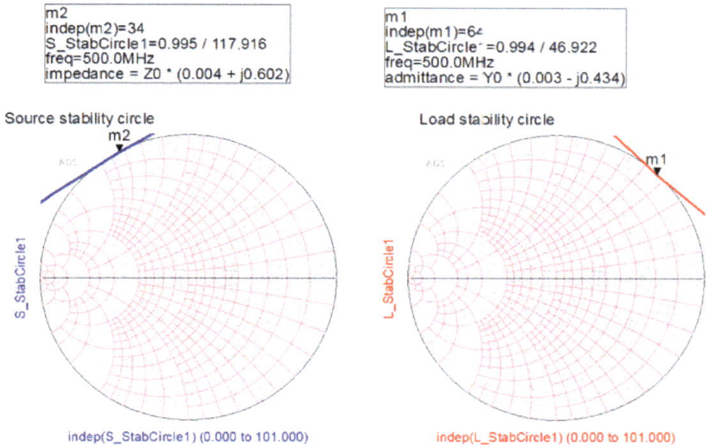

Figure 6.19 Simulation output at 500 MHz.

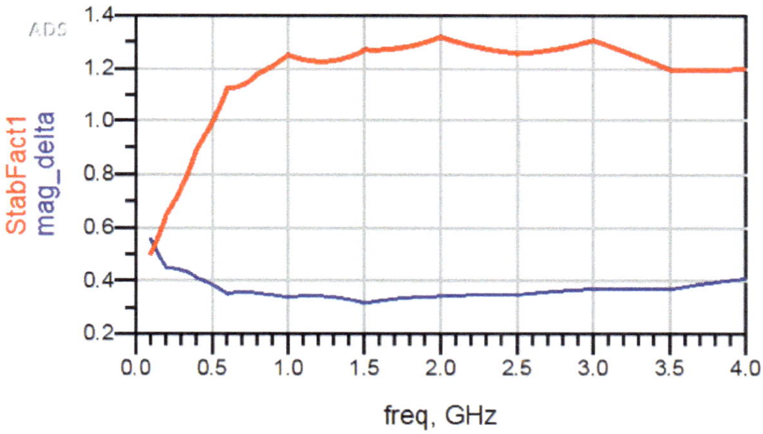

Figure 6.20 Frequency sweep of the k-factor.

The low frequency potential instability can also be addressed when designing the bias network. This approach can avoid the disadvantages above and will be discussed later.

The stabilization could have also been achieved at the input with a series or shunt resistance. Referring again to Figure 6.15, the normalized real part of the source stability circle is 0.18. If a series resistance of $0.18 \times 50 = 9$ ohms were added, a similar result would be obtained. However, adding extra resistance at the input of an amplifier will degrade its noise figure.

6.3 Gain Circles

We wish to describe the variation in gain with Γ_S and Γ_L in a graphical form. Let's assume that $|S_{11}| < 1$ and $|S_{22}| < 1$ [2] .

- Values of Γ_S and Γ_L that produce constant gain lie on circles in the Γ plane.
- Maximum gain occurs when $\Gamma_S = S_{11}^*$ and $\Gamma_L = S_{22}^*$. These are points on Γ_S and Γ_L planes respectively. The centers of the circles lie on the line connecting these points with the origin. Gain is relative to $|S_{21}|^2$.

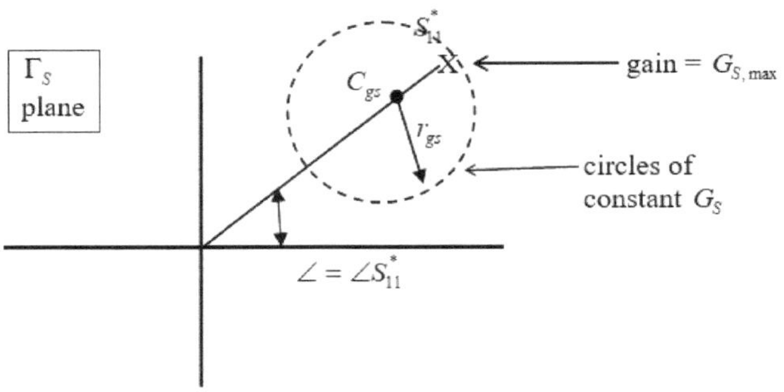

Figure 6.21 Definition of gain circles on the Γ_s plane.

In Figure 6.21 we see that the maximum available gain is obtained at S_{11}^*. Therefore a Γ_S value less than S_{11}^* reduces the G_S gain factor [2] .

- By necessity, a 0 dB circle will always pass through the origin ($\Gamma_S = 0$ or $\Gamma_L = 0$). This comes about because $G_S = 1$ and $G_L = 1$ when $\Gamma_S = \Gamma_L = 0$, i.e. matched to Z_0.
- Circles of constant G_L can be similarly drawn on the Γ_L plane.

From [2] ,

1. Draw line from origin to S_{11}^* (for Γ_S plane) or S_{22}^* (for Γ_L plane).
2. Determine gain steps of interest and calculate normalized gain factor

$$g_i = \frac{G_i}{G_{i,\max}} \tag{6.29}$$

where $0 \le g_i \le 1$. $i = S$ or L.

3. Calculate

$$r_{gs} = \frac{\sqrt{1 - g_S}\left(1 - |S_{11}|^2\right)}{1 - |S_{11}|^2 (1 - g_S)}. \tag{6.30}$$

4. Calculate

$$r_{gs} = \frac{\sqrt{1-g_S}\left(1-|S_{11}|^2\right)}{1-|S_{11}|^2\left(1-g_S\right)}. \tag{6.31}$$

or, use ADS to plot the circles:
G_S Cir
G_LCir
} icons found in Simulation-S-param palette (see Equation (6.19) for definitions of G_S and G_L.

Gain circles show you where Γ_S or Γ_L must be to achieve a certain gain from the device,

$$G_{TU} = G_S G_0 G_L \tag{6.32}$$

G_0 remains constant $= |S_{21}|^2$; G_S, G_L depend on Γ_S, Γ_L respectively

Since this unilateral case was defined to be unconditionally stable, ($|S_{11}| < 1$ and $|S_{22}| < 1$), we do not need to base our selection of Γ_S and Γ_L on stability but rather on design convenience, or other factors such as VSWR or bandwidth or reproducibility.

When stability is an issue, as it is whenever $|S_{12}| > 0$, both gain and stability circles must be considered in choosing the optimum G_S and G_L.

6.3.1 Bilateral case

Given that it is highly unlikely for any useful microwave device to be truly unilateral, a procedure to use the gain circles to guide us in a bilateral design is necessary. Again, we will employ ADS to cut through the boring math, and we can proceed with designing the input and output matching networks.

In the bilateral case($|S_{12}| > 0$), we need a systematic design method, because changes in Γ_S will affect Γ_{OUT} and changes in Γ_L will affect Γ_{IN}. Γ_L is the load reflection coefficient presented to the device. This would produce an unfortunate iterative solution chasing both input and output matches. The operating power gain, G_P, provides a graphical design method suitable for bilateral amplifiers. Gain circles can be calculated that show contours of constant operating power gain. G_P is useful since it is independent of the source impedance; the gain circle represents the gain that would be obtained if a magic genie adjusted $\Gamma_S = \Gamma_{IN}{}^*$ for each value of Γ_L on the circle. Then, $G_P = G_T$, i.e. the operating power gain equals the transducer gain.

$$G_P = \frac{P_L}{P_{\text{in}}} = \frac{\text{power delivered to load}}{\text{power delivered to input}} \tag{6.33}$$

$$G_P = \frac{1}{1 - |\Gamma_{in}|^2} |S_{21}|^2 \frac{1 - |\Gamma_L|^2}{|1 - S_{22}\Gamma_L|^2} \tag{6.34}$$

Note independent of Γ_S (Equation (6.10))

G_P is independent of the source match, whereas the transducer gain G_T

$$G_T = \frac{P_L}{P_{AVS}} \tag{6.35}$$

includes the gain term between P_{AVS} and P_{in} as shown in Equation (6.7).

Power gain circles can be plotted in the Γ_L plane. They will be independent of Γ_S : so the iterative design problem is cured.

If conditionally stable or unconditionally stable,

- choose Γ_L
- calculate Γ_{in}
- set $\Gamma_S = \Gamma_{in}^*$.

Under this condition, since the input is conjugately matched, *VSWR*$_{in}$. The output may have significant *VSWR* because it is intentionally mismatched in order to make the design less sensitive to component variations and provide an extra margin of stability.

Example 6.1: Let us use the S-parameter model of the transistor and the ADS schematic in Figure 6.14. We established that the device is not unilateral, and that it can be made unconditionally stable at the design frequency of 500 MHz with resistive loading on the output (see Figures 6.18 and 6.19). MaxGain1 (MSG in this case), StabFact1 (*k*), and mag_delta are calculated and shown in the table of Figure 6.22.

It is often more convenient to calculate the operating power gain circles on the data display rather than on the schematic. On the data display, you can change the gain values without having to re-simulate the amplifier. Use the Eqn function to write gain circle equations. The syntax is: gp_circle(S, gain, # points on circle) where S is the S-parameter matrix. Let's illustrate.

The gain circles are at MSG (MaxGain1 in this case), and 1 and 2 dB below MSG. A marker (m1) is placed on the –2 dB circle. Γ_L can be read off the display as magnitude = 0.314 with angle = 69.0°. If the input is conjugately matched (and stable), then the gain should be MSG – 2 = 20.4 dB.

The marker is on the –2 dB operating power gain circle and represents our choice of Γ_L. Notice that we have many options for Γ_L as long as we keep a respectable distance from the load stability circle. As shown above, if we

freq	mag_delta	StabFact1	MaxGain1
500.0MHz	0.386	0.999	22.442

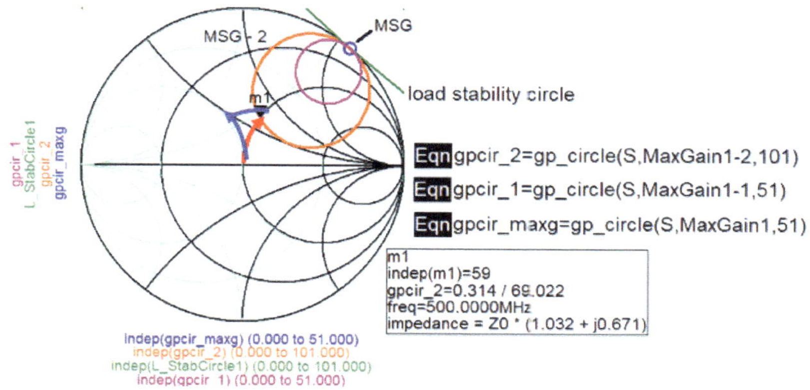

Figure 6.22 Operating power gain circles and the load stability circle are plotted on the load plane at 500 MHz.

choose Γ_L at the location of m1, we are nearly on the unit constant resistance circle (r = 1). Adding an inductance (red arrow) in series with the 50 ohm load will place us at Γ_L. The series inductance could be implemented with a lumped inductor (if feasible) or a series transmission line of high impedance. This approach will require an RFC or a $\lambda/4$ high Z_0 line for biasing.

Example 6.2: Design a series transmission line that will provide the required inductance.

Plotting Γ_L on a Smith chart, we find that an inductive reactance x of $+j0.6$ or $X = j30\Omega$ is required. Referring back to Chapter 5, Figure 5.15 explains how an inductor can be made from a series high impedance line. Let's choose $Z_0 = 80\Omega$, not so high that it would be difficult to implement on typical dielectric materials. From Equation (5.24),

$$L = Z_0\tau = \frac{30}{\omega} = 10.6 \text{ nH}$$

and therefore, $\tau = 0.13$ ns and electrical length $E = \tau f_{\text{ref}}\, 360 = 21.5°$.

Alternately, a series–shunt matching network could be used. This approach is shown in blue arrows in Figure 6.22. A series 50 ohm transmission line connects the collector to a shunt shorted stub or an inductor. This stub can also serve as our bias port if we are careful to provide appropriate bypass capacitance for the power supply and DC blocking where needed.

Next, calculate the input reflection coefficient for your choice of Γ_L and make sure it is stable. To do this, plot the source plane stability circle and compare with Γ_S. We can see in Figure 6.23 that this choice will be stable at the design frequency.

Eqn gamma_L=m1[0] Eqn gamma_S=conj(gamma_IN)

Eqn gamma_IN=S11+(S12*S21*gamma_L)/(1-S22*gamma_L)

gamma_IN	gamma_S	gamma_L
0.462 / -104.136	0.462 / 104.136	0.307 / 69.739

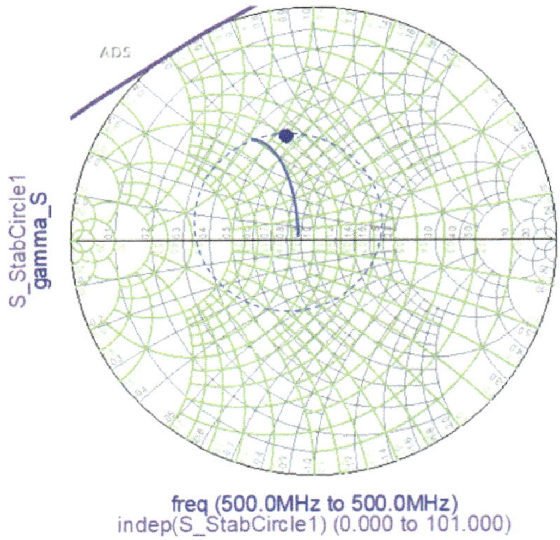

Figure 6.23 Source plane showing the calculated Γ_s , the blue dot. This is the complex conjugate of Γ_{IN} .

Now we can implement the input matching network with a shunt–series transmission line L network. Starting from $z = 0$, we add inductive susceptance until it intersects the circle at Γ_s, approximately $b = -j1.0$. From the angle scale on the Smith chart, a shorted transmission line of 45°, an eighth-wave line, will provide this. Remember to divide the angle by 2 when converting to electrical length.

Then add a series transmission line rotating clockwise on the circle to shift impedance to Γ_S. A length of 7° can again be found from the angle scale on the chart.

Question: How do you plot multiple lines and equations in the data display? Figure 6.24 is the explanation for Figure 6.22.

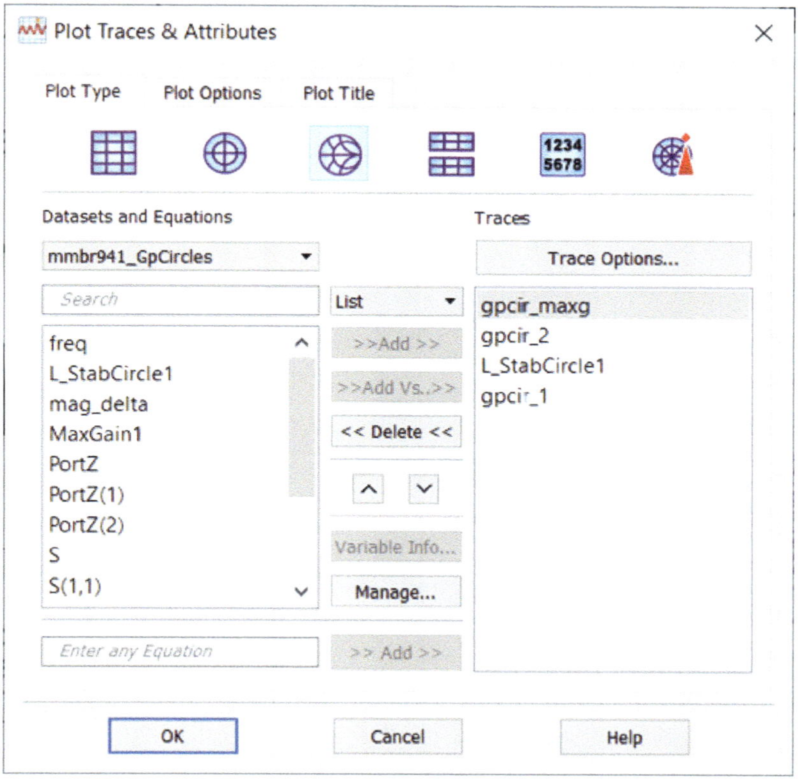

Figure 6.24 Plot traces and attributes panel in which variables to be plotted can be selected.

Place a Smith chart on the data display. Double click to open up Plot Traces and Attributes. Any variable in the left-hand list can be added – highlight it and click >> Add >>. Equation names must be typed in to the Enter any Equation box and then >> Add >>. All variables thus selected show up in the right-hand list.

Question: How do you select colors and widths for the display?

In the right-hand list, highlight the variable you want to edit. Then select Trace Options (Figure 6.25).

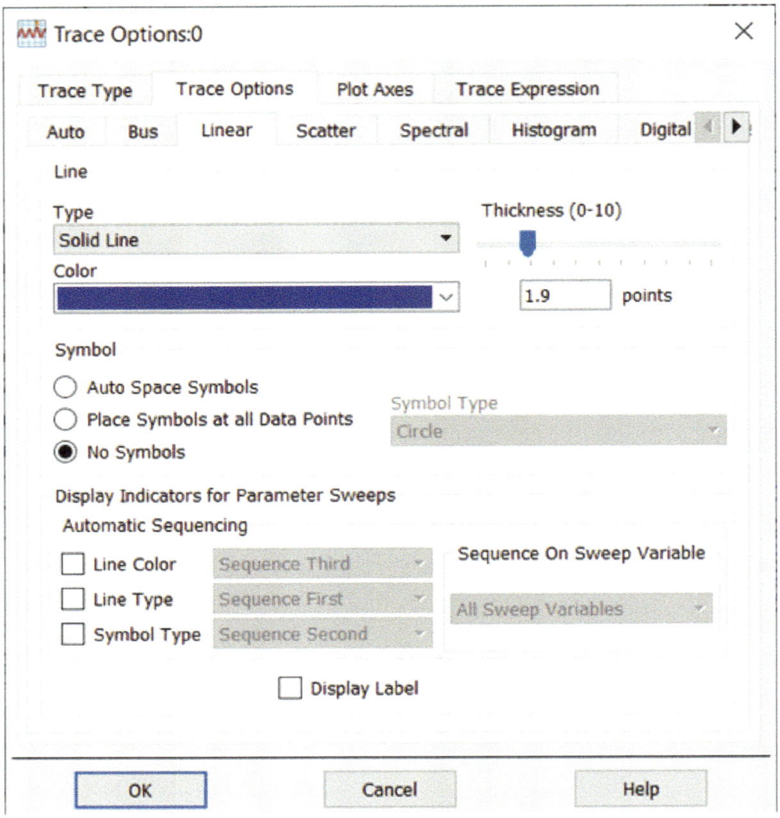

Figure 6.25 Trace options panel. Color, line style, and line thickness can be selected.

Example 6.3: Now add the matching networks as per Figures 6.22 and 6.23. The schematic diagram that includes these transmission lines and other components is shown in Figure 6.26.

The input network consists of shunt–series transmission lines. The output is a high impedance series line which provides series inductance. The arrow points to the series-RLC icon used to represent blocking and bypass capacitors as described in Chapter 1. Note that the transistor model is a small-signal S-parameter model evaluated at a specific bias condition (V_{CC} = 6 V, I_C = 10 mA). No DC sources are required for the simulation.

Checking the results, we find that it does meet the expected gain of 20 dB but is potentially unstable at frequencies below 500 MHz (Figure 6.27).

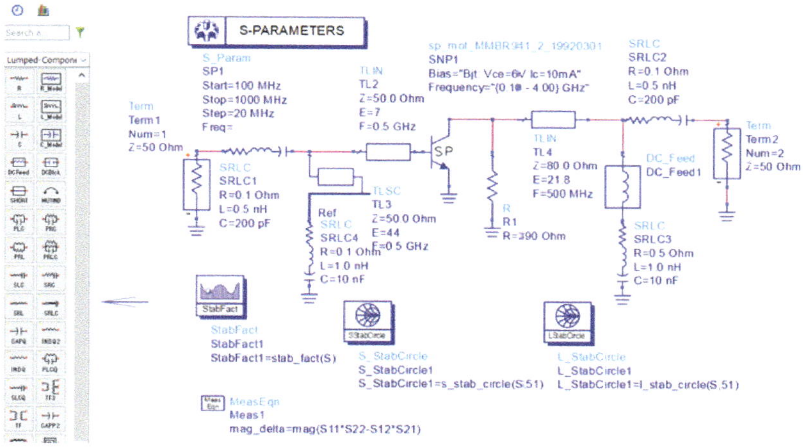

Figure 6.26 Amplifier design with matching networks on input and output.

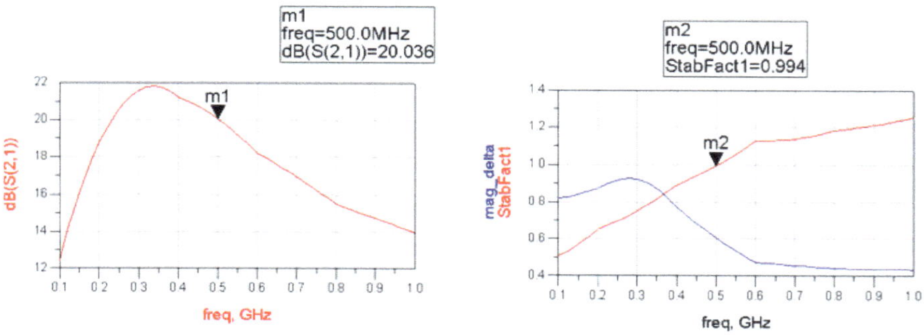

Figure 6.27 Transducer gain S(2,1) and stability factors over the 100 MHz to 1 GHz frequency range.

Note that the capacitors needed for bypassing and DC blocking have been modeled with equivalent networks. The DC blocking capacitors at the input and output have been designed to be series-resonant at the operating frequency. This isn't always the best practice. In our implementation, however, at low frequencies where the DC blocking capacitors have high impedance, the shunt stub at the input is bypassed with a much larger capacitor providing low impedance to ground over a wide frequency range. In the next section, we will add some series resistance in order to kill low frequency gain. The output will see the shunt stability resistor in parallel with the bypass capacitor

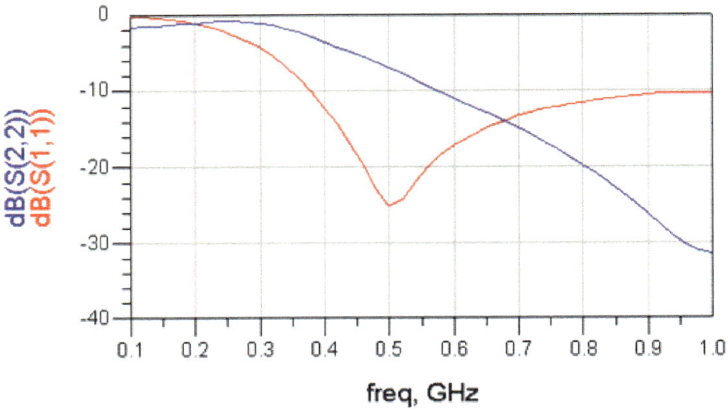

Figure 6.28 Magnitude of S(1,1) and S(2,2) is less than 1 over this frequency range.

network. The DC feed will need to be replaced with either an RF choke or a high impedance series transmission line.

Further improvements in accuracy for design purposes could be obtained by using the MLIN microstrip physical line models with bends and tees as needed. When RF chokes are used for biasing instead of the matching network approach, you should model the RFC as an equivalent RLC network and determine stability of the amplifier over a wide frequency range as well.

If you want to have a successful design on the first attempt, you must include equivalent circuit models for all of the components which will sensitively affect the performance.

Wideband stability: We must also guarantee that the amplifier is stable not only at the design frequency but everywhere else as well (Figure 6.27). To do this, we will sweep the frequency over a wide range (100 MHz to 1 GHz) and look at k, S_{11} and S_{22} (Figure 6.28). We noted already, in Figure 6.20, that $k < 1$ below 500 MHz. The magnitude of S_{11} and S_{22} are <1, thus stability is not in great danger. Nevertheless, we must modify the circuit to prevent $k < 1$ to any significant extent.

We could try a different Γ_L and test again. However, we may be able to solve the problem directly by adding losses to the input match. Notice that the input matching network has a shunt line that will become a short circuit at low frequencies. The device may become unstable as we approach $\Gamma_S = 1$. So, add some small series resistance to the shunt stub to modify the low frequency behavior (Figure 6.29). Fortunately, this works in this case. Adding

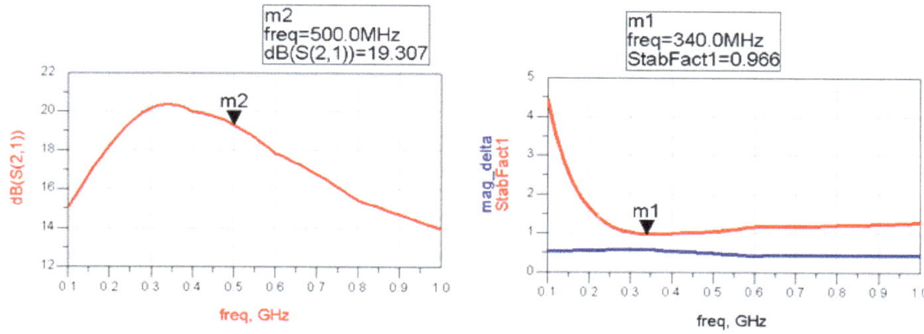

Figure 6.29 This shows the effect produced by adding loss to the shunt transmission line on the input of the amplifier. Minimal loss of power gain in exchange for greatly enhanced low frequency stability as revealed by the plot of StabFact1 (k-factor).

10 ohms to SRLC4 in series with the input stub costs about 0.7 dB of gain, but keeps k acceptable at all frequencies. The worst case is now k = 0.966 at 340 MHz.

Finally, the S-Parameter model should be replaced with a nonlinear device model and bias insertion circuitry added. The device can be biased to the desired value and stability evaluated once again. The bias network potentially can add an undesired feedback loop. Extra levels of bypassing may be needed. If significant changes in the prediction of k-factor are observed, then more in-depth stability evaluation using the WS Probe described below may be required.

An additional comment about gain circles: Device data sheets often plot G_T on the Γ_S plane with the output matched. This condition provides the available power gain $= G_A$.

$$G_A = \frac{P_{AVN}}{P_{AVS}} \tag{6.36}$$

Since output is always matched, G_A is independent of Γ_L.

$$G_A = \frac{1 - |\Gamma_S|^2}{|1 - S_{11}\Gamma_S|^2} |S_{21}|^2 \frac{1}{1 - |\Gamma_{out}|^2} \tag{6.37}$$

since $\Gamma_L = \Gamma_{out}^*$

This gain definition depends upon the input match because the actual power absorbed in the input is not necessarily the same as P_{AVS} (available power from the source which only occurs when there is a conjugate match at the input). G_A is helpful for design of low noise small-signal amplifiers, as will be discussed in Chapter 9.

If input is conjugately matched, then

$$\begin{aligned} G_A = & G_{A,\max} = MAG = G_{T,\max} \\ & = \frac{1}{\left(1 - |S_{11}|^2\right)} |S_{21}|^2 \frac{1}{1 - |\Gamma_{\text{out}}|^2}. \end{aligned} \tag{6.38}$$

6.3.2 Limitations of Rollett's stability factor

The original paper defining this stability factor included a proviso: that the circuit will have only left-hand plane poles when terminated with either open or short circuit [3, 5]. In other words, the circuit is known to be stable when ideally terminated. Because the method applies at the external input and output ports of an amplifier, multistage amplifiers might have unstable internal stages that would not be detected by k-factor calculations. Or internal resistances could mask negative resistances that would lead to instability. While the k-factor method is useful for simple amplifiers because of the design guidance insights given though stability and gain circles, it is not always sufficient to guarantee success.

The developers of Pathwave Keysight ADS have introduced another stability probe, the WS_Probe, that is based on traditional and more recent feedback analysis. When a circuit includes a feedback path, one must be careful to ensure that the magnitude of the loop gain af is never equal to one [4] :

$$A_{CL} = \frac{a(s)}{1 - a(s)f(s)} \tag{6.39}$$

Here, A_{CL} is the closed loop gain, a, the open loop gain, and f the feedback factor. Traditional feedback analysis using Bode plots and the Nyquist criteria can be used to identify right-hand plane poles. The WS_Probe enables this analysis under the following limitations.

1. There must be an identifiable feedback loop in which to insert the probe. This may be less than obvious in some circuits because it can also depend on layout of the chip or PCB. EM simulations would be needed to extract an equivalent network corresponding to layout.
2. Unless the large or small signal device model is accessible in an equivalent circuit form, then the device(s) internal feedback loop can't be probed. So, using simple S-parameter or equation-based device models

(Gummel-Poon, Curtice, etc.) does not provide this access. In that case, the WS Probe analysis is not that much superior to using the k-factor because extra feedback paths are being ignored.

3. Keysight has provided videos [7] and application notes [8] that describe how to use the probe and what it will reveal regarding stability. Nyquist plots can show whether there are right-hand plane poles.

6.4 Summary of Amplifier Design Methodology

6.4.1 Bilateral case $|S_{12}| > 0$ (potentially unstable)

A. Check for stability: If $k < 1; |\Delta| < 1$, then: *potentially unstable.*
B. Plot stability and G_P gain circles. ($G_{TU,max}$ isn't available – oscillator!).
C. Use resistive stabilization to force $k > 1$ and avoid Γ_L near the stability circle.
D. Design for low sensitivity to Γ_S and Γ_L at the desired gain.

6.4.2 Bilateral case for k $>$ 1 (conditionally stable)

In all cases, also do a wideband frequency sweep to verify stability at all frequencies of interest. Kill low frequency instability with additional resistive loading, typically less harmful to gain when it can be included in the bias network.

6.4.3 WS Probe

If the device model is an equivalent circuit and there are known feedback paths around the device, then the WS Probe can be used to plot Nyquist loops. This will reveal right-hand plane poles that can cause oscillation.

References

[1] Gonzalez, G. (2007) *Microwave Transistor Amplifiers, Analysis and Design*, 2nd edn., Prentice-Hall, Section 2.6. This explains how to use signal flow graphs to derive S-parameter relationships.

[2] Gonzalez, G. (2007) *Microwave Transistor Amplifiers, Analysis and Design*, 2nd edn., Prentice-Hall, Sections 3.2.1 – 3.2.4, 3.3, Appendix B.

[3] Rollett, J.M. (1962) Stability and power gain invariants of linear two-ports *IRE Trans. Circuit Theory* CT-9 (3) 29–32.

[4] Gray, P.R. et al. (2009) *Analysis and Design of Analog Integrated Circuits*, 5th edn. Wiley, ch. 9.
[5] Jackson, R.W. (2006) Rollett proviso in the stability of linear microwave circuits, a Tutorial *IEEE Trans. Microwave Theory Tech.* 54 (3) 993–1000.
[6] Pozar, D. M. (2005) *Microwave Engineering* 3rd edn. Wiley, Section 11.3.
[7] https://youtube.com/playlist?list=PLtq84kH8xZ9Gz-1_VvqWsWeGwzMnKjRwY
[8] "Designing for Stability in High Frequency Circuits", www.keysight.com, 2021.

Homework

1. a. Consider the source Γ_S plane in the figure below. $|S_{11}| < 1$ and $|S_{22}| < 1$. Which region of the plane is stable? Justify your answer.

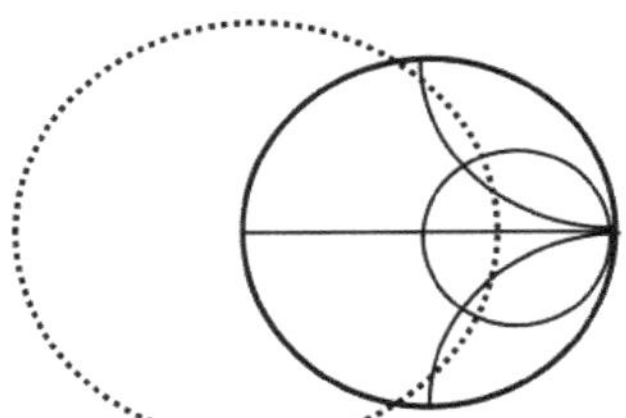

b. Now, $|S_{11}| < 1$ and $|S_{22}| > 1$. Which is the stable region of the figure below? Explain.

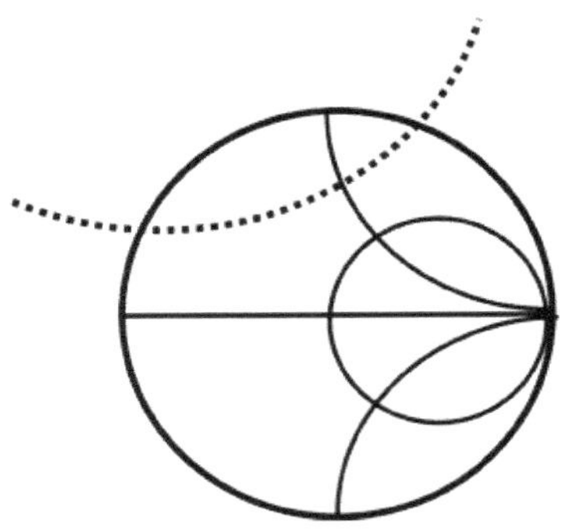

2. Refer to the plot in the figure below. This is an X-band GaAsFET.

a. There is a potential problem with this choice of Γ_L and Γ_S. Explain.

b. Does MaxGain represent MSG or MAG in this case? Why?

c. How much transducer gain would you obtain if the input and output were simply terminated in 50 ohms?

d. For the unstabilized device as shown, design a one-element matching network for Γ_L.

e. Design an L network to present Γ_S to the device input.

f. Find as many ways as possible to make the device unconditionally stable with $k = 1$. Evaluate the components required for each configuration.

g. How much error in gain would you encounter if you used the unilateral approximation for this device?

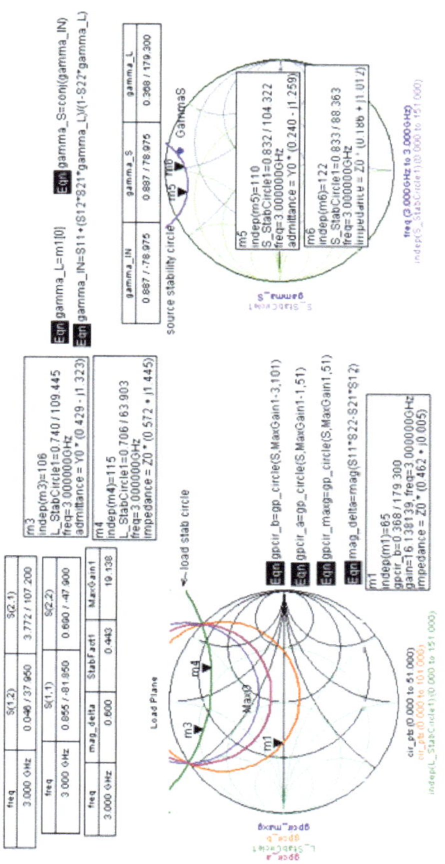

3. Design an amplifier at 450 MHz using the Infineon BFR93A BJT.

Background. Bandpass RF amplifiers are often used in receiver front ends to improve sensitivity and noise performance. Design this amplifier for unconditional stability while achieving close to maximum available gain and good return loss at the input and output. Note that the device is bilateral and potentially unstable.

Specifications:

S_{11}, S_{22}	Better than −10dB @ 450 MHz
Gain	Greater than 15dB
$V_{CE} = 8$ V	$I_C = 5$ mA
3 dB bandwidth	At least 50 MHz
Frequency	450 MHz

a. Use Keysight ADS to plot MaxGain1, StabFact1 over the 100 MHz to 2 GHz range. Use the sp_sms_BFR93A_1 S parameter model. (Insert > Component > Component Library)
b. Plot the source and load stability circles at 450 MHz.
c. Modify the design by adding resistive stabilization element(s) as needed to provide unconditional stability at 450 MHz.
d. Add G_P circles on the load plane. You can sacrifice gain in order to make the matching network easier to implement. Select a value for Γ_L and, using the marker, calculate the corresponding value of Γ_s Plot $\Gamma_s$on the source plane.
e. Design distributed or lumped element matching networks at the input and output.
f. Verify that the device is stable over the full 100 MHz to 2 GHz range. If not, add additional loss to produce stability at lower frequencies as needed.
g. Show that your design meets the specifications.

4. a. Derive the voltage gain and input impedance of this circuit. Use a simple model of the device including only Cgs and the gm Vgs dependent source.

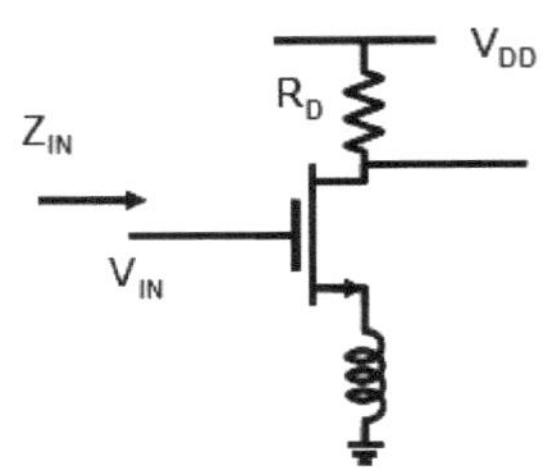

b. If the source inductor were replaced by a capacitor, what effect would this have on the Z_{IN}? Under what conditions could you observe a negative resistance? What might this do to circuit stability?

7

Bias Circuit Design and Wideband Microwave Amplifiers

Goals

1. Learn to provide stable DC bias conditions for active devices
2. Design wideband amplifiers using shunt – series feedback
3. Use ADS for DC analysis using large-signal component models
4. Add more accurate transmission line models to a design.

Recall in the last chapter, design techniques for a narrowband high frequency amplifier were presented:

- Stability and gain circles with S-parameter models for the device were used for evaluation of bilateral devices.
- Resistive loading was employed to achieve stability both at the design frequency and over wide bandwidth.
- DC biasing was not discussed because the S-parameter models are measured with specific bias conditions.

1. In the actual implementation, the active devices must be presented with a stable bias condition over a range of expected temperatures and anticipated device variation. A large signal device model, either equation based or equivalent circuit based is needed. Feedback methods are often suitable for achieving this.
2. There are also applications in which a wideband amplifier is necessary. Shunt-series feedback is one way to accomplish this.
3. DC analysis is straightforward using ADS. It also requires large signal device models.
4. Layout of a circuit also must use more accurate models for passive components such as transmission lines. Keysight ADS also provides such models.

7.1 Biasing

We need to provide a stable bias condition for our device in any amplifier application.

Bipolar transistors: Must force the DC (average) value of V_{CE} and I_C to desired values and keep them constant using feedback techniques.

Never fix V_{BE}: $I_C = I_{SE}e^{VBE/VT}$. I_C varies exponentially with temperature.

Never fix I_B: $I_C = \beta\ I_B$. β varies tremendously from device to device and also increases with temperature as well (0.7%/°C).

Field Effect Transistors: Force V_{DS} and I_D to desired values and keep them constant. The main weakness of microwave FETs is the variation in threshold voltage, V_T, and the transconductance g_m from device to device and with temperature.

In some cases, bias stabilization may be accomplished with passive circuit elements. An emitter or source resistor provides negative feedback to stabilize bias current. For example, as seen in Figure 7.1, R_S is a self-bias resistor. $V_{GS} = -I_D R_S$ in order to provide a negative gate-source voltage. If I_D increases, V_{GS} decreases to compensate. However, wiring inductance is introduced in the source circuit, even with a bypass capacitor across R_S. This will become significant when $\omega L = 1/(10g_m)$, which is generally a very small inductance.

Referring again to Figure 7.1, the BJT circuit uses the conventional four-resistor bias approach where the emitter resistor provides negative feedback stabilization against drift of the bias point with temperature or device parameter variation from batch to batch of devices.

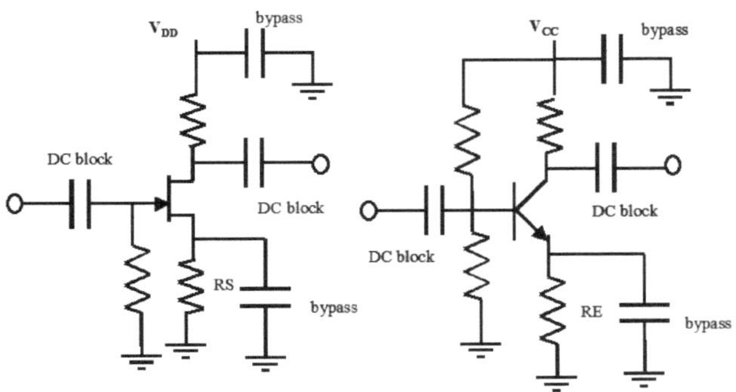

Figure 7.1 Passive bias circuits for MESFET (usually depletion mode) and BJT.

However, these circuits are not often used for RF applications because the biasing resistances also load the circuit and reduce the gain. Therefore, circuit techniques that permit use of a directly grounded source or emitter connection are preferred for high frequency amplifiers when implemented using discrete components on PC boards. In RFIC implementations, more flexibility is possible. One can use CC, CB, CG, CD connections as well as choosing device areas to optimize circuit performance.

The MESFET (or JFET or PHEMT) circuit (Figure 7.2) uses two power supplies when the source is grounded on a PCB or RFIC because the threshold voltage of a typical microwave FET is negative. Microwave FETs are always n-channel.

So, in some cases, at VHF/UHF frequencies, RFCs can be used for biasing, as shown in Figure 7.2. They provide high Z at the design frequency and so will not usually sacrifice gain. However, they always have resonances, so a complete RFC equivalent circuit model such as that discussed in Chapter 1 is essential if the circuit performance is to be accurately predicted.

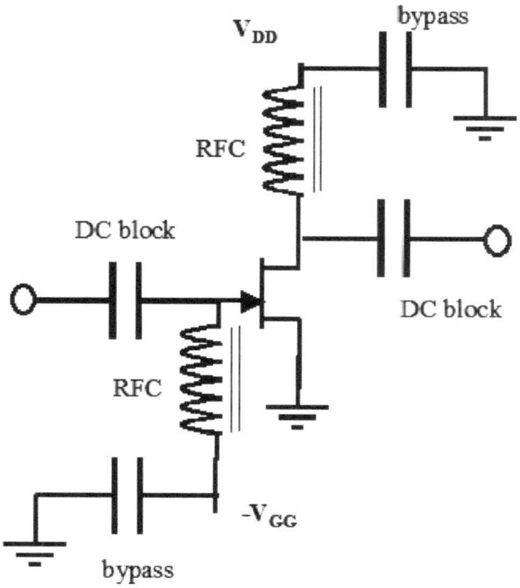

Figure 7.2 Example of GaAs or InP FET bias insertion using two power supplies and RFC.

When possible, a better approach to bias insertion than resistors or RFCs is to use the matching network itself to insert bias voltages. This is possible when a shunt shorted stub or shunt inductor is connected to the input or output terminals of the device without any series DC blocking or impedance matching capacitor. The grounded end is bypassed with a grounded capacitor with low impedance over the full range of frequencies. See Figure 7.3.

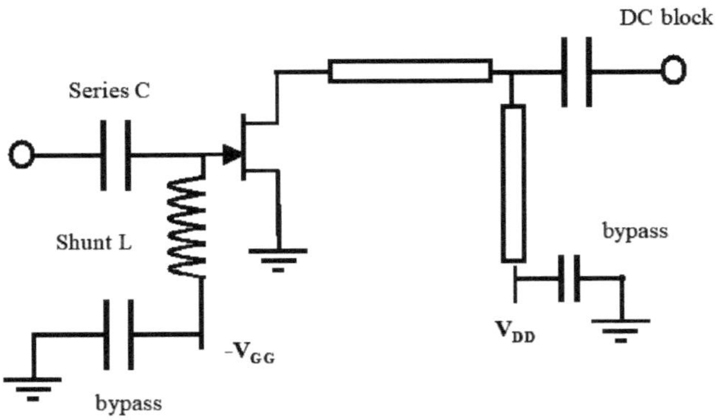

Figure 7.3 Biasing the device through the matching networks. Examples of lumped and distributed element matching are shown.

Another bias insertion method that is widely used and is more suitable for microwave circuits utilizes high impedance lines that are one-quarter wavelength at the design frequency. This is shown in Figure 7.4 as a bias feed for the drain. If a low impedance bypass capacitor is used to short the bias feed end, the $\lambda/4$ line transforms the short into an open at f_0. An added bonus

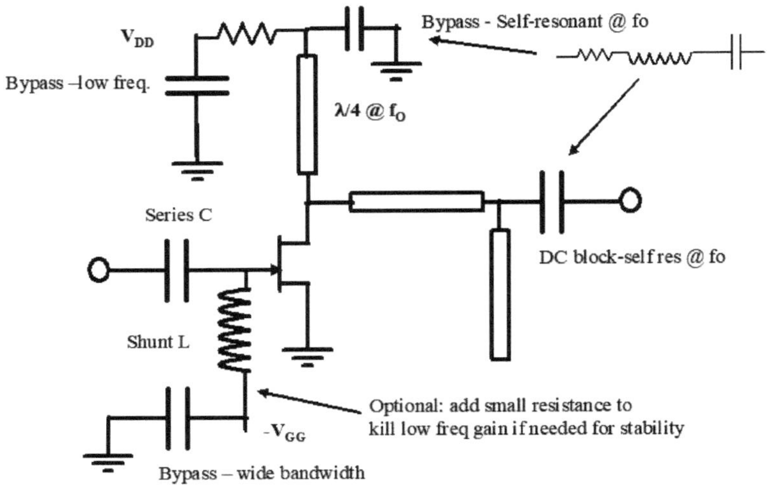

Figure 7.4 Bias insertion with quarter wave high impedance line.

comes from short circuiting the even harmonics because the line is also $\lambda/2$ at $2f_0$, λ at $4f_0$, etc. This can improve efficiency of some power amplifiers or can be used with a resistor to force the stability factor $k > 1$ at $2f_0$ by killing high frequency gain with lossy loading.

7.1.1 Active bias circuit

We have shown in Figure 7.1 that passive bias circuit approaches using resistors can load the amplifier, creating extra losses and add source or emitter inductance. The best practice is to directly ground the emitter or source for microwave amplifiers. However, grounding the emitter or source leaves the devices wide open to DC bias problems such as thermal runaway on the BJTs or temperature drift of the bias point for FETs. So, an active feedback circuit such as that shown in Figure 7.5 can be used for biasing. While an op amp could be used, this requires an extra power supply, so a simpler approach using a PNP transistor is frequently more efficient.

Q2 can be nearly any lower frequency PNP with sufficient current capability. V_{DS} is set by the voltage divider

$$V_1 = V_{DD}\left(\frac{R2}{R1 + R2}\right) - V_{D1} \quad \text{and} \quad V_{D1} = -V_{BE,Q2} = 0.7V. \quad (7.1)$$

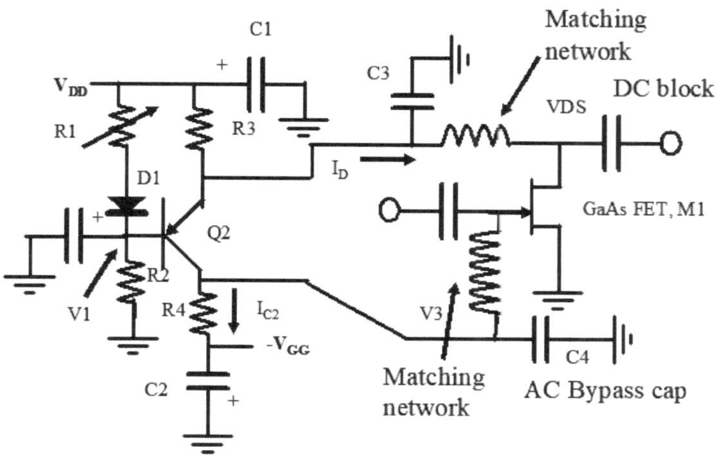

Figure 7.5 Active bias generator (on left) provides gate and drain voltages to the GaAs FET.

So,

$$V_{DS} = V_1 - V_{BE,Q2} = V_{DD}[R2/(R1 + R2)]. \quad (7.2)$$

$$I_D + I_{C2} = (V_{DD} - V_{DS})/R3 \qquad I_{C2} << I_D \quad (7.3)$$

$$I_{C2} = (V3 + V_{GG})/R4. \quad (7.4)$$

D1 is provided for temperature compensation and can be a diode-connected PNP of the same type as Q2.

The DC feedback can be seen as follows: Suppose that:

1. I_D increases (maybe due to temperature variation or change in device).
2. Then, the voltage drop across R3 will increase, reducing $V_{BE,Q2}$.
3. I_{C2} drops, reducing the voltage drop across R4. This reduces V_{GS} and I_D to correct for the change in step 1.

C1 and C2 are low frequency bypass capacitors. These are typically polarized tantalum or electrolytic caps that are used for decoupling and stability of the active bias circuit. C3 and C4 are high frequency bypass caps. These must NOT be tantalum or electrolytic types, but must be high quality ceramic chip caps with very low series inductance. Any series L must be included as part of the matching network.

When choosing values for C3 and C4, we must recognize that the active bias generator is a feedback system. R3C3 and R4C4 add two low frequency poles in the feedback loop. A second order FB system can be underdamped (ringing) or oscillatory if the phase margin is insufficient. For stability, create a dominant pole with one of the two RC networks with one time constant at least 10 times larger than the other. This should provide nearly a 90° phase margin.

A similar active feedback approach can be used for biasing a BJT (Figure 7.6).

7.2 Wideband Shunt-series Feedback Amplifier

One well-known technique for achieving wide bandwidth, controlled gain and input–output impedance over a moderate range of frequencies is the shunt-series feedback design. The example shown (Figure 7.7) uses a BJT, however FET implementations are also widely used. The resistors set gain and impedances and reduce dependence on transistor parameters [1].

First, determine R_E, R_F, R_S and R_L to establish gain and R_{IN}. Typically for the RF application of this type of feedback amplifier, R_{IN} = R_{OUT} = R_L

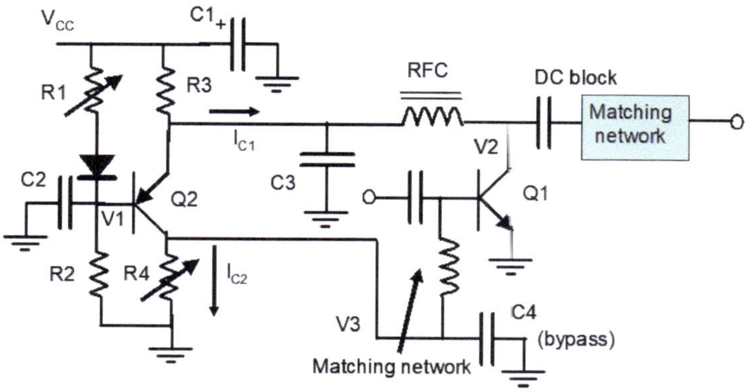

Figure 7.6 An active feedback approach for biasing a BJT.

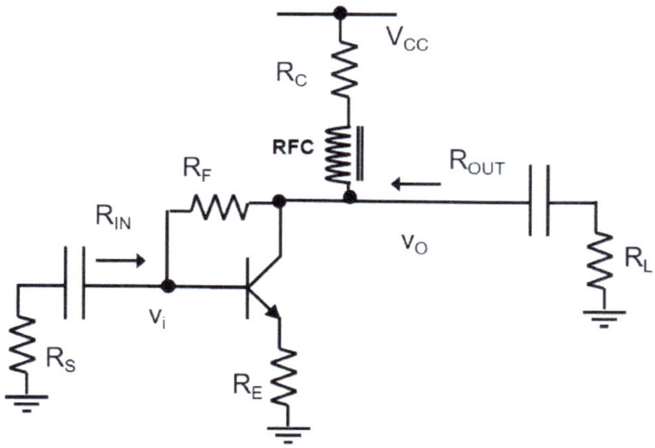

Figure 7.7 Shunt-series feedback wideband amplifier.

= R_S = R_O. R_O is defined to be equal to all input and output parameters. We assume that the RFC isolates R_C from the AC analysis.

Using the analysis in [1] , voltage gain can be estimated by

$$|A_v| + 1 = \frac{v_o}{v_i} \cong \frac{-R_O}{(R_E + r_e)} \tag{7.5}$$

where, r_e is the emitter resistance of the transistor,

$$r_e = kT/qI_E \cong \frac{1}{40I_C}. \tag{7.6}$$

This gain estimate is approximately true when the open loop gain is large. Then, the gain specification can be used to estimate R_E:

$$R_E \cong \frac{R_O}{|A_V| + 1} - r_e. \tag{7.7}$$

Then if we assume that the input resistance of the device is much higher than R_{IN} = R_O, then the feedback resistor sets the input impedance:

$$R_F \cong (|A_V| + 1)\, R_O. \tag{7.8}$$

Thus, we have design equations for the shunt-series feedback amplifier. Of course, it is possible to have different input and output resistances, but this condition requires a more exact analysis.

7.2.1 DC biasing with shunt-series feedback

Biasing resistors R_F and R_C are shown in Figure 7.7. Once R_E is determined by the gain requirement, the DC drop due to $I_C R_E \cong I_E R_E$ can simply be added to the base voltage V_B. You will also find that the required value of R_F for the AC analysis is much smaller than is needed for proper biasing. In that case, a capacitively bypassed resistor must be added in series with R_F to satisfy both requirements.

An approximate solution can be made using a few typical design rules.
For a BJT design, make these assumptions: β = 100,
then base current

$$I_B = \frac{I_E}{\beta} \cong \frac{I_C}{\beta}. \tag{7.9}$$

Finally, assume $V_{BE} \cong 0.75$ V. Then, knowing the desired V_{CC}, V_{CE} and I_C, and the R_E value determined from Equation (7.7), then,

$$V_B = V_{BE} + I_E R_E \tag{7.10}$$

$$R_C = \frac{(V_{CC} - V_{CE})}{I_C} \tag{7.11}$$

$$R_F = \frac{V_{CE} - V_B}{I_B} \tag{7.12}$$

7.3 DC Simulation with ADS

Next let's illustrate how to perform a DC simulation on Keysight ADS in order to verify or modify the biasing circuit parameters. We can use the shunt feedback bias approach in Figure 7.7.

In order to do a DC simulation, a large-signal device model must be selected.

Insert > Component Library. In this example, the RF Transistor library was selected, and the model for the BFR193 BJT chosen. Double-click on the transistor model, then click in the schematic window to place it.

You must add a DC source. These are found in the Sources-freq domain menu.

Next, insert a current probe to monitor supply current. Probe components menu.

Then, a simulation controller is needed. Simulation-DC menu.

Nodes must be labelled in order to display voltages. Use the icon below to do this.

Then, the result can be displayed on the schematic as seen in Figure 7.8 using the Simulate > DC Annotation menu to annotate voltage and annotate pin current. These results can also be presented in the data display window as a table, Figure 7.9.

7.4 Implementation: Microstrip Line Modeling in ADS

You have used the ideal transmission line TLIN and open and shorted stubs in the homework assignments. You should be aware that there is a more accurate set of microstrip line models and elements in another menu (TLines-Microstrip) and palette. These elements take into account some of the nonidealities of the line: corners, tees, and losses. Msub is used to describe the line and dielectric. MLIN is a series line section. MLOC is an open circuit stub; MLSC is a short circuit stub. Mcorn represents a right angle corner without a miter bend. MBend represents a mitered corner. MTEE is a

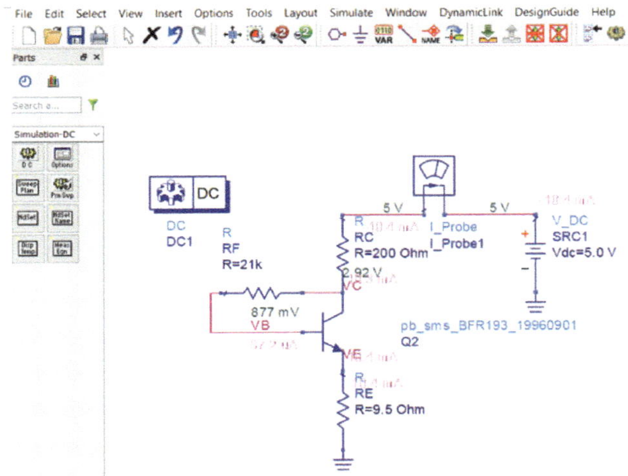

Figure 7.8 Schematic for DC simulation of the biasing of the BJT using shunt-series feedback. A V_{CC} of 5 V was assumed. The design aimed for V_{CE} = 3 V, I_C = 10 mA.

freq	I_Probe1.i	VB	VC	VE
0.0000 Hz	-10.41 mA	876.7 mV	2.918 V	98.92 mV

Figure 7.9 Table from the data display window for the above simulation.

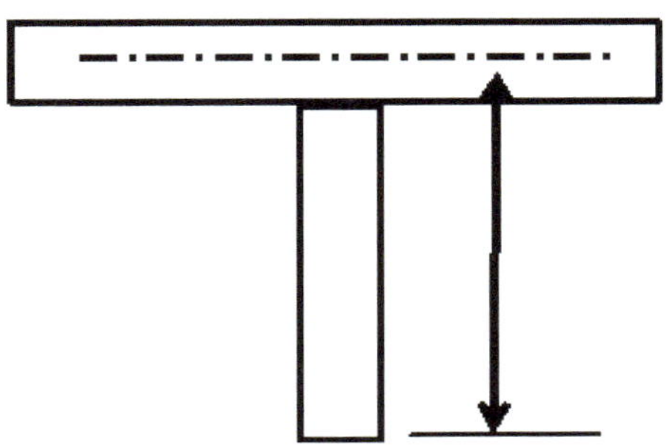

Figure 7.10 T-connection showing length to be computed from midpoint of the thru-line.

T junction. Note that the length of the stub is measured from the center of the transmission line at the T junction (Figure 7.10).

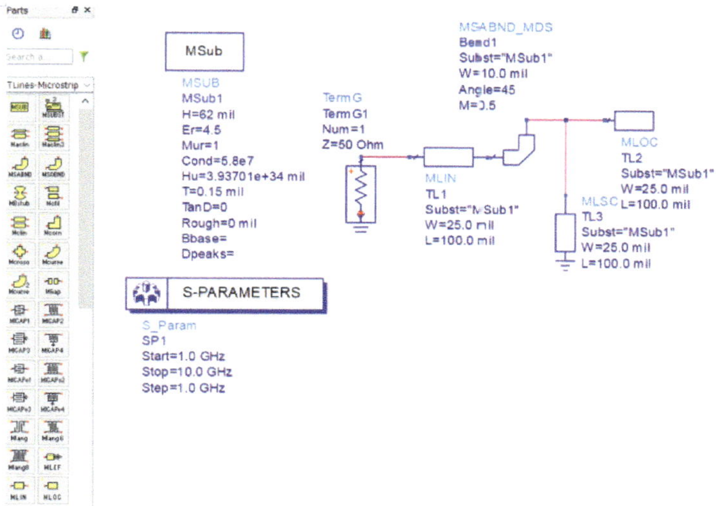

Figure 7.11 Illustration of various microstrip line models available for accurate physical implementations.

Also, recall that ADS Linecalc can be used to simulate the parameters of a transmission line. Use models such as these when implementing a hardware design. An example is shown in Figure 7.11.

References

[1] McWhorter, M. et al. (1995) EE344 High Frequency Laboratory, Stanford University, Ch. 2.

Homework

1.a. Design a shunt-series BJT feedback amplifier (Figure 7.7) using the approximate design equations presented in this chapter. I_C = 12 mA. Design for a power gain of 10 dB with R_O = 50 Ω.

b. Complete the DC design for a bias condition of V_{CE} = 4 V and I_C = 12 mA. V_{CC} = 6 V. Using the large signal device model for the BFR193 BJT, perform a DC analysis on ADS to confirm that the bias condition is nearly correct.

c. Use the S-parameter analysis with the large signal model to show whether the design estimates led to the correct gain and input/output resistances.

8

Performance Limitations of Amplifiers – Distortion and Noise

Goals

1. Understand the origins and consequences of nonlinearity in RF systems
2. Evaluate and calculate gain compression and intermodulation distortion
3. Understand the sources of noise in RF systems
4. Calculate noise factor, noise figure, and signal-to-noise ratio
5. Dynamic range of a radio frequency system can be evaluated by the spurious-free dynamic range (SFDR)
6. Use of harmonic balance simulator to evaluate IMD and distortion

Thus far, our analysis of amplifiers has considered only gain and stability. While these parameters are essential, other factors must be considered. In this chapter, we will consider two such limitations. Performance is limited under large signal conditions by nonlinear behavior causing distortion and spurious signal generation. At the weak signal end of the operating range, noise, inherent in all components and devices, limits the minimum detectable signal, especially important for satellite receivers. The design for low noise amplifiers will be presented in the next chapter.

8.1 Distortion in Nonlinear Systems

The upper limit of useful operation is limited by distortion [1, 2, 4]. All analog systems and components of systems (amplifiers and mixers for example) become nonlinear when driven at large signal levels. The nonlinearity distorts the desired signal. This distortion exhibits itself in several ways:

1. Gain compression or expansion (sometimes called AM–AM distortion).
2. Phase distortion (sometimes called AM–PM distortion).
3. Unwanted frequencies (spurious outputs or spurs) in the output spectrum. For a single input, this appears at harmonic frequencies,

creating *harmonic distortion* or HD. With multiple input signals, in-band distortion is created, called *intermodulation distortion* or IMD.

Simulation of non-linearity requires use of the Harmonic Balance (HB) mode of ADS. Use of HB will be described in Appendix 8.1.

When these spurs interfere with the desired signal, the S/N ratio or SINAD (signal to noise plus distortion ratio) is degraded.

8.1.1 Gain compression

The nonlinear transfer characteristic of the component shows up in the grossest sense when the gain is no longer constant with input power. That is, if P_{out} is no longer linearly related to P_{in}, then the device is clearly nonlinear and distortion can be expected (Figure 8.1).

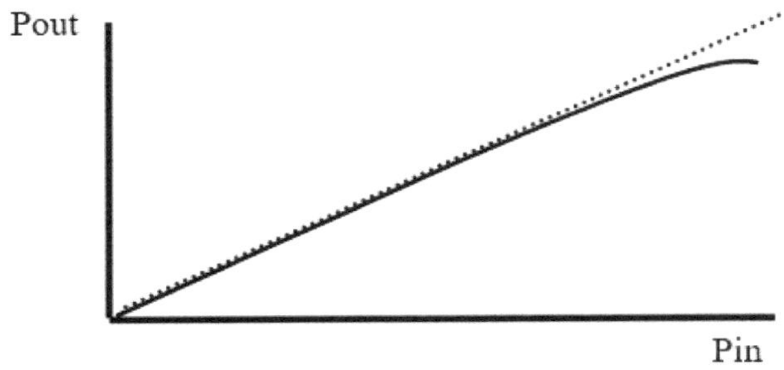

Figure 8.1 Transfer characteristic of a realistic amplifier.

P_{1dB}, the input power required to compress the gain by 1 dB, is often used as a simple to measure index of gain compression. An amplifier with 1 dB of gain compression will generate severe distortion.

Distortion generation in amplifiers can be understood by modeling the amplifier's transfer characteristic with a simple power series function:

$$V_{out} = a_1 V_{in} + a_2 V_{in}^2 + a_3 V_{in}^3 + \cdots \tag{8.1}$$

Of course, in a real amplifier, there may be terms of all orders present, but the simple cubic nonlinearity is easy to visualize. The coefficient a_1 represents the linear gain; a_3 the third-order distortion. When the input is small, the cubic term can be very small. At high input levels, much nonlinearity is present. This leads to gain compression among other undesirable things.

Suppose an input $V_{in} = A\sin(\omega t)$ is applied to the input.

$$V_{out} = A\underbrace{\left[a_1 - \frac{3a_3A^2}{4}\right]}_{\text{Gain compression}}\sin(\omega t) + \underbrace{\frac{1}{4}a_3A^3\sin(3\omega t)}_{\text{Third order distortion}} \tag{8.2}$$

Gain compression is a useful index of distortion generation. It is specified in terms of an input power level (or peak voltage) at which the small signal conversion gain drops off by 1 dB.

The example above assumes that a simple cubic function represents the nonlinearity of the signal path. When we substitute $Vin(t) = A\sin(\omega t)$ and use trig identities, we see a term that will produce gain compression: $A(a_1 - 3a_3A^2/4)$.

If we knew the coefficient a_3, we could predict the 1 dB compression input voltage. Typically, we obtain this by measurement of gain vs. input voltage.

8.1.2 Harmonic distortion

We also see square and cubic terms in Equation (8.1) that represent the second- and third-order *harmonic distortion* (HD) that also is caused by the nonlinearity of the signal path. Harmonic distortion is easily removed by filtering; it is the *intermodulation distortion* that results from multiple signals that is far more troublesome to deal with.

Note that in this simple example, the fundamental is proportional to A whereas the third-order HD is proportional to A^3. Thus, if P_{out} vs. P_{in} were plotted on a dBm scale, such as in Figure 8.2, the HD power will increase at three times the rate that the fundamental power increases with input power. This is often referred to as being *well behaved*, although given the choice, we could easily live without this kind of behavior!

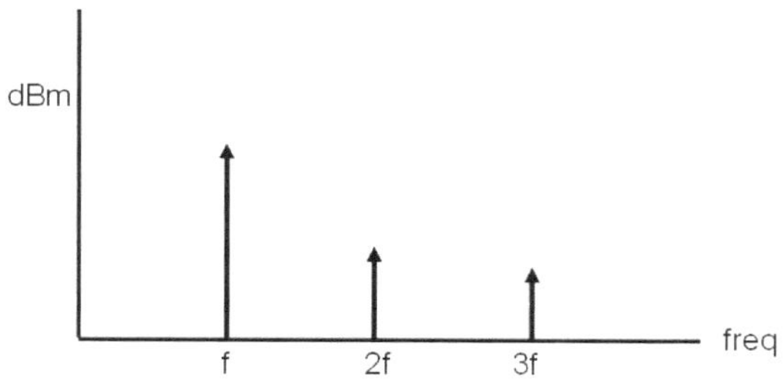

Figure 8.2 Spectral plot showing the harmonic distortion produced with a strong fundamental signal and nonlinear coefficients a_2 and a_3.

8.1.3 Intermodulation distortion [1, 2, 4]

Let's consider again the simple cubic nonlinearity $a_3 Vin^3$. When two inputs at $\omega 1$ and $\omega 2$ are applied simultaneously to the RF input of the system, the cubing produces many terms, some at the harmonics and some at the IMD frequency pairs. The trig identities show us the origin of these nonidealities:

$$V_{in}(t) = V_1 \sin(\omega_1 t) + V_2 \sin(\omega_2 t) \tag{8.3}$$

Then, taking V_{in} to the third power,

$$\begin{aligned} &V_{out3} = a_3 \left[V_1^3 \sin^3(\omega_1 t) + V_2^3 \sin^3(\omega_2 t)\right] + \\ &a_3 \left[\underline{3V_1^2 V_2 \sin^2(\omega_1 t) \sin(\omega_2 t)} + 3V_1 V_2^2 \sin(\omega_1 t) \sin^2(\omega_2 t)\right] \end{aligned} \tag{8.4}$$

Expanding further the underlined term,

$$\frac{3V_1^2 V_2 a_3}{2} \left\{\sin(\omega_2 t) - \frac{1}{2}\left[\sin(2\omega_1 - \omega_2)t - \sin(2\omega_1 + \omega_2)t\right]\right\}. \tag{8.5}$$

8.1.4 Cross modulation

The first term in Equation (8.5) represents *cross-modulation*. The amplitude of one signal (say ω_1) influences the amplitude of the desired signal at ω_2 through the coefficient $3V_1^2 V_2 a_3/2$. A slowly varying modulation envelope on V_1 will cause the envelope of the desired signal output at ω_2 to vary as well since this fundamental term created by the cubic nonlinearity will add to the linear fundamental term. If this term is strong, it can cause serious interference to the desired output at ω_2.

8.1.5 Intermodulation distortion

The second term describes *third-order intermodulation distortion (IMD3)*. We will be mainly concerned with IMD3. (Actually, any distortion terms can create in-band signals – we will discuss this later.) IMD is especially troublesome since it can occur at frequencies within the signal bandwidth. For example, suppose we have two input frequencies at 899.990 and 900.010 MHz. Third order products at $2f1 - f2$ and $2f2 - f1$ will be generated at 899.980 and 900.020 MHz. These IM products may fall within the filter bandwidth of the system and thus cause interference to a desired signal. The spectrum would look like Figure 8.3, where you can see both third and fifth order IM.

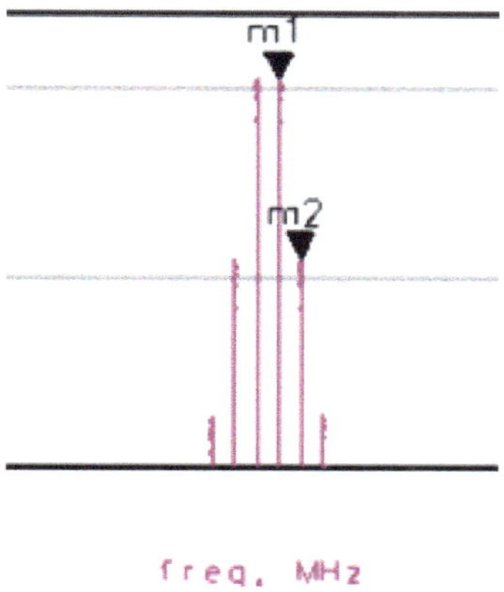

Figure 8.3 Spectrum showing third- and fifth-order IMD products.

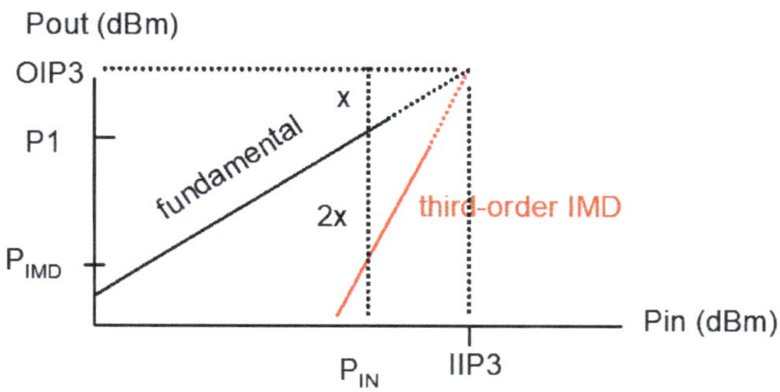

Figure 8.4 Log plot (dBm) of P_{in} vs. P_{out}.

In Figure 8.4 , the red line represents the amplitude of one of the two IMD3 distortion products. From Equation (8.5), you can see that IMD3 is proportional to $V_1^2V_2$ or $V_1V_2^2$, hence the slope of 3 . The vertical dotted line at P_{in} intersects both the fundamental and IMD lines at P_1 and P_{IMD} respectively.

IMD power, just as harmonic distortion power, will have a slope of 3 on a P_{out} vs. P_{in} (dBm) plot. A widely used figure of merit for IMD is the *third-order intercept* (TOI) point. This is a fictitious signal level at which the fundamental and third-order product terms would intersect. In Figure 8.4, input power IIP3 is usually defined as the TOI. In reality, the intercept power is 10 to 15 dBm higher than the $P1$dB gain compression power, so the circuit does not amplify or operate correctly at the IIP_3 input level. The higher the TOI, the better the large signal capability of the system. If specified in terms of input power, the intercept is called IIP_3. Or, at the output, OIP_3. This power level can't be actually reached in any practical amplifier, but it is a calculated figure of merit for the large-signal handling capability of any RF system.

It is common practice to extrapolate or calculate the intercept point from data taken at least 10 dBm below $P1$dB. One should check the slopes to verify that the data obeys the expected slope = 1 or slope = 3 behavior. The TOI can be calculated from the following geometric relationship:

$$IIP_3 = P_{IN} + \frac{1}{2}\left(P_1 - P_{IMD}\right) \tag{8.6}$$

Also, the input and output intercepts (in dBm) are simply related by the gain (in dB):

$$OIP_3 = IIP_3 + \text{power gain}. \tag{8.7}$$

Other higher odd-order IMD products, such as the fifth and seventh, are also of interest, and can also be defined in a similar way, but may be less reliably predicted in simulations unless the device model is precise enough to give accurate nonlinearity in the transfer characteristics up to the $(2n-1)$th order.

Example 8.1: Calculate the P_{IMD} and IIP_3 of an amplifier given V_{in} = 100 mV, gain G = 10, R = 50 Ω and a_3 = 0.33. Recall that dBm = 10 log (power in milliwatts).

$$P_{IN} = \frac{V_{in}^2}{2R} = 10^{-4}\ W = -10dBm$$

$$P_1 = P_{IN}G = 10^{-3}\ W = 0dBm$$

$$P_{IMD} = \frac{\left(\frac{3}{4}a_3V_{\text{in}}^3\right)^2}{2R} = 6.1 \times 10^{-10} = -62.2dBm$$

$$IIP_3 = P_{IN} + \frac{1}{2}\left(P_1 - P_{IMD}\right) = +21dBm.$$

8.1.6 Second order nonlinearity

In the simplified model above, we have neglected second order nonlinear terms in the series expansion. In many cases, an amplifier or other RF system will have some even-order distortion as well. The transfer function then would look like this, repeated from Equation (8.1):

$$V_{out} = a_1 V_{in} + a_2 V_{in}^2 + a_3 V_{in}^3$$

If we once again apply two signals at frequencies ω_1 and ω_2 to the input, we obtain:

$$V_{out2} = a_2 \left[V_1^2 \sin^2(\omega_1 t) + V_2^2 \sin^2(\omega_2 t) + 2V_1 V_2 \sin(\omega_1 t)\sin(\omega_2 t)\right]. \tag{8.8}$$

The $\sin^2$ terms expand into:

$$\frac{1}{2}a_2 V_1^2 \left[1 - \cos(2\omega_1 t)\right] + \frac{1}{2}a_2 V_2^2 \left[1 - \cos(2\omega_2 t)\right]. \tag{8.9}$$

From this, we can see that there is a DC term and a second harmonic term present for each input. The DC term is proportional to the square of the voltage, therefore power. This is one use of second-order nonlinearity – as a power sensor. The HD term is also proportional to the square of the voltage. Thus, on a power out vs. power in plot, it has a slope of 2.

When the next term is expanded, the product of two sine waves is seen to produce the sum and difference frequencies.

$$a_2 V_1 V_2 \left[\cos(\omega_2 - \omega_1)t - \cos(\omega_2 + \omega_1)t\right]. \tag{8.10}$$

This can be both a useful property and a problem. The useful application is as a frequency translation device, often called a mixer, a downconverter, or an upconverter. The desired output is selected by inserting a filter at the output of the device.

Second order distortion, if generated by out-of-band signals, can also lead to interference in-band. When the difference or sum of two out-of-band signals lands in-band, there can be interference to the desired signals. Preselection filtering can generally suppress this in narrowband amplifiers, but it can be a big problem for wideband circuits.

An SOI, or second-order intercept can also be defined as shown in Figure 8.5:

The second-order IMD slope $= 2$. The separation $P_1 - P_{\text{IMD}} = x = OIP2 - P_1$. Then, $IIP2$ can be calculated from measurement by:

$$IIP2 = P_{in} + (P_1 - P_{IMD}) \tag{8.11}$$

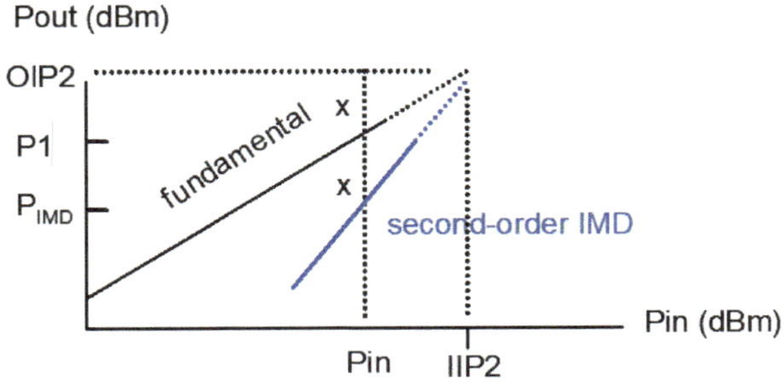

Figure 8.5 Log plot (dBm) of P_{in} and P_{out}.

$$OIP2 = IIP2 + \text{ Gain } = 2P_1 - P_{IMD} \tag{8.12}$$

How is the intercept point (Figure 8.6) affected by cascaded stages?

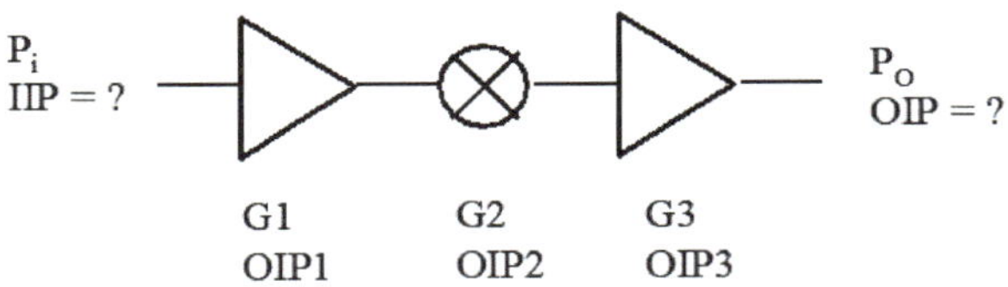

Figure 8.6 Cascaded stages. The intercept point must be calculated at the same reference plane.

Gains multiply in a cascade: $P_O = P_i G_1 G_2 G_3$ (or add them if in dB).

Individual intercept points must be referred to the same reference plane. It can be either at the input or the output. In this example, the output IP, OIP, is specified for each stage.

1. Convert all OIPs from dBm to mW and gains from dB to a power ratio.
2. Let's refer all of these OIPs to the output plane.
   OIP_3
   $G_3 OIP_2$
   $G_2 G_3 OIP_1$.
3. The *third order intercept* cascading relationship is:

$$\frac{1}{OIP} = \frac{1}{G2G3\, OIP1} + \frac{1}{G3\, OIP2} + \frac{1}{OIP3} \tag{8.13}$$

$$IIP_3 = \frac{OIP_3}{G_1 G_2 G_3} \tag{8.14}$$

4. Convert the results back to dBm if desired.

Second order intercept cascading is accomplished by the following equations:

$$\frac{1}{\sqrt{OIP}} = \frac{1}{\sqrt{G2G3OIP1}} + \frac{1}{\sqrt{G3OIP2}} + \frac{1}{\sqrt{OIP1}} \quad (8.15)$$

$$IIP = \frac{OIP}{GIG2G3} \quad (8.16)$$

Example 8.2: Receiver front end (Figure 8.7).

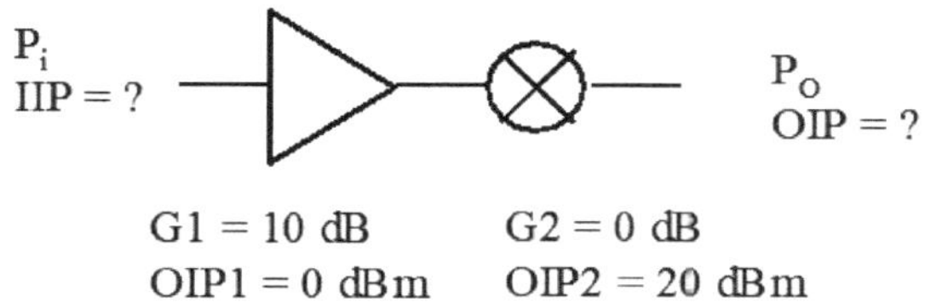

Figure 8.7 Two-stage system includes an amplifier and a mixer.

1. Convert dBm to mW: OIP_1 = 1 mW, OIP_2 = 100 mW
   Convert dB to a power ratio: G_1 = 10, G_2 = 1
2. Refer to the output plane using Equation (8.13):
   $1/OIP = 1/G_2OIP_1 + 1/OIP_2 = 1 + 1/100 = 1.01$ $OIP = 1$ (0 dBm)
3. $IIP = OIP/10 = 0.1$ (−10 dBm).

We can see that the first stage completely dominates the *IIP* in this example. IF we eliminated this stage, then $OIP = OIP_2$ = 20 dBm and *IIP* = 20 dBm:

a 30 dB improvement!

What do we lose by eliminating the first stage, often a low-noise amplifier (LNA)?

8.1.7 Measuring IMD performance

Figure 8.8 represents a typical test bench setup for measuring IMD. Two sources, a combiner, variable attenuator and spectrum analyzer are needed for this test.

1. Set the amplitude of the two sources to be equal.
2. Start with a very low input power using the variable attenuator. Then, increase power in steps until you begin to see the IMD signals on the

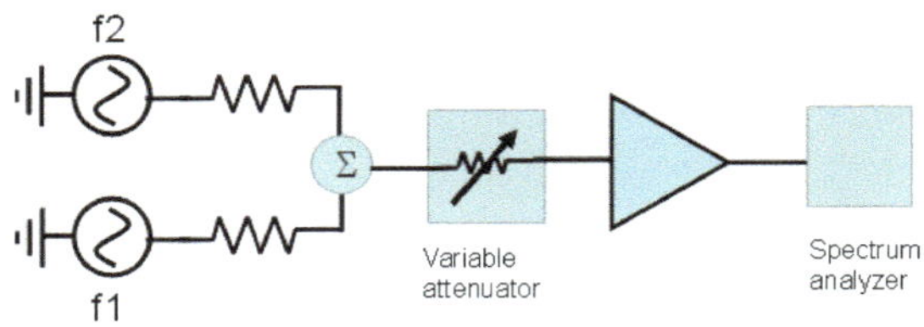

Figure 8.8 Measurement setup for evaluation of intermodulation distortion.

spectrum analyzer. The resolution bandwidth should be narrow to reduce the noise floor in order to increase visibility of the IMD at lower power levels.

3. Keep P_{in} at least 10 dB below the 1 dB compression power, P_{1dB}.
4. Plot the IMD power vs. input power and verify that the slope is close to 3.

Note that at higher input power levels the slope may depart from 3. This is due to the contribution of fifth and higher order IMD power.

Lab Exercise 8.1: P_{1dB} and harmonic distortion

Measure gain compression and HD on a commercial LNA. Minicircuits[1] has many such amplifiers available [5] . The ZFL-500LN, for example, covers 0.1 to 500 MHz.

A. P_{1dB} and harmonic distortion. Figure 8.9 illustrates a test configuration for this exercise.

1. Measure small signal gain at a frequency within the useful range of the amplifier that is low enough in frequency that you can measure at least a second harmonic on the spectrum analyzer. (Maximum input +5 dBm with no damage.)
2. Increase input power and determine the 1dB compression input power level.
3. At this power level, determine the output power at all harmonic frequencies that can be seen on the spectrum analyzer. List the HD/fundamental Pout ratios in dB in a table. The span should be wide enough to view several harmonics.

Be sure to add attenuation to the input of the spectrum analyzer to avoid overloading it and causing distortion in its front end.

B. Intermodulation distortion and IIP_3. Refer to Figure 8.8.

[1] Minicircuits.com

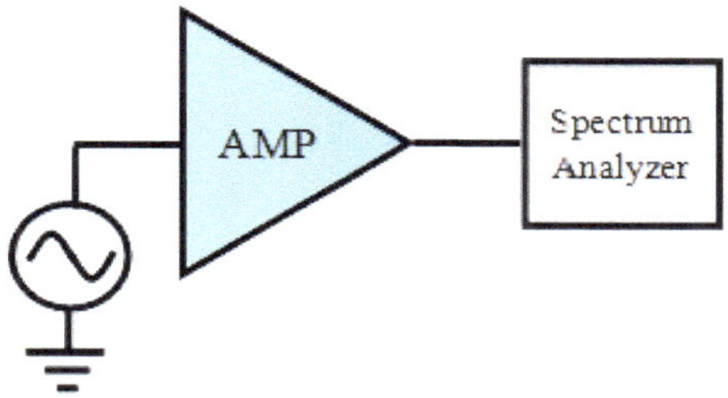

Figure 8.9 Measurement setup for evaluation of intermodulation distortion.

Use two generators to test for intermodulation behavior. Set both generator powers equal and sum together with an isolating power combiner (Minicircuits ZSC-2-1 for example –see datasheet on their website). Use a step attenuator to decrease or increase the total power level at the input of the amplifier.

1. Choose f_1 and f_2. The third order IMD will be visible at $2f_1 - f_2$ and $2f_2 - f_1$. Set the total power of both signals at least 20 dB below P_{1dB} and reduce the resolution bandwidth and span of the spectrum analyzer so that the IM_3 is clearly visible.
2. Measure the ratio of output power P1 at f_1 or f_2 to IMD power as a function of input power in 1dB steps. Make sure the slope of the IMD power vs. input power is approximately 3. Calculate the IIP_3 from this measurement. If the slope differs much from 3, reduce the input power level further and try again. Does your calculated IIP_3 vary with input power? Compare IIP_3 with P_{1dB}.

8.2 Next Topic: Noise

Noise determines the minimum signal power (minimum detectable signal or MDS) at the input of the system required to obtain a signal to noise ratio of 1. An S/N = 1 is usually considered to be the lower acceptable limit except in systems where signal averaging or processing gain is used. The noise figure is a figure of merit used to describe the amount of degradation in the S/N ratio that the system introduces as the signal passes through.

For some applications, the minimum signal power that is detectable is important.

- Satellite receiver
- Terrestrial microwave links
- 802.11.

Noise limits the minimum signal that can be detected for a given signal input power from the source or antenna.

We will identify sources of noise, and define related quantities of interest:

- S/N = signal to noise ratio
- MDS = minimum detectable signal
- F = noise factor
- NF = 10 × log(F) = noise figure.

8.2.1 Noise basics: sources of noise

What is noise? How is it evident to us? Why is it important?[3, 4]

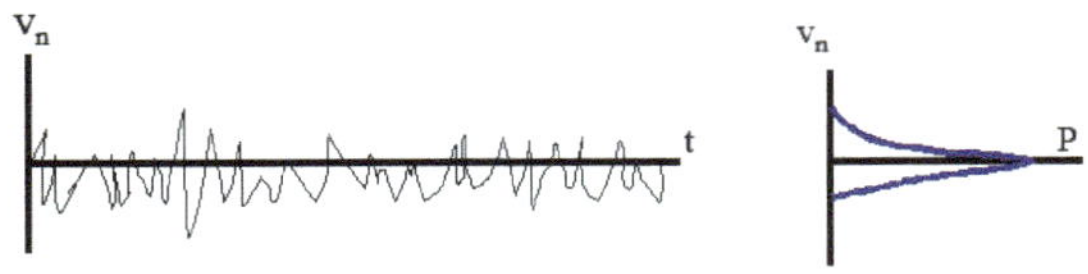

Figure 8.10 Voltage variations due to noise. Probability distribution.

What is noise. We see in Figure 8.10 that noise is represented in the time domain and frequency spectral density. It comes about due to:

1. Any unwanted random disturbance
2. Random carrier motion produces a current. Frequency and phase are not predictable at any instant in time
3. The noise amplitude is often represented by a Gaussian probability density function.

The cumulative area under the curve represents the probability of the event occurring. Total area is normalized to 1.

Because of the random process, the average value is zero:

$$\bar{v}_n = \lim_{T\to\infty} \frac{1}{T} \int_{t_1}^{t_1+T} [v_n(t)]\, dt = 0. \tag{8.17}$$

We cannot predict $v_n(t)$, but the variance (standard deviation) is finite:

$$\bar{v}_n^2 = \lim_{T\to\infty} \frac{1}{T} \int_{t_1}^{t_1+T} [v_n(t)]\, dt = \sigma^2. \tag{8.18}$$

Often we refer to the rms value of the noise voltage or current:

$$v_{n,rms} = \sqrt{\bar{v}_n^2}. \tag{8.19}$$

Sources of noise in circuits:

- Shot noise — forward-biased junctions
- Thermal Noise — any resistor
- Flicker (1/*f*) noise — trapping effects

Shot noise: This is due to the random carrier flow across a pn junction. Electrons and holes randomly diffuse across the junction producing noise current pulses that occur randomly in time. The steady state current measured across a forward biased diode junction is really a large number of discrete current pulses.

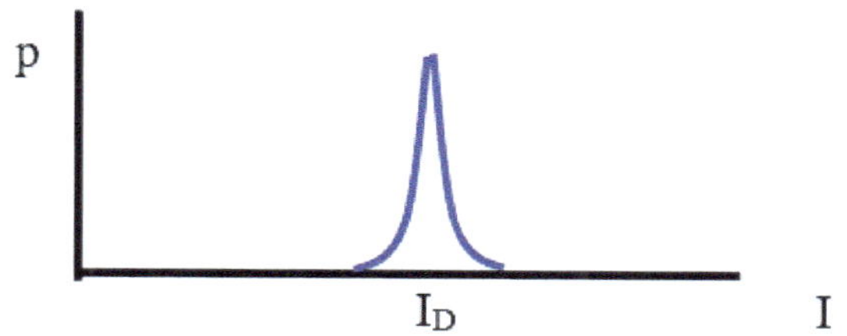

Figure 8.11 Probability distribution of shot noise in a pn junction.

In Figure 8.11, I_D represents the DC bias current through the pn junction. The variance of this current:

$$\bar{i}^2 = \lim_{T\to\infty} \frac{1}{T} \int_0^T (I - I_D)^2\, dt = \sigma^2 \tag{8.20}$$

It can be shown that this mean square noise current can be predicted by

$$\bar{i}^2 = 2qI_D B \tag{8.21}$$

where
q = charge of an electron = 1.6×10^{19}
I_D = diode current
B = bandwidth in Hertz (sometimes called Δf).
The noise current spectral density:

$$\boxed{\bar{i}^2/B = 2qI_D} \tag{8.22}$$

- Independent of frequency (white noise)
- Independent of temperature for a fixed current
- Proportional to the forward bias current

- Gaussian probability distribution

1 mA of current corresponds to a noise current spectral density of

$$\boxed{18\ \mathrm{pA}/\sqrt{\mathrm{Hz}}}$$

Read: 18 picoamp per root Hertz

Thermal noise: Thermal noise, sometimes called Johnson noise, is due to random motion of electrons in conductors. It is unaffected by DC current and exists in all conductors. Its spectral density is also frequency independent, but is directly proportional to temperature. The noise probability density is Gaussian.

$$\boxed{\overline{v}^2 = 4kTRB} \quad (8.23)$$

$$\boxed{\overline{i}^2 = 4kTB/R} \quad (8.24)$$

$$k = \text{Boltzmann constant} = 1.38 \times 10^{-23}\ \text{J/K}$$

A 50 ohm resistor produces a noise voltage spectral density of

$$\boxed{0.9\ \mathrm{nv}/\sqrt{\mathrm{Hz}}}$$

or a Norton equivalent noise current spectral density of

$$\boxed{18\ \mathrm{pA}/\sqrt{\mathrm{Hz}}}$$

Flicker or 1/f noise: This noise source is most evident at very low frequencies. It is hard to localize its physical mechanisms in most devices. There is usually some $1/f$ noise contribution due to charge traps with long time constants. The trap charge then is randomly released after some relatively long period of time. $1/f$ noise is modeled by:

$$\overline{i}^2/B = K\frac{I}{f} \quad (8.25)$$

- K is a fudge factor. It can vary wildly from one type of transistor to the next or even from one fabrication lot to the next.
- I is the current flowing through the device.
- B is the bandwidth.

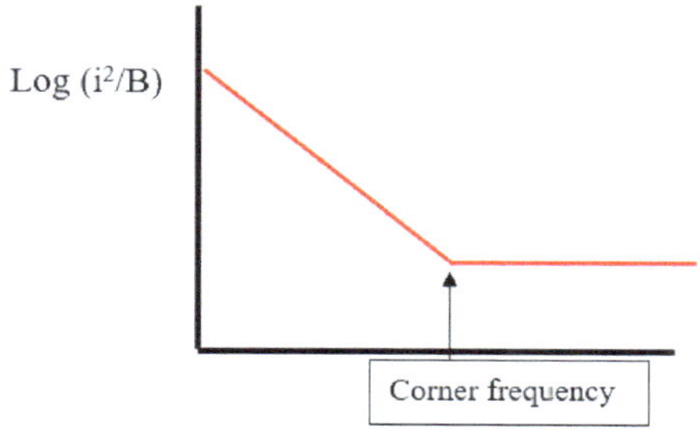

Figure 8.12 Spectral distribution of 1/*f* noise. Plotted vs. log frequency.

- 1/*f* noise can be described by a corner frequency, as shown in Figure 8.12 where the 1/*f* noise intersects the background noise floor that is independent of frequency.
- Carbon resistors exhibit 1/*f* noise; metal film resistors do not.

Noise can be modeled as a Thevenin equivalent voltage source or a Norton equivalent current source, as in Figure 8.13. The noise contributed by the resistor is modeled by the source, thus the resistor is considered noiseless.

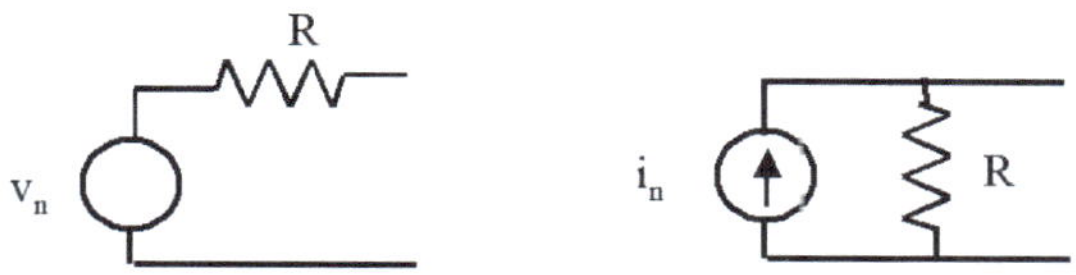

Figure 8.13 Equivalent circuit noise generators.

It is important to note that noise sources:

- Do not have polarity (the arrow is just tc distinguish current from voltage).
- Do not add algebraically, but as RMS sums calculated as in Equation (8.26) and shown schematically in Figure 8.14.

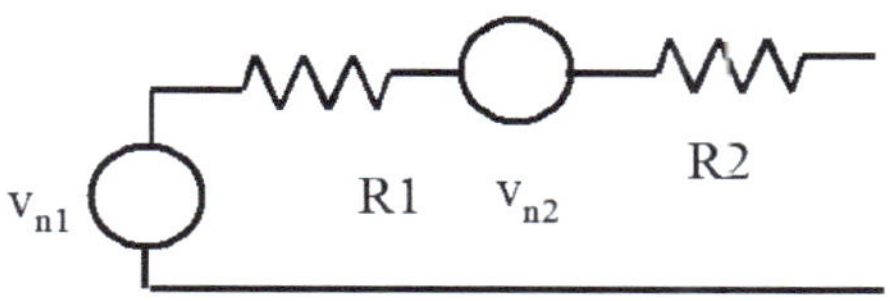

Figure 8.14 Noise sources add as RMS sums.

$$v_{n,total}^{-2} = v_{n1}^{-2} + v_{n2}^{-2} = 4kTBR_1 + 4kTBR_2 \tag{8.26}$$

If the sources are correlated (derived from the same physical noise source), then there is an additional term:

$$\bar{v}_{n,total}^2 = \bar{v}_{n1}^2 + \bar{v}_{n2}^2 + 2Cv_{n1}v_{n2} \tag{8.27}$$

C can vary between –1 and 1.

The *available noise power*, P_{av}, can be calculated from the RMS noise voltage or current:

$$\boxed{P_{a\nu} = \frac{\bar{v}_n^2}{4R} = \frac{\bar{i}_n^2 R}{4} = kTB.} \tag{8.28}$$

That is, the available noise power from the source is

- independent of resistance
- proportional to temperature
- proportional to bandwidth
- has no frequency dependence

$$P_{av} = 4 \times 10^{-21} \text{ watt}$$

in a 1 Hz bandwidth at the standard noise room temperature of 290 K. If converted to dBm = 10log(P/10^{-3}), this power becomes

$$\boxed{-174 \text{ dBm/Hz.}}$$

We are generally interested in the noise power in other bandwidths than 1 Hz. It's easy to calculate: $P = kTB$, where $kT = -174$ dBm.

To convert bandwidth in Hertz to dB: 10log$B$.

Example 8.3: Suppose your B = 1000 Hz. $P = kTB$.

In dBm, $P = -174 + 10\log(1000) = -174 + 30 = -144$ dBm.

Question: Can a resistor produce infinite noise voltage (refer to Figure 8.15)?

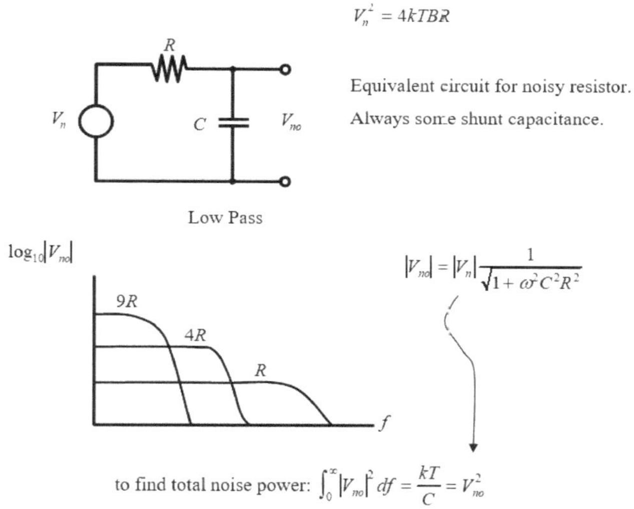

Figure 8.15 Any real resistive network includes some capacitance. Hence the low pass characteristic. So, total noise power is independent of *R*.

8.2.2 Noise equivalent bandwidth

An amplifier or filter has a nonideal frequency response. Noise power transmitted through is determined by the bandwidth (Figure 8.16).

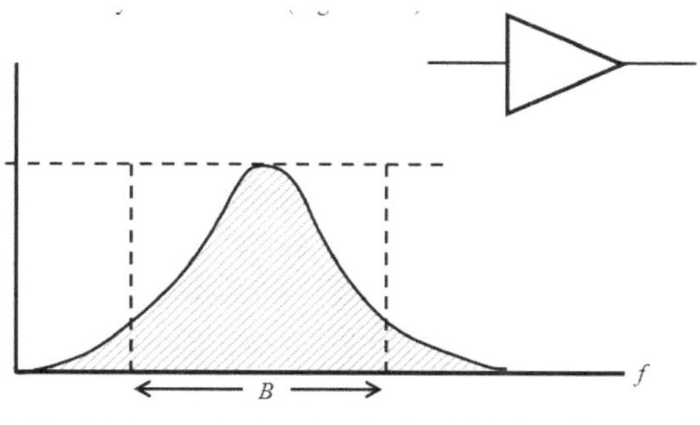

Figure 8.16 Noise equivalent bandwidth. Brick wall approximation.

Noise power $\propto$ (mean square voltage):

$$\bar{v}_i^2 |A(f)|^2 = \bar{v}_o^2 / Hz \text{ in a 1 Hz interval.} \tag{8.29}$$

Finding the total noise power at the output requires summation over the entire frequency band

$$\int_0^\infty \bar{v}_o^2(f)df = \bar{v}_i^2 \int_0^\infty |A(f)|^2 df. \tag{8.30}$$

We choose an equivalent BW, B, with a rectangular profile whose area is the same.

$$A_m^2 B = \int_0^\infty |A(f)^2| df. \tag{8.31}$$

$$\boxed{B = \frac{1}{A_m^2} \int_0^\infty |A(f)^2| df.} \tag{8.32}$$

This is the definition of bandwidth that we will assume in subsequent calculations.

8.2.3 Signal-to-noise ratio

Having determined the sources of noise and how to determine noise power in a given bandwidth, we can now define an important figure-of-merit, the *signal-to-noise ratio*, which will be used to determine a minimum detectable signal in a receiver:

$$SNR = \frac{P_S}{P_N} = \frac{S}{N}. \tag{8.33}$$

Most noise calculations use available power for signal and for noise. Available noise power was defined in Equation (8.28) which refers to Figure 8.11 . In some cases, the ratio can be defined with noise power in the numerator (Equation (8.34)). Also, distortion power, *D*, can also be considered when large signal behavior limits performance.

$$\frac{S+N}{N} \text{ and } \frac{\mathrm{S+N+D}}{\mathrm{N}} \text{ or } \mathit{SINAD}. \tag{8.34}$$

Why is the S/N important? It affects the error rate when receiving information.

8.2.4 Noise factor, F

In Figure 8.17 we see an amplifier or passive component with signal and noise powers represented at input and output. It has an available gain of G_a. Noise factor *F* is a measure of how much noise is added by a component such as an amplifier.

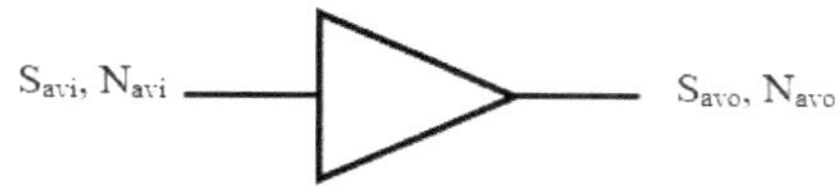

Figure 8.17 A component with available gain G_a. Signal and noise available power is shown at input and output.

$$G_a = \frac{S_{avo}}{S_{avi}} \tag{8.35}$$

$$F = \frac{S_{avi}/N_{avi}}{S_{avo}/N_{avo}} > 1 \tag{8.36}$$

$$F = \frac{N_{avo}}{N_{avi}G_{av}}. \tag{8.37}$$

F is always greater than one because *S/N* at input will always be greater than *S/N* at output. The component can only add noise. Therefore, noise factor represents the extent that the S/N is degraded by the system.

$$F = \frac{\text{total noise power available at output}}{\text{noise power available at output due to source @290}k}$$

In terms of dB, this called the *noise figure*, NF.

$$NF = 10\log_{10} F. \tag{8.38}$$

Let's look at an example:

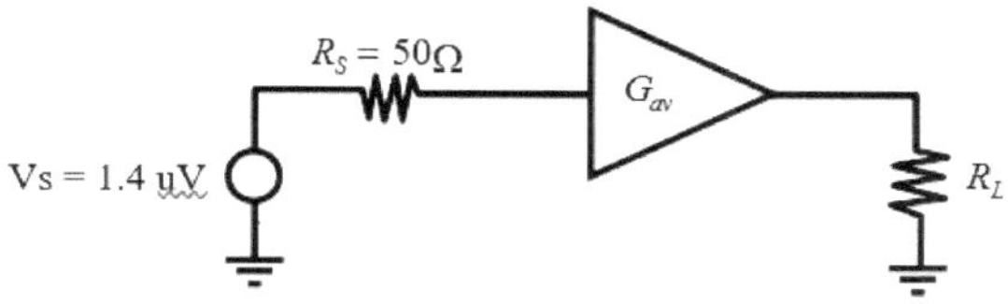

Figure 8.18 Amplifier with input from a 50 ohm source.

Calculate the signal-to-noise ratio of the amplifier shown in Figure 8.18 with the following specifications:

$$G_{av} = 10 \text{ dB}$$

$$NF = 3\text{ dB}$$

$$B = 10^6\text{Hz}$$

Signal available power:

$$S_{avi} = \frac{v_S^2}{8R_S} = \frac{2 \times 10^{-12}}{400} = 5 \times 10^{-15}\text{ W} = -113\text{ dBm}$$

Noise average power = N_{avi} = kTB = $-174 + 60 = -114$ dBm.
Because noise power increases with B $10\log_{10}B$ = 60 dB (in this example)

$$10\log\left(\frac{S_{avo}}{N_{avo}}\right) = 10\log\left(\frac{S_{avi}}{N_{avi}}\right) - NF$$
$$= -113 - (-114) - 3$$
$$= 1\text{ dB} - 3\text{ dB} = -2\text{ dB (not very good).}$$

How can S_o/N_o be improved?

1. Reduce F. Slight room for improvement.
2. Reduce B. Major improvement if application can tolerate reduced B.
3. Increase antenna gain. Lots of room for improving S_i/N_i.

say $B = 10^5$

$$N_{avi} = -174 + 50 = -124\ dBm$$

$$\frac{S_{avi}}{N_{avi}} = 11dB \text{ and } \frac{S_{avo}}{N_{avo}} = 8dB$$

Example 8.4: Noise floor of spectrum analyzer

Typical $NF \cong 25$ dB

$N_{avo} = N_{avi} \times F \times G_{av}$

$N_{avi} = (-174\text{ dBm/Hz}) + 10\log B$

$NF = 25$ dB — B: resolution bandwidth (RBW)

G_{av} = 1 (0 dB)

RBW	N_{avo}
1 KHz	−119 dBm
10 kHz	−109
100 kHz	−99
etc.	

We will see later how this can be improved.

8.2.5 Noise temperature

The excess noise added by an active circuit such as an amplifier can also be modeled by an extra resistor at an effective input noise temperature, T_e .

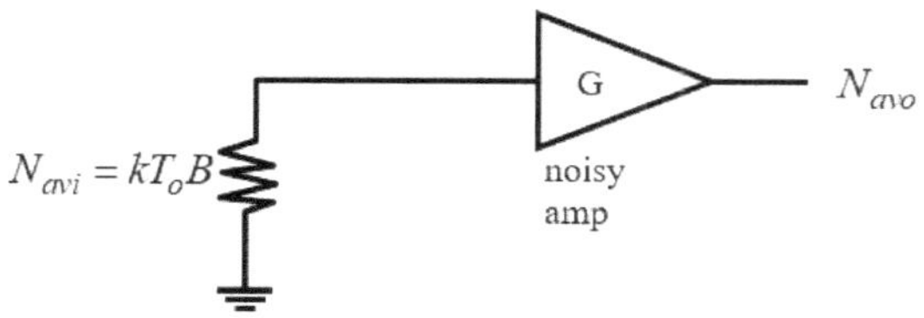

Figure 8.19 Noisy amplifier with input available noise power N_{avi}.

The circuit shown in Figure 8.19 can be represented in figure 8.20 as a separate noise source and signal source at the input.

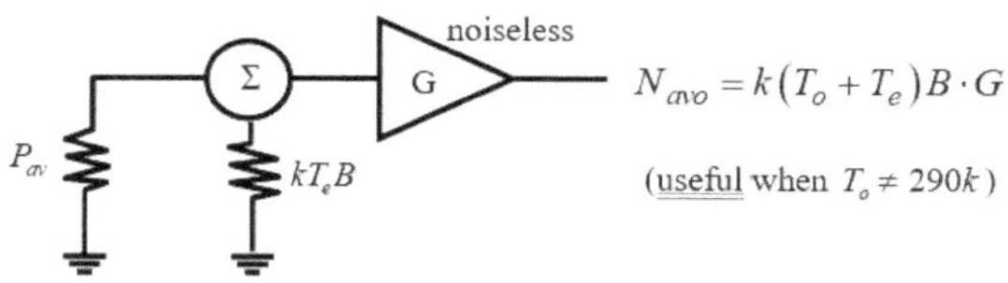

Figure 8.20 Noise contribution of the amplifier is modelled by an equivalent resistor at temperature T_e.

In terms of noise factor:

$$F = \frac{\text{noise out due to DUT } + \text{ noise out due to source}}{\text{Noise out due to source}} \tag{8.39}$$

$$= \frac{kT_eBG + kT_oBG}{kT_oBG} = 1 + \frac{T_e}{T_o}$$

$$\text{or } T_e = 290(F - 1) \tag{8.40}$$

(where F is a number, not dB).

Significance of T_e: The equivalent noise temperature can be used to model excess noise due to system components. The impact of the excess noise depends on the application. For terrestrial applications, 290 K source temperature is typical. But space applications (satellite receivers for example) excess noise is much more critical.

$$N_{avo}(total) = kBG(T_o + T_e) \tag{8.41}$$

Example 8.5:

$$NF = 1 \quad dB \Rightarrow F = 1.26$$
$$= 1 + T_e/T_o = 1 + T_e/290$$
$$\text{so } T_e = 75 \text{ K}$$

Total output noise $\Rightarrow 290 + 75 = 365$ K equivalent source temperature.

So what? Not a major increase in noise power. Further reduction in F may not be justified.

But, for space applications, a much lower background noise temperature is normally visible. As low as is possible.

In that case,

$$T = T_o + T_e = 20 + 75 = 95K$$

The amplifier noise temperature of 75K will produce a major degradation in noise temperature and consequently signal-to-noise ratio. Noise factor or noise figure at room temperature doesn't reveal this so clearly.
$F = 1 + 75/20 = 4.5$ ($NF = 7$ dB).

8.2.6 Noise figure of cascaded stage

Use available gain. Why available gain?

Noise power is defined as available power. Cascading of noise is more convenient when G_A is used. Also, it assumes that the output network is matched for any input impedance presented to the amplifier.

8.2.6.1 Second stage noise contribution

Let the noise contribution of each amplifier be defined by its respective noise temperature. Then we will define an input equivalent noise power that includes the influence of the second stage. Refer to Figure 8.21. Here noise factors, noise power and gain are attributed to each stage.

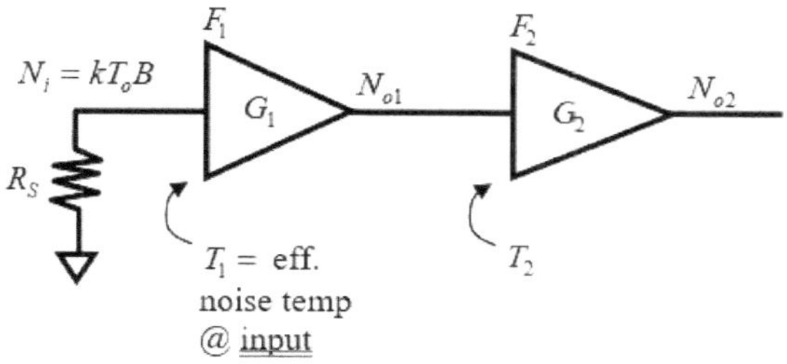

Figure 8.21 Block diagram of a cascaded two-stage amplifier.

$$\begin{aligned} N_{o1} &= k\left(T_o + T_1\right) BG_1 \\ N_{o2} &= k\left(T_o + T_1\right) BG_1G_2 + kT_2BG_2 \end{aligned} \tag{8.42}$$

To get *total input referred noise* power we take into account the gain of each stage and find what noise at the input would represent each of the stage contributions.:

$$\frac{N_{o2}}{G_1G_2} = N_i \text{ (equivalent)} = k\left(T_o + T_1\right) B + kT_2B/G_1. \tag{8.43}$$

Therefore, the excess noise at the input is:

$$kT_1B + kT_2B/G_1 \tag{8.44}$$

$$\text{Recall that } F = 1 + \frac{T_e}{T_o}, \text{ therefore } T_e = T_1 + T_2/G_1. \tag{8.45}$$

Then, the cascading relationship for F is given by:

$$F_{TOTAL} = \underbrace{1 + \frac{T_1}{T_0}}_{F_1} + \underbrace{\frac{T_2}{T_0G_1}}_{\frac{F_2-1}{G_1}} + \tag{8.46}$$

Then, we can by extension include the noise contribution of a third stage:

$$+\frac{F_3 - 1}{G_1G_2}. \tag{8.47}$$

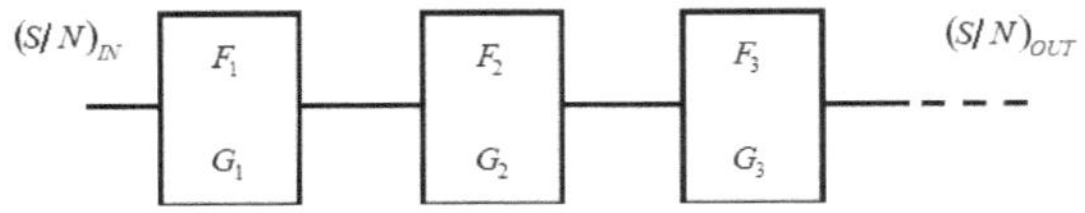

Figure 8.22 Noise figure of cascaded stages.

It is important to recall that the parameters F and G are not in dB, but are numbers.

$$\left.\begin{array}{rl} F_i = & \text{Noise Factor} \\ \mathrm{G_i} = & \text{Available Gain} \end{array}\right\} \underline{\text{not}} \text{ in } dB$$

$$\boxed{F_{TOTAL} = F_1 + \frac{F_2 - 1}{G_1} + \frac{F_3 - 1}{G_1 G_2} + \ldots} \tag{8.48}$$

= Input Total Noise Factor

Then, the impact on signal-to-noise ratio can be seen.

$$\frac{(S]N)_{IN}}{(S/N)_{OUT}} = F_{TOTAL} \tag{8.49}$$

Or, in terms of dB:

$$(S/N)_{OUT}\ \mathrm{dB} = (S/N)_{IN}\ \mathrm{dB} - NF_{TOTAL}. \tag{8.50}$$

If the noise figure is important in a receiver, it is standard procedure to design so that the first stage sets the noise performance. This will require a large enough first stage gain to diminish the noise contribution of the second stages.

8.2.7 Minimum detectable signal

How is the *minimum detectable signal* or MDS defined?

* At a given B (very important)

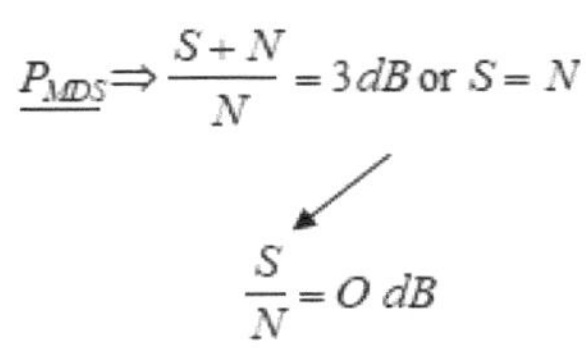

$$P_{MDS} = 10\log(kTB) + NF(dB). \tag{8.51}$$

Substitute the thermal noise power in a 1 Hz bandwidth at the standard noise room temperature of 290 K. If converted to dBm = 10 log(P/10^{-3}), this power becomes −174 dBm/Hz. Then, the influence of the bandwidth and noise figure on MDS can be clearly seen

$$P_{MDS} = -174dBm/Hz + 10\log B + NF(dB) \tag{8.52}$$

We can see now how bandwidth B affects the MDS. The larger the requirement for bandwidth, the higher the noise floor becomes. In many cases the bandwidth is determined by the signal spectral requirements, so is not an independent variable.

8.2.8 Noise figure of passive networks

Why is this important? For example, a component that introduces attenuation, whether due to cable losses, resistive networks, filters, or matching networks, will add to the noise factor. Here, we assume that there are no active components, only resistors and reactances.

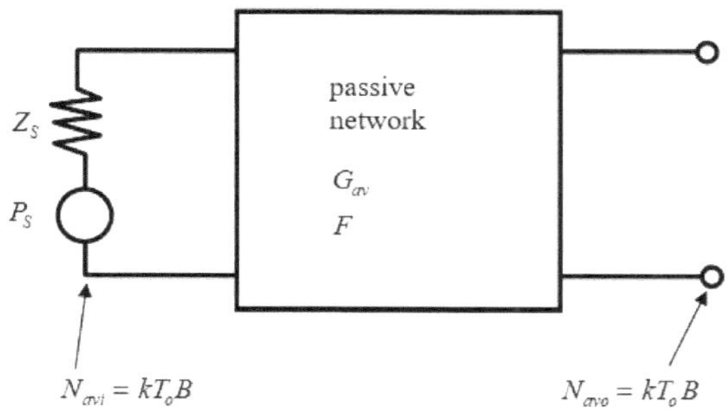

Figure 8.23 A passive network containing only resistances and reactances.

The noise available power does not depend on resistance; however, the signal power is degraded by the attenuation. P_S is the available power from the source. Thus, G_{av} is negative.

Based on Equation (8.49) and Figure 8.23,

$\frac{S_{avo}}{S_{avi}} = G_{av}$

so, determining the signal-to-noise ratio at the input and output of the network, an overall noise factor, F, can be determined as in Equation (8.55).

$$(S/N)_i = \frac{P_S}{kT_oB} \tag{8.53}$$

$$(S/N)_o = \frac{G \cdot P_S}{kT_oB} \tag{8.54}$$

$$F = \frac{(S/N)_i}{(S/N)_o} = \frac{1}{G} \tag{8.55}$$

Therefore, the noise factor is just the inverse of gain.

Or in terms of dB, the noise figure, $NF = -G$ (dB).

Example 8.6: A 10 dB attenuator introduces 10 dB of loss in signal power: $G_{av} = -10$ dB.

Thus, $NF = 10$ dB. You don't want this at the input of your receiver unless it is intended for large signal applications.

8.2.9 How to measure the noise figure of an amplifier

Three methods.

Method 1: Use the spectrum analyzer as a noise receiver.

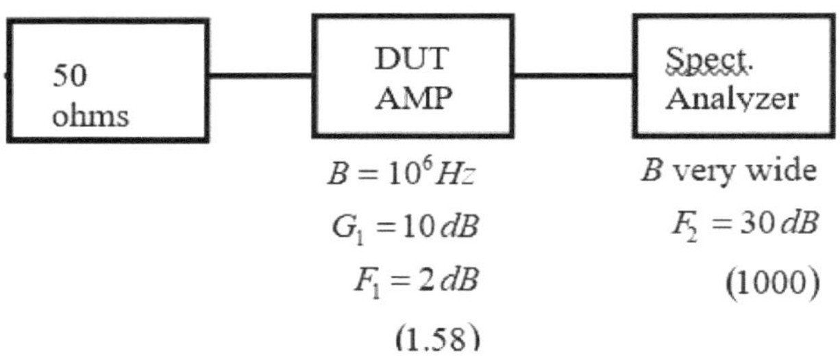

Figure 8.24 Block diagram of noise measurement with a spectrum analyzer.

One must be aware that accurate measurement of noise power is affected by the spectrum analyzer noise floor. The high noise figure of the spectrum analyzer completely dominates the measurement.

$$F_{TOTAL} = 1.58 + \frac{999}{10} = 101.5(20 \text{ dB})$$

(completely dominated by second stage).

Now add a low-noise, high gain preamp ahead of SA (Figure 8.25).

$$F_{TOTAL} = 1.58 + \frac{(1.26-1)}{10} + \frac{999}{(10)(1000)} = 1.71$$
$$(2.3dB)$$

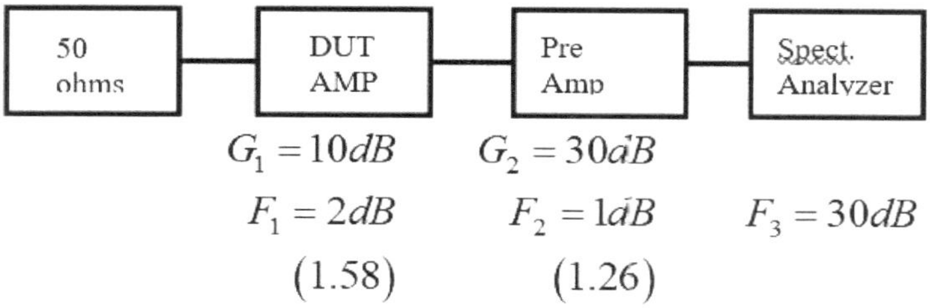

Figure 8.25 Add a low-noise, high gain preamplifier in front of the spectrum analyzer.

With the addition of a low-noise preamp, the spectrum analyzer noise contribution can be kept small enough that the front end noise figure can be determined with accuracy. Otherwise, it is rather hopeless.

Method 2

Use a calibrated signal source, matched correctly to the amplifier under test.

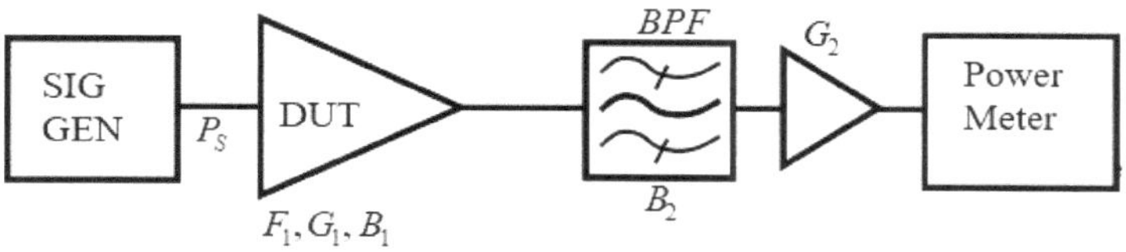

Figure 8.26 Measurement of noise figure using a signal generator and power meter.

The power meter will allow us to calculate V_{rms} across the known load resistance at its input, R_L.

$*B_2 << B_1$ Figure 8.26 shows how the noise equivalent BW is set by a bandpass filter, (BPF).

Two measurements are required.

1. Generator inactive, but still properly terminating the amplifier input. The generator *must* have the correct source impedance.

$$P_S = P_{avs} = N_{avi} = kT_o B_1 \tag{8.56}$$

$$P_1 = FkT_0 A_t B_2 \tag{8.57}$$

2. Next, set the generator on. $P_S = \frac{V_S^2}{8R_S}$ is the available power from the generator in excess of kTB. Then measure P_2 with the power meter.

$$P_2 = FkT_o A_t B_2 + P_S A_t \tag{8.58}$$

3. We see now that the gain can be removed from the calculation. Therefore, the measurement becomes independent of gain. But we do need to know bandwidth B_2. Finally, solve for F.

$$F = \frac{1}{P_2/P_1 - 1} \cdot \frac{P_S}{kT_oB_2} \tag{8.59}$$

The ratio $Y \equiv \frac{P_2}{P_1}$ is often referred to as the Y factor.

Method 3: Hot-cold NF

You can also use a calibrated noise source for measuring NF.

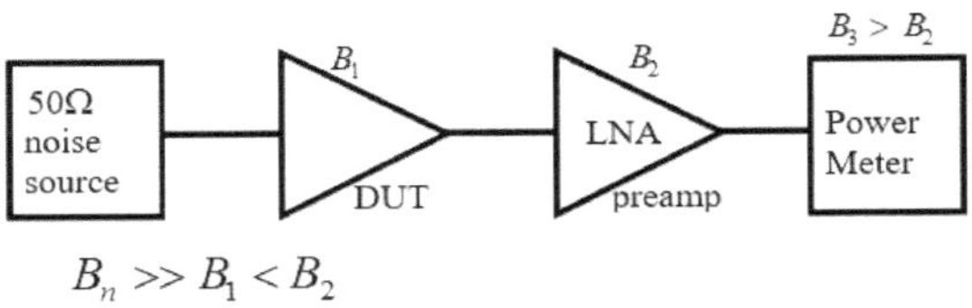

Figure 8.27 Hot-cold noise figure measurement.

The advantage here is that we don't need to know noise equivalent *BW* accurately. The noise source shown in Figure 8.27 has *very wide BW* compared with system under test.

1. "Hot" measurement

$$\begin{aligned} P_H &= \text{ noise power with source on } = kT_HB \\ T_H &= \text{ effective noise temperature of source.} \end{aligned}$$

2. "Cold" measurement.

$$P_o = kT_oB = \text{noise power with source off}$$
$$T_o = 290 \text{ K.}$$

$$\text{Excess noise ratio } = ENR = \frac{P_H - P_o}{P_o} = \frac{T_H}{T_o} - 1 \tag{8.60}$$

$\text{ENR}(dB) = 10\log_{10}\left(\frac{T_H}{T_o} - 1\right)$

Y factor for noise source:

$$Y_S = \frac{P_H}{P_o} = \frac{T_H}{T_o}. \tag{8.61}$$

So, we can use the noise source instead of the signal generator.

1. Source off. Noise power at meter:

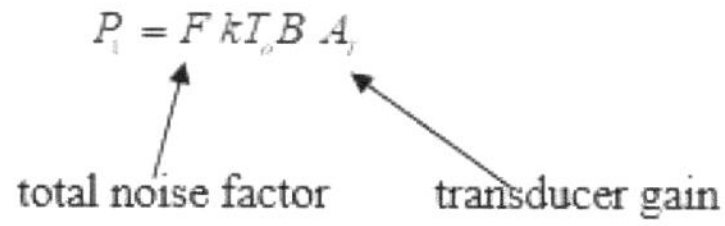

2. Source on:
   $P_2 = P_1 + Y_S kT_0 B A_T$

Divide: $\frac{P_2}{P_1} = Y = 1 + \frac{Y_S}{F}$

Again, the transducer gain cancels, and now B cancels too. We can solve for F from the measured P_2/P_1.

$$F = \frac{Y_S}{Y - 1} \tag{8.62}$$

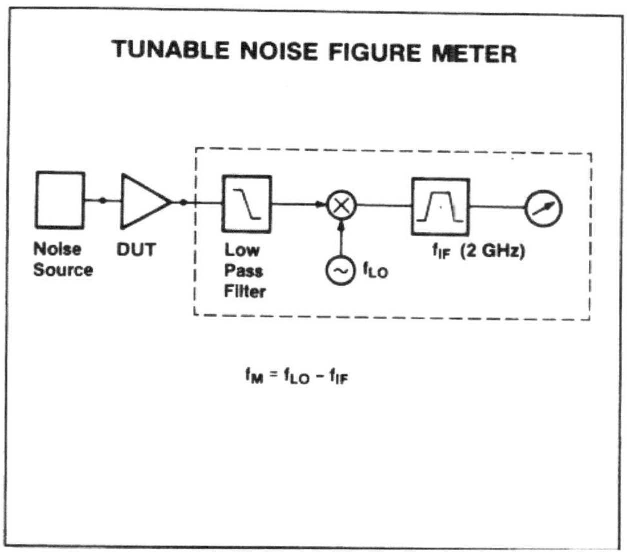

Figure 8.28 Block diagram of a typical tunable noise figure meter.

Figure 8.28 shows that the tunable noise figure meter is a receiver. The mixer block upconverts the input noise signal to a 2 GHz power meter.

$$f_{in} = 2GHz - f_{LO}$$

Thus, by choosing the local oscillator frequency f_{LO}, we measure the noise power within the bandwidth of the IF filter. The noise figure meter also applies a square wave to turn the noise source on and off, obtaining the HOT/COLD input noise condition needed to determine F. As an added bonus, the meter also measures the gain of the device under test.

8.2.10 Spurious free dynamic range

The spurious free dynamic range (SFDR) measures the ability of a receiver or any other RF system to operate between noise limits and interference limits. It is a commonly used figure of merit. The maximum input signal power is limited by distortion, which we describe by IIP_3. The noise limits are described by the minimum detectable signal power, MDS, the signal power at the input that will produce . $(S/N)_{out} = 1$ These two limits on performance are defined in Figure 8.29.

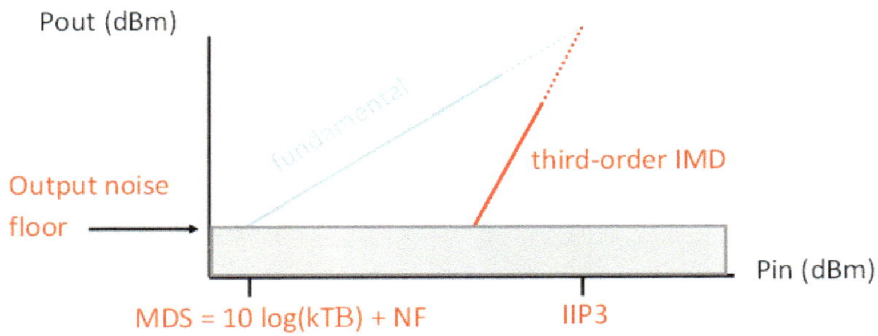

Figure 8.29 Definition of SFDR.

$$SFDR = \frac{2}{3}(IIP3 - MDS) \tag{8.63}$$

If the signal power is increased beyond the point where the third-order IMD rises above the noise floor, then the signal-to-distortion ratio dominates and degrades by 3 dB (slope = 3) for every 1 dB increase in signal power. SFDR is calculated from the geometric 2/3 relationship between the IIP_3 and the *IMD*. It is important to note that the *SFDR* depends directly on the bandwidth, *B*. It has no meaning without specifying the bandwidth of the system under consideration.

8.2.11 Noise and distortion example

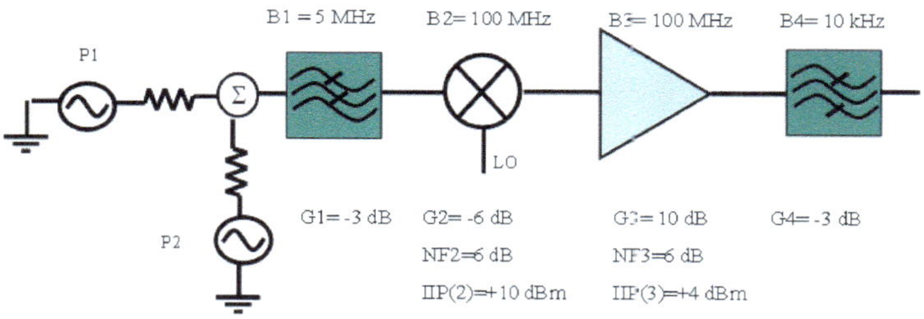

Figure 8.30 Calculate the *SFDR* of this system consisting of two sources, a combiner, a low-pass filter, a mixer, an amplifier and a bandpass filter.

The mixer component is often found in receiver systems and is used to translate frequency between its input and a local oscillator source (LO) either up or down. Mixers will be discussed in much detail in Chapter 11.

First calculate the system noise figure. Convert NF and G in dB to numerical ratios. Recall that for a passive lossy network,

$$F = \frac{1}{G}$$

$$\begin{aligned} F_{TOTAL} &= F_1 + (F_2 - 1)/G_1 + (F_3 - 1)/(G_1G_2) + (F_4 - 1)/(G_1G_2G_3) \\ &= 2 + 6 + 24 + 0.8 = 32.8 \\ &\quad (15.1 \text{ dBNF}) \end{aligned}$$

Minimum detectable signal level (MDS) is defined as that signal power at the input that will produce $(S/N)_{out} = 1$. First find the minimum bandwidth in the chain: in this example, stage 4; $B_4 = 10$ kHz. The NF in the sum below is the total noise contribution found from F_{TOTAL}

$$MDS = -174 \text{ dBm/Hz} + 10 \log B_4 + NF = -119 \text{ dBm}.$$

Could we improve the noise figure? The third stage is the major contributor. We do not need such a wide band IF amplifier for a 10 KHz bandwidth, so this stage could be redesigned for minimum noise figure. Even then, the total NF is high due to the losses in stages 1 and 2. The first stage filter should be replaced with one with lower loss, since its noise figure adds directly to the receiver total NF. The best way to improve NF is to add an LNA, but this will have an impact on the IIP_3.

Next, we must calculate the input third-order intercept for the chain.

Refer the intercepts of stages 2 and 3 to the input of the filter at stage 1:

$IIP(2)' = IIP(2) + 3$ dB = +13 dBm (20 mW)

$IIP(3)' = IIP(3) + 6 + 3 = +13$ dBm

$(IIP_3\text{total})^{-1} = 1/20 + 1/20 = 1/10$

$IIP_3\text{total} = +10$ dBm.

Now, we can calculate *SFDR* using Equation (8.63):

$SFDR = \frac{2}{3}(IIP3 - MDS)$

$$SFDR = \frac{2}{3}(+10 - (-119)) = 86dB$$

Also, we can define another receiver figure of merit: the *receiver factor* [1]:

$$RxF = IIP_3 - NF = 10\text{ dBm} - 15.1\text{ dB} = -5.1\text{dBm}$$

The receiver factor also takes into account both noise and intermodulation properties of the system. It is independent of bandwidth.

8.3 Summary

Two phenomena limit the performance of amplifiers and by extension to receivers.

1. **Distortion**: Limits the large-signal usefulness of the device. Gain compression, harmonic distortion (HD) and intermodulation distortion (IMD) were described.
2. **Noise**: Limits the small-signal usefulness. Sources of noise, noise factor F and noise figure NF, minimum detectable signal (MDS) were identified and quantified. Noise measurement methods also were discussed.
3. **Noise and Distortion**: Spurious free dynamic range (SFDR) and receiver factor.

References

[1] Hayward, W. (1994) *Introduction to Radio Frequency Design* American Radio Relay League, Section 6.3.

[2] Pozar, D. (2005) *Microwave Engineering, 3rd Edn* Wiley, Section 10.2.

[3] Pozar, D. (2005) *Microwave Engineering, 3rd Edn* Wiley, Section 10.1.

[4] Minicircuits.com

[5] Razavi, B. (1998) *RF Microelectronics* Prentice Hall, Ch. 2,. Also, 2nd Edn, 2012.

Homework

1. Assume that there are two input sources to the receiver with equal power levels. How much input power will be required to produce a third order intermodulation component of –60 dBm at the output?

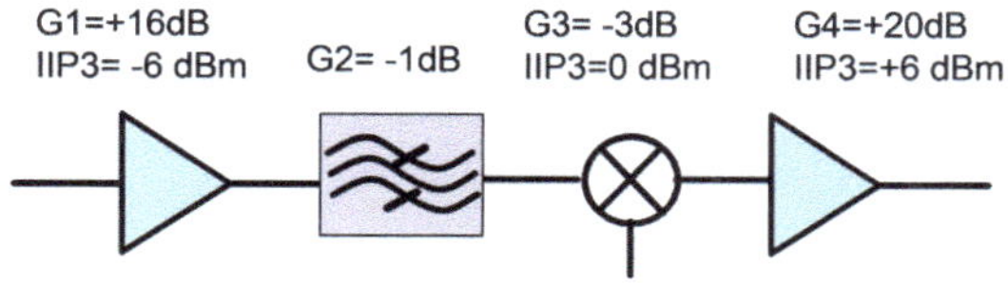

b. If the noise figure of the first amplifier stage is 1 dB, the mixer NF = 6 dB, and IF amplifier NF = 3 dB, calculate the spurious free dynamic range of this receiver front end. The IF bandwidth is 10 kHz, limited by a channel selection filter following the IFA.

2. Derive the cascading relationship

$$\frac{1}{OIP_{total}} = \frac{1}{G2G3OIP(1)} + \frac{1}{G3OIP(2)} + \frac{1}{OIP(3)}$$

for the third order output intercept point. The voltages seen in RL should include the OIP of each stage as scaled to the output. Assume that all voltages add in phase – (worst case).

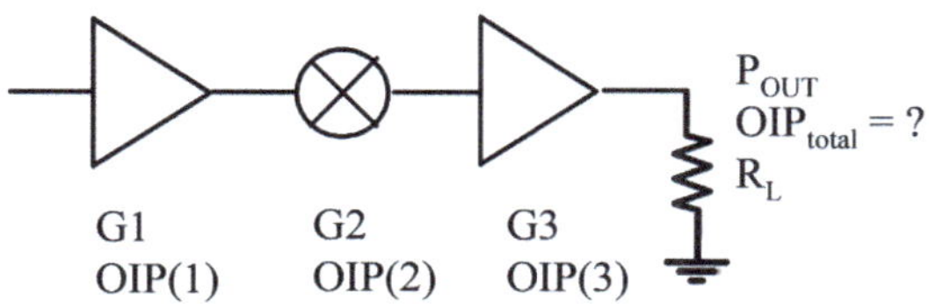

3. The amplifier below is being tested for intermodulation distortion. The measurement data is in the table below.

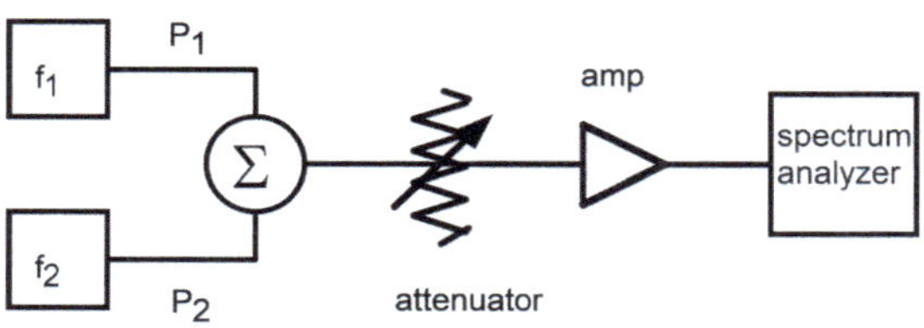

f_1 = 100 MHz; f_2=101 MHz; P_1 = P_2 = 5 dBm

a. At what frequencies will the third order IMD components be generated?
b. At what frequencies will the second order IMD components be generated?
c. Assuming typical losses in the splitter, determine the amplifier gain.
d. Determine the *input* intercept points for second and third order IMD.

IIP2 = dBm IIP3 = dBm

Attenuator dB	Pout @ fundamental	Pout @ second order	Pout @ third order
50	−28		
40	−18	−90	
36	−14	−82	
33	−11	−76	
30	−8	−70	−88
27	−5	−64	−79
24	−2	−58	−70

4. The amplifier manufacturer, stikit2u.com, claims that their model DOA-1 LNA has an IIP_3 of 0 dBm under two-tone measurement conditions (f_1 = 999.9 MHz; f_2 = 1000.1 MHz). The gain of the amplifier is 10 dB, and its noise figure is 3 dB. Given the shady reputation of the manufacturer, you decide to perform a measurement to verify their claim.

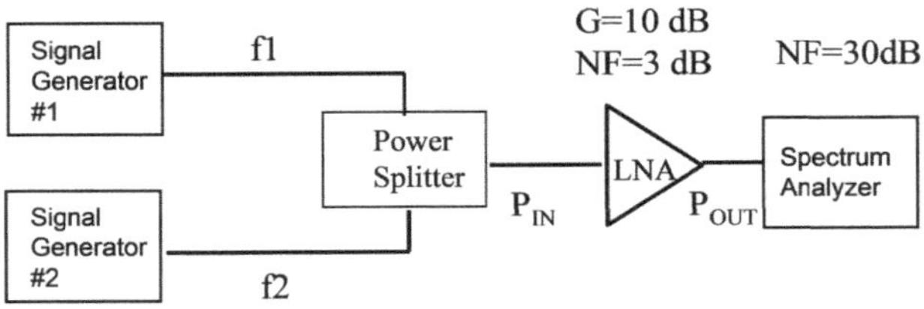

a. If your spectrum analyzer has NF = 30 dB, what is the total noise figure at the input of the system (at P_{in})?
b. Assume your spectrum analyzer has a 70 dB dynamic range. Find the minimum P_{in} that could be used for the carrier-to-IMD measurement on the DOA-1. How much intermodulation power P_{IMD} would you measure?
c. What is the maximum resolution bandwidth that could be used and still measure an P_{IMD}-to-noise ratio of 0 dB under this condition.

5. Consider the wireless LAN receiver front end shown below.
a. If the system is at room temperature, find the noise figure of the overall system.
b. What is the resulting S/N ratio at the output if the input signal power is –90 dBm?
c. Can the components be rearranged to give a better noise figure?
d. Compare the output intercept point and SFDR for both configurations. Which has the better receiver factor?

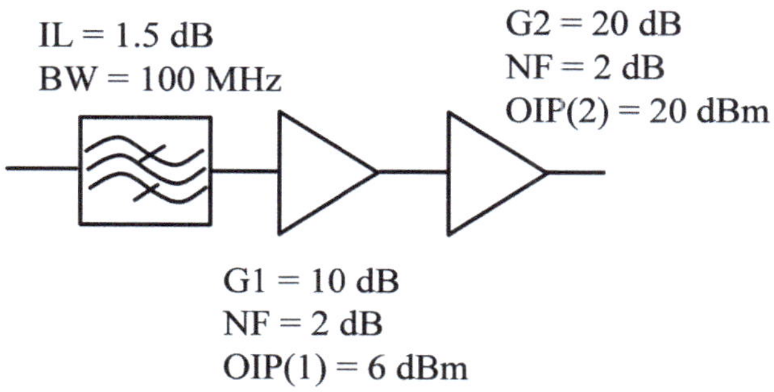

6. Use the three amplifiers described in the table to design a three-stage cascaded amplifier.

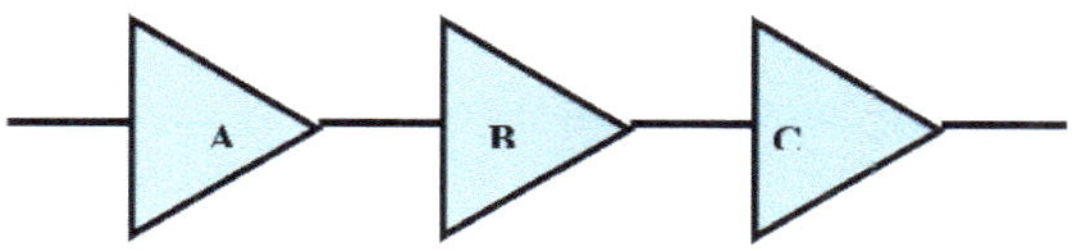

Amplifier	Third-order input intercept (dBm)	Gain (dB)	Noise Figure (dB)
A	+3	10	2
B	-10	13	3
C	+13	16	4

Determine the optimum sequence of amplifiers to obtain the highest possible spurious free dynamic range on the three-stage cascade. What is the SFDR for that combination?

Appendix 8.1. Harmonic Balance Simulation on ADS:

General Description of Harmonic Balance in Keysight ADS[2]

Harmonic balance is a frequency-domain analysis technique for simulating nonlinear circuits and systems. It is well-suited for simulating analog RF and microwave circuits, since these are most naturally handled in the frequency domain. Circuits that are best analyzed using HB under large signal conditions are:

- Power amplifiers
- Frequency multipliers
- Mixers
- Oscillators
- Modulators.

Harmonic balance simulation calculates the magnitude and phase of voltages or currents in a potentially nonlinear circuit. Use this technique to:

- Compute quantities such as P1dB, third-order intercept (TOI) points, total harmonic distortion (THD), and intermodulation distortion components
- Perform power amplifier load-pull contour analyses
- Perform nonlinear noise analysis
- Simulate oscillator harmonics, phase noise, and amplitude limits.

In contrast, S-parameter or AC simulation modes do not provide any information on nonlinearities of circuits. Transient analysis, in the case where there are harmonics and or closely spaced frequencies, is very time and memory consuming since the minimum time step must be compatible with the highest frequency present while the simulation must be run for long enough to observe one full period of the lowest frequency present.

Harmonic balance simulation makes possible the simulation of circuits with multiple input frequencies. This includes intermodulation frequencies, harmonics, and frequency conversion between harmonics. Not only can the circuit itself produce harmonics, but each signal source (stimulus) can also produce harmonics or small-signal sidebands. The stimulus can consist of up to twelve non-harmonically related sources. The total number of frequencies in the system is limited only by such practical considerations as memory, swap space, and simulation speed.

---

[2] From Keysight ADS Circuit Simulation Manual, Chap. 7, Harmonic Balance.

8A.1. The Simulation Process[2] (FYI – skip to next section if you want to get started now).

The harmonic balance method is iterative. It is based on the assumption that for a given sinusoidal excitation there exists a steady-state solution that can be approximated to satisfactory accuracy by means of a finite Fourier series. Consequently, the circuit node voltages take on a set of amplitudes and phases for all frequency components. The currents flowing from nodes into linear elements, including all distributed elements, are calculated by means of a straightforward frequency-domain linear analysis. Currents from nodes into nonlinear elements are calculated in the time-domain. Generalized Fourier analysis is used to transform from the time-domain to the frequency-domain.

A frequency-domain representation of all currents flowing away from all nodes is available. According to Kirchoff's current law (KCL), these currents should sum to zero at all nodes. The probability of obtaining this result on the first iteration is extremely small.

Therefore, an error function is formulated by calculating the sum of currents at all nodes. This error function is a measure of the amount by which KCL is violated and is used to adjust the voltage amplitudes and phases. If the method converges (that is, if the error function is driven to a given small value), then the resulting voltage amplitudes and phases approximate the steady-state solution.

- Designers are usually most interested in a system's steady-state behavior. Many high-frequency circuits contain long time constants that require conventional transient methods to integrate over many periods of the lowest-frequency sinusoid to reach steady state. Harmonic balance, on the other hand, captures the steady-state spectral response directly.
- The applied voltage sources are typically multitone sinusoids that may have very narrowly or very widely spaced frequencies. It is not uncommon for the highest frequency present in the response to be many orders of magnitude greater than the lowest frequency. Transient analysis would require an integration over an enormous number of periods of the highest-frequency sinusoid. The time involved in carrying out the integration is prohibitive in many practical cases.
- At high frequencies, many linear models are best represented in the frequency domain. Simulating such elements in the time domain by means of convolution can results in problems related to accuracy, causality, or stability.

- Analysis of harmonic and intermodulation distortion and gain compression requires the use of nonlinear models for active devices.

8A.2. Harmonic Balance Setup

The HB method depends on calculating currents and voltages at many harmonically related frequencies for each fundamental signal under consideration. Because we are interested in the steady state solution of a nonlinear problem, we must allow the HB simulator to use enough harmonics so that a Fourier series constructed from these harmonic amplitudes and phases can reproduce a reasonable replica of the time domain solution.

Figure 8.31 illustrates a very basic HB simulation setup. The HB controller specifies several key simulation parameters. In the example below, one fundamental frequency, Freq[1]=RF_Freq, is specified as an input. The index [1] shows that only one fundamental frequency is being considered. Order[1] specifies the number of harmonic frequencies to be calculated (9) for the first (and only) frequency in this case. One of the most common errors in HB simulation setup is to use too low of an order. You can determine what order is optimum if you first simulate your circuit with a small order then increase the order in steps of 1 or 2 harmonics. When the solution stops changing within a significant bound, you have reached the optimum order. Using too high of an order is wasteful of memory, file size and simulation time, so it is not efficient to just clobber the problem with a very high order. Some user discretion is advised.

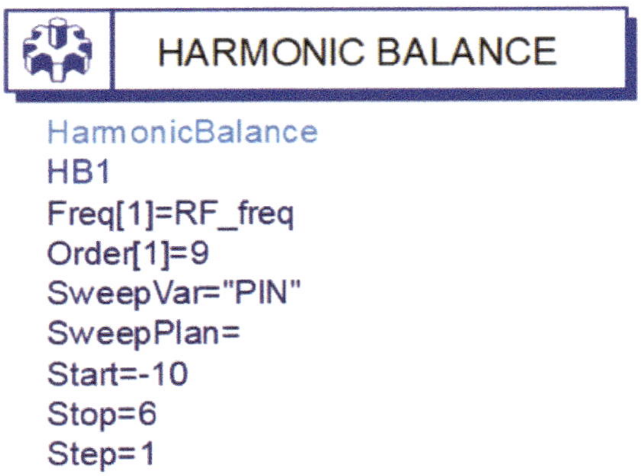

Figure 8.31 Example of the HB controller used for a very simple single tone (frequency) simulation. In addition, power PIN is being swept from −10 to 6 dBm. PIN and RF_freq are defined in Figure 8.32.

8A.2.1. Input power sweep

Note that we are sweeping the input power variable PIN. The variables that you define must be declared using the VarEqn icon shown in Figure 8.32 in order for the controller to recognize them.

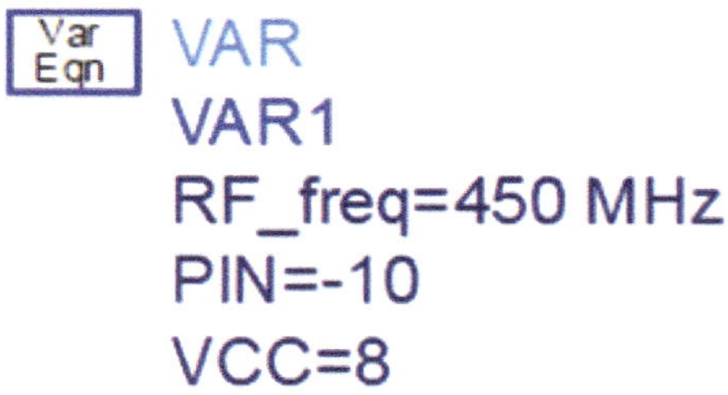

Figure 8.32 We also see independent variables, PIN, VCC and RF_freq, defined here.

To do a parameter sweep, the sweep limits and variable must be selected in the HB controller in Figure 8.33. Double click the controller and select Sweep from the menu options.

Selecting the Sweep tab allows you to sweep one independent variable. Choose the parameter to be swept, the sweep type (Linear, Log), and the Start, Stop, and Step variables (or number of points instead of step-size). In this example, we are sweeping the input power to the amplifier to determine the gain compression behavior. The more sweep points chosen, the longer the simulation time and the greater the data file size.[3]

The fundamental frequency of the input source must be the same as specified on the controller. The *Sources-Freq Domain* palette includes many sources suitable for use with HB. The single tone source *P_1Tone* is illustrated below. This source provides a single frequency sinusoid at a specified available power. Here we see that the internal source resistance (50 ohms) is included. The available source power is provided as PIN (in dBm, which will be converted to Watts by the dbmtow function) and 0 degrees of phase.

Frequency, source impedance, and available power must be specified (Figure 8.34).

You could also have selected a voltage source, *V_1Tone*, or for multiple frequency simulations, there are *V_nTone* and *P_nTone* sources. These are often used for intermodulation distortion simulations. Voltage or current

---

[3] If two variables are to be swept, a ParamSweep controller icon from the HB menu must be added to the schematic.

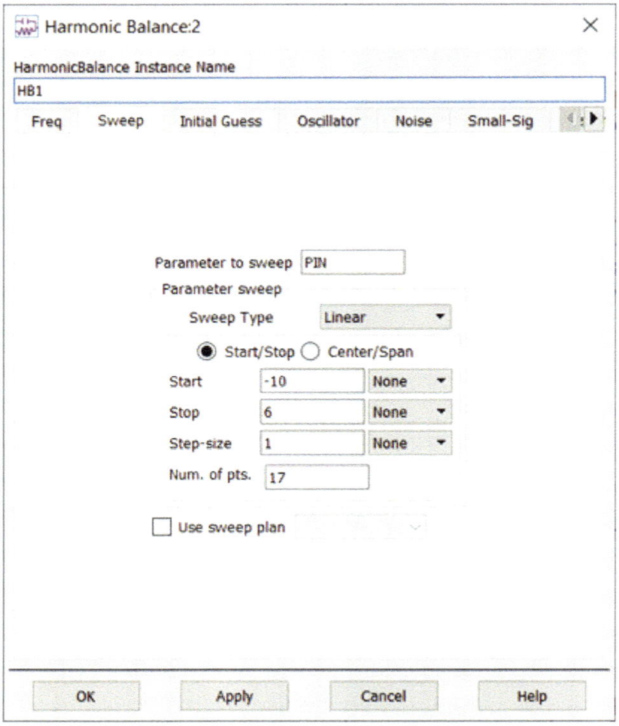

Figure 8.33 Set up the simulation to sweep a variable by naming the variable and setting the limits and step size.

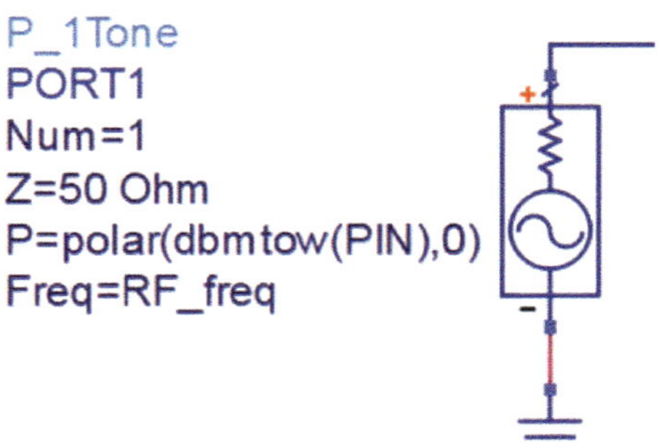

Figure 8.34 Single frequency source.

sources require an external source resistance or impedance whereas the power sources include an internal source resistance or impedance, *Z*.

Nodes must be labeled in the harmonic balance simulation in order to transfer their voltages to the display. If currents are to be used in calculations as well, a current probe must be inserted from the *Probe Components* menu. An example of a power sweep is shown in the next figure. Node Vload is labeled using the Insert pulldown menu: *Insert > Wire/Pin Label.* This opens a text box where you can enter the node name you want. I-DC and I-load were measured with the current probes as shown. No input and output matching networks are included in this example. The schematic in Figure 8.35 is configured to determine the 1dB compression input power and harmonic distortion generated by nonlinear device characteristics. A large-signal nonlinear device model has been used.

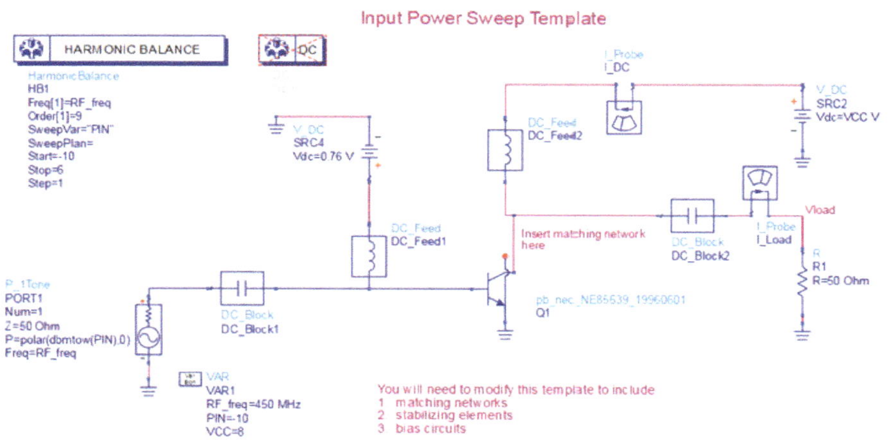

Figure 8.35 HB schematic for a basic power sweep.

8A.2.2. Displaying results

Once the simulation has been run, you will be able to see the results on the display panel. The output voltages and currents calculated by the HB analysis will contain many frequency components. You can display all of them in a spectral display by just plotting the voltage or current on an X-Y plot (Figure 8.36). Markers can be used to read out the spectral line amplitudes or powers at each value of PIN.

You can see the 9 harmonics according to the *Order[1]* specified for the HB controller. Each spectral line displays 17 different power levels.

Often you will want to plot power in dBm. If your load impedance is real, you can use the dbm function in an equation as above. If the load has a complex impedance, then use the definition for sinusoidal power.

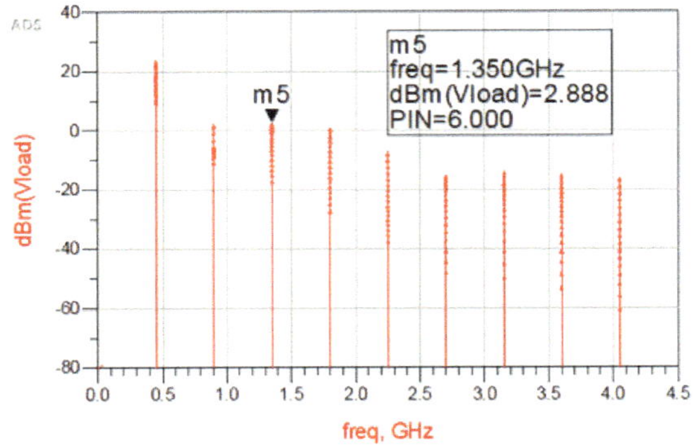

Figure 8.36 Spectrum in dBm is plotted for Vload.

$$P_{out\_dBm} = 10^{*} \log(0.5^{*} real\ (Vload^{*} conj\ (I\_load.i))) + 30$$

This will give you the power in dBm in all cases. This is the preferred method. Note that calculated quantities much below –100 dBm are probably not very reliable due to the limited precision of the device models.

8A.2.3 Harmonic Distortion

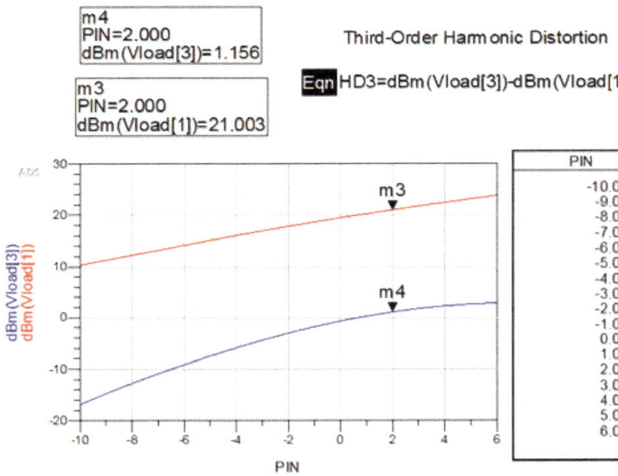

PIN	dBm(Vload[1])	HD3
-10.000	10.414	-27.157
-9.000	11.377	-26.013
-8.000	12.337	-24.997
-7.000	13.291	-24.093
-6.000	14.236	-23.284
-5.000	15.166	-22.555
-4.000	16.078	-21.901
-3.000	16.968	-21.321
-2.000	17.832	-20.822
-1.000	18.667	-20.415
0.000	19.474	-20.110
1.000	20.252	-19.917
2.000	21.003	-19.847
3.000	21.728	-19.905
4.000	22.431	-20.097
5.000	23.115	-20.425
6.000	23.781	-20.893

Figure 8.37 Several ways to display harmonic distortion.

In this case, shown in Figure 8.37, third-order was selected. The harmonic index was selected by the number in the square brackets after the variable.

To perform calculations in the display panel, you will want to be able to select specific frequency components. The harmonic index (harmindex) can be used to do this. If you plot your output variable in a table format (Figure 8.38), you will see a list of frequencies.

freq	Pout	Pout_dBm
0.0000 Hz	0.0000	<invalid>
450.0MHz	134.5m	21.29
900.0MHz	11.47u	-19.41
1.350GHz	336.8n	-34.73
1.800GHz	219.8p	-66.58
2.250GHz	2.250p	-86.48
2.700GHz	9.389p	-80.27
3.150GHz	1.465p	-88.34

Figure 8.38 Table showing the value of Pout and Pout_dBm at several harmonic frequencies.

The frequencies are printed in order and can be designated by an index, ranging from 0 for DC to Order[1] – 1 for the highest harmonic frequency.

The first frequency in the table is DC and has index 0. Fundamental is index 1. So, to select the voltage at the fundamental frequency, for example, you could write Vload[1] or to select power, Pout[1] or Pout_dBm[1] in this example. The second harmonic would be Pout[2]. Of course, we do not need to draw a table to use the index. For example, the DC component of the power supply voltage can be extracted by using the 0 index: VCC[0]. Then, if the supply voltage and current were measured and passed to the output display, you could calculate DC input power by

```
Eqn P_DC=real(VCC[0]*I_DC.i[0])
```

To display the results of equations such as this, you use the table or rectangular plot features in the display panel. The data set must be changed to Equations as shown below in Figure 8.39 in order to find the result of the calculations.

The equations that you have defined will be displayed on the left panel. Highlight an equation and click on >> Add >> and it will appear as a trace on the right panel.

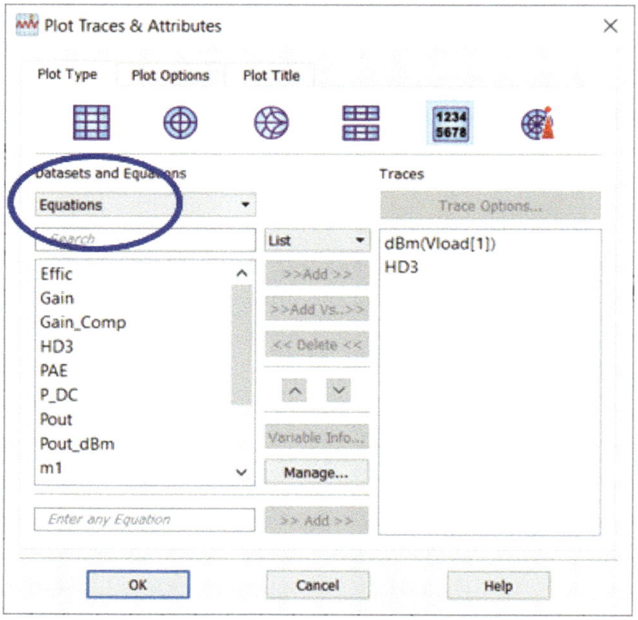

Figure 8.39 To plot the results of an equation in the data display, select Equations in the data set.

8A.2.4. Gain compression

We did the power sweep in Figure 8.35 so that the P1dB gain compression point could be determined. Here, we see the result of the sweep. Gain was defined using the equations shown.

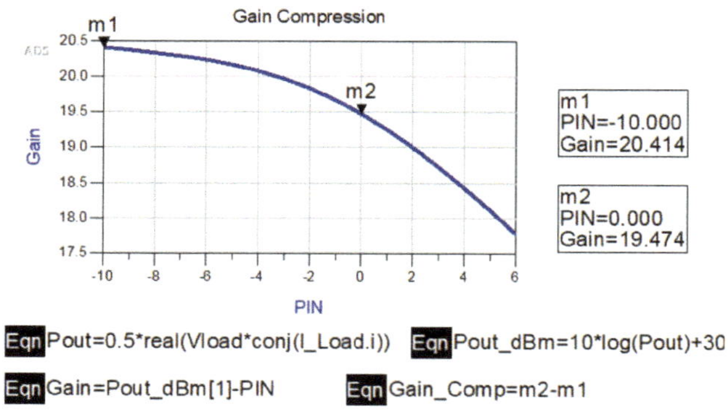

Figure 8.40 Plot of data showing the gain compression that happens at higher PIN levels.

Note the use of equations and indices once again in Figure 8.40 to select the fundamental component. The 1 dB compression power is about 0 dBm for this case.

8A.3 Multiple frequency simulations

Multiple frequencies or "tones" (mainly two-tone) are widely used for evaluation of intermodulation distortion in amplifiers or mixers. In Figure 8.41, you can see that now two frequencies have been selected, Freq[1] and Freq[2]. Each frequency must also declare an order (number of harmonic frequencies to be considered).

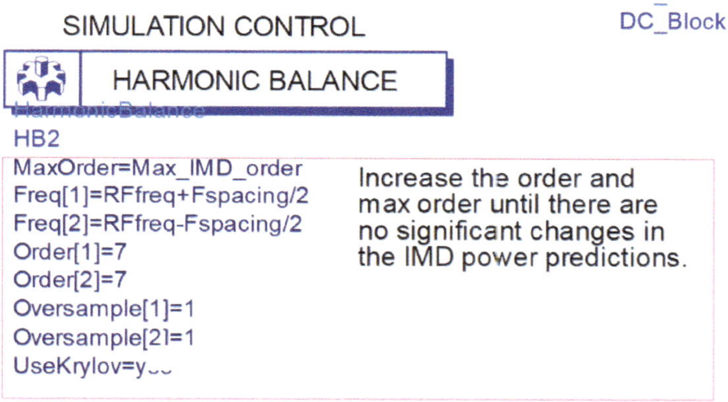

Figure 8.41 HB controller example for a two-tone PA simulation.

Intermodulation distortion in a nonlinear system occurs when more than one input frequency is present in the circuit under evaluation. Therefore, additional frequencies need to be specified when setting up for this type of simulation. Two-tone simulations are generally performed with two closely spaced input frequencies. In this example, the two inputs are at 899.9 and 900.1 MHz for an RF frequency of 900 MHz for a spacing of 200 KHz shown in Figure 8.42. The frequency spacing must be small enough that the two tones are well within the signal bandwidth of the circuit under test.

Maximum order corresponds to the highest order mixing product (n + m) to be considered (n*freq[1] $\pm$ m*freq[2]). There will be a frequency component in the output file corresponding to all possible combinations of n and m up to the *MaxOrder* limit. The simulation will run faster with lower *MaxOrder* and fewer harmonics of the sources, but may be less accurate. Often accurate IMD simulations will require a large maximum order. In this case, a larger number of spectral products will be summed to estimate the time domain waveform and therefore provide greater accuracy. This will increase the size of the data file and time required for the simulation. Increase the orders and *MaxOrder* in increments of 2 and watch for changes in the IMD output power. When no further significant change is observed, then the order is large enough. If large asymmetry is noted in the intermodulation components, higher orders are indicated.

Sometimes, increasing the *oversampling ratio* for the FFT calculation (use the *Param* menu of the HB controller panel) can reduce errors. This oversampling controls the number of time points taken when converting back from time to frequency domain in the harmonic balance simulation algorithm. A larger number of time samples increases the accuracy of the transform calculation but increases memory requirements and simulation time. Both order and oversampling should be increased until you are convinced that further increases are not worthwhile.

For multiple frequency simulations, the simulation time will be reduced substantially by using the Krylov option which can be selected on the *Display* tab of the HB controller.

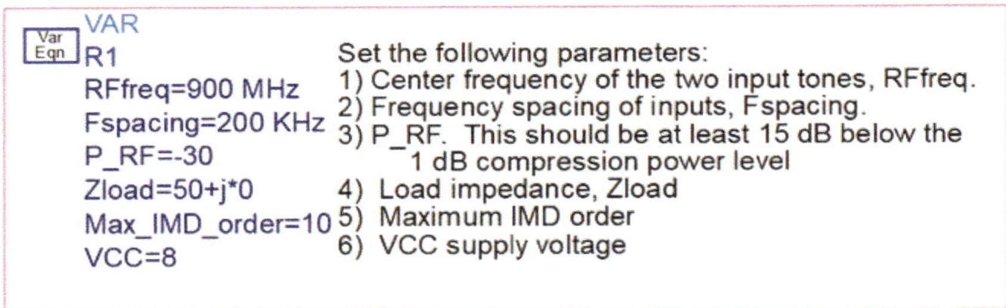

Figure 8.42 Definition of the variables used in the IMD simulation.

The two-tone source frequencies are provided with a *P_nTone* generator from the Sources – Freq Domain menu. The two frequencies are specified in the *Var* block in Figure 8.42. The same approach is used to specify frequencies in the HB controller so that the effect caused by changes in P_RF spacing could be evaluated by changing only one variable. The available power, PIN, is specified in dBm for each source frequency. A P_nTone source is required at the input (Figure 8.43).

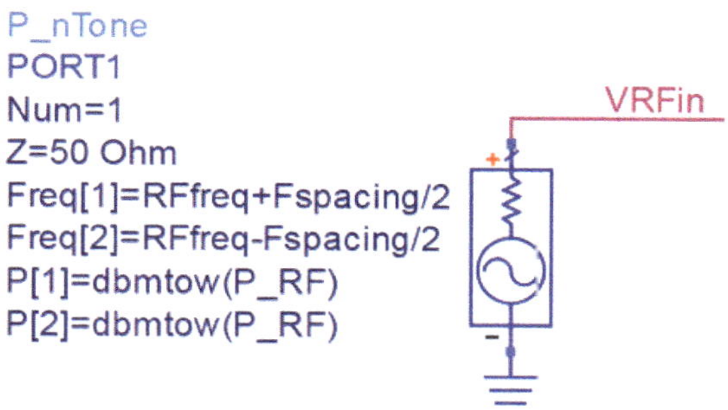

Figure 8.43 Two-tone source example.

8A.3.1. Displaying Results of Multitone Simulations

You can view the result around the fundamental frequencies by disabling the autoscale function in the plot and specifying your own narrow range. The display below in Figure 8.44 shows intermodulation products up to the seventh order (MaxOrder specified on the HB controller).

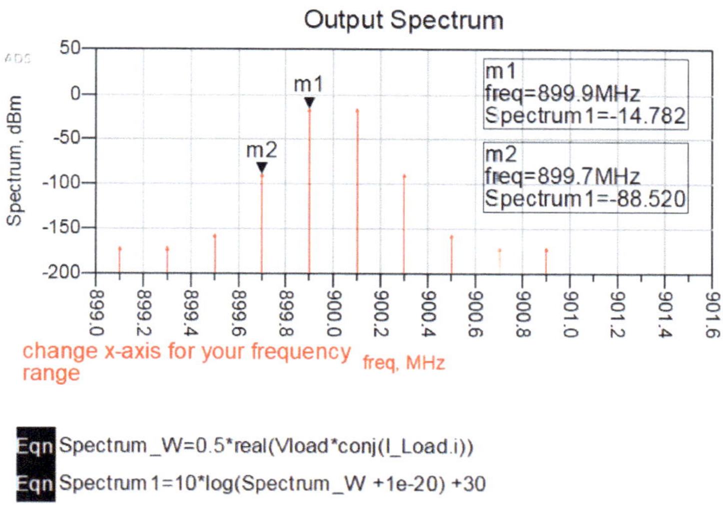

Figure 8.44 Output spectrum (Spectrum1) calculated by the two-tone IMD simulation.

We would like to study the output voltage at the fundamental frequencies and the third-order IMD product frequencies. This can be selected from the many frequencies in the output data set by using the *mix* function. The syntax of the mix function is shown in the equations extracting the output power in watts and then converting to dBm for the two fundamental frequencies.

```
Eqn LSB_W=mix(Spectrum_W,{0,1},Mix)
Eqn USB_W=mix(Spectrum_W,{1,0},Mix)
Eqn LSB=10*log(LSB_W+1e-20) + 30
Eqn USB=10*log(USB_W+1e-20) + 30

Eqn IM3_LSB=mix(Spectrum1,{-1,2},Mix)
Eqn IM3_USB=mix(Spectrum1,{2,-1},Mix)
```

Figure 8.45 Use of the mix function to select output powers at certain frequencies.

The respective indices used with the mix function to select these frequencies are shown in the equations in Figure 8.45. The curly brackets and indices are ordered according to the HB fundamental analysis frequencies. Looking at the HB controller, we see that the Freq index 1 corresponds to the upper fundamental frequency, whereas index 2 the lower. So, {0,1} will extract the lower fundamental.

To extract the third-order output power, the mix function uses indices 2 and 1. Recall that the third-order intermodulation frequencies are given by $2f_1 - f_2$ and $-f_1 + 2f_2$. The result is shown in the table below. We can see in Figure 8.46 that the IMD powers are quite close in value. This indicates that the order used for the calculation was sufficiently large.

LSB	USB	IM3_LSB	IM3_USB
-14.782	-14.784	-88.520	-88.518

Figure 8.46 Output power in dBm for fundamental ({0,1} and {1,0}) and third-order IMD components ({-1,2} and {2,-1}).

Having these output powers available now, the output third-order intercept (TOIout or OIP3) can be calculated from

$$OIP3 = TOI_{out} = P_{OUT} + \frac{1}{2}\left(P_{OUT} - P_{IMD}\right) = \frac{3}{2}P_{OUT} - \frac{1}{2}P_{IMD}$$

In our example, Figure 8.47 shows equations for P_{out} = LSB or USB and P_{IMD} = IM3_LSB or IM3_USB.

Eqn TOIoutput_USB=1.5*USB-0.5*IM3_USB

Eqn TOIoutput_LSB=1.5*LSB-0.5*IM3_LSB

TOIoutput_USB	TOIoutput_LSB
22.083	22.087

Figure 8.47 Equations used to calculate the third-order output intercept (OIP3) and the calculated values.

8A.4. Convergence Woes

Any user of the harmonic balance simulator will eventually encounter convergence problems. Unfortunately, when this happens, no useful information is provided by the simulator. Problems with convergence generally arise when the circuit under simulation is or becomes highly nonlinear. In the case of mixers, there are inherent nonlinearities that are required for the mixing process, but these are usually not so bad unless you are seriously overdriving one of the inputs. If the simulation fails, check the biasing of the transistors. HB doesn't do well with BJTs driven into their saturation region. If that is not the problem, then try decreasing either LO power or the RF power sweep range. You may be driving the mixer or an amplifier well beyond saturation when using default power levels in amplifier or mixer Design Guide templates. As a last resort, you can try using the Direct solver rather than Krylov, but this will increase simulation time by a large factor.

8A.5. Time Domain Simulation from Harmonic Balance

While there is a transient analysis (simulation-transient) controller in ADS, there may be some occasions where a time domain plot from harmonic balance would be more efficient. This would especially be the case when more than one frequency is used to drive a circuit.

In harmonic balance, if you want to see the time domain version of a voltage or current, the display can perform the inverse Fourier transform while plotting. Select the *time domain signal* option using the *Plot Traces & Attributes* panel in the display (Figure 8A.18).

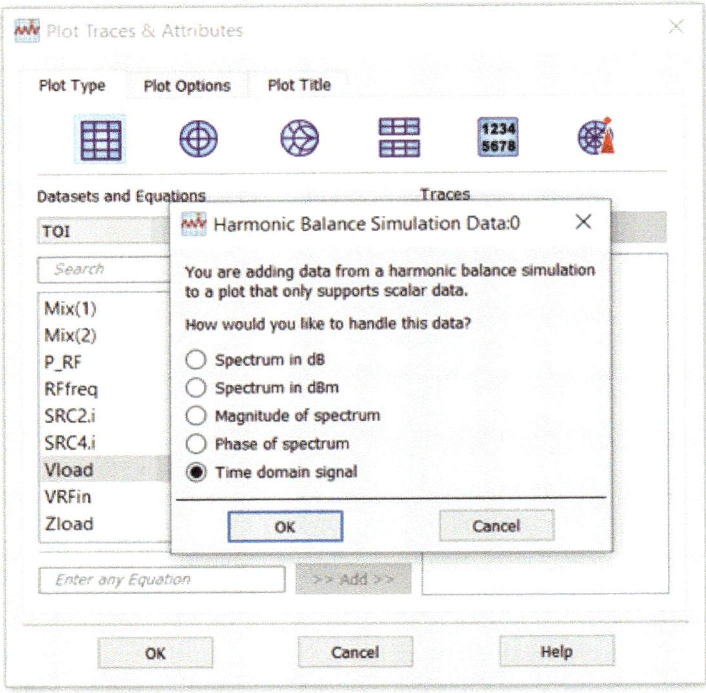

Figure 8.48 When plotting HB data, you must convert it to a scalar quantity (dB, dBm, magnitude or phase).

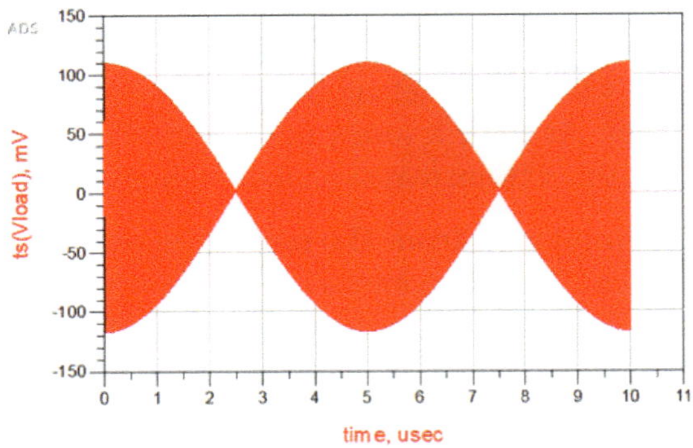

Figure 8.49 Time domain plot of the two-tone IMD simulation.

Notice in Figures 8.48 and 8.49 that a time domain conversion is being performed by an FFT if requested. The ts function can be used to make the conversion.

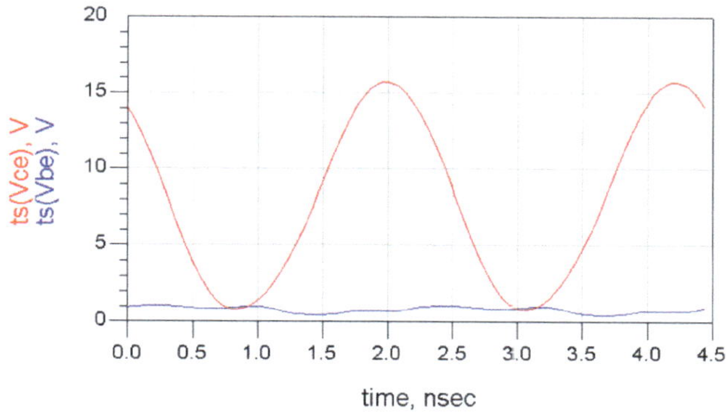

Figure 8.50 Example of a time domain plot from a HB simulation.

Note that the period of the signal, 5 microseconds, corresponds to the 200 kHz separation of the input frequencies.

In Figure 8.50, the collector-emitter and base-emitter voltages are converted into time domain functions. This can be helpful to see if a device is being driven into saturation. You can also uncheck Auto in the x-axis scaling and zoom in on a smaller time period of your own choosing.

9

Design of Low Noise Amplifiers

In the previous chapter, the trade-off between noise and intermodulation distortion was shown to determine the useful dynamic range (SFDR) of an RF system. Recall that we have already studied amplifier design methods to provide stability and gain. Now we will consider how to design for lowest noise.

Recall $F_{total} = F_1 + \frac{F_2-1}{G_1} + \frac{F_3-1}{G_1 G_2} + \cdots$

- The noise factor of the first stage, F_1, dominates the overall noise performance if G_1 is sufficiently high. Recall that these are numerical ratios, not dB. Gains are available gains.
- We will see that maximum gain and minimum noise cannot be obtained at the same time. Similarly, one generally can't achieve minimum noise and large IIP3 concurrently. Noise performance is controlled by Γ_S. So, the design winds up being a compromise between gain and noise and distortion.

Goals

1. Understand why the two-port device model (S-parameters and noise parameters) is superior for the design of low noise microwave amplifiers.
2. Use ADS to plot noise figure circles and available gain circles.
3. See how tradeoffs between noise and available gain are needed for low noise amplifiers (LNA). Distortion mitigation is a secondary goal which may be affected by the choice of device and its bias conditions.
4. Calculate path loss on link applications.
5. Evaluate the signal-to-noise ratio for terrestrial and space applications.

9.1 Device Noise Models

There are two techniques widely used to model the noise contribution of an amplifier.

9.1.1 Input-referred noise voltage and currents

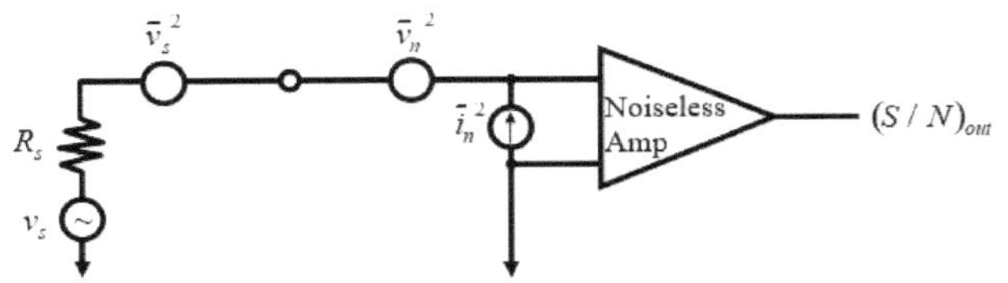

Figure 9.1 All noise contributions of the amplifier are modelled at the input by noise voltage and noise current sources.

All noise sources in the amplifier (devices, resistors) are combined to form input-equivalent voltage and current sources at the input, as shown in Figure 9.1.

The procedure involves choosing components and biasing in the amplifier to minimize $\bar{v}_n^2$ and $\bar{i}_n^2$. Then, select R_S (or Z_S) to provide the highest $(S/N)_{OUT}$.

This technique is good for encouraging intuition. Similar to the use of device models consisting of equivalent circuits. This procedure is widely used in low frequency circuits where $\bar{v}_n^2$ and $\bar{i}_n^2$ can be measured by the application of input short and open circuits.

However, at high frequencies, pure short and open circuits are not possible due to parasitic inductances and capacitances. Because of this, a more accurate technique for low noise design has been developed that makes use of two-port noise parameters.

9.1.2 Two-port noise parameter representation

We will derive the optimum source admittance Y_s which when presented to the input of the transistor leads to the minimum degradation of S/N.

The total output noise power is proportional to $\bar{i}_{sc}^2$, the mean-square input port short circuit current. Noise voltages and currents add as mean-square values (Figure 9.2).

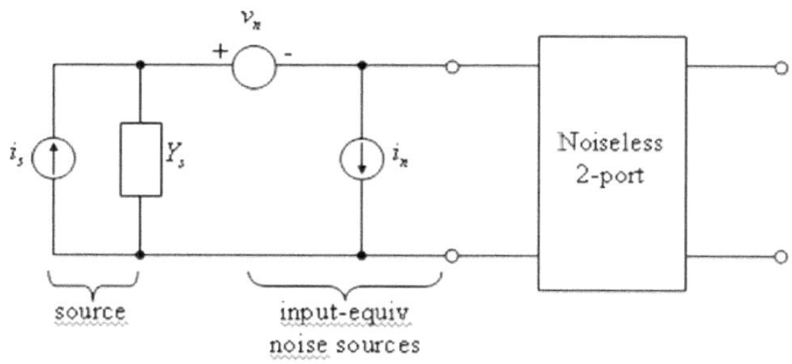

Figure 9.2 Two-port noise model equivalent network.

$$F = \frac{\bar{i}^2_{sc}}{\bar{i}^2_s} = \frac{\text{noise power due to source and amp}}{\text{noise power of source alone}} \tag{9.1}$$

$$i_{sc} = -i_s + i_n + v_n Y_s$$

$$\bar{i}^2_{sc} = \bar{i}^2_s + \overline{\left(i_n + v_n Y_s\right)}^2 - \underbrace{\overline{2i_s\left(i_n v_n Y_s\right)}}_{=0 \text{ since } i_s \text{ not correlated with } i_n \text{ or } \mathrm{v}_n} \tag{9.2}$$

so:

$$F = 1 + \frac{\overline{\left(i_n + v_n Y_S\right)}^2}{\bar{i}^2_s} \tag{9.3}$$

$$\bar{i}^2_s = 4kT_o G_s B \tag{9.4}$$

where G_s is the source conductance, and

$$\bar{v}^2_n = 4kT_o R_N B \tag{9.5}$$

R_N is a fictitious equivalent noise resistance that represents $\bar{v}^2_n$, and i_n can be represented by a correlated (with v_n) and uncorrelated part.

$$i_n = i_{nu} + i_{nc} \tag{9.6}$$

$$i_{nc} = Y_c v_n \tag{9.7}$$

(Y_c is a fictitious correlation admittance)

$$i_{nu} = 4kT_o G_u B \tag{9.8}$$

(G_u is a fictitious equivalent noise conductance).

To continue, we must minimize F. (see appendix L in Gonzalez [1] for a detailed derivation).

This leads to the following:

$$F = F_{\min} + \frac{R_n}{G_s}\left[(G_s - G_{opt})^2 + (B_s - B_{opt})^2\right] \tag{9.9}$$

where $Y_s = G_s + jB_s$ and $Y_{opt} = G_{opt} + jB_{opt}$, and

$F_{\min}$ = lowest possible noise factor

Y_{opt} = optimum source admittance for minimum noise

Normalizing everything to Z_o leads to:

$$\boxed{F = F_{\min} + \frac{r_n}{g_s}\left|y_s - y_{opt}\right|^2.} \tag{9.10}$$

In terms of reflection coefficients:

$$y_S = \frac{1 - \Gamma_S}{1 + \Gamma_S}, \quad y_{opt} = \frac{1 - \Gamma_{opt}}{1 + \Gamma_{opt}}$$

$$\boxed{F = F_{\min} + \frac{4r_n\left|\Gamma_S - \Gamma_{opt}\right|^2}{\left(1 - \left|\Gamma_S\right|^2\right)(1 + \Gamma_{opt})^2}.} \tag{9.11}$$

In Equation (9.11), $F_{\min}, r_n, \Gamma_{\text{opt}}$ are *noise parameters* given by the manufacturer or that are measured.

So what? Well, we see from the above that there is a minimum noise factor possible for a device, F_{min}, that is achieved only when a particular reflection coefficient, Γ_{opt} is presented to the input. So, $\Gamma_s = \Gamma_{opt}$ leads to the minimum noise figure for the amplifier built with this transistor.

9.1.3 Measuring the two port noise parameters

1. Use a tuner to fiddle with Γ_s until minimum F is observed. This is the matching network (MN) shown in Figure 9.3. Read from the *NF* meter, and disconnect the amplifier and use the network analyzer to measure $\Gamma_{opt} = \Gamma_s$.
2. Set at the amplifier input with a termination in Z_o. Measure F.

$$r_n = (F_{\Gamma_s=0} - F_{\min})\frac{\left|1 + \Gamma_{opt}\right|^2}{4\left|\Gamma_{opt}\right|^2} \tag{9.12}$$

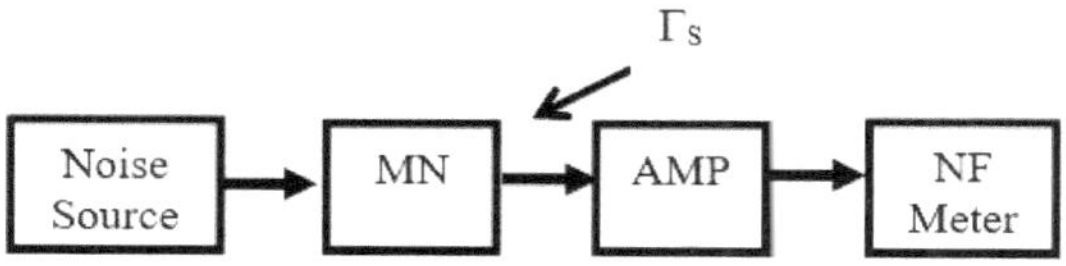

Figure 9.3 Block diagram of the noise parameter measurement setup. MN is an adjustable matching network, a tuner appropriate for the frequency of interest.

Ok. But, this is equivalent to using equivalent input-referred noise voltages and currents (method 1), isn't it?Yes, but the noise parameters in method 2 are *measurable* whereas in method 1 they are not at high frequencies.

Note that the noise figure is often dependent on the bias point selection for the transistor. The manufacturer should provide noise parameters for several bias conditions, as well as noise circle plots such as described below and shown in Figure 9.4.

9.2 Noise Figure Circles

We can also use Equation (9.11) to predict noise figure vs.Γ_s . The equation can be reconfigured:

$$\frac{|\Gamma_S - \Gamma_{opt}|^2}{1-|\Gamma_S|^2} = \underbrace{\frac{F_i - F_{\min}}{4r_n} \mid 1+\Gamma_{opt}\mid^2}_{\text{constant for each } F_i} \equiv N_i \tag{9.13}$$

Then, circles can be drawn on the source Γ_s Smith chart that correspond to a particular noise figure (see Figure 9.4). When these noise circles are plotted with available gain circles, you can see the tradeoff between minimum noise and available gain.

To calculate noise circles [1] :

Define:

$$N_i = \frac{F_i - F_{\min}}{4r_n}|1+\Gamma_{opt}|^2 = noise\ figure\ parameter. \tag{9.14}$$

Rearrange the equation so that

$$|\Gamma_S - C_{Fi}| = r_{Fi} \tag{9.15}$$

where:

$$C_{F_i} = \frac{\Gamma_{opt}}{1+N_i} \quad \text{(center)} \tag{9.16}$$

$$r_{Fi} = \frac{1}{N_i + 1}\sqrt{N_i^2 + N_i\left(1 - |\Gamma_{opt}|^2\right)} \quad \text{(radius).} \tag{9.17}$$

Note that $N_i = 0$ and $r_{Fi} = 0$ when $\Gamma_S = \Gamma_{\text{opt.}}$. Therefore, the circle Γ_{opt} is just a dot.

(Or use ADS to plot these as described in Section 9.4).

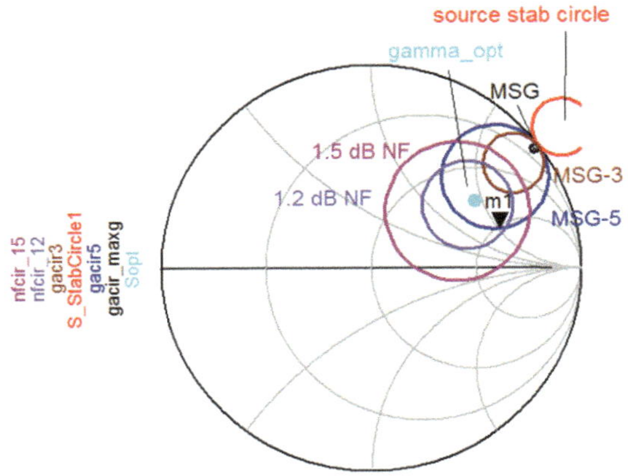

Figure 9.4 Γ_S Smith chart showing noise figure circles and available gain circles generated by ADS. Maximum stable gain (MSG) is the maximum gain allowed by stability.

9.3 Available Gain Circles

Device data sheets often plot G_T on the Γ_S plane under the assumption that the output is matched ($\Gamma_L = \Gamma_{out}{}^*$). This is the *available power gain* $= G_A$

$$G_A = \frac{P_{AVN}}{P_{AVS}} \tag{9.18}$$

Since output is always matched, and Γ_{out} depends on Γ_S, G_A is independent of Γ_L.

$$G_A = \frac{1 - |\Gamma_S|^2}{|1 - S_{11}\Gamma_S|^2}\,|S_{21}|^2\,\frac{1}{1 - |\Gamma_{\text{OUT}}|^2} \tag{9.19}$$

(since $\Gamma_L = \Gamma_{OUT}^*$).

The first term depends upon input match because actual power absorbed in the input is not necessarily the same as Γ_{AVS} (unless conjugately matched at input, almost never true for a low noise design). Circles of available gain or MSG can be plotted on the same Γ_S plane Smith chart as described in Chapter 6.

If input were to be conjugately matched, then we would get the maximum available gain:

$$G_A = G_{A \cdot \max} = MAG = G_{T,\max}$$
$$= \frac{1}{\left(1 - |S_{11}|^2\right)} |S_{21}|^2 \frac{1}{1 - |\Gamma_{\text{OUT}}|^2}. \tag{9.20}$$

When not possible for stability purposes or for achieving the minimum noise figure, then the MSG would be plotted.

9.4 Using ADS to Simulate the Noise Figure

Pathwave/Keysight ADS can be used to design low noise amplifiers much in the same way you have already used it for MAG or MSG designs in Chapter 6. Noise circles and available gain circles are the tools that give the most guidance on design tradeoffs. Refer to chapter 4 and appendices K and L of Gonzalez [1] for the theory behind these analyses.

Here are three cases that you might encounter with device models when analyzing a low noise amplifier.

1. The ADS large signal transistor model is used to represent the device. This is the ideal case, but, unfortunately, the large signal models sometimes do not produce accurate S-parameters because they are equation based. If you use this model, you should check the simulated S-parameters against the manufacturer's data sheet to verify that it provides reasonably accurate results. It should be useful for DC simulations however.
2. The ADS S-parameter transistor model is used to represent the device. This is the most accurate case. Of course, no DC simulations will be possible with this model, but it will represent the S-parameters and noise parameters accurately. Be sure to select the model that represents the actual bias condition to be used in your analysis. In most cases, the manufacturer's data will provide noise figure data at various bias conditions. This can be used to select the model for minimum noise when other considerations don't apply.
3. No S-parameter model is available for the device in the ADS library. In this case, check the manufacturer's web site and download an S2P file

for the device. Place an S2P block from the Data Items menu in your schematic and identify the file name. ADS will look for the file in the ADS project's data directory. See Appendix 9.1 for more information on S2P files. In this case, you may have to enter the noise parameters as well using equations on the data display panel.

9.4.1 Using ADS large signal model library

Let's walk through this as a detailed design example. First you must bias the transistor. Let's say V_{CE} = 5 V and I_C = 5 mA. This bias condition might be selected from the plot of noise figure vs. bias condition provided by the manufacturer such as seen in Figure 9.5.

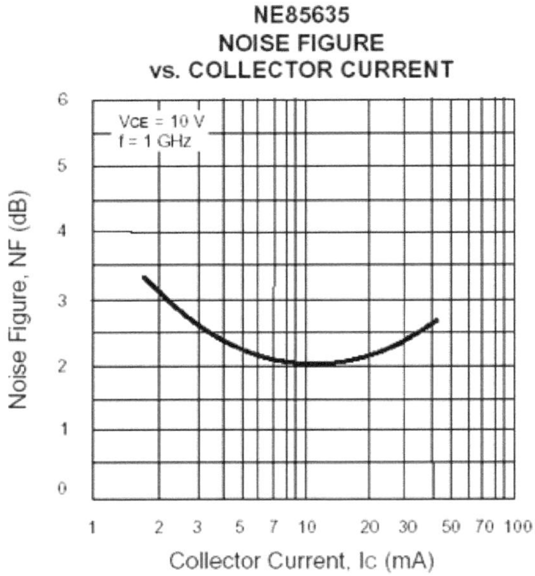

Figure 9.5 Plot of noise figure vs. collector current for a microwave BJT. Note that VCC = 10 V for this plot.

Set up a biasing circuit such as the one in Figure 9.6. Select a large signal device model from the Analog/RF–RF Transistor/Packaged BJT library. Then perform a DC simulation. To see the results of the DC simulation, you go to the Simulate Menu > Annotate DC, Annotate voltage (V), Annotate pin current (ctrl+I). Sometimes it is helpful to move component text aside so that the annotation is easier to read (use F5 to move text on the schematic).

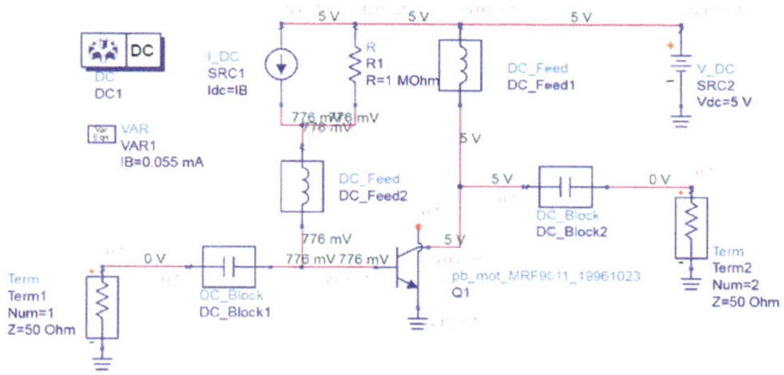

Figure 9.6 DC simulation of a microwave BJT.

This is the result. We have V_{CE} = 5 V and I_C = 5.06 mA. Of course, in a complete design, we would need to apply bias through real components as part of or in addition to the matching networks.

Next, compare the S-parameters from the large signal model with those from the small signal measured S-parameter model (Figures 9.7 and 9.8). This model can be found in the Analog/RF S Parameter component library. Simulate over the specified frequency range.

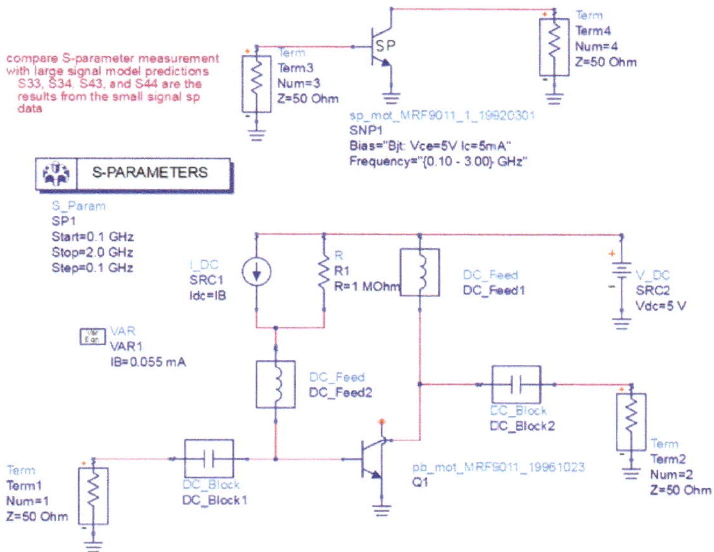

Figure 9.7 Schematic of S-parameter simulations comparing the large signal model (S(1,1), S(1,2), S(2,1), S(2,2)) with the small-signal model (S(3,3), S(3,4), S(4,3), S(4,4).

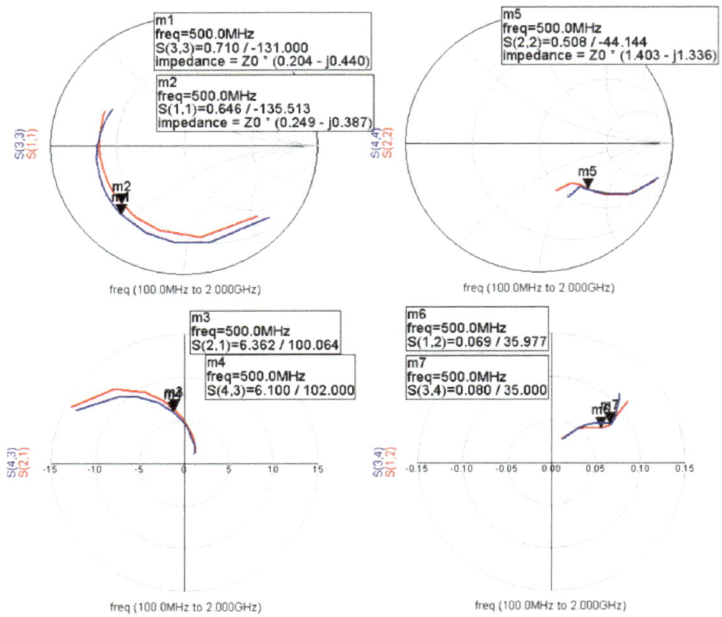

Figure 9.8 Comparison of simulated S-parameters from a large-signal model with the small-signal S-parameter model for the MRF901 microwave device.

From this simulation, we can see that the agreement at our design frequency of 500 MHz is fairly good. S(1,2) is off by about 20%, which will have an effect on the predictions of the MSG, but the other parameters fit pretty well at this frequency. Let's assume this is sufficient. Now we must test for stability (Figures 9.9 and 9.10).

We can see that the device is potentially unstable at the design frequency, but if we are careful and can give up some gain, we may be able to find a stable solution and still retain low noise figure. This may give us lower noise figure than if we first make the device unconditionally stable with resistive loading. To evaluate this possibility, we need to look at the available gain and noise figure circles. It is more convenient to calculate the gain and noise circles (available gain for noise calculations) on the display panel rather than the schematic panel so you can change the noise figure without having to re-simulate the circuit.

If the device model includes noise data as this one does in the example above, enable the noise simulation in the S-parameter controller

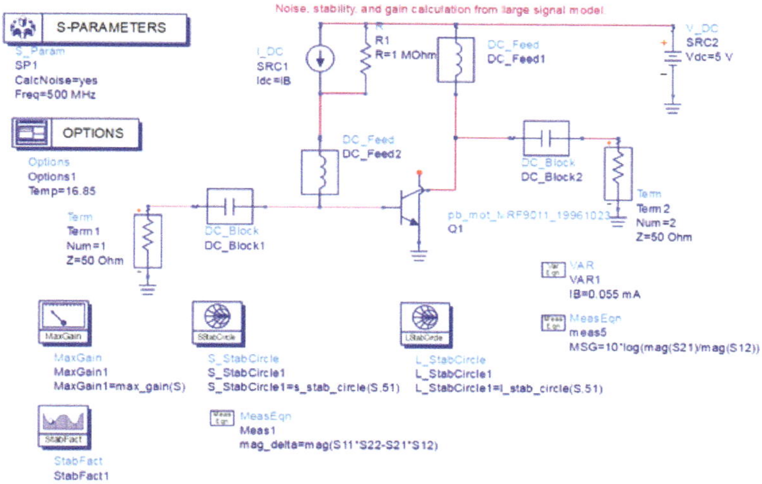

Figure 9.9 Schematic file for simulation of stability, available gain and noise figure.

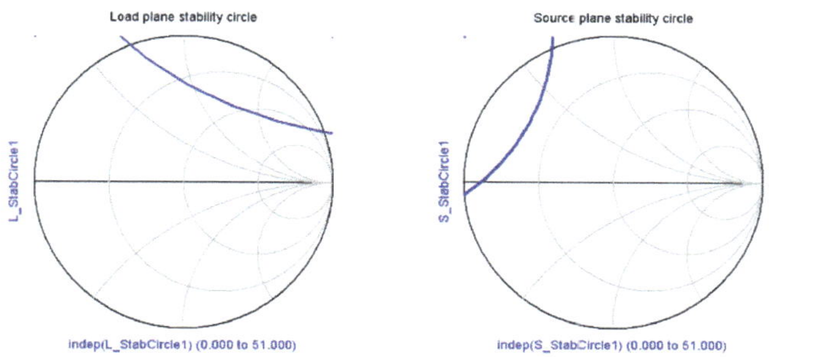

freq	mag_delta	StabFact1
500.0MHz	0.299	0.485

freq	MSG	MaxGain1
500.0MHz	19.769	19.769

Figure 9.10 Simulation result for stability factor and circles.

(Figure 9.11). You can display this function using the *Display* setting in this controller. This will then calculate NFmin, Γ opt (called Sopt), and Rn (un-normalized). Also include on the schematic panel the maximum gain (MaxGain1) equation (calculates either Maximum Available Gain MAG or Maximum Stable Gain, MSG) and stability circle equations. You must first determine stability before proceeding with the noise analysis. Since this is a low noise design, we will try first for a solution that does not require resistive stabilization which would degrade the noise figure.

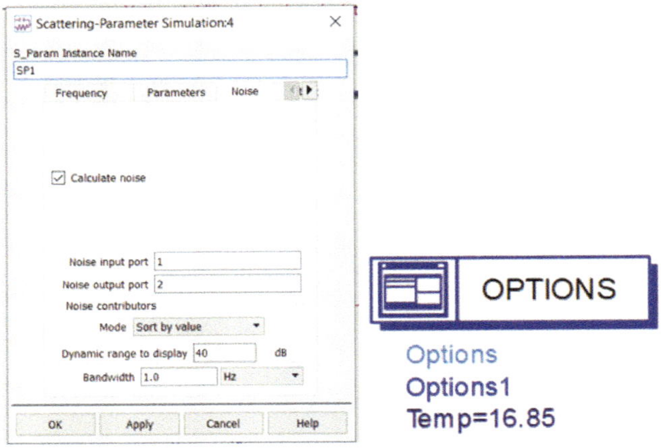

Figure 9.11 Set S-parameter controller for noise calculation. Standard temperature of 290 K is set by inserting the Options icon from the Simulation-S_Param pulldown menu.

You will generally want to calculate noise parameters at one frequency, so use the Single Point Sweep option on the frequency panel. Also choose Options and set the temperature to 16.85° (standard temperature for noise is 290 K). Then Sopt, NFmin, and Rn are determined.

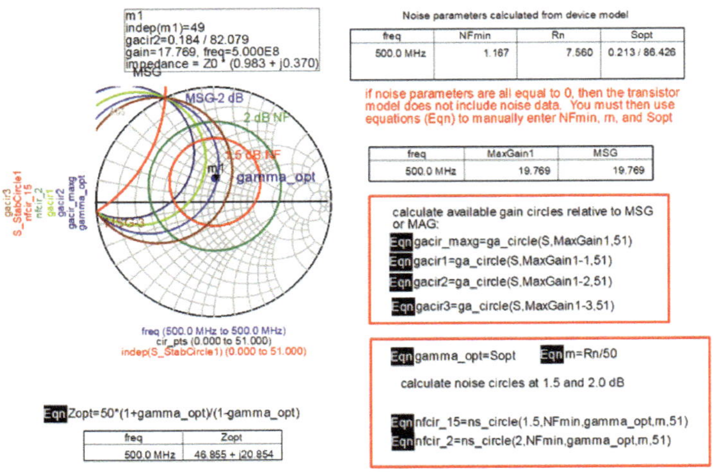

Figure 9.12 Results obtained from the simulation of MSG and Noise Figure.

You can see in Figure 9.12 that the MSG is high (19.8 dB), and if we give up 2 dB in gain, a perfect noise source match can be achieved, as seen by the

MSG-2 circle (blue) passing over Gamma Opt. The noise parameters for the transistor under the given bias condition are shown in the table. These were calculated from the large signal model.

The syntax for calculating available gain circles (ga_circle equation) is the same as that for power gain circles. It is usually convenient to plot the gains relative to the MaxGain1 value. The Gav circles must be plotted on the source plane. They assume that the load is conjugately matched to the output for all Γ_S values. For example, from the circles display in Figure 9.16,

$$gacir\_maxg = ga\_circle(S, MaxGain1, 51) \tag{9.21}$$

S is the S-parameter matrix, MaxGain1 is the MSG, and 51 chooses the number of points that you want to use for plotting the circle. In the example in Figure 9.12, gain circles are plotted for MSG, MSG – 1, MSG – 2, and MSG – 3.

The noise circle parameters (for the ns_circle equation) are defined as:

$$ns\_circle(nf, NF\min, Sopt, rn, 51) \tag{9.22}$$

where nf = noise figure of the device represented by Γ_S values that fall on the circle, NFmin = minimum possible noise figure of device at this bias and frequency, Sopt = Γ opt = the optimum input match for best noise figure, rn = the noise resistance parameter (normalized to 50 ohms), and 51 = number of points plotted on the circle.

If the transistor model does not include noise data (not all of them do), you must enter the NFmin, gamma_opt, and rn from the transistor data sheet manually using equations (Eqn) or an S2P file in order to calculate noise circles.

In Figure 9.13 it is clear that the calculated value for Γ_L is quite close to the load stability circle. This is not an acceptable design; therefore, some resistive stabilization will be needed. As was presented in Chapter 6, a shunt resistance on the output can be added to move the load stability circle away from Γ_L. Don't add resistance to the input because this will degrade the noise figure. Because this is a simulation with DC bias, a blocking capacitor will be used to prevent DC current from being shunted by the resistor. A resistance of 160 ohms was calculated from the closest point on the circle to Γ_L. Figure 9.14 shows the modified schematic, and Figures 9.15 and 9.16 the results.

In Figure 9.15 we can see that the MSG was not affected by the resistive loading. The minimum noise figure was only slightly increased, presumably

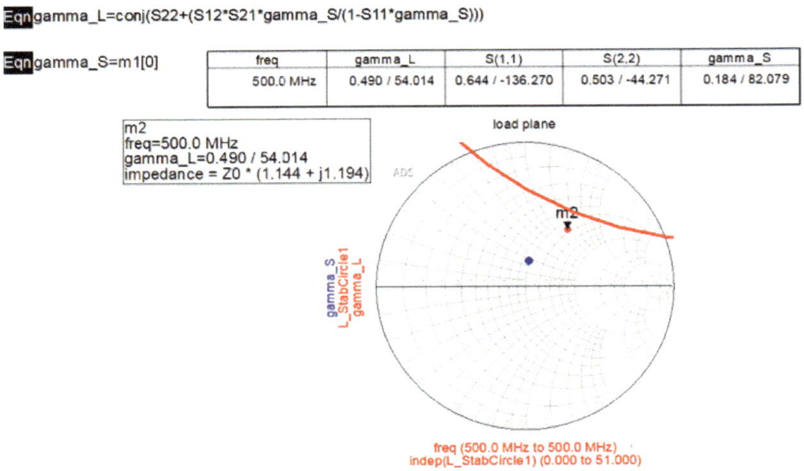

Figure 9.13 The Eqn function on the display panel can also calculate and . Note that the value is determined by the location of marker m1 on the gain circle.

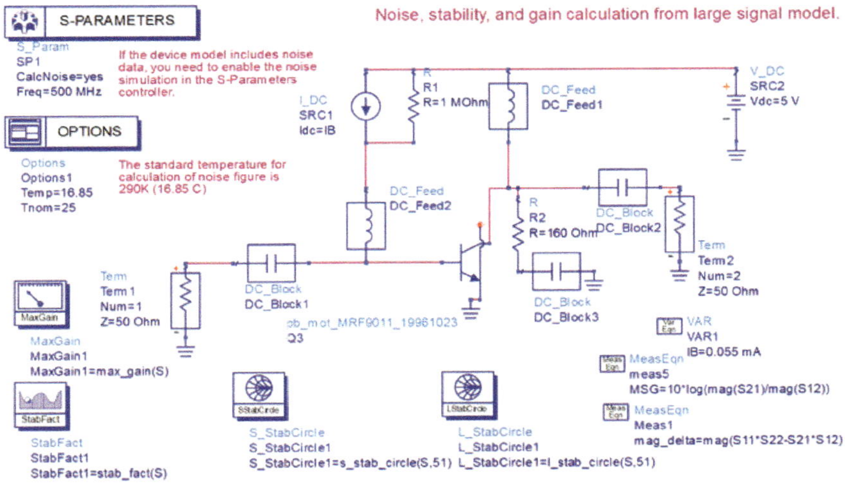

Figure 9.14 Schematic diagram of the simulation including a stabilizing resistor on the output.

because the amplifier has high enough available gain to mask the added resistive noise. 4 dB of available gain is sacrificed by relocating m1 to the MSG-4dB circle. Figure 9.16 shows that this modified design pushes the load stability circle well away from Γ_L at m2.

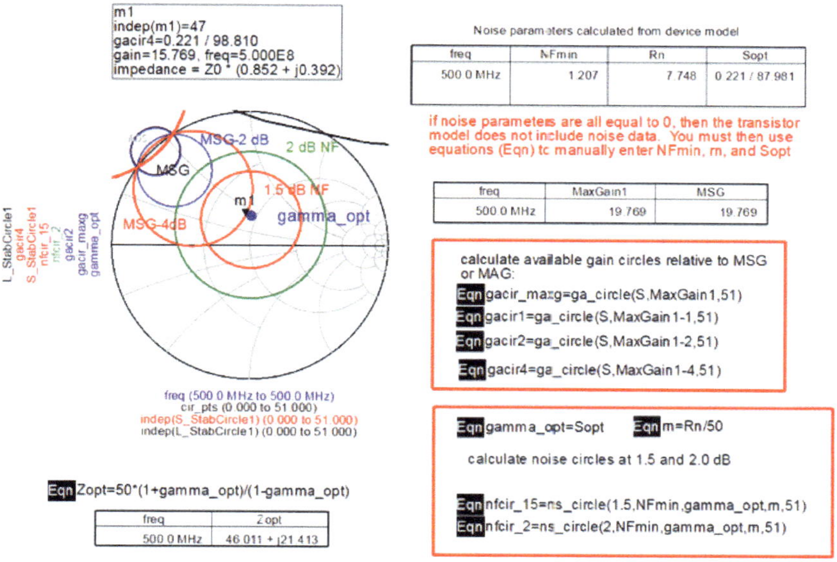

Figure 9.15 Gain and noise circles for the modified amplifier.

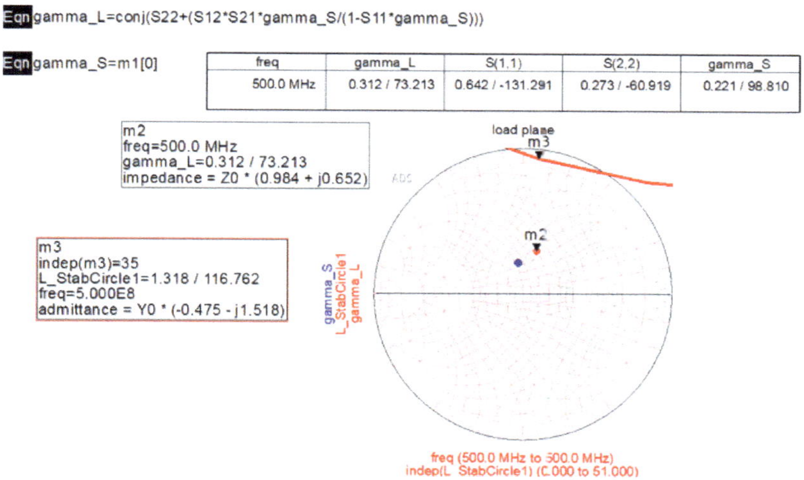

Figure 9.16 Improved stability of the modified design.

Now, you could go back and vary the bias to determine whether you could improve on gain or noise. If satisfied with the result, then complete the design by adding matching networks (Chapter 5), biasing circuitry (Chapter 7) and simulate the amplifier over a range of frequencies to verify stability out of band just as you did in the Stability and Gain tutorial (Chapter 6).

This exercise used the large signal model because it agreed well with the S-parameter model. If there had been substantial disagreement, then the S-parameter model would be preferred.

9.5 Summary

1. Select device model; S-parameter model is preferred.
2. Evaluate stability at design frequency. Don't use stabilizing resistors at input – these will greatly increase noise figure.
3. Plot available gain (G_A) and noise figure circles. Tradeoff between gain and noise to facilitate stability and reduce sensitivity to the match at input and output.
4. Design matching networks.
5. Include bias insertion.
6. Simulate over wide frequency range. Must have stability at all frequencies. Modify circuit if necessary with frequency-dependent stabilizing circuits.

References

[1] Gonzalez, G. (1997) *Microwave Amplifiers: Analysis and Design*, 2nd edn. J. Wiley, Chapter 4.
[2] Pozar, D. M. (2005) *Microwave Engineering*, 3rd edn. J. Wiley, Chapter 13.
[3] Leeson, D. (2005) Stanford University, personal communication.

Homework

1a. The S-parameters at 1.2 GHz and the load stability circle are shown below.

- Identify the stable region.
- Show how you would resistively stabilize this amplifier (estimate the resistance required).

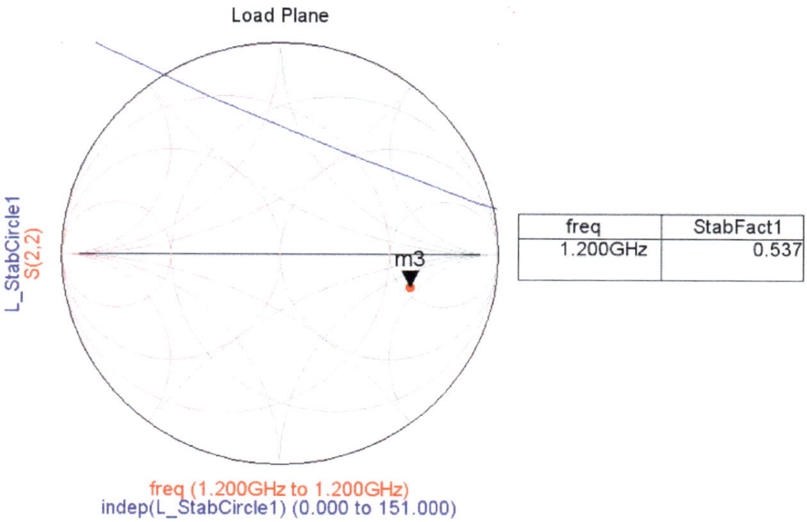

	S11	S21	S12	S22
1.200GHz	$0.862\angle -42.8$	$7.844\angle\, 137.2$	$0.036\angle\, 70.6$	$0.615\angle -16.9$

Available gain (MAG, MAG – 3 dB and MAG – 5 dB) and noise figure circles (1.0, 1.5 and 3 dB) are plotted on the source plane as shown below.

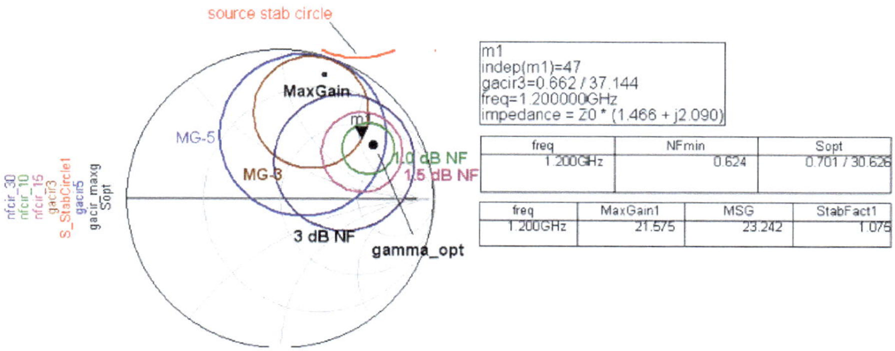

b. Explain why the maximum available gain is less than the maximum stable gain (MSG).

c. Determine the noise parameters for the device at this frequency.

F_{min} =

Γ_{opt} =

r_n =

d. Estimate the available gain and noise figure that would be obtained at marker m1
e. Design a distributed matching network that provides the Γ_s indicated by m1.

2. Create an S2P file on ADS using the S-parameters in problem (a). Perform the simulation to reproduce the stability and circles figures shown above.

3. A particular transistor data sheet provides the available gain, stability and noise circles for a microwave BJT at 500 MHz. At this bias point and frequency, the S-parameters of the transistor were:

$$S11 = 0.51\angle-122^\circ \quad S12 = 0.098\angle 45^\circ$$
$$S21 = 4.24\angle 98^\circ \quad S22 = 0.55\angle-39^\circ$$
$$\text{Also, } \Gamma_{opt} = 0.39\angle 52^\circ$$

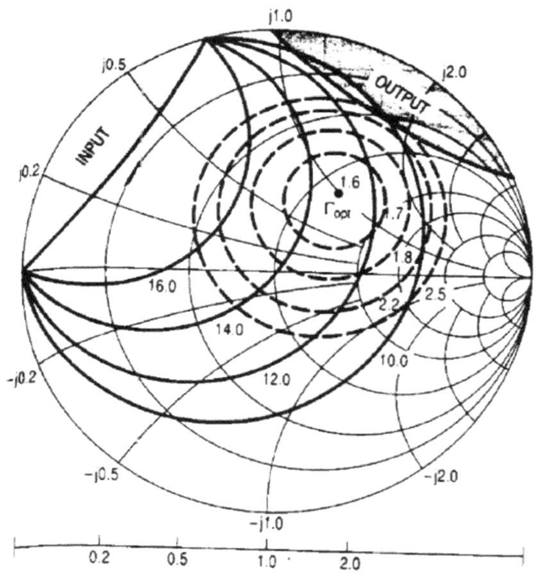

a. If the input and output are terminated with 50 ohms, calculate the transducer gain in dB.
b. Calculate the maximum stable gain at this frequency.
c. Resistively stabilize the device in two different ways. In a low noise application, which would be best? Explain why.

d. Assuming you choose not to stabilize the device, show a point on the Smith chart which would be your best choice for Γ_S that would give 1.8 dB noise figure. Justify your choice. Estimate the available gain at that Γ_S.
e. Estimate or calculate the three noise parameters (NFopt, Zopt, Rn) given the chart and the S parameter information.
f. Design an input matching network that would provide NFopt. Specify component values at 500 MHz.
g. How much will the noise figure be degraded if the input network has a loss of 1 dB?
h. Design a bias circuit that would provide stable biasing at VCE = 3 V and IC = 3 mA.

4.Use the NE85639 S-parameter model in the ADS components library – biased at 2.5 V and 3 mA – for this exercise.
a. Use ADS to determine stability using the stability factor *k* and delta = det(S) from 50 MHz to 3 GHz. Plot the source and load stability circles at 500 MHz on separate Smith chart displays. Identify clearly on the plots which region is stable and which is unstable.
b. Show how to make the transistor unconditionally stable at 500 MHz on the load side by adding a stabilizing resistor. Explain how you predict the resistor value from (a) and show that it produces an unconditionally stable device.
c. Using the load side stabilization from part (b), plot available gain circles for this device on the Γ_S plane. Also, plot noise figure circles on the same display. Determine suitable Γ_S and Γ_L to achieve at least 12 dB of gain at 500 MHz with noise figure less than 2 dB.
d. Design lumped element matching networks that will provide the selected Γ_S and Γ_L. Simulate the complete amplifier gain and stability at 500 MHz to verify your design.
e. Modify the matching network design adding low frequency resistive loading so that the amplifier will also be unconditionally stable from 50 to 500 MHz. Verify that the gain and noise figure at 500 MHz still meet specifications and that the amplifier will also be stable up to 3 GHz.

5. Use the NE34018 S2P S-parameter model biased at V_{DS} = 2 V and I_D = 10 mA for this exercise.
a. Use ADS to determine stability using the stability factor *k* and delta = det(S) from 50 MHz to 5 GHz. Plot the source and load stability circles at 1.4 GHz on separate Smith chart displays. Identify clearly on the plots which region is stable and which is unstable.

b. Show how to make the transistor unconditionally stable at 1.4 GHz on the load side by adding a stabilizing resistor. Explain how you predict the resistor value from (a) and show that it produces an unconditionally stable device.
c. Using the load side stabilization from (b), plot available gain circles and noise figure circles for this device. Determine suitable Γ_S and Γ_L to achieve at least 15 dB of gain and noise figure less than 1.6 dB at 1.4 GHz.
d. Design lumped element or distributed matching networks which will provide the selected Γ_S and Γ_L. Simulate the amplifier gain and stability at 1.4 GHz to verify your design.
e. Modify the matching network design adding low frequency resistive loading so that the amplifier will also be unconditionally stable from 50 MHz to 1.4 GHz. Verify that the gain and noise figure at 1.4 GHz still meet specifications and that the amplifier will also be stable up to 5 GHz.

Appendix 9.1: Representing devices as an S2P file

For example, in the case of the NE34018, the ADS component library does not contain a set of S-parameters for the device. Instead, we can use an S2P file provided by NEC. The file for this device was measured at the V_{DS} = 2 V; I_D = 5 mA bias point. An S2P file is simply a tab delimited ASCII text file containing S parameters measured on a two-port device. Figure 9.17 shows how it is inserted into ADS. The format is:

Frequency	S11	S21	S12	S22
In GHz	mag ang	mag ang	mag ang	mag ang

Example file:
GHZ placeS MA R 50

0.500	0.978	–17.0	6.806	162.8	0.018	80.1	0.723	–7.5
0.600	0.969	–20.3	6.731	159.7	0.022	78.7	0.719	–9.1
0.700	0.960	–23.6	6.691	156.5	0.025	77.3	0.712	–10.5
0.800	0.949	–26.9	6.624	153.4	0.028	75.5	0.706	–12.0

! NOISE PARAMETERS ADDED......
! NOTE:The following data was just taken from the pdf file of NE34018 data sheet,
! For the Bias of VDS=2 V and IDS=10 mA.
! For Syntax, refer to the ADS S2P file format
! Ghz NFmin mag(gammaopt) phase(gammaopt) Rn/50

!NOISE DATA

0.9	0.56	0.76	30	0.45
2.0	0.63	0.61	41	0.28
2.5	0.68	0.49	51	0.18
3.0	0.70	0.39	49	0.16

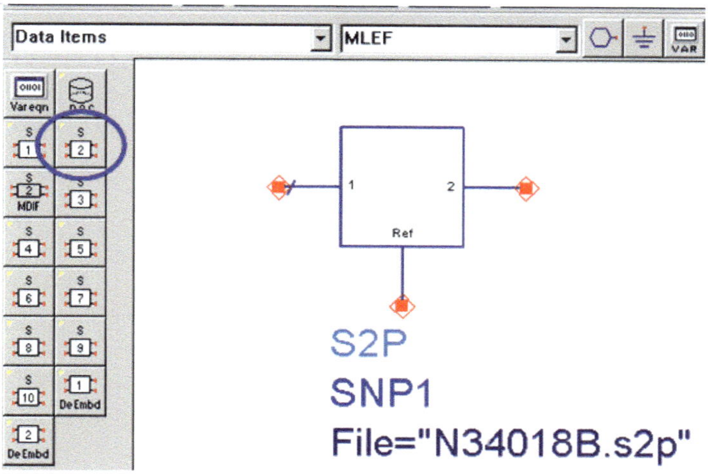

Figure 9.17 Place an S2P block from the Data Items menu and identify the file name. ADS will look for the file in the ADS project's data directory.

Appendix 9.2. Free Space Propagation Model

We have devoted considerable discussion on how to design an amplifier for reduced noise contribution. In this appendix, we will see for one application why this could be quite important.

The free-space propagation model is used to calculate path loss between a transmitter and receiver. This can be used for terrestrial or satellite applications [2] .

Consider the following.

Radiated power density:

$$S = \left(\frac{P_T}{4\pi r^2}\right) D \tag{9.23}$$

where D = directivity = G_T = peak radiation power/average radiation power, and P_T = transmitter power.

A receiving antenna is described by its effective area A_e.

$$A_e = \frac{P_r}{S} = \frac{\text{power delivered to input}}{\text{incident power at antenna aperture}}. \tag{9.24}$$

This relates to the directivity, *D*:

$$A_e = \frac{\lambda^2}{4\pi} D = \frac{\lambda^2 G_R}{4\pi} \tag{9.25}$$

(wavelength enters into the calculation because A_e must scale with λ^2 to maintain a given directivity or gain).

The power at the receiver is given by the Friis equation:

$$P_R = SA_e = \left(\frac{P_T G_T}{4\pi r^2}\right)\left(\frac{\lambda^2 G_R}{4\pi}\right) = \frac{P_T G_R G_T \lambda^2}{(4\pi r)^2 L} \tag{9.26}$$

where,

$P_R =$ received power

$P_T =$ transmit power

$G_T =$ transmit antenna gain

$G_R =$ receive antenna gain

$\lambda =$ wavelength (meters)

$L \geq 1$(system loss factor) due to transm ission line losses, filter attenuation

$r =$ distance from transmitting antenna to receiving antenna.

P_R falls off as square of distance, (20 dB/decade rate). Then, path loss can be defined by the ratio of transmit power to the received power in dB.

$$\text{Path loss } = 10 \log \frac{P_T}{P_R} = -10 \log \left[\frac{G_t G_r \lambda^2}{(4\pi)^2 r^2}\right] \tag{9.27}$$

(by convention, path loss is represented as a positive number)

Example 9.1. Terrestrial Microwave Link

How much transmit power is needed to get $(S/N)_{OUT}$ = 0 dB?

Assume antenna sees standard terrestrial background temperature of 290 K.

Frequency = 2.4 GHz (λ = 12.5 cm), r = 50 km.

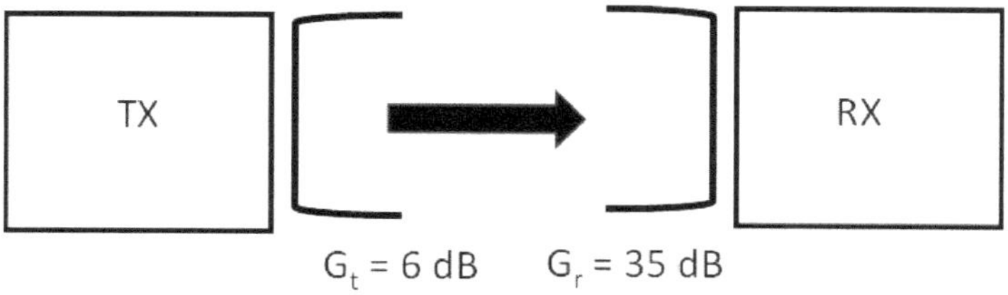

Figure 9.18 Terrestrial microwave link example.

Assume $L = 1$ (no loss – not a good assumption, but we will address this later).

Figure 9.18 represents such a terrestrial link with antenna gains labeled. In addition, P_t = transmit power, P_r = power received at input of receiver, receiver specifications: NF = 3 dB; bandwidth B = 20 MHz.

1. Calculate path loss using Equation (9.27):

$$10\log\left(\frac{P_r}{P_t}\right) = -10\log\left[\frac{G_t G_r \lambda^2}{(4\pi)^2 r^2 L}\right] = -134dB.$$

2. Find power required at receiver input accounting for signal and noise.

$S_{avi} = P_t$ – path loss + antenna gains = P_t (dBm) – 134 dB + 6 + 35 = P_t – 93dB

$N_{avi} = kT_oB$ = –174 dBm + 73 dB = –101 dBm

$S_{avi} = N_{avi} + NF = N_{avi}$ – 3 dB = – 98 dBm.

3. Calculate transmitter power required.

P_t (dBm) = –98 dBm + 93 dB = –5 dBm

More realistically, we need much higher S/N than 0 dB!

A9.2.1. Antenna noise model.[1]

The antenna acts as a transducer for all incident radiation.

Objects in the view of the antenna radiate blackbody noise at their surface temperature.

In the example above, the microwave link was point-to-point terrestrial. Thus, the antenna sees a noise temperature of 290K, and Navi = k 290 B.

If the antenna is pointed toward space, but at a low angle to the horizon, part of the field of view will be the earth and part will be space (Figure 9.19). You can calculate an equivalent noise temperature for the antenna by the relative areas subtended by the solid angle of view of the antenna. Here, γ is the area of view of the earth (Figure 9.19).

[1]Examples courtesy of D. Leeson, Stanford University [3]

$$\mathrm{Ta} = \gamma \mathrm{T}_{\mathrm{earth}} + (1 - \gamma)\mathrm{T}_{\mathrm{sky}} \tag{9.28}$$

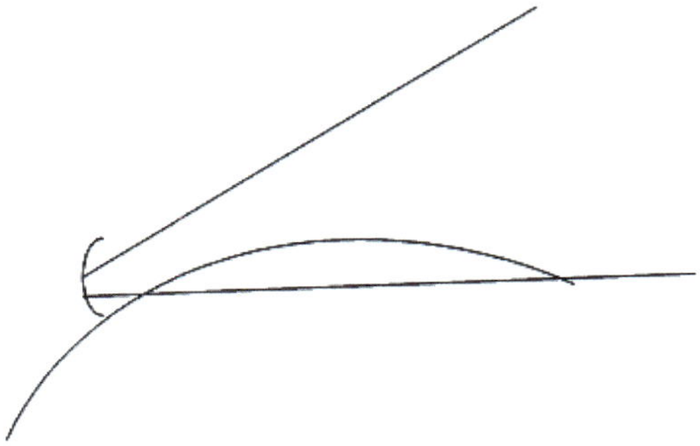

Figure 9.19 Antenna sees partially earth and partially space.

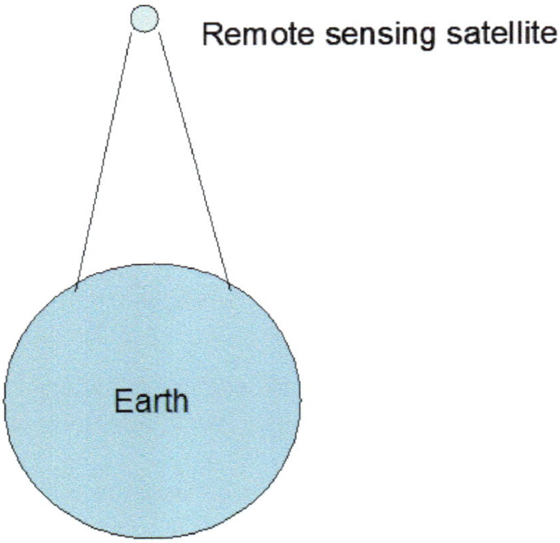

Figure 9.20 View of remote sensing satellite.

In the case of the remote sensing satellite in Figure 9.20, its field of view is earth, so the antenna temperature will be 290 K once again.

If it were much farther away from the earth so that the antenna also was viewing partially earth and part space, the antenna temperature would be lower in proportion to the solid angles of each.

A9.2.2. Earth satellite receiver

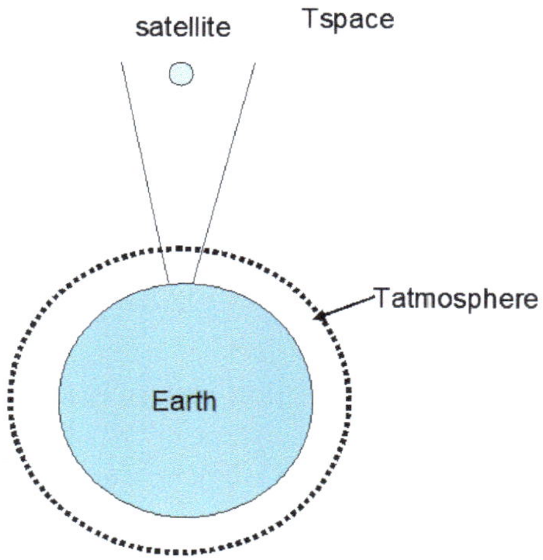

Figure 9.21 View of space for a satellite receiver.

Now, the antenna sees only space (Figure 9.21). Atmospheric attenuation reduces the noise from space but also replaces it with noise from a hotter atmosphere.

η_{atm} = transmission coefficient of atmosphere.

$$T_{antspace} = \eta_{atm} T_{space} + (1 - \eta_{atm})\, T_{atm}$$

For a vertical pass on a clear day at 12 GHz: η_{atm} = 0.98, T_{atm} = 150 K, T_{space} can vary depending on what is in view of the antenna between 3 K and 100 K. Of course, you want to avoid pointing the antenna at the sun!

Let's say T_{space} = 10 K, then,

$$T_{antspace} = T_{space} \eta_{atm} + (1 - \eta_{atm})\, T_{atm} = 12.8 \text{ K}.$$

Now, we must consider other noise sources that add to the system noise temperature.

1. Antenna efficiency (ohmic loss)
2. Sidelobe noise
3. Feedline loss
4. LNA excess noise.

A9.2.2.1. Antenna efficiency

This consists of the ratio of radiation resistance to total resistance:

$$\eta_{ant} = \frac{R_{rad}}{R_{rad} + R_{loss}} \tag{9.29}$$

Since the antenna is at 290 K, if η_{ant} = 0.98, then

$$T_{ant\ loss} = T_{earth} * (1 - \eta_{ant}) = 5.8 \text{ K}.$$

This adds to the effective antenna noise temperature.

A9.2.2.2. Sidelobe noise

Antennas are not perfect. The radiation pattern also includes sidelobes with reduced gain, but still significant noise or interference can be added by these from undesired directions. Suppose we have a sidelobe gain of say –14 dB, then the sidelobe gain

$$G_S = 0.04.$$

The sidelobes may see the earth, so suppose

$$T_{sidelobe} = G_S T_{earth} = 0.04 * 290K = 11.6K.$$

We can now add the temperatures to obtain a total antenna temperature.

$$T_{ant} = 12.8 + 5.8 + 11.6 = 30.2K.$$

Now we will include system losses between the antenna and receiver.

A9.2.2.3. Feedline loss

This might also include filter losses. T = 290 K. If the loss is 0.3 dB, then the feedline "gain", G_F = 0.94. This reduces antenna noise but adds much more thermal noise.

$T_{ant\_withfeed}$ = G_F * T_{ant} + (1 – G_F)T_{earth} = 0.94 * 30.2 + 0.06 * 290 = 45.8 K.

This is a lot of extra noise. This is why the LNA is usually located on the antenna right at the feed point.

A9.2.2.4. LNA and system

Now, we consider the system noise: T_{system} = 35 K.

Adding all contributors together give us a total noise temperature of:

T_{ant_space} = 12.8K
T_{ant_loss} = 5.8K
$T_{sidelobe}$ = 11.6K
so, T_{ant} =30.2 K.
Then, including feedline noise: $T_{ant\_withfeed}$ = 45.8 K.
Including system noise:

$$T_{TOTAL} = 45.8 + 35 = 80.8 \text{ K}.$$

Converting to available noise power:

$N_{avi} = kT_{TOTAL} = 1.38 \times 10^{-23} * 80.8 = 1.12 \times 10^{-21}$ W/Hz or –179.5 dBm/Hz.

Now, let's apply this to the satellite example:
Suppose the path loss is –180 dB
Transmitter power = 10 W (40 dBm)
Transmit antenna gain = G_T = 10 dB
Receive antenna gain = G_R = 30 dB
Signal bandwidth = 10 MHz.
Signal available power:
P_{SIG} = 40 dBm – 180 + 10 + 30 = $-$1C0 dBm.
Available noise power:
$P_N = -179.5$ dBm + 10 log B = -109.5 dBm.
This gives us an S/N ratio of 9.5 dB.

What if the system noise temperature had been 70 K rather than 35 K. How much would that degrade S/N?

Our total noise temperature now increases to: T_{TOTAL} = 45.8 + 70 = 115.8 K

$N_{avi} = kT_{TOTAL} = 1.38 \times 10^{23} * 115.8 = 1.60 \times 10^{21}$ W/Hz or –178 dBm/Hz.

Available noise power:
$P_N = -178$ dBm + 10 log B = -108 dBm.
This reduces the S/N ratio by 1.5 dB.

The transmitter power on the satellite would need to increase from 10 W to 14.1 W. This can be very expensive on a satellite as it increases the system power and cooling requirements.

Compare the noise figure of the system in these two cases. Remember that noise factor is defined in terms of noise temperature by:

$F = 1 + T_{system}/290$
$F_{case1} = 1 + 35/290 = 1.12$ (0.5 dB NF)
$F_{case2} = 1 + 70/290 = 1.24$ (0.93 dB NF).

We see that going from 70 to 35 K improves the noise figure by only 0.43 dB but improves S/N by 1.5 dB. This discrepancy is due to the fact that we are looking out into a cooler background (12.8 K) rather than 290 K. The excess noise contribution at 290 K really hurts the S/N.

The S/N improvement would be even greater if some of the other noise contributors (sidelobe noise, feedline loss) were smaller.

10

Introduction to Receivers

Goals

Understand:

1. What constitutes a receiver
2. Two dominant receiver architectures
   (a) Superheterodyne and direct conversion
   (b) Pros and cons of each
3. Receiver dynamic range
4. Images and image rejection
5. Benefits of multiple conversion architectures.

10.1 Receiver Architectural Concepts

We have described the analysis and design of key RF components and their limitations in previous chapters. These would include:

- Small-signal amplifiers:
  - stability
  - matching networks
  - distortion
  - noise.
- Free-space propagation models.

In this chapter, receivers are used to represent an RF system example that makes use of these categories of analysis. Many receivers must be capable of handling a very wide range of signal powers at the input while still producing the correct output (Figure 10.1). This must be done in the presence of noise and interference which occasionally can be much stronger than the desired signal.

The purpose of a receiver is to translate RF signals to baseband where they can be demodulated.

Noise sets the threshold for minimum detectable signal power – MDS. MDS was defined in Chapter 8 (Equation (8.52)).

Distortion sets the maximum signal power level. The third order input intercept (IIP3) is a figure of merit that is directly related to the intermodulation distortion produced by a particular design. (Figure 8.4 and Equation (8.6))

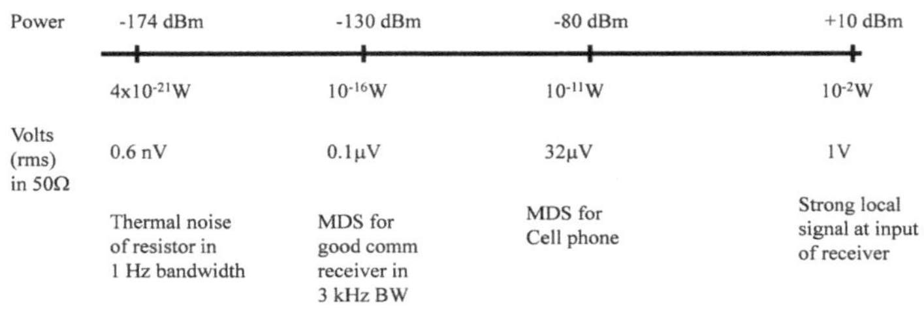

Figure 10.1 Range of signal powers that a receiver might encounter.

Why frequency translation? To bring modulation to the baseband where it can be demodulated, there are two dominant architectures for frequency translation [1, 2]:

- Superheterodyne
- Direct conversion.

Figures 10.2 and 10.3 represent generic functions needed for all receivers. The original superheterodyne concept in 1917 addressed the current technology of that day. The vacuum tubes were not capable of providing any gain above 1 or 2 MHz. By using the nonlinearity of a vacuum tube for AM detection along with adequate gain at low *intermediate* frequencies (IF, a few hundred kHz typically), receivers could be built that were sensitive in the MHz range. This enabled the power level of radio transmitters to be greatly reduced.

Today, gain is cheap, but the superheterodyne architecture has lived on and has much broader use. It allows the designer to optimize the receiver performance through a clever choice of intermediate frequencies and filtering.

Direct conversion is less common but has become recently more popular in single chip radios. It can eliminate off-chip bandpass filters, replacing them with on-chip DSP lowpass filters.

Note that both architectures use mixers for either up or down conversion.

The superheterodyne or superhet architecture uses an intermediate frequency (IF) following the mixer. This is selected such that amplifiers and channel selection bandpass filters are available with suitable performance. Image rejection also plays a role in selecting IF frequencies, as will be seen

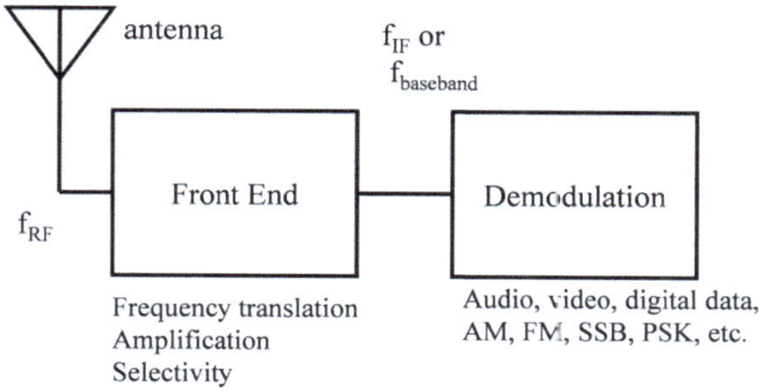

Figure 10.2 The front end of the receiver performs the frequency translation, channel selection and amplification of the signal.

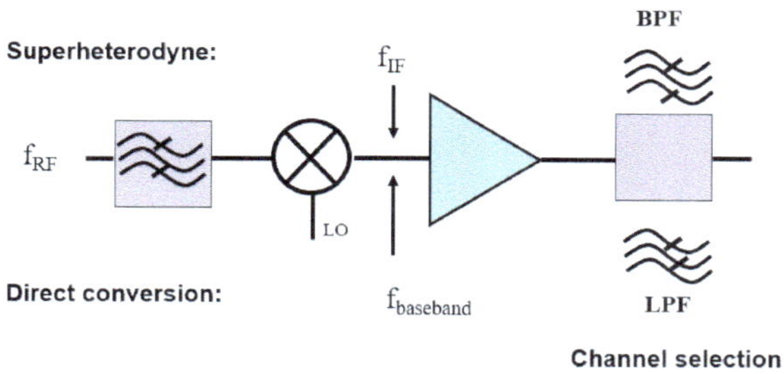

Figure 10.3 Block diagram of both receiver architectures. They differ by either mixing to an IF frequency or to baseband.

later. Today, A/D converters can have enough speed to directly convert IF to digital.

The direct conversion mixes down to DC. The advantage is that low-pass filters can be integrated on a chip using active or digital filter design approaches. However, local oscillator leakage causes a DC offset. Also, the mixer in most cases must be a complex image rejecting design because the signal and image fold over onto the same frequency.

Both utilize a mixer to produce the frequency translation. So, what is a mixer?

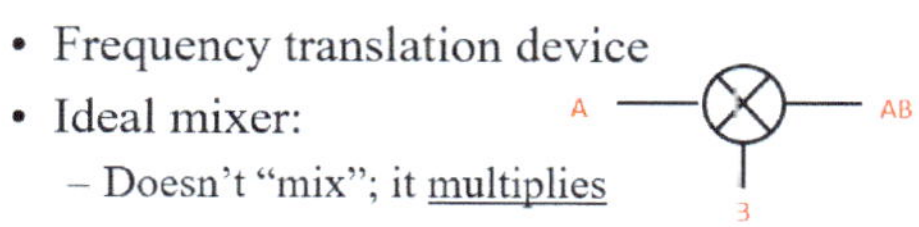

Figure 10.4 Mixers are multipliers.

A mixer doesn't really "mix" or sum signals; it multiplies them (Figure 10.4).

$$(A \sin \omega_1 t)(B \sin \omega_2 t) = \frac{AB}{2}[\cos(\omega_1 - \omega_2)t - \cos(\omega_1 + \omega_2)t] \quad (10.1)$$

Note that both sum and difference frequencies are obtained by the multiplication of the two input sinusoidal signals with amplitudes A and B. A mixer can be used to either *downconvert* or *upconvert* the RF input signal. The designer must provide a way to remove the undesired output, usually by filtering. Mixers are discussed in much more detail in Chapters 11 and 12.

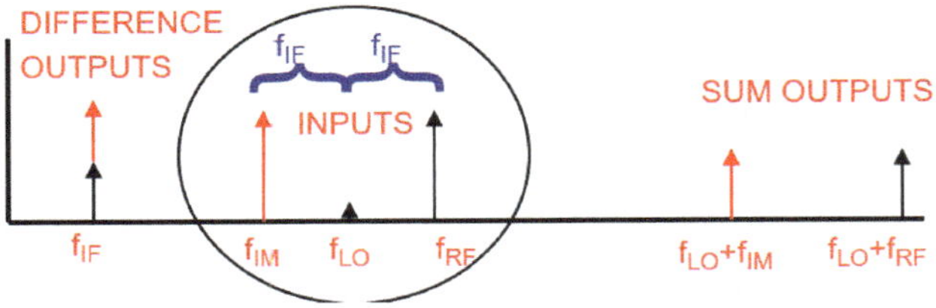

Figure 10.5 Two outputs are produced with the multiplication for downconversion and upconversion.

10.2 Image Rejection

Even in an ideal multiplier, there are two RF input frequencies (F_{RF} and F_{IM}) whose second-order product has the same difference IF frequency.

$$F_{RF} - F_{LO} = F_{LO} - F_{IM} = F_{IF}. \quad (10.2)$$

The two results are equally valid. One is generally referred to as the "*image*" and is undesired. In the example in Figure 10.5, the lower input frequency is designated the image in this case.

10.2.1 What is the source of the image signal?

Both signals enter through the antenna. While federal and international agencies regulate spectral usage through frequency allocation, these other users may be transmitting in bands that coincide with our image frequency. Our job as designers, is to choose LO and IF frequencies to avoid potential interferers in the image band.

In a receiver front end, out-of-band inputs at the image frequency could cause interference when mixed to the same IF frequency. Also, the noise present at the image would also be translated to the IF band, degrading the signal-to-noise ratio.

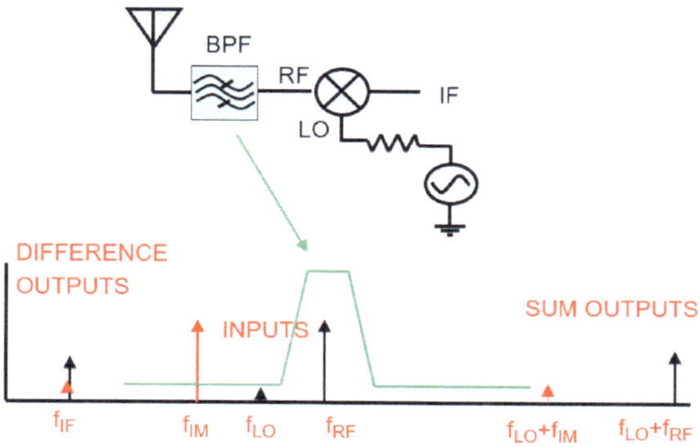

Figure 10.6 Bandpass filter at the receiver front-end can suppress the signal and noise at the image frequency.

In Figure 10.6 we see that a *bandpass preselection filter* is often used ahead of the mixer to suppress the image signal. The IF and LO frequencies must be carefully selected to avoid image frequencies that are too close to the desired RF frequency to be effectively filtered. Alternatively, an *image-rejection mixer* could be designed which suppresses one of the input sidebands by phase and amplitude cancellation. This approach requires two mixers and some phase-shifting networks.

So far, the spectrum exhibited by the ideal multiplier is free of harmonics and other spurious outputs (*spurs*). The RF and LO inputs do not show up in the output. While accurate analog multiplier circuits can be designed, they do not provide high dynamic range mixers since noise and bandwidth often are sacrificed for accuracy.

Figure 10.7, **case 1:** LO frequency is higher than RF frequency. This places the image frequency above the RF frequency. A sharp cutoff lowpass filter (LPF) or bandpass filter (BPF) could be used to attenuate the image.

$$f_{IF} = f_{RF} - f_{LO} \tag{10.3}$$

Case 2: RF frequency is higher than LO frequency. This places the image frequency at below the RF frequency – now in-band for an LPF. A sharp

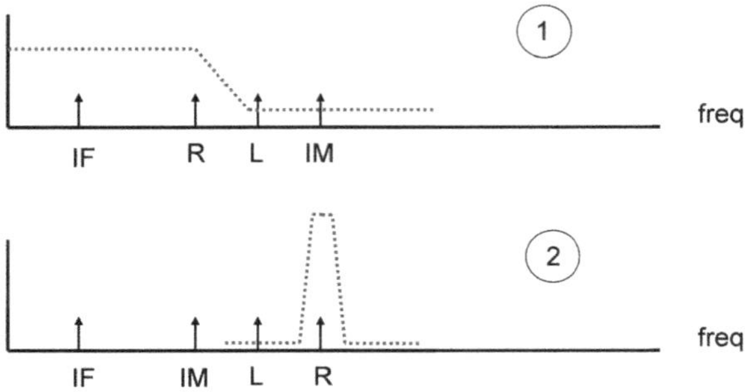

Figure 10.7 There are two choices of LO vs. RF frequency for downconversion – IF frequency lower than RF.

cutoff bandpass filter (BPF) must be used to attenuate the image.

$$f_{IF} = f_{LO} - f_{RF} \tag{10.4}$$

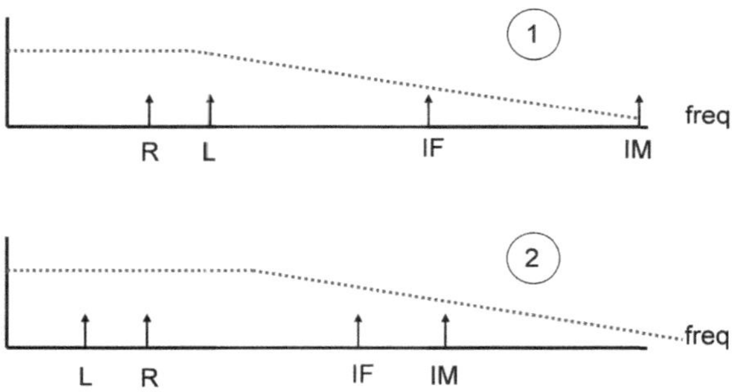

Figure 10.8 Upconversion $F_{IF} > F_{RF}$.

. The upconversion cases often can use an LPF for image rejection (Figures 10.8 and 10.9). In fact, the whole reason for upconverting in a receiver is to make image rejection more effective. However, we see that for the same f_{RF}, the two cases give much different results.

Case 1: Here the LO is higher than RF. Two input frequencies produce the same IF

$$f_{RF} + f_{LO} = f_{IF} \tag{10.5}$$

$$f_{IM} - f_{LO} = f_{IF} \tag{10.6}$$

The image frequency is much higher than the RF frequency. This makes it easy to use a simple LPF to get significant image rejection.

Case 2: Same equations, but now the LO is lower than RF. This places the IF and IM frequencies lower, making it more demanding for the LPF to provide significant image rejection.

An IF filter is often used here to block potentially interfering spectral inputs from creating distortion downstream in the receiver where amplification is provided. This function is often called a "roofing filter".

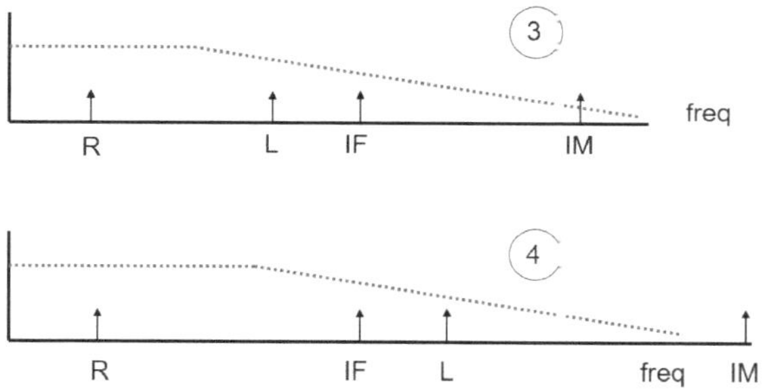

Figure 10.9 Cases 3 and 4 maintain the same IF frequency.

Case 3:

$$f_{RF} + f_{LO} = f_{IF} \tag{10.7}$$

$$f_{IM} - f_{LO} = f_{IF} \tag{10.8}$$

Case 4:

$$f_{LO} - f_{RF} = f_{IF} \tag{10.9}$$

$$f_{IM} - f_{LO} = f_{IF} \tag{10.10}$$

Once again, the high LO injection leads to a higher image frequency and better image rejection.

10.2.2 Channel selection

Or, alternatively, if we chose to keep the same IF frequency, probably a common choice since IF filters are available at only certain frequencies, the picture changes slightly from cases 1 and 2.

In Figure 10.10, you can see that a narrow band, fixed frequency filter (crystal, SAW, ceramic) is often used for channel selection. It is easier to

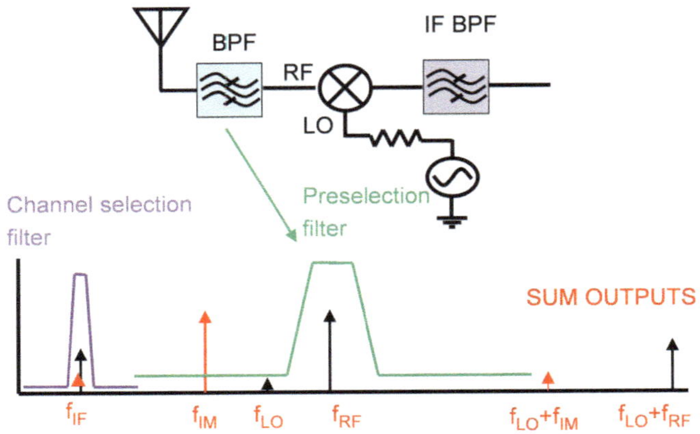

Figure 10.10 An additional filter is used with a superhet architecture to select the desired channel. This filter is typically narrow band.

build a high Q narrowband fixed frequency filter at a lower frequency than to build a tunable high Q high frequency filter. The local oscillator tunes the front end to select the input frequency .

The example shown in Equation (10.3) downconverts to a lower intermediate frequency. This is the *superhetrodyne* approach invented by Armstrong who applied for a patent on the concept in 1918.

We also have a cost consideration. IF filters are available only at certain frequencies if we want inexpensive mass-produced filters. Here are some common ones: 455 kHz, 10.7 MHz, 21.4 MHz, 45 MHz, 70 MHz.

There are also filters available in the VHF/UHF range.

Another choice, the *direct conversion* architecture, downconverts directly to baseband (zero IF). Then, a simple lowpass filter is used for anti-aliasing, an A/D converter and DSP is used for demodulation.

10.2.3 Example. FM broadcast receiver

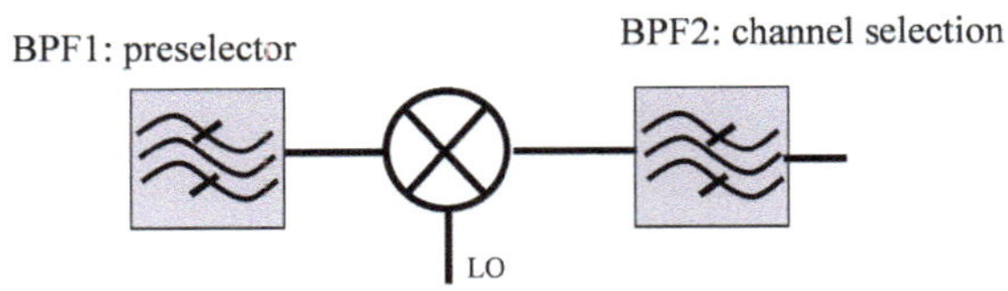

Figure 10.11 Block diagram of an FM Broadcast receiver front end.

- FM broadcast band: 88 MHz to 108 MHz.
- IF frequency: 10.7 MHz.

$$f_{IMAGE} = f_{RF} + 2f_{IF} \tag{10.11}$$

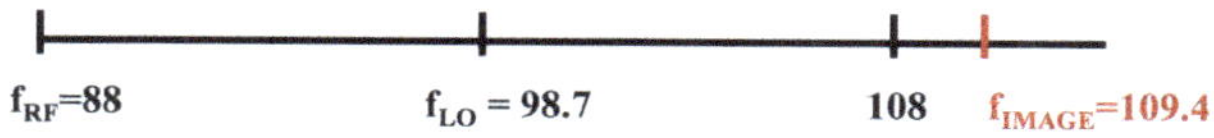

Figure 10.12 Spectrum showing RF, LO, and image frequencies.

Figure 10.11 represents the radio frequency "front end" of an FM broadcast receiver. Figure 10.12 shows the frequency plan. Both the RF and IMAGE frequency will be translated to the same IF frequency. With a 10.7 MHz IF frequency, the image is always outside of the FM broadcast band. Therefore, strong in-band FM signals are never to be found at the image frequency. Equation (10.11) is derived from Figure 10.7, case 1.

A preselection filter at RF frequency can be used to reject the image that is 21.4 MHz away from the desired RF signal. In the usual implementation, this filter is a bandpass filter with narrow bandwidth, and is tuned, tracking the LO frequency.

Question: Why does it use LO injection on the high side (above the RF in frequency)?

10.2.4 Example: AM broadcast band

Refer again to Figure 10.11. In the AM broadcast band case,

1. AM broadcast band: 530 kHz to 1700 kHz
2. IF frequency: 455 kHz.

In this case, for historic reasons, a low IF frequency was used. This was because gain at higher frequencies with vacuum tubes was insufficient. However, this selection creates a serious image problem.

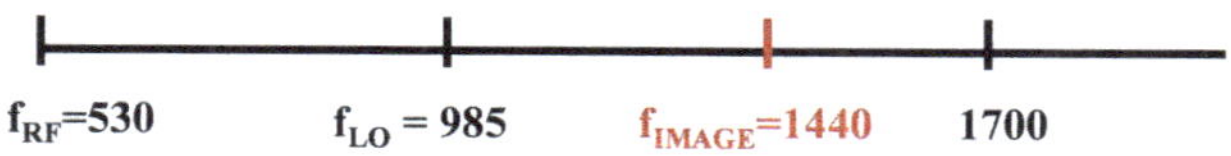

Figure 10.13 AM broadcast band spectrum showing the image falling in-band.

From the frequency plan shown in Figure 10.13, you can see that the image frequency can fall within the band according to Equation (10.11). The solution required a high-Q tunable preselector bandpass filter at the front end. For example, when f_{RF} = 530 kHz, f_{IMAGE} = 1440 kHz with high LO frequency.

The choice of lower LO injection frequency has other problems. LO frequency selection: we always have two choices. Image rejection and oscillator implementation affect the choice.

1. $F_{LO1} = F_{RF} - F_{IF}$
   530 − 455 = 75 KHz
   1700 − 455 = 1245 KHz.
2. $F_{LO2} = F_{RF} + F_{IF}$
   530 + 455 = 985 KHz
   1700 + 455 = 2155 KHz.

LO case #1 requires a 16.6 to 1 tuning range for the LO; #2 only requires 2.2 to 1. The oscillator will be much easier to implement.

Can we redesign the receiver to use fixed low-pass preselector?

- Upconvert: Use higher $F_{IF} >> F_{RF}$
- Preselector admits entire AM band
- No tuning allowed.

10.2.5 Calculate image rejection

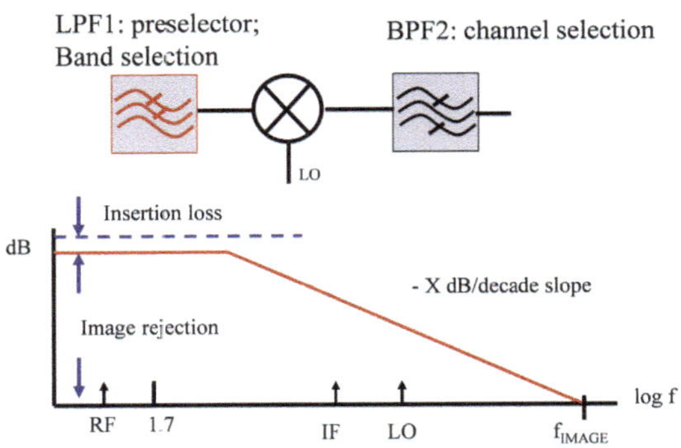

Figure 10.14 Another example. AM receiver. Calculate image rejection provided by a known low-pass filter slope.

In Figure 10.14 we will use an upconversion approach to achieve a high image frequency. Let's make the preselection filter simple and cheap: two poles give −40 dB/decade.

We will design according to two requirements:

- Minimum of 40 dB image rejection ratio
- Inexpensive IF filter: try 10.7 MHz IF frequency.

$f_{LO} < f_{IF}$

Frequency	Low end	High end
f_{RF}	0.53 MHz	1.7 MHz
f_{LO}	10.17	9.0
f_{IM}	20.87	19.7

$f_{RF} + f_{LO} = f_{IF}$
$f_{IM} - f_{LO} = f_{IF}$

$f_{LO} > f_{IF}$

Frequency	Low end	High end
f_{RF}	0.53 MHz	1.7 MHz
f_{LO}	11.23	12.4
f_{IM}	21.93	23.1

$f_{LO} - f_{RF} = f_{IF}$
$f_{IM} - f_{LO} = f_{IF}$

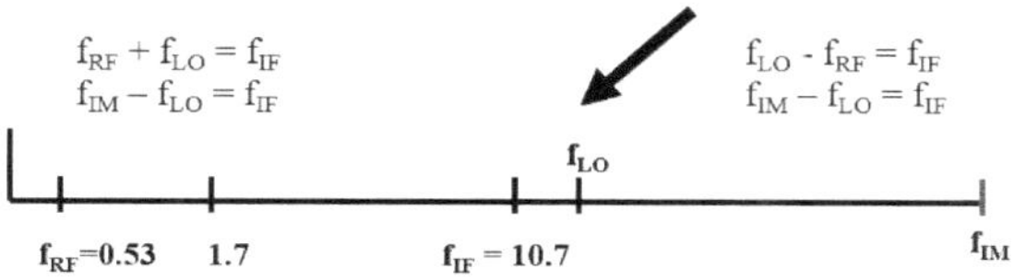

Figure 10.15 Calculation of image frequency for two different LO choices.

In Figure 10.15 we see again two choices of LO frequency, $f_{IM} - f_{LO} = f_{IF}$.

The worst-case image frequency with low LO injection would be for f_{RF} = 1.7 MHz. In this case, f_{IM} = 19.7 MHz.

The worst-case image frequency with high LO injection would be for f_{RF} = 0.53 MHz. In this case, f_{IM} = 21.93 MHz.

As we have seen previously, the higher LO frequency will give us better image rejection:

LPF filter cutoff frequency must be at 1700 KHz to cover entire AM band, so check image rejection to see if it meets our spec. With −40 dB/decade, we will beat the spec. The filter will be 40 dB down at 17 MHz.

So, at 21.93 MHz: 40 log(21.93/1.7) = 40 × 1.11 = 44.4 dB.

10.3 Dual Conversion Architecture

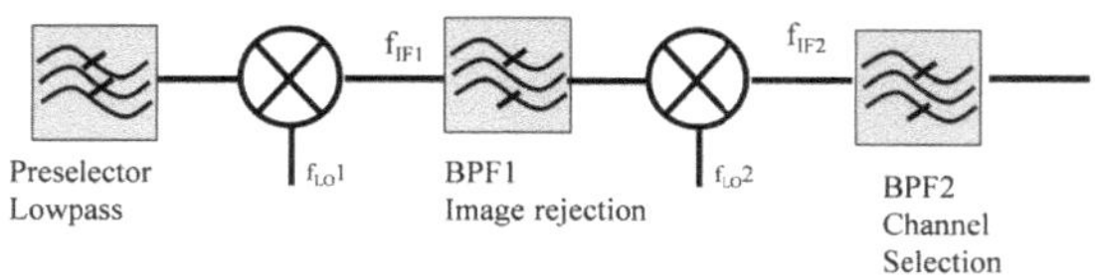

Figure 10.16 Dual conversion receiver.

A high first IF frequency, as shown in the previous example, places the image frequency well away from the desired signal. Then, a simple lowpass filter can be used for preselection in some cases (Figure 10.14).

However, this high first IF may present problems for channel selection. If a narrow modulation bandwidth is used, the filter bandwidth of BPF1 will be small. Then, a high loaded Q is required, with the associated high losses. In order to gain added flexibility in managing images and spurs, as well as providing for a lower Q channel selection filter, a second mixer is often used to downconvert to a much lower second IF frequency, as seen in Figure 10.16. With this architecture, we avoid having to trade off selectivity for sensitivity.

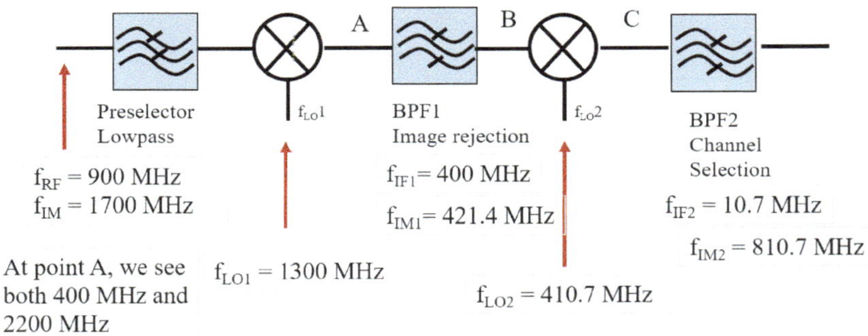

Figure 10.17 Example of a dual-conversion cellphone receiver.

Figure 10.17 shows that the output of the first mixer contains both the down and upconversion terms:

400 MHz and 2200 MHz.

The higher frequency is easily removed by BPF1 with frequency 400 MHz.

BPF1 must also prevent an image from passing through the second mixer. At point B, the IF frequency is 400 MHz, but the image frequency would be 421.4 MHz. So, the bandwidth of BPF1 must be small enough to reject signals at 421.4 MHz.

At C, we have both 10.7 MHz and 810.7 MHz. $f_{IF2} = f_{LO2} - f_{IF1}$. But we also get the sum term. The higher frequency is easily removed by BPF2. This can be a narrow bandwidth filter for channel selection.

10.4 Automatic Gain Control (AGC)

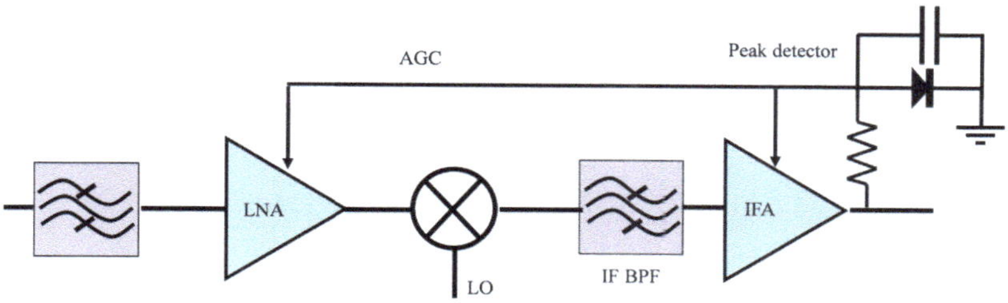

Figure 10.18 Block diagram of an AGC feedback loop.

Automatic gain control (AGC or RSSI) is used as a low frequency feedback loop within a receiver (Figure 10.18). The signal amplitude is measured with a peak detector and rectified. This control voltage can then be used to control the gain of amplifier stages so that the signal path can remain linear and avoid distortion from large amplitude input signals.

In some cases, the LNA can be switched out of the system or attenuation switched ahead of the LNA to handle strong signals.

The AGC path must accommodate the delay found in the filters. This can make the loop unstable unless the AGC voltage to the LNA and other early stages (pre-filtering) are suitably delayed.

10.5 Compare Superheterodyne and Direct Conversion

Superhet:

Benefits:

- Low cost, high quality fixed frequency IF bandpass filters are available
- $1/f$ noise at IF is negligible
- Good dynamic range with AGC.

Challenges:

- Image and spurious signal control
- Off-chip filters consume power, area
- Power dissipation
- Simple image control solutions (LPF for example) may create a strong signal overload problem.

Direct conversion:
Benefits:

- Simplest receiver architecture
- Baseband filtering can be done digitally or with active filters.

Challenges:

- $1/f$ noise
- DC offset can be caused by LO to RF leakage at mixer input
- Requires image rejecting mixer – precision components needed
- Second order distortion. If there is a strong input signal, the second order nonlinearity creates a signal at $2f$. This mixes with the LO at frequency f producing another source of DC offset.

The susceptibility to DC offset from LO feedthrough and second-order distortion can be reduced by careful design. The local oscillator is often set to twice the frequency and divided by 2 to avoid LO leakage. Balanced circuits in the mixer and amplifier will help to suppress second-order distortion. Finally, many have opted for a low frequency IF rather than a DC IF to avoid offset problems. This has its own hazards with regard to image rejection.

Image rejection mixers can also address these problems using complex signals. This will be discussed in Chapter 12 on mixers.

10.6 Summary

A brief introduction to receiver architectures was presented. Superheterodyne (or simply heterodyne) and direct conversion. The topic of frequency planning for image rejection was shown to be dependent on availability of suitable IF filters. FM and AM radio receivers were used as examples. Single and dual conversion concepts were presented.

The brief introduction to mixers will be expanded in a subsequent Chapters 11 and 12 along with mixers based upon image rejection using complex signals.

References

[1] Razavi, B. (1998) *RF Microelectronics* Prentice Hall, ch.5. Also 2nd Edn. 2012.

[2] Lee, T.H. (2004) *The Design of CMOS Radio-Frequency Integrated Circuits*, 2nd Edn. Cambridge, ch. 19.

Homework

1 a. Consider the receiver front end below. You are to determine f_{IF} and f_{LO} such that the image rejection for the FM broadcast band will be at least 30 dB. The minimum RF frequency is 88 MHz and maximum is 108 MHz. The preselector low pass filter has a cutoff frequency of 108 MHz and has a 40 dB/decade slope after cutoff.

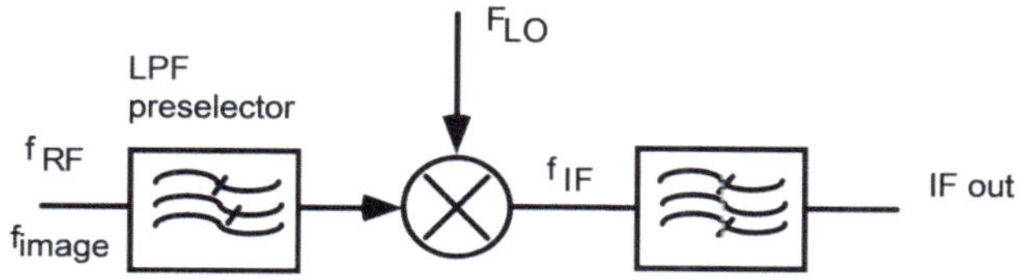

b. Now add a second mixer with a 10.7 MHz channel selection IF filter at the output. Specify the image filter bandwidth needed to suppress the image at the output of mixer 1 by 30 dB. Assume BPF1 has a 3 to 60 dB bandwidth ratio of 5. Also specify f_{LO2}.

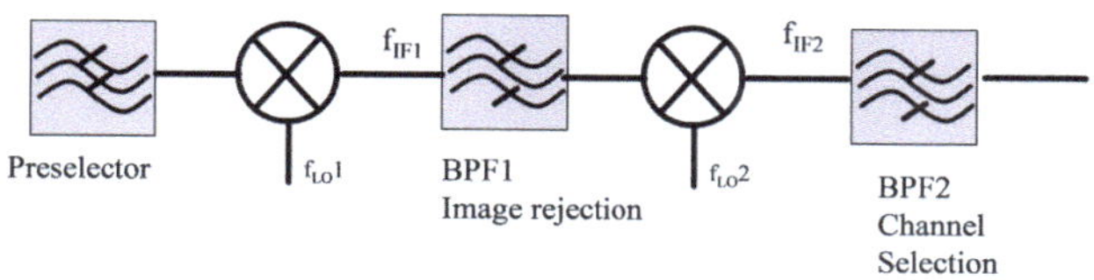

2. A receiver front end that uses a high first IF frequency is an effective tool for improving image rejection while using simple fixed input low-pass preselecting filters. However, it does leave the receiver vulnerable to second order IMD.

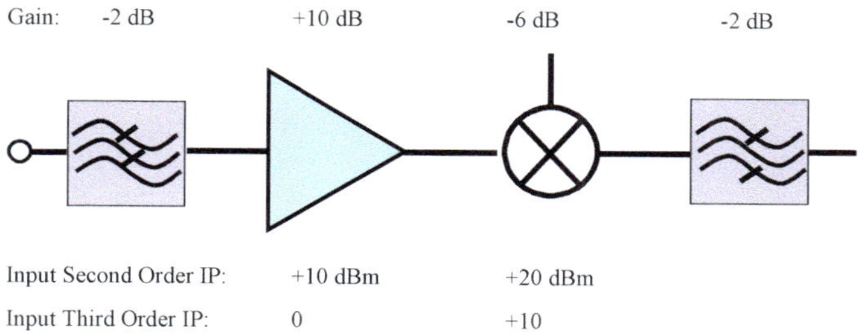

a. Find the second order and third order intercept points at the input of the receiver.
b. Explain why second order IMD could be a problem with this design.
c. There will be an input signal power level where the second and third order distortion powers are equal. Determine this level and note the region where the second order distortion dominates.

3. a. Choose an LO frequency that will provide the best image rejection for this combination of RF and IF frequencies. Explain why. What is the image frequency in this case? If the input bandpass filter has a 40 dB/decade slope outside of the passband, how much image rejection will be obtained in this case?

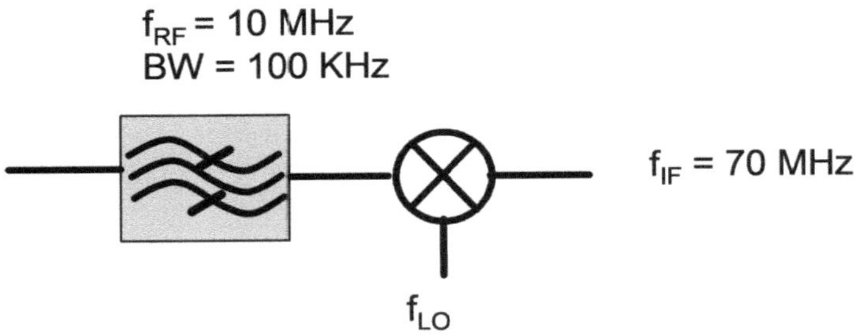

b. Dual conversion RF front end. The second mixer can also produce image problems if not properly filtered. Use your LO frequency from part (a). If the 70 MHz IF bandpass filter has a narrow bandwidth but 60 dB/decade slope outside of the passband, calculate the second IF image rejection in dB.

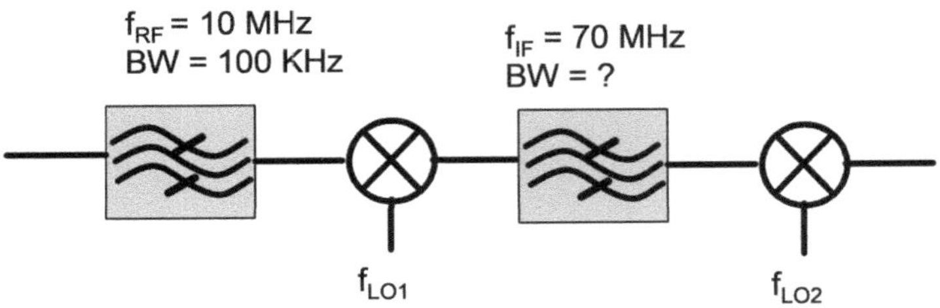

4. a. Assume that there are two input sources to the receiver with equal power levels. How much input power will be required to produce a third order intermodulation component of –60 dBm at the output?

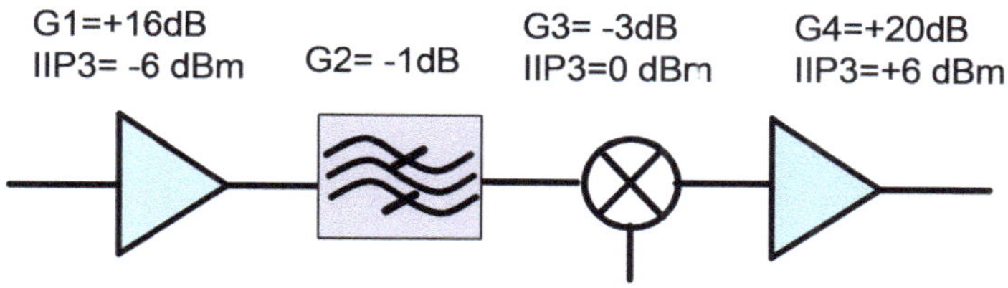

b. If the noise figure of the first amplifier stage is 1 dB, the mixer NF = 6 dB, and IF amplifier NF = 3 dB, calculate the spurious free dynamic range of this receiver front end. The IF bandwidth is 10 kHz, limited by a channel selection filter following the IFA.

5. The receiver front end shown below is intended to operate over the 1 to 10 MHz input (RF) frequency range. The preselector lowpass filter has a 15 MHz cutoff frequency and a 60 dB/decade rolloff. Choose the IF such that the worst case image rejection is 50 dB. Also indicate the LO frequency range corresponding to this choice of IF frequency.

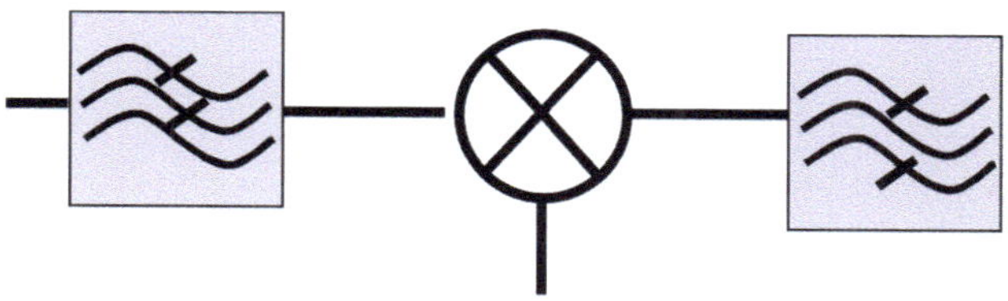

Comment. This is an upconversion mixer. The required IF frequency is higher than the signal band of interest. There are two solutions, depending on whether you consider 1 MHz or 10 MHz to be the worst case for the image and whether the LO frequency is higher than or lower than the IF frequency. Find both solutions.

11

Mixers

We have seen in the previous chapter the important role that the mixer plays in RF systems. Frequency translation is necessary in almost any application, so it is important to understand the design options and tradeoffs.

Goals:

- Understand the operating principles of a mixer
- Understand what makes a good mixer
- Choices: Nonlinear/switching mode; single/double balance; active/ passive
- Specify performance: Gain, NF, P_{1dB}, TOI, SFDR, isolation, image rejection
- Review some mixer design examples
- Use ADS to evaluate mixer performance.

Chapter 12 will extend the discussion to include image-rejecting mixers using complex signals.

11.1 Types of Mixer

There are many different mixer circuit topologies and implementations that are suitable for use in receiver and transmitter systems. How do you select the best one for a particular application? Why does the choice depend on the application and technology available?

Here is breakdown of typical mixer functions.

Receivers:

- Up or downconversion
- Demodulation of suppressed carrier SSB or DSB
- Input must support a large dynamic range
- AGC is often required.

Transmitters:

Up-conversion

Modulation: amplitude and phase

Input requires choosing the optimum signal level for high performance.

The superheterodyne (or heterodyne) receiver architecture often has several frequency translation stages (IF frequencies) to optimize image rejection, selectivity, and dynamic range. Direct conversion receiver architectures such as used in pagers use mixers at the input to both downconvert and demodulate the digital information. Mixers are thus widely used in the analog/RF front end of receivers. In these applications, often the mixer must be designed to handle a very wide dynamic range of signal powers at the input.

The mixer can be used for demodulation, although the trend is to digitize following a low IF frequency and implement the demodulation function digitally. They can also be used as analog multipliers to provide gain control. In this application, one input is a DC or slowly varying RSSI signal which when multiplied by the RF/IF signal will control the degree of gain or attenuation.

In transmitter applications, the mixer is often used for upconversion or modulation. In this application, the input signal level can be selected to optimize the overall signal-to-noise ratio at the output.

High performance RF mixers use nonlinear or time-varying characteristics to generate the multiplication. Thus, they also generate lots of undesired output frequencies. Three techniques have proven to be effective in the implementation of mixers with high dynamic range:

1. Use a device that has a known and controlled nonlinearity.
2. Switch the RF signal path on and off at the LO frequency.
3. Sample the RF signal with a sample-hold function at the LO frequency.

11.1.1 Non-linear mixer

The *non-linear mixer* (Figure 11.1) can be applicable at any frequency where the device presents a known nonlinearity. It is the main approach available at the upper millimeter-wave frequencies.

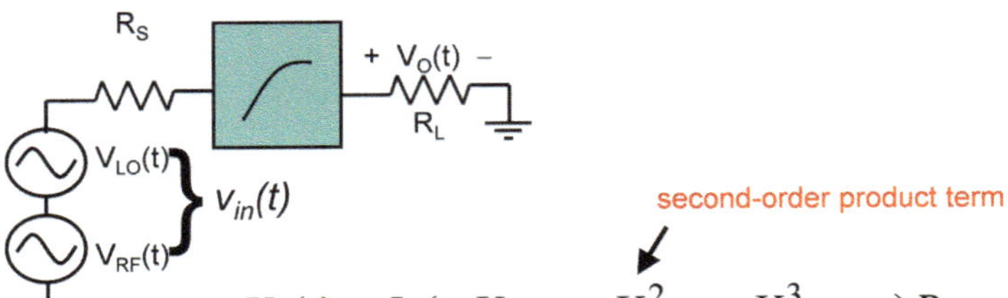

$$V_o(t) = I_s(a_1 V_D + a_2 V_D^2 + a_3 V_D^3 + \cdots) R_L$$

Figure 11.1 Non-linear mixer operation.

Any diode or transistor will exhibit nonlinearity in its transfer characteristic at sufficiently high signal levels.

Only the second-order product term produces the desired output. Let's suppose that V_R and V_L are the fraction of V_{RF} and V_{LO} that appear across the nonlinear device (possibly a diode),

$$a_2 V_D^2 = a_2 \left[V_R^2 \sin^2(\omega_R t) + V_L^2 \sin^2(\omega_L t) + 2V_R V_L \sin(\omega_R t)\sin(\omega_L t)\right]. \tag{11.1}$$

The product term produces the desired mixer output at the difference frequencies. Substituting the trig identity for the product of two sine functions,

$$V_R V_L \left[\cos(\omega_R - \omega_L)t - \cos(\omega_R - \omega_L)t\right]. \tag{11.2}$$

From the $\sin^2$ term, DC and second harmonic terms are also present. Recalling,

$$V_R^2 \sin^2(\omega_R t) = V_R^2 \left[1 - \cos(2\omega_R t)\right]. \tag{11.3}$$

If the application is frequency multiplication rather than mixing, the harmonic term can be useful. The DC term is proportional to RF input power, so it can often be used as a power meter.

In addition, when V_{RF} consists of multiple carriers, the power series also will produce cross-products that make the desired output products dependent on the amplitude of other inputs. Spurious output signal strengths can be decreased when devices that are primarily square-law, such as FETs with longer gate lengths, are used in place of diodes or bipolar transistors.

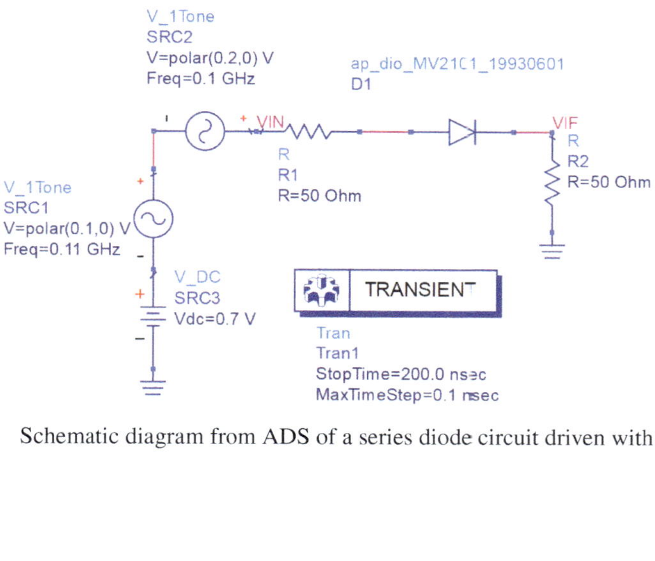

Figure 11.2 Schematic diagram from ADS of a series diode circuit driven with two signals.

As an illustration, a simple diode circuit (Figure 11.2) can be simulated on ADS to reveal the mixing behavior.

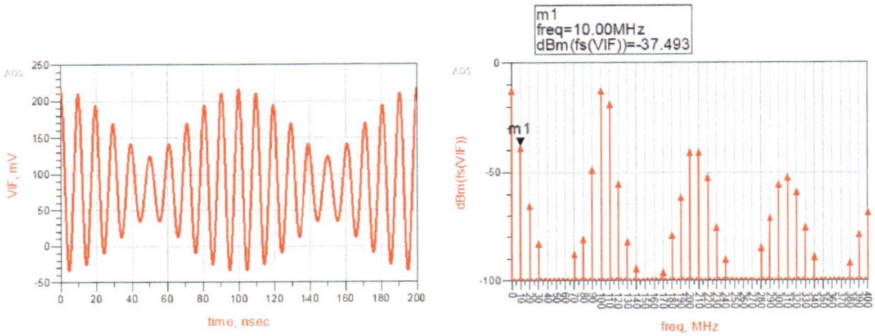

Figure 11.3 Time domain and frequency domain output from the non-linear diode mixer.

In Figure 11.3, we see the diode mixer with inputs

$$f_{RF} = 110 \text{ MHz}$$
$$|V_{RF}| = 0.1 \text{ V}$$

$$f_{LO} = 100 \text{ MHz}$$
$$|V_{LO}| = 0.2 \text{ V}.$$

In addition, the diode was biased at its DC threshold voltage of 0.7 V. The spectral output reveals RF and LO feedthrough and a DC term, as predicted. We can also see that there are a lot of spurious outputs (*spurs*) generated. Ideally, we would like to see outputs only at 10 MHz and 210 MHz. The output at 10 MHz is relatively small, −37 dBm.

ADS tip: In this example, the transient analysis mode of ADS was used to simulate the circuit containing nonlinearities. S-Parameter analysis only applies with linear components. While the output for VIF is a time-domain signal, the function dBm(fs(VIF)) performs a Fourier Transform in order to display the spectrum.

The StopTime setting must be long enough that one full period of the lowest frequency component of interest would be calculated. In this example, twice the period of a 10 MHz signal was used. The MaxTimeStep should be small enough to include the highest frequency input period and to assure that the calculation converges. It will give an error message if the time step is too large. Too small a MaxTimeStep will require excessive simulation time.

While transient analysis is efficient for relatively simple circuits, the typical mixer with multiple RF and LO signals is more efficiently simulated with harmonic balance (HB), especially for intermodulation evaluation. Refer to the tutorial in Chapter 8 on how to use HB.

11.1.2 Switching mixer [1–3]

We prefer the switching-type mixer (Figure 11.4) when the RF and LO frequencies are low enough that we can make decent switches. This takes us up through much of the millimeter-wave spectrum. The reasons will be described in some depth.

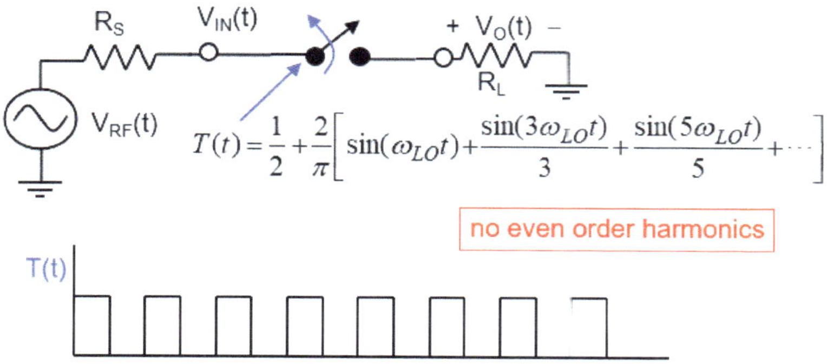

Figure 11.4 A very basic switching mode mixer. Ideally, 50% duty cycle is needed for LO.

This simple switch is operated by the local oscillator (LO). If the LO is a square wave with 50% duty cycle, it is easily represented by its Fourier Series. The symmetry causes the even-order harmonics to drop out of the LO spectrum. When multiplied by a single frequency cosine at ω_{RF}, the desired sum and difference outputs will be obtained as shown in Equation (11.4) and Figure 11.5. Note that everything is single-ended; there is no balancing on this design.

$$V_o(t)=\underbrace{\frac{V_R}{2}\cos(\omega_{RF}t)}_{\text{RF feedthrough}}+\frac{2V_R}{\pi}\left[\underbrace{\cos(\omega_{RF}t)\sin(\omega_{LO}t)}_{\text{2nd-order product}}+\underbrace{\frac{\cos(\omega_{RF}t)\sin(3\omega_{LO}t)}{3}}_{\text{4th-order spurs}}+\cdots\right] \tag{11.4}$$

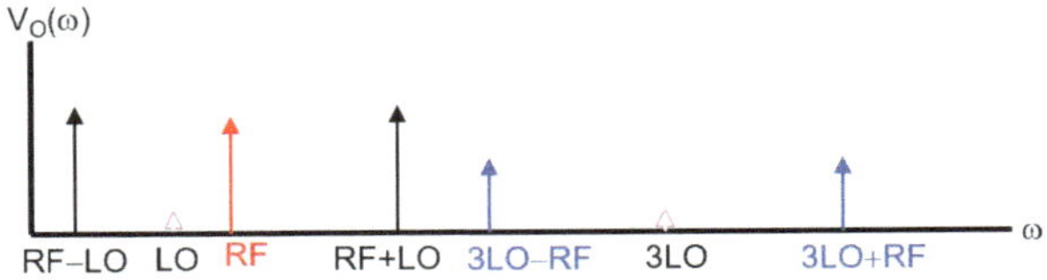

Figure 11.5 Output spectrum of the simple switching mixer.

The product of $V_{RF}(t)T(t)$ produces the desired output frequencies at

$$\omega_{RF} - \omega_{LO} \text{ and } \omega_{RF} + \omega_{LO}, \tag{11.5}$$

from the second order product. Both downconversion (RF – LO) and upconversion (RF + LO) are present. We can see that there is a significant feedthrough of the RF signal present in the output. There will also be harmonics of the LO present at $3\omega_{LO}$, $5\omega_{LO}$, etc. that will mix to produce "*spurs*" as seen in Figure 11.5. These harmonics also convert broadband noise that is generated internal to the mixer (or that is allowed into the mixer input in the absence of a preselection filter) into the IF output band.

Odd harmonics of the LO frequency are also present since we have a square wave LO switching signal. These produce spurious 4th, 6th, ... order products with outputs at

$$n\omega_{LO} \pm \omega_{RF} \tag{11.6}$$

where n is odd.

None of the LO signal should appear in the output if the mixer behaves according to this equation. However, if there is a DC offset on the RF input, there will be a LO frequency component in the output as well. This requirement is not unusual, since many mixer implementations require some bias current which leads to a DC offset on the input.

Exercise 11.1: Use the diode in the nonlinear diode mixer simulation as a switch. Put a square wave LO in series with the RF generator and simulate the output spectrum using transient analysis.

11.1.3 Single-balanced mixer

The RF feedthrough can be eliminated by using a differential IF output and a polarity reversing LO switch.

Active or *passive implementations* can be used for the mixer [4]. Each has its advantages and disadvantages. The passive implementations using diodes as nonlinear elements or switches or FETs as passive switches always exhibit conversion loss rather than gain. This can impact the overall system noise performance, so if noise is critical, an LNA is usually added before the mixer.

The polarity reversing LO switching function is shown in Figure 11.6.

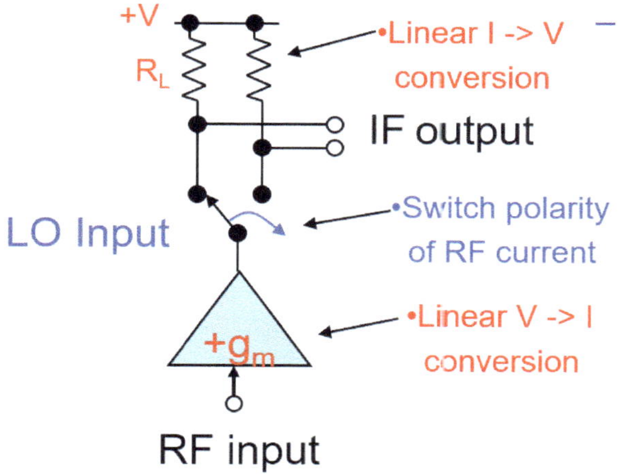

Figure 11.6 Single-balanced switching mixer uses differential output.

The single-balanced LO function consists of the equivalent of two square waves, as shown in Figure 11.7.

$$T_1(t) = \frac{1}{2} + \frac{2}{\pi}\left[\sin(\omega_{LO}t) + \frac{1}{3}\sin(3\omega_{LO}t) + \ldots.\right], \tag{11.7}$$

$$T_2(t) = -\frac{1}{2} + \frac{2}{\pi}\left[\sin(\omega_{LO}t) + \frac{1}{3}\sin(3\omega_{LO}t) + \ldots.\right]. \tag{11.8}$$

When added together, the DC terms (1/2 and $-1/2$) cancel. The DC term was responsible for the RF feedthrough in the unbalanced mixer since the term was multiplied only by $T_1(t)$.

$$V_{IF}(t) = g_m R_L V_R \cos(\omega_{RF}t)\,\frac{4}{\pi}\left[\sin(\omega_{LO}t) + \frac{1}{3}\sin(3\omega_{LO}t) + \frac{1}{5}\sin(5\omega_{LO}t) + \ldots\right]. \tag{11.9}$$

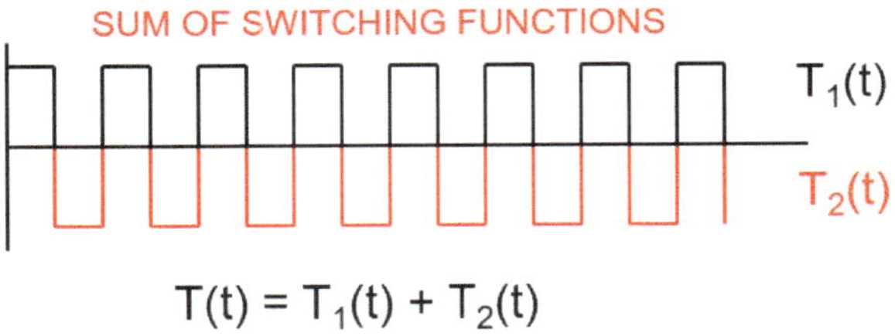

Figure 11.7 Sum of switching functions.

Second-order term:

$$\frac{2g_m R_L V_R}{\pi}\left[\sin\left(\omega_{RF}+\omega_{LO}\right)t+\sin\left(\omega_{RF}-\omega_{LO}\right)t\right]. \tag{11.10}$$

Here we see that the ideal conversion gain $(V_{IF}/V_R)^2 = (2/\pi)^2$ is 6 dB greater than for the unbalanced design (if $g_m R_L = 1$).

However, we can still get LO feedthrough if we take a single-ended output or if there is a DC current in the signal path. There is often DC present since the output of the transconductance amplifier will have a DC current component. This current shows up as a differential output.

$$\begin{aligned} V_{lF}(t) &= R_L\left[I_{DC}+g_m V_k\cos\left(\omega_{RF}t\right)\right]\times \\ &\frac{4}{\pi}\left[\sin\left(\omega_{LO}t\right)+\frac{1}{3}\sin\left(3\omega_{LO}t\right)+\frac{1}{5}\sin\left(5\omega_{LO}t\right)+\ldots\right]. \end{aligned} \tag{11.11}$$

$$\begin{aligned} &= \frac{4R_L}{\pi}\left\{I_{DC}\sin\left(\omega_{LO}t\right)+\right. \\ &\left.\frac{1}{2}g_m V_R\left[\sin\left(\omega_{RF}+\omega_{LO}\right)t+\sin\left(\omega_{RF}-\omega_{LO}\right)t\right]\right\}. \end{aligned} \tag{11.12}$$

As you can see, the output spectrum of the single-balanced switching mixer is much less cluttered than the nonlinear mixer spectrum. This was simulated with transient analysis using an ideal switch (Figure 11.9). The behavioral switch model has an on-threshold of 2 V and an off-threshold of 1 V. The LO was generated with a 4 V pulse function and the duty cycle was set to 50%. F_{LO} = 1 GHz. F_{RF} = 0.2 GHz. F_{IF} can be seen at 800 and 1200 MHz. The output is taken differentially.

Note the strong LO feedthrough component in the output. This is present because of the DC offset on the RF input which produces a differential LO voltage component in the output.

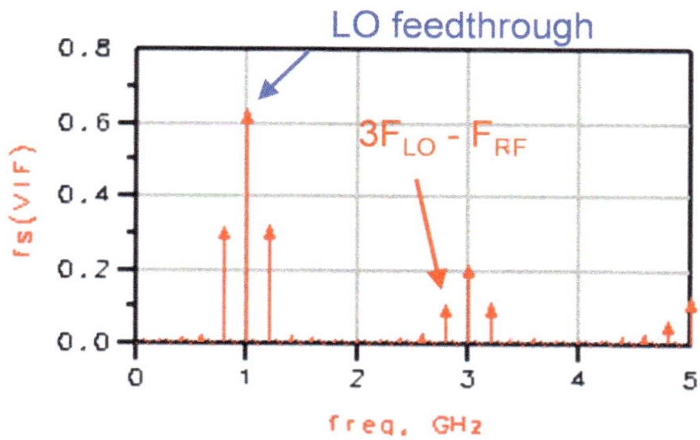

Figure 11.8 Output spectrum of a single-balanced mixer using a simple SPDT switch as a mixer.

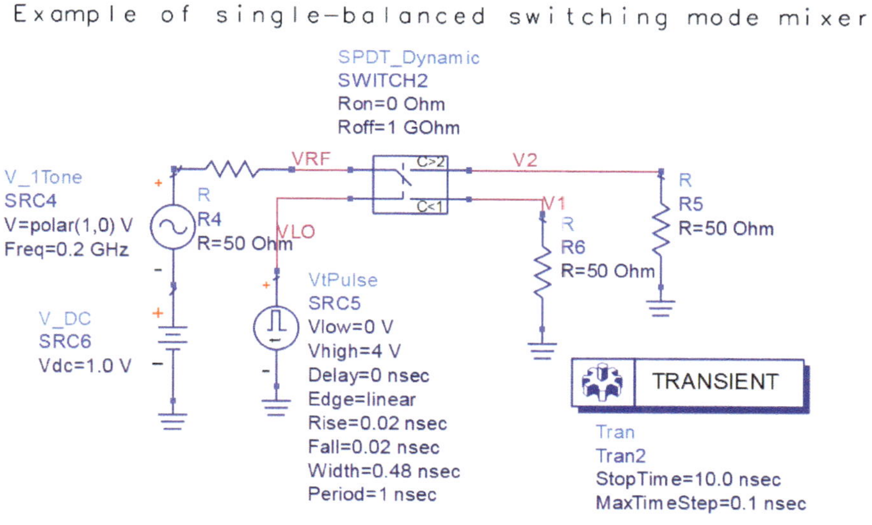

Figure 11.9 Example of how to simulate the single-balanced mixer using an ideal switch.

The VtPulse and V_DC sources are found in the Sources-TimeDomain pulldown menu. The switch component can be found in the System-Switch and Algorithmic menu. The total transient simulation time must be at least equal to the period of the lowest frequency present.

Exercise 11.2: Set the DC offset voltage to 0 and resimulate. Observe that the LO feedthrough is gone. Compare V1 and V_{IF} vs. time with and without the DC offset. Use markers to measure the IF output power and calculate the conversion gain.

This LO component is highly undesirable because it could desensitize a mixer post-amplifier stage if the amplification occurs before IF filtering. Eliminating the LO component when a DC current is present requires *double-balancing*.

11.1.4 Double-balanced mixer

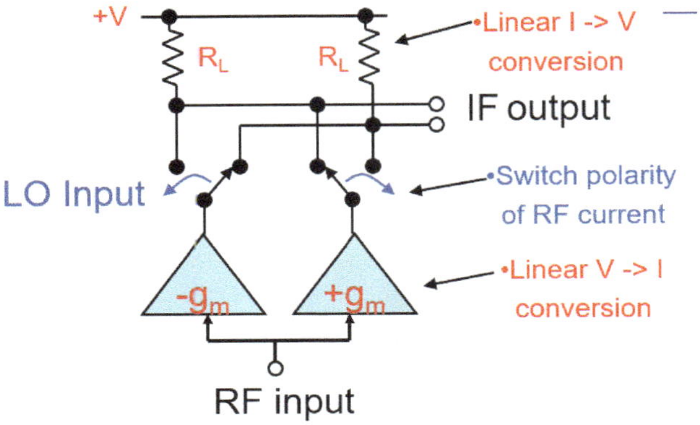

Figure 11.10 Block diagram of an ideal double-balanced switching mixer.

An ideal double balanced mixer (Figure 11.10) consists of a switch driven by the local oscillator that reverses the polarity of the RF input at the LO frequency [1] and a differential transconductance amplifier stage [1, 5]. The polarity reversing switch and differential IF cancels any output at the RF input frequency since the DC term cancels as was the case for the single balanced design. The double LO switch cancels out any LO frequency component, even with currents in the RF to IF path. The LO is typically suppressed by 50 or 60 dB if the components are well matched and balanced.

Two single-balanced mixers – difference cancels LO feedthrough

$$V_o = R_L\left[I_{DC} + I_{RF}\cos(\omega_{RF}t)\right]\frac{4}{\pi}\left[\sin(\omega_{LO}t) + \frac{1}{3}\sin(3\omega_{LO}t) + \cdots\right] - R_L\left[I_{DC} - I_{RF}\cos(\omega_{RF}t)\right]\frac{4}{\pi}\left[\sin(\omega_{LO}t) + \frac{1}{3}\sin(3\omega_{LO}t) + \cdots\right] \quad (11.13)$$

From - gm side of mixer (desired output adds)

The LO cancellation due to double balancing can be seen from these equations. It is equivalent to two single-balanced mixers combined differentially. The DC terms cancel and the RF terms add.

An IF balun, either active (a differential amplifier) or passive (a transformer or hybrid), is often used, however, so that the conversion gain will be maximized.

To get the highest performance from the mixer we must make the RF to IF path as linear as possible and minimize the switching time of the LO switch. The ideal mixer above would not be troubled by intermodulation distortion (IMD) at the high end of the operating signal range since the ideal transconductors and resistors are linear and the switches are ideal. Active components, however, will experience non-linearity and the consequent distortion at large signal levels.

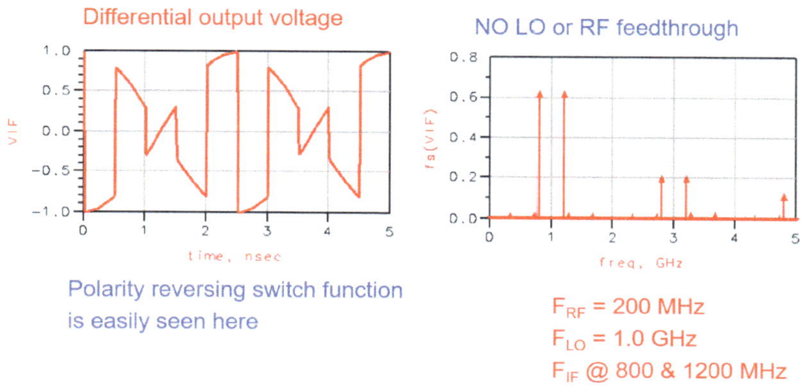

Figure 11.11 Time-domain and spectrum of the ideal double-balanced mixer output.

The differential output voltage and frequency spectrum are simulated using a transient analysis in ADS. The polarity switching action can be clearly seen in the output voltage in Figure 11.11. There is no LO or RF feedthrough in this ideal DB mixer.

In real mixers, there is always some imbalance. Transistors and baluns are never perfectly matched or balanced. These nonidealities will produce some LO to IF or RF to IF feedthrough (thus, isolation is not perfect). This is usually specified in terms of a power ratio relative to the desired IF output power: dBc

Secondly, the RF to IF path is not perfectly linear. This will lead to intermodulation distortion. Odd-order distortion (typically third and fifth order are most significant) will cause spurs within the IF bandwidth or cross-modulation when strong signals are present. Also, the LO switches are not perfectly linear, especially while in the transition region. This can add more distortion to the IF output and will increase loss due to the resistance of the switches.

11.2 Mixer Performance: Conversion Gain.

Conversion gain (CG) is usually defined as the ratio of the IF output power to the available RF source power. In order to be compatible with the ADS output format, in these equations, the voltages are amplitudes, not RMS. If the source and load impedances are different, the power gain must account for this, as shown in Equation (11.14). Voltage gains are also useful, especially in RFIC implementations of mixers,

$$\frac{output@F_{IF}}{P_{AVS}@F_{RF}} = \frac{V_{IF}^2/2R_L}{V_{RF}^2/8R_S} \tag{11.14}$$

$$\begin{aligned} A_V &= \frac{2}{\pi}\cos(\omega_{RF}t)\sin(\omega_{LO}t) \\ &= \frac{1}{\pi}\{\sin[(\omega_{RF}-\omega_{LO})t]+\sin[(\omega_{RF}+\omega_{LO})t]\} \\ CG &= \frac{1}{\pi^2} = 0.1 \quad \text{or} \ -10dB, \end{aligned} \tag{11.15}$$

if $R_S = R_L$. Otherwise,

$$CG = 10log(A_V R_S/R_L). \tag{11.16}$$

We see that the simple switching mixer has low conversion gain because the voltage gain A_V is only $1/\pi$. Also, in the RF feedthrough problem, and in most instances, an LO feedthrough problem exist. All of these deficiencies can be improved by the use of balanced topologies that provide some cancellation of RF and LO signals as well as increased conversion gain.

Referring back to Equation (11.13), we observe that in the double-balanced mixer, the differential connection not only cancels the LO feedthrough, but also doubles the IF signal. If,

$$I_{RF} = g_m V_{RF}/2 = g_m V_{IN}, \tag{11.17}$$

then,

$$\frac{V_{IF}}{V_{IN}} = \frac{2}{\pi} g_m R_L \left[\sin(\omega_{LO}+\omega_{RF})t + \sin(\omega_{LO}-\omega_{RF})t\right]. \tag{11.18}$$

Therefore, the conversion transducer gain $CG = (\frac{2}{\pi})^2$ has increased by a factor of 4 from -10 dB to -4 dB (assuming $g_m R_L = 1$). We see then that the extra circuitry required for double balancing has multiple benefits. We will consider such implementations later in this chapter.

11.3 Gain compression

Mixer conversion gain also exhibits the gain compression seen in amplifiers that was described in Chapter 8. Figure 11.12 shows a simulation schematic using a generic mixer in ADS. The mixer can be found in the System-Amps & Mixers pull-down menu. It allows the user to specify isolation, conversion gain, Third-order intercept (TOI) and sidebands (down and/or up conversion). An RF power sweep is needed in order to reveal conversion gain compression behavior. This is best accomplished using Harmonic Balance simulation. Here we see that the LO source uses odd harmonics up to the seventh order to provide a square wave like local oscillator. In the mixer application, the LO port is always labeled as PORT3.

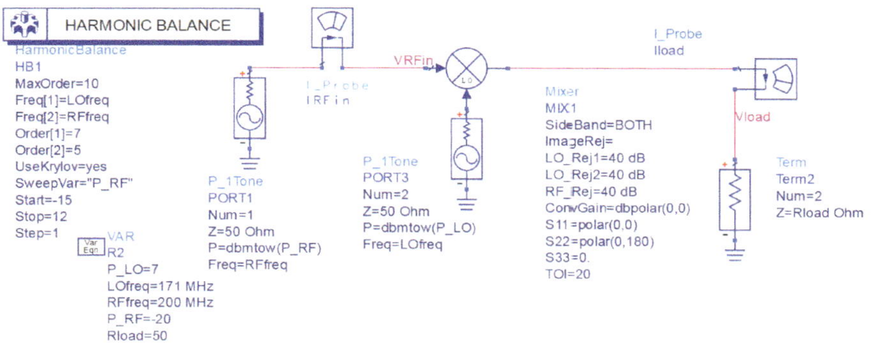

Figure 11.12 Schematic used to perform a power sweep of the RF input power.

Conversion gain compression is evident in the output in Figure 11.13.

If one simply wants to find the P1dB compression level for a mixer, the Mixer Design Guide provides templates and sample mixers. The harmonic balance XDB controller can be used to find this parameter. In the Design Guide under single-ended mixer analysis select Mix_SE_PNdB.

The XDB controller shown in Figure 11.14 is an easy way to find the P1dB. A template for this simulation (and others) can be found in the Mixer Design Guide. It produces two results: inpwr and outpwr. Inpwr is the input

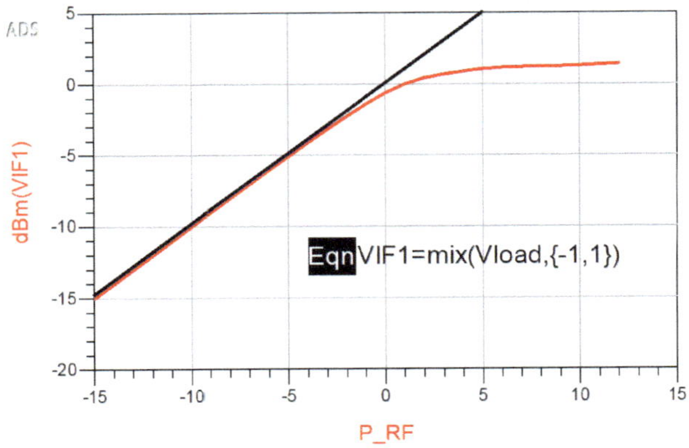

Figure 11.13 Input power sweep reveals significant gain compression above −2 dBm. We see the conversion gain = 1 up to that input power.

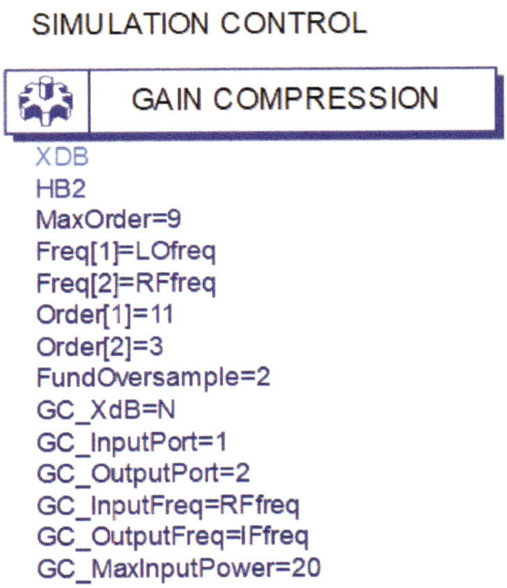

Figure 11.14 Gain compression controller for harmonic balance simulation.

power required for N dB of gain compression. Outpwr is the corresponding output power. Thus, the conversion gain at the N dB compression point is provided.

LOfreq, RFfreq, IFfreq, LO power, N dB, and Zload must be defined. Figure 11.15 shows the output from the gain compression simulation.

1.0 dB compression input power
and associated conversion gain

RF frequency	RF Input voltage	1.0 dB gain compression input power level (dBm)	Conversion gain
900.0 MHz	0.077	-15.135	1.646

LO voltage @ LOfreq	LO Power (dBm - 50 ohms)
0.316 / 0.000	0.000

LO frequency
1.100E9

Load Impedance R + jX
200.000

IF Output Voltage

freq	V_dwnconv	freq	V_upconv
200.0 MHz	0.134 / 49.383	2.000 GHz	0.069 / -170.7...

Mix_SE_PNdB

Figure 11.15 Simulation result from the XDB gain compression example.

In receiver applications, a mixer is often exposed to several signals within its preselected input bandwidth. It is important to understand that it is the peak signal voltage, not average signal power, that dictates when distortion becomes excessive in an amplifier or mixer.

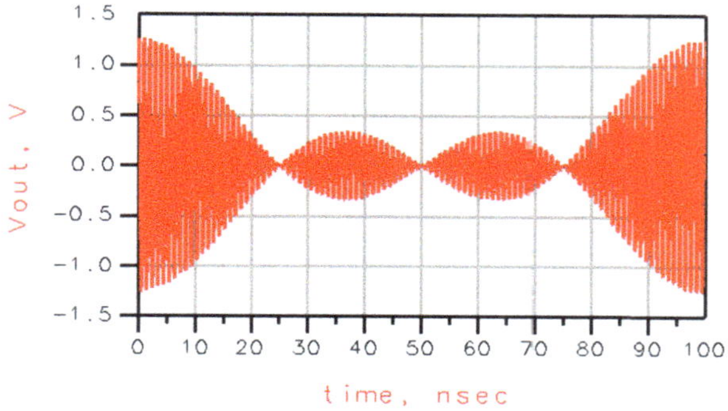

Figure 11.16 Four signals, each with 0 dBm average power applied to an input.

In Figure 11.16, four carriers, each with 0 dBm average power, are applied to the input. Each signal is separated by 10 MHz and all four are in phase at $t = 0$. As can be seen above, these signals will appear in phase periodically, with four times the peak signal voltage of a single carrier. While we are fond of expressing the large-signal performance in terms of an input power in dBm, let's remember that the time domain instantaneous signal peaks are what stress the system. Real signals are likely to be much more complex than this, so the probability of having a large peak like this is less likely in a real application. However, even infrequent overdrive and distortion generation can degrade bit error rates.

11.4 Mixer Performance: Intermodulation Distortion

IMD consists of the higher order signal products that are generated when *two* RF signals are present at the mixer input. Different from an amplifier IMD, the mixer IMD will be down and up converted by the LO as will the desired RF signal. The LO input is provided as before. These two signals can interact with the nonlinearities in the mixer signal path (RF to IF) to generate unwanted IMD products (distortion) which then get mixed down or up to IF. Absolute accuracy of the calculation is highly dependent on the accuracy of the device model, but the relative accuracy is valuable for optimizing the circuit parameters for best IMD performance.

The gain compression power characterization provides a good indication of the signal amplitude that the mixer will tolerate before really bad distortion is generated. You should stay well below the P_{1dB} input level. Intercept calculation and simulation at the input and output (IIP_3 and OIP_3) were described in Chapter 8. Harmonic balance is the better choice for such simulations. Refer to the tutorial in Chapter 8. Beyond that, mixer simulation is a little more complicated due to the dependence on LO amplitude. Some additional ADS guidance on this topic is included in the Section 11.6.

We will be mainly concerned with the third-order IMD. This is especially troublesome since it can occur at frequencies within the IF bandwidth. For example, suppose we have two input frequencies at 899.990 and 900.010 MHz. Third order products at $2f_1 - f_2$ and $2f_2 - f_1$ will be generated at 899.980 and 900.020 MHz. Once multiplied with the LO frequency, these IMD products may fall within the filter bandwidth of the IF filter and thus cause interference to a desired signal. IMD power, just as HD power, will ideally have a slope of 3 on a dB plot. Recall the third-order IM equations from Chapter 8.

$$
\begin{aligned}
V_{out3} = {} & a_3[V_1^3 \sin^3(\omega_1 t) + V_2^3 \sin^3(\omega_2 t) + \\
& 3V_1^2 V_2 \sin^2(\omega_1 t)\sin(\omega_2 t) + 3V_1 V_2^2 \sin(\omega_1 t)\sin^2(\omega_2 t)]
\end{aligned}
$$

$$
\frac{3V_1^2 V_2 a_3}{2}\left\{\sin(\omega_2 t) - \tfrac{1}{2}\left[\sin(2\omega_1 - \omega_2)t - \sin(2\omega_1 + \omega_2)t\right]\right\} \tag{11.19}
$$

Cross-modulation   Third-order IMD

Third-order intermodulation distortion is illustrated by the equations in (11.19).

In addition, the cross-modulation effect can also be seen. The amplitude of one signal (say ω_1) influences the amplitude of the desired signal at through the coefficient $3{V_1}^2 V_2 a_3/2$. A slowly varying modulation envelope on V_1 will cause the envelope of the desired signal output at ω_2 to vary as well since this fundamental term created by the cubic nonlinearity will add to the linear fundamental term. This cross-modulation can have annoying or error generating effects at the IF output.

Other higher odd-order IMD products, such as fifth and seventh, are also of interest, but may be less reliably predicted unless the device model is precise enough to give accurate nonlinearity in the transfer characteristics up to the $(2n - 1)$th order.

11.5 Mixer Performance: Isolation

In any practical mixer there will be feedthrough from port-to-port, as shown in Figure 11.17. This is in addition to whatever is produced by an imperfect switch waveform. Isolation can be quite important for certain mixer applications. For example, LO to RF leakage can be quite serious in direct conversion receiver architectures because it will remix with the RF and produce a DC offset. Large LO to IF leakage can degrade the performance of a mixer postamp if it is located prior to IF filtering.

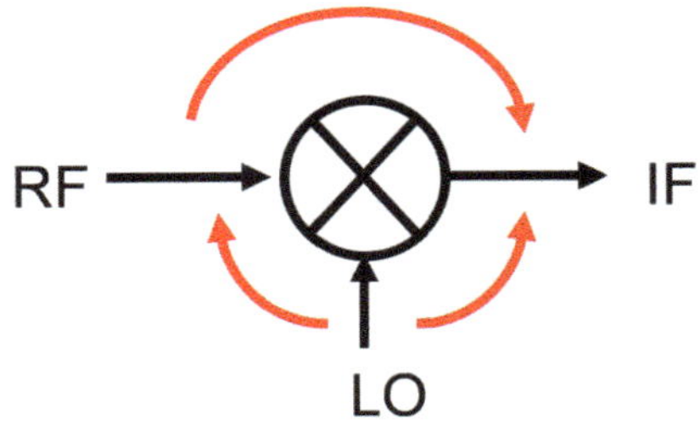

Figure 11.17 Leakage paths for RF – IF, LO – RF and LO – IF feedthrough.

$$\text{LO to IF isolation} = \text{LO power at IF/LO power at LO} \tag{11.20}$$

$$\text{RF to IF isolation} = \text{RF power at IF/RF power at RF} \tag{11.21}$$

$$\text{LO to RF isolation} = \text{LO power at RF/LO power at LO.} \tag{11.22}$$

A quick look at specifications for double-balanced diode mixers from Mini-Circuits [6] gives typical LO – RF isolation at 45–50 dB; LO – IF at 40–45 dB midband.

You can determine the magnitude of these leakage components at the IF and RF ports using harmonic balance. Use the mix function to select frequencies.

11.6 Harmonic Balance Simulation for Mixer Intermodulation

Harmonic balance is the method of choice for simulation of mixers. By specifying the number of harmonics to be considered for the LO and RF input frequencies and the maximum order (highest order of sums and differences) to be retained, you get the frequency domain result of the mixer at all relevant frequencies. To get this information using SPICE or other time domain simulators can often require a very long simulation time since at least two complete periods of the lowest frequency component must be generated in order to get accurate FFT results. This becomes a serious problem with two-tone input simulations. Concurrently, the time step must be compatible with the highest frequency component to be considered.

Maximum order corresponds to the highest order mixing product (n + m + k) to be considered $(nf[1] + mf[2] + kf[3])$. The simulation will run faster with lower order and fewer harmonics of the sources, but may be less accurate. You should test this by checking if the result changes significantly as you increase order or number of harmonics.

The frequency with the highest power level (the LO) or the most nonlinear is always the first frequency to be designated in the harmonic balance controller. Other inputs follow sequencing from highest to lowest power. More detail can be found in the help files on ADS, *harmonic balance simulation-basic setup* (Figure 11.18).

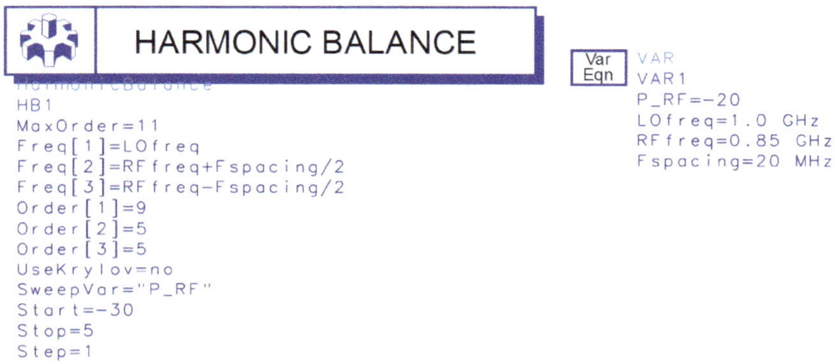

Figure 11.18 Harmonic balance controller setup for IMD simulation. The variables used must be defined in a Var Eqn icon as shown.

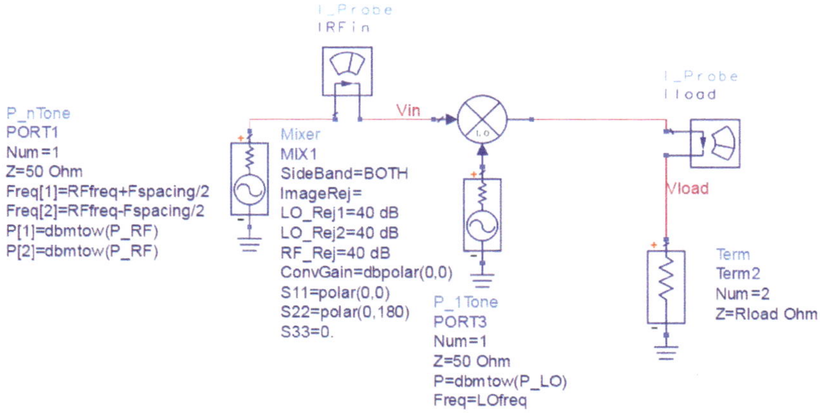

Figure 11.19 Schematic for IMD simulation using a generic mixer. Parameters can be set inside the mixer for gain compression and TOI.

In most cases, you will replace this generic mixer with your own design. Information on creating a subnetwork for your own circuit that can be inserted hierarchically into a simulation is described briefly in Appendix 11.1. For HB simulation, sources must be selected from the Sources – Freq Domain palette. Time domain sources do not work in HB. In order to determine the third-order intercept, a power sweep is necessary.

The IMD simulation (Figure 11.19) is performed with a two-tone generator (P_nTone) at the RF input. The frequency spacing should be small enough so that both fall within the IF bandwidth. You should keep in mind that both of the generator tones are in phase, therefore the peak voltage will add up periodically to twice the peak of each source independently. Because of this, you will expect to see some reduction in the P_{1dB} on the order of 6 dB. Also, choose frequencies that do not create overlapping output tones.

Often accurate IMD simulations will require a large *maximum order* and *LO harmonic order* when using harmonic balance. In this case, a larger number of spectral products will be summed to estimate the time domain waveform and therefore provide greater accuracy. This will increase the size of the data file and time required for the simulation. Increase the orders in steps of two and watch for changes in the IMD output power. When no further significant change is observed, then the order is large enough. Simulation of very low power levels is subject to convergence errors and numerical noise.

Selection of specific output frequencies uses the *mix* function. In the case of mixers, for IMD simulations, we have three input frequencies to consider,

LOfreq,
RFfreq + Fspacing/2, and
RFfreq – Fspacing/2.

The IF downconversion lower sideband frequency (LSB) then would be selected by

mix(Vout,{1,-1,0}), or 140 MHz in this example.

The upper sideband (USB) would be

mix(Vout, {1,0,-1}).

A third-order IM frequency would be given by

mix(Vout,{1,-2,1}) or mix(Vout,{1,1,-2}).

freq	Mix(1)	Mix(2)	Mix(3)
P_RF=-30.000			
0.0000 Hz	0	0	0
20.00 MHz	0	1	-1
40.00 MHz	0	2	-2
60.00 MHz	0	3	-3
100.0 MHz	1	-3	2
120.0 MHz	1	-2	1
140.0 MHz	1	-1	0
160.0 MHz	1	0	-1
180.0 MHz	1	1	-2
200.0 MHz	1	2	-3
260.0 MHz	2	-3	1

Figure 11.20 Frequency indices can be used to select any of the frequencies generated in the simulation.

The index assignments with frequency can be displayed in a table format using, in this example, Mix(1), Mix(2) and Mix(3). We see part of this table in Figure 11.20.

The Simulation-HB palette contains a function IP3out which when given Vout and the above indices, will calculate OIP3 (Figure 11.21).

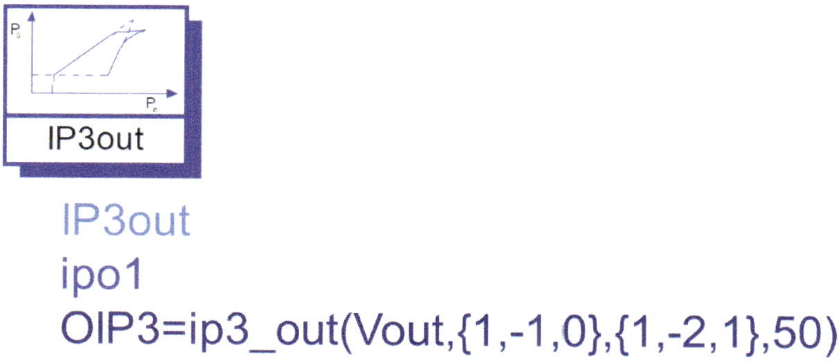

Figure 11.21 Output third-order intercept function.

11.7 Mixer Performance: Noise Figure

The *noise figure* is defined as the ratio between the input and output S/N ratio.

$$\text{NF (dB)} = 10 \log[(\text{S/N})\text{in}]/[(\text{S/N})\text{out}]. \tag{11.23}$$

Any real mixer or amplifier will degrade S/N because noise is added to the signal. The *minimum detectable signal* (MDS) power is determined by noise and corresponds to a signal whose strength just equals the noise. The thermal noise power in bandwidth is where k is Boltzmann's constant and T is absolute temperature. Thus,

$$\text{MDS (dBm)} = 10\log(kT\Delta f) + \text{NF}. \tag{11.24}$$

In addition, the system generated noise adds to the thermal noise ambient.

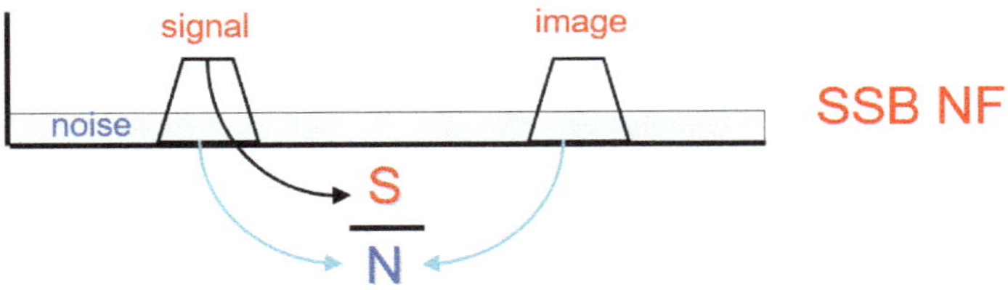

Figure 11.22 Single-sideband (SSB) noise figure.

There are two definitions used for noise figure with mixers – often a source of confusion. SSB NF assumes signal input from only one sideband, but noise inputs from both sidebands, as shown in Figure 11.22.

Measuring the SSB noise figure is relevant for heterodyne receiver architectures in which the image frequency is removed by filtering or cancellation. Noise figure is generally measured with a wideband noise source that is switched on and off. The NF is then calculated from the "Y factor" [3] and gain does not need to be known. With a SSB measurement, the mixer internal noise shows up at the IF output from both signal and image inputs, but the excess noise is only introduced in the signal frequency band.

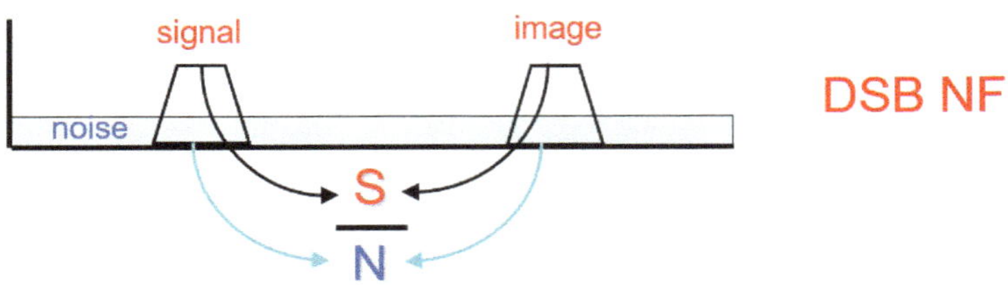

Figure 11.23 Double-sideband (DSB) noise figure.

The DSB NF shown in Figure 11.23 includes both signal and noise inputs from both sidebands. A DSB NF is easier to measure; wideband excess noise is introduced at both the signal and image frequencies. It will be 3 dB less than the SSB noise figure in most cases. This is perhaps more relevant for direct conversion receivers where the image cannot be filtered out from the signal.

Either type of measurement is valid so long as you clearly specify what type of measurement is being made. Once MDS and IIP3 are known, then the spurious free dynamic range can be calculated for the mixer just as it was for an amplifier.

11.8 Lab Exercise. Mixer Characterization

Goal: Characterize a double balanced mixer for large signal performance and isolation.

In this exercise, you could use a double-balanced diode mixer (ZAD-1) from Mini-Circuits[1]. Refer to their website for the data sheet [6]. See Figure 11.25 for a schematic. Here is a short summary:

Absolute maximum ratings for Mini-Circuits ZAD-1 mixer

LO Power:	+10 dBm
RF Input Power:	+13 dBm

Be very careful about synthesizer power adjustment so that these maximum levels are not exceeded. The mixer is expensive!

Frequency range:

RF/LO	0.5 to 500 MHz
IF	DC to 500 MHz

Mixer Evaluation. Apply a 250 MHz, +7 dBm sine wave to the local oscillator input (L) of the Mini-Circuits ZAD-1 mixer. Measure this signal on the spectrum analyzer or a power meter first to account for cable loss and verify its power level and spectral content. Then, connect the spectrum analyzer to the IF output of the mixer.

On all of these measurements: record the spectrum analyzer settings (RBW, Ref Level, Span) with the data.

a. Connect the second generator to the RF input of the mixer. Set the mixer input RF signal amplitude to –20 dBm. Select an RF input frequency that will produce a downconversion output at 50 MHz. Determine the conversion loss at both down- and upconversion output frequencies.

b. Change the RF input to the image frequency. Record frequencies and show that the conversion losses are essentially the same.

[1] Minicircuits.com

c. Measure the 1 dB gain compression point for the mixer.

d. Measure the conversion loss of the mixer as a function of LO input power starting from 7 dBm and reducing the LO power in 3 dBm increments. Can you explain why the loss increases when the LO drive is reduced?

e. Restore the LO power to +7 dBm. Measure the LO to IF and RF to IF isolation. Then, connect the spectrum analyzer to the RF port and measure LO to RF isolation.

f. Reconnect the spectrum analyzer to the IF port. With the RF input power close to the 1 dB compression point, set the reference level of the spectrum analyzer so that the largest output from the mixer is close to the top of the screen. Reduce the resolution bandwidth so that the noise level is less than −70 dBm. Record and identify the frequency and amplitude of all IF output frequencies up to the fifth order. Explain how each frequency was generated (for example, a fourth order product would be $3f_{LO} + f_{RF}$).

11.9 Simulation of Mixer Noise Figure

Harmonic balance simulation provides for determination of mixer NF. It takes into account any nonlinearities and harmonics that could mix noise into the IF band.

If P_RF << P_ LO, either a 1-tone generator or a passive termination can be used at the input with equal accuracy. The noise figure is calculated with an ideal filter (centered on F_RF) added in simulation. Thus, SSB noise figure is calculated. Noise contributors within the mixer are listed by value.

For passive switching mixers, NF $\cong$ −ConvGain

There are examples of mixer noise figure simulation on ADS in the mixer design guide (Figure 11.24).

Note in Figure 11.25 the many parameters in the HB controller that must be set in order to simulate noise figure. Also note that the options must set Temp = 16.85 for all noise simulations. Figure 11.26 presents the output, SSB NF at the designated IF frequency.

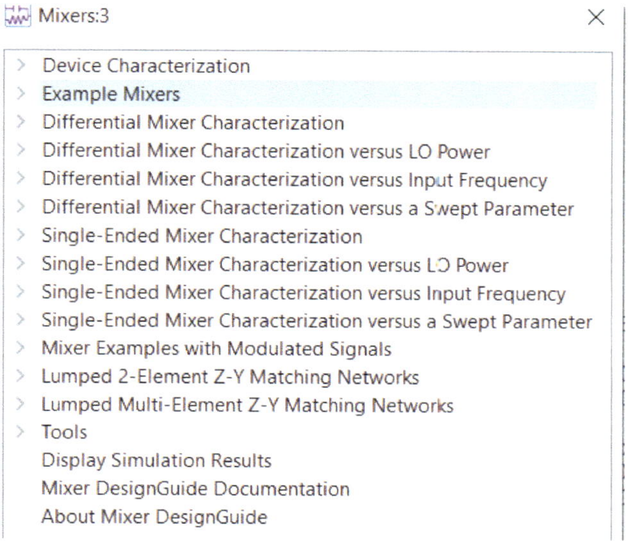

Figure 11.24 The mixer design guide provides simulation schematic and display templates and example mixer circuits.

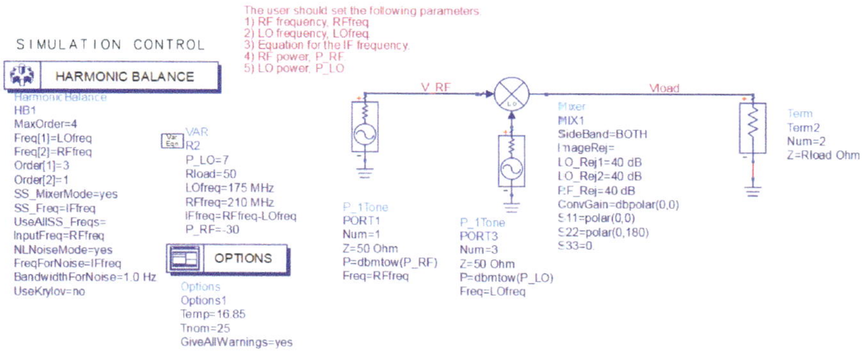

Figure 11.25 Setup for a noise simulation for a mixer.

Display for Mixer Noise Figure Simulation

noisefreq	NFssb
35.00 MHz	9.011

Figure 11.26 SSB noise figure was calculated for the generic mixer in Figure 11.25.

11.10 Mixer Circuit Examples

Now let's look at some circuit examples of how mixers are often implemented.

1. Diode DB quad:
   - familiar and widely used
   - wide bandwidth, limited by baluns.
2. FET DB quad:
   - not as well known, but good performance.
3. Gilbert multiplier
   - Very widely used active mixer.

11.10.1 Diode DB quad

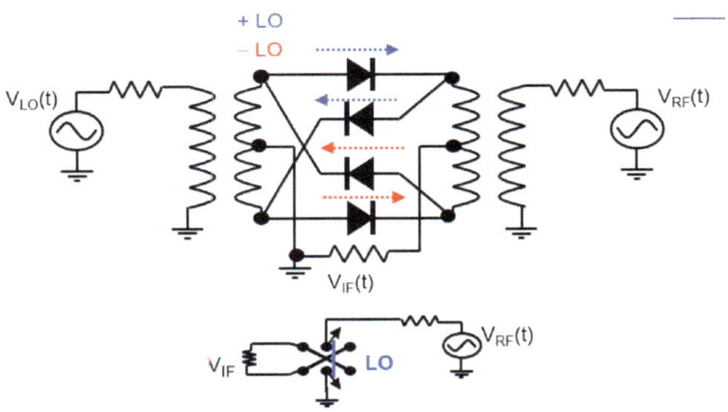

Figure 11.27 Diode double-balanced quad mixer.

The diode double-balanced quad mixer shown in Figure 11.27 is a very popular design and available in a wide variety of frequency bandwidths and distortion specs [1, 7]. The Mini-Circuits™ catalog [6] is full of these. The diodes act as a polarity reversing switch as seen in the bottom of the figure. When the top of the LO transformer is positive, the blue path is conductive and will ground the top of the RF transformer. When the top is negative, the red path is conducting and the RF polarity reverses. Both LO and RF feedthrough are suppressed by the symmetry and balancing provided by the transformers. The LO signal at the RF and IF ports appear to be a virtual ground for either LO polarity.

Since the LO signal must switch the diodes on and off, a large LO power is required, typically 7 dBm when one diode is placed in each leg, 17 dBm with two diodes per path! With this much LO power, even with good isolation, there may still be significant LO in the IF output.

When the diodes are conducting with LO current >> RF current, the mixer should behave linearly. At large RF signal powers, the RF voltage modulates the diode conduction, so lots of distortion will result in this situation. The diodes are also sensitive to RF modulation when they are biased close to their threshold current/voltage. For both reasons, we prefer high LO drive with a fast transition (high slew rate – a square wave LO is better than sine wave) between on and off. The IMD performance is very poor with small LO power.

- The RF and LO impedances at the diode ring theoretically should be determined and matched at all of the relevant harmonics [7]. For most designs, optimizing the transformer ratios with the IF port connected to 50 ohms should be sufficient, since we cannot select impedances at each frequency independently, and this approach would not be possible for a broadband design such as this.
- Transformer ratios can be swept using HB or XDB. A diode DB mixer example can be seen in the design guide and used for simulation of mixer performance.
- Sweep the LO power to maximize conversion gain and gain compression
- Or, buy one from Mini-Circuits^{TM}!

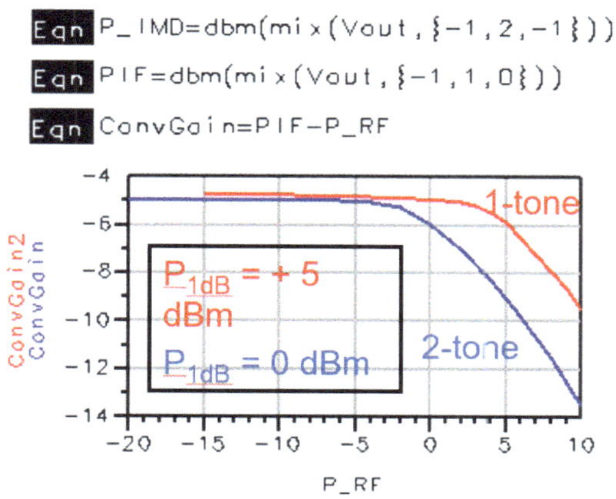

Figure 11.28 ADS simulation of conversion gain for a diode double-balanced mixer.

A harmonic balance simulation can be used to estimate the mixer performance. With the double-balanced diode mixer shown in Figure 11.27, the currents in the diodes are half wave, so a high number of LO harmonics and maximum order and some oversampling of the FFT operation are necessary to reproduce this waveform and therefore get reasonable accuracy on the third-order product. Gain compression is not as sensitive to the LO order.

Gain compression behavior depends strongly on the signal statistics as discussed earlier (see Figure 11.16). There is about 5 dBm difference in P_{1dB} between single and two-tone simulations.

When plotting IMD power vs. RF power, we see that the third order IMD predictions are not "well behaved"; the slope is 33.2 dB/decade instead of 30 in this example. This puts our TOI calculation in doubt, but it is still useful for design optimizations.

The noise figure of these passive switched mixers is usually close to

$$\mathrm{NF} = -\mathrm{ConvGain}$$

Thus, for a −5 dB conversion gain, a noise figure of about 5 dB is expected.

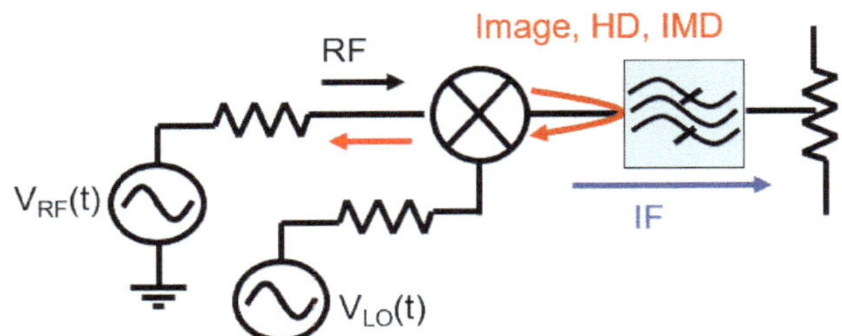

IF term	P1dB	TOI
50Ω	0 dBm	15 dBm
BPF	−2	−2

Figure 11.29 The passive DB diode mixer is bilateral, hence the match at each port is critical.

Since passive diode switch mixers are bilateral (Figure 11.27), that is, the IF and RF ports can be reversed, the performance of the mixer is very sensitive to the termination impedances at all ports. A wideband resistive termination is needed to absorb not only the desired IF output but also any images, harmonics, and IMD signals. If these signals are reflected back into the mixer, they will remix and show up at the RF port and again at the IF port.

The phase shifts associated with the multiple replicas of the same signals can seriously deteriorate the IMD performance of the mixer.

A simulation was carried out using a bandpass filter in the IF port as shown above. The P_{1dB} was degraded by 2 dB and the third-order IMD power was not well behaved. A calculated TOI showed nearly 17 dB degradation.

Thus, it is important to terminate. Terminations can consist of:

1. Attenuator. Obviously not a good idea if the NF is important.
2. Wideband amplifier with good S_{11} or S_{22} return loss.
3. Diplexer. A passive network that separates frequencies but provides Z_0 termination for all components.

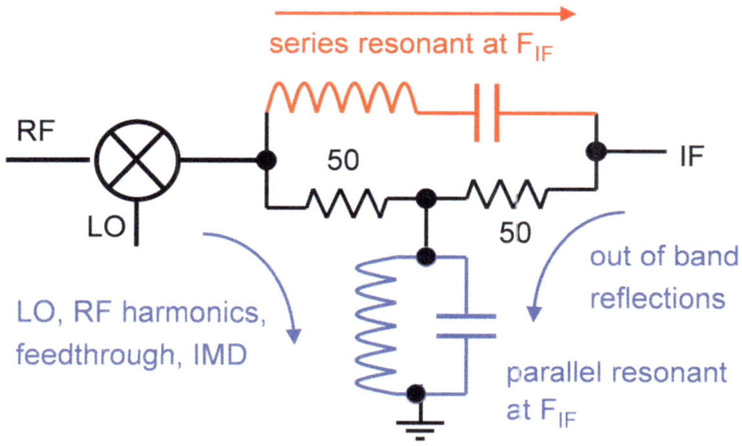

Figure 11.30 Passive diplexer.

The passive diplexer shown in Figure 11.30 provides a low-loss forward path through the series resonant branch. At F_{IF}, the parallel resonant branch has a high impedance and does not load the IF. Outside of the IF band (you need to set the Q for the design to control the bandwidths) the series resonant branch presents a high impedance to the signals and the parallel resonant branch a low, but reactive impedance. At these frequencies, above (through C) and below F_{IF} (through L), the resistors terminate the output. The farther away from F_{IF} you are, the better the match.

The series LC branch (C1 and L1) reactances can be determined by

$$X_{C1} = Z_0/Q \text{ and } X_{L1} = Z_0 Q \tag{11.25}$$

The parallel LC branch (C2 and L2) reactances:

$$X_{C2} = Z_0/Q = X_{L2} \tag{11.26}$$

and $R = Z_0$.

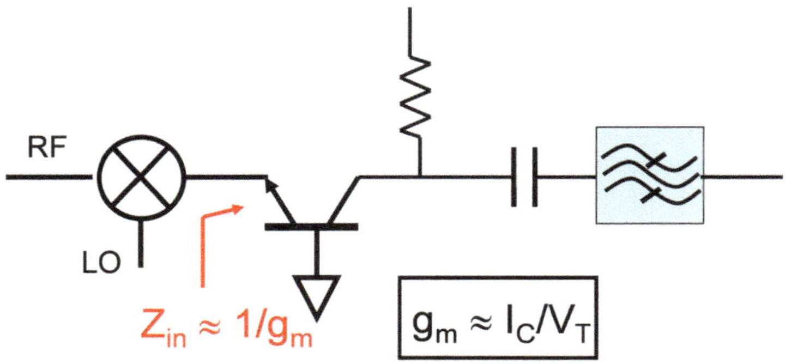

Figure 11.31 Active wideband termination.

The common base stage shown in Figure 11.31 provides a wideband resistive impedance provided the maximum frequency at the input is well below the f_T of the transistor. The bias current can be set to provide a 50 ohm input impedance. Alternatively, one can bias the device at higher current levels and add a series resistor at the input. Of course, this degrades noise, but will improve IMD performance. The amplifier must be capable of handling the complete output power spectral density of the mixer without distorting.

11.10.2 Mixer Circuit Examples: Double-balanced Switching FET Mixer

Figure 11.32 presents a schematic of a double-balanced FET mixer. LO voltage must be large enough to drive the FETs into their triode region. The channel resistance of a large FET when in its triode region (below saturation) can be quite low as shown in Figure 11.33 and is not as current dependent as the diode. Therefore, switching configurations using FETs can be more linear in the RF to IF path than diode switching mixers:

1. The conversion loss will be similar to the diode mixer.
2. Large LO drive voltage is needed (1 to 5 volts)

With the FET ring mixer, devices alternate polarity between the RF input and IF output. If the devices were ideal switches, then the input and output would be directly connected. So, transformers with identical turns ratios

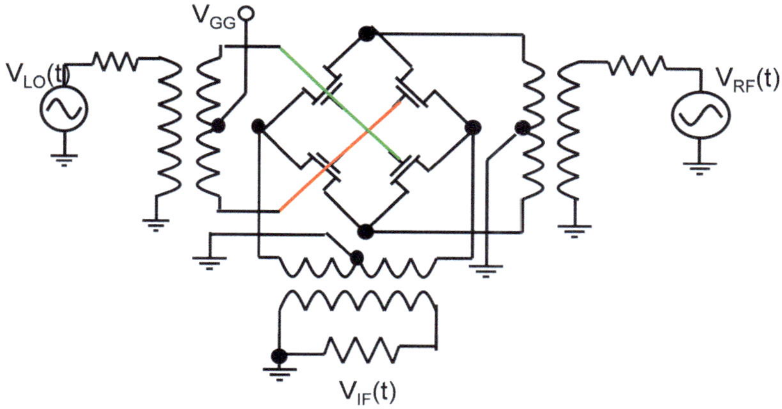

Figure 11.32 Double-balanced switching FET mixer.

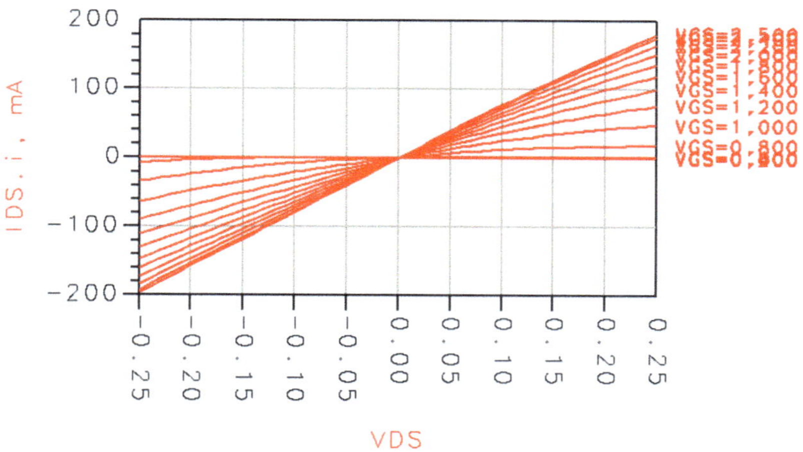

Figure 11.33 ID vs. VDS characteristic of a MOS transistor.

should be used. The impedance level at the FET ring should be much higher than the series resistance of the FETs in order to reduce conversion losses. This may also help with linearity. If the impedance is too high, however, LO feedthrough may be higher and frequency response more limited. So, some optimization is needed.

The mixer RF to IF path will be quite linear if the total drain voltage (VDS) remains small. As can be seen from the DC simulation in Figure 11.33, the MOSFET exhibits quite linear channel resistance up to at least a VDS of + and −0.25V.

11.11 Passive vs. Active Mixers?

Passive nonlinear devices or switches:

- Conversion loss, not gain
- High tolerance to IMD
- External baluns or transformers needed.

Active mixers:

- Can provide conversion gain
- Active baluns - better for IC implementation
- More difficulty in achieving good IMD performance.

Passive mixers are widely used because of their relative simplicity, wide bandwidth, and good IMD performance. The transformers or baluns generally limit the bandwidth. They must introduce some loss into the signal path, however, which can be of some concern for the noise figure. In this case, an LNA can be introduced ahead of the mixer, usually with some degradation in IMD performance.

Active mixers are preferred for RFIC implementation. They can be configured to provide conversion gain, and can use differential amplifiers for active baluns. Because of the need for additional amplifier stages in the RF and IF paths with fully integrated versions, it is often difficult to obtain really high third-order intercepts and 1 dB compression with active mixers.

The mixer designs shown previously are passive. The devices are acting as switches and are not active – therefore we have conversion loss, not gain. Also, if the balancing involves transformers, then integrating onto an IC is not usually possible. So, other implementations that provide gain and are more amenable to integration are frequently used in IC front end chips.

The design objectives are generally the same however:

1. Maximize linearity in signal path
2. Idealize switching in LO path
3. Minimize the noise contribution due to thermal and shot noise.

11.11.1 Mixer examples: differential pair is the basis for an active mixer

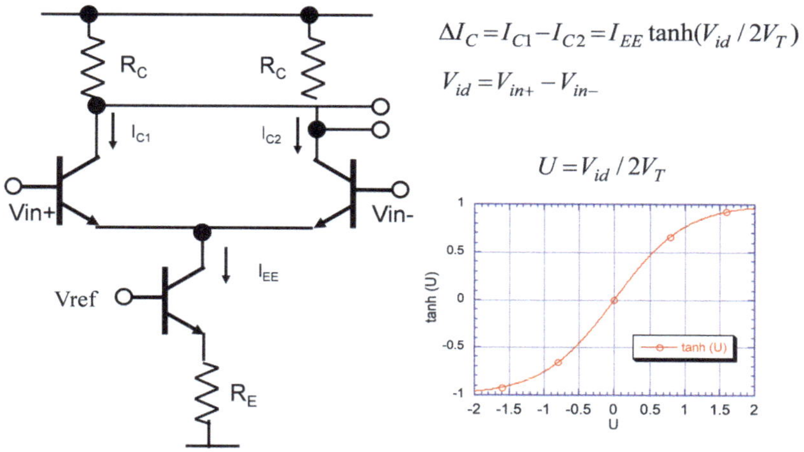

Figure 11.34 Differential pair of BJTs. Transfer characteristic follows a hyperbolic tangent function (Equation (11.28)).

The emitter–coupled pair seen in Figure 11.34 can be used to provide multiplication if the input range limitations are observed [2, 5]. Here we have a differential transconductance stage with voltage V_{id} as input and $\Delta\ I_C$ as output. If $V_{id} << 2V_T$, then

$$\Delta I_C \gg I_{EE}\left(V_{id}/2V_T\right) = g_m V_{id}. \tag{11.27}$$

Since the transconductance $I_{EE}/2V_T$ can be varied with I_{EE}, a second input shown in Figure 11.35 can be added which will produce a two-quadrant multiplication,

$$\tanh(u) = u - \frac{1}{3}u^3 + \frac{2}{15}u^5 - \cdots \tag{11.28}$$

However, $V_{i2} > V_{BE}$ and $V_{id} = V_{in+} - V_{in-}$ must be small, $<< V_T$. Rather limited use.

In the terminology of mixers, this is a single balanced mixer. If we apply V_{LO} to V_{id} and keep V_{i2} constant, our output is full amplitude at the LO frequency. NOT DESIRABLE – will desensitize the IF amplifier. So, we really should go one step further and design a doubly balanced mixer to suppress the LO feedthrough.

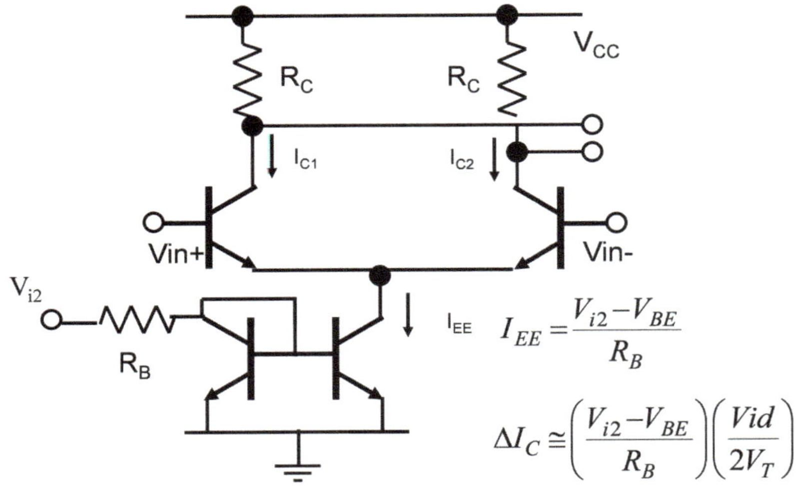

Figure 11.35 Single-balanced active differential mixer.

11.29

The double-balanced BJT active mixer shown in Figure 11.36 was first described in the 1960s. Barrie Gilbert was credited with the mixer design [2]. Of course, implementation with MOSFETs is also used today.

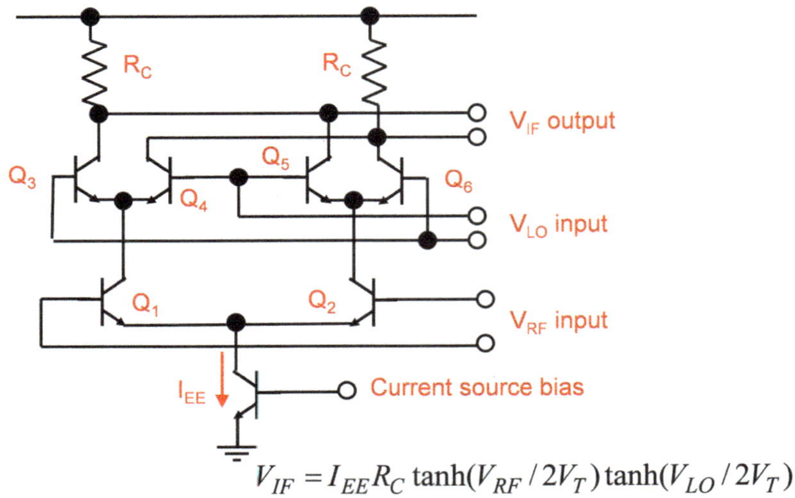

Figure 11.36 Double-balanced BJT (Gilbert) mixer [2].

11.30

We can make several observations on this remarkably useful circuit:

1. Q3–Q6 provide a fully balanced, phase reversing switch.
   When V_{RF} = 0, I_{OUT} also is 0 regardless of the status of V_{LO}. This is because I_{C3-5} and I_{C4-6} each will add up to $I_{EE}/2$ in this condition.
2. When $V_{LO} \neq 0$, the same condition applies. Therefore, there is never any LO component in the output differential current. The upper tier of BJTs only provides the phase reversal of the RF signal as controlled by the LO voltage.
3. The mixer can provide conversion gain – depending on the load impedance presented to the collectors and I_{EE}.
4. The signal handling capability is still limited below V_T, but we can use emitter degeneration to improve linearity.
5. The distortion is entirely odd-order for perfectly matched transistors.
   The IF output voltage is the product of the RF differential pair current passed through to R_C and the LO switching function (second term),

$$V_{IF} = I_{EE} R_C \tanh\left(V_{RF}/2V_T\right) \left[\frac{4}{\pi} \cos\left(\omega_{LO} t\right)\right] \tag{11.31}$$

Linearity of RF → IF path: We have seen that the RF input differential pair converts the input voltage into a differential current. The RF → IF path is supposed to be very linear. We can analyze the transfer function for linearity and distortion:

$$\Delta I = I_{EE} \tanh\left(V_{LO}/2V_{T}\right) \tanh\left(V_{RF}/2V_{T}\right). \tag{11.32}$$

The LO input is intentionally overdriven so that it is just acting as a polarity switch. So, we only need to investigate the behavior of the RF port.

Let's normalize: $U = V_{RF}/2V_T$. Then, the gain is the derivative of tanh$U$:

$$gain = I_{EE}\,\mathrm{sech}^2\,U \tag{11.33}$$

To evaluate the linearity, we must remember that it is the instantaneous peak voltage at the input that drives the stage into nonlinearity, not the average voltage. From the graph in Figure 11.37, we can see that the tanh function is not very linear.

The gain drops 10% for inputs as small as $U = \pm 0.33$. This corresponds to an amplitude of only 17 mV and is approximately the 1 dB compression input voltage. This is –19 dBm in a 50 ohm system, not a very good P1dB.

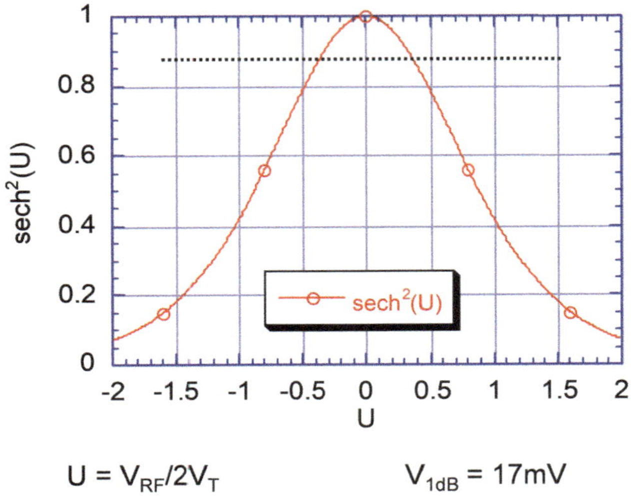

Figure 11.37 Gain of the DB Gilbert mixer as a function of input RF voltage.

Please note that V_{1dB} does *not* imply any particular input or available power without specifying a reference impedance Z_0! Unless Z_0 is specified, do not assume that P_{1db} (dBm) and V_{1dB} are related. V_{1dB} is simply the voltage at the input of the mixer that compresses the conversion gain by 1 dB (a ratio).

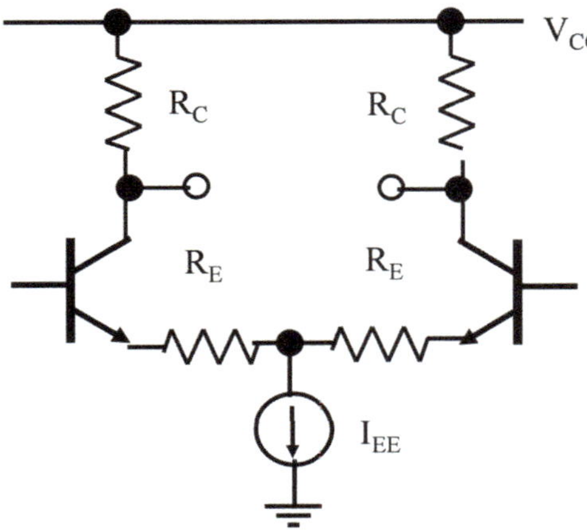

$$g_{mo} = \frac{I_{EE}}{2V_T}$$

$$g_m = \frac{g_{mo}}{(1+\sigma)}$$

where

$$\sigma = g_{mo} R_E \tag{11.34}$$

Figure 11.38 The input range can be extended by emitter or source degeneration.

Emitter degeneration, as seen in Figure 11.38, can be used to extend the input range and linearize the incremental gain of the amplifier.

The emitter-base $r_e = 2V_T/I_{EE}$ leads to $g_{mo} = 1/r_e$. The factor of 2 comes from having the emitter current source divided between Q1 and Q2. At small signal, each DC emitter current is $I_{EE}/2$.

When an external resistor R_E is added, the gain is reduced because of the local negative feedback. Emitter currents flow in these resistors and subtract from the input voltage.

If we define a dimensionless parameter σ to indicate the amount of degeneration,

$$\sigma = g_{mo}R_E \tag{11.35}$$

Then,

$$g_m = g_{mo}/(1+\sigma) \tag{11.36}$$

This will lower the conversion gain, but raise the P_{1dB} and the third-order intercept.

The Table 11.1 shows that the incremental gain is reduced but made more linear. Here Z_0 = 50 ohms.

Table 11.1 Incremental gain reduction and P_{1dB} extension with emitter degeneration.

$\sigma = I_{EE}/2V_T$	V1dB (mV)	P1dB (dBm)	PTOI (dBm)	Gain reduction (dB)
0	36	-18.7	-5	0
0.5	64	-13.8	+0.4	-3.5
2.0	158	-6.0	+9.4	-9.6
5.0	361	+1.1	+18.3	-15.6

Next, calculate the conversion gain. Refer back to Figure 11.38.

The input gain is determined by the lower differential pair and the emitter degeneration R_E.

$$V_{IF} = V_o^+ - V_o^- = \left(\frac{g_{mo}R_C}{1+\sigma}\right)\left(V_{in}^+ - V_{in}^-\right)\cos\left(\omega_{RF}t\right) \tag{11.37}$$

Next add the commutating switch function $T(t)$ (Figure 11.39).

$$T(t) \cong \frac{4}{\pi}\sin\left(\omega_{LO}t\right) \tag{11.38}$$

If we neglect higher order terms in the switch function, it is easy to estimate the voltage gain. Simply multiply the switch function $T(t)$ and the

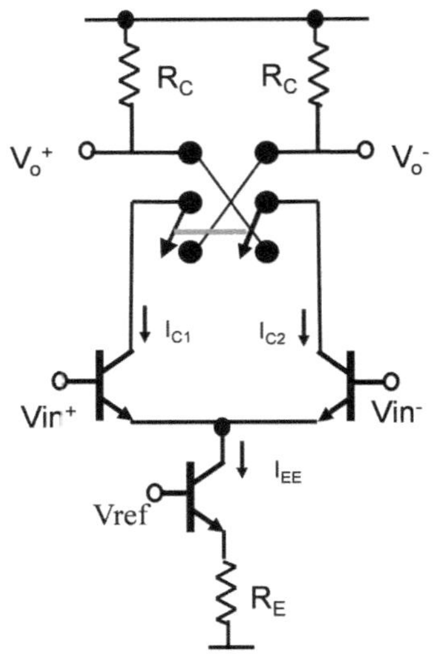

Figure 11.39 Adding the commutating switch function $T(t)$.

small signal gain of the diff pair using the differential input and output voltages.

$$V_{IN} = V_{IN}^{+} - V_{IN}^{-} \tag{11.39}$$

$$V_{IF} = \left(\frac{g_{m0}R_C}{1+\sigma}\right)\left(\frac{4}{\pi}\right)V_{IN}\left[\sin\left(\omega_{LO}t\right)\cos\left(\omega_{RF}t\right)\right] \tag{11.40}$$

$$V_{IF} = \left(\frac{g_{mo}R_C}{1+\sigma}\right)\left(\frac{2}{\pi}\right)V_{IN}\left[\sin\left(\omega_{RF}-\omega_{LO}\right)t + \sin\left(\omega_{RF}+\omega_{LO}\right)t\right] \tag{11.41}$$

Now we can calculate the conversion gain (transducer gain) which is the power delivered to the load over the available power from the source,

$$CG = \frac{V_{IF}^2/2R_C}{V_{IN}^2/2R_S} \tag{11.42}$$

And, since we want the available source power, $V_{IN} = \frac{V_S}{2}$.

11.11.2 Noise analysis of the differential active mixer [2, 5]

We have explored one dimension: distortion. But noise will determine the full dynamic range of the mixer. Recall that spurious free dynamic range

(Chapter 8) was defined as

$$SFDR = (2/3)(IIP3 - MDS) \tag{11.43}$$

For our purposes, let's define dynamic range = $P_{1dB} - P_{MDS}$.

Then, a noise analysis of the BJT mixer will be needed. Assume the noise of the BJT is dominated by shot noise. We will consider the effect of emitter degeneration later.

1. Assume that the conversion gain of the mixer is high enough to make the second stage contribution of any collector load devices small, and
2. If we neglect the effect of the base resistance of the BJTs, we can make a simple estimate of the noise behavior of this mixer.

Upon neglecting these sources, we are left only with the shot noise associated with the base-emitter forward biased pn junction. Shot noise is caused by random flow of carriers across a junction. Each carrier is emitted individually and so the current arriving across the junction consists of tiny pulses randomly distributed in time. This produces a noise current whose mean square variance is given by

$$i_N^2 = 2qI_C B. \tag{11.44}$$

Here, q is the electron charge and B is the noise equivalent bandwidth. Typically, we normalize to a 1 Hz bandwidth centered on some "spot" frequency. Then this is a noise spectral density.

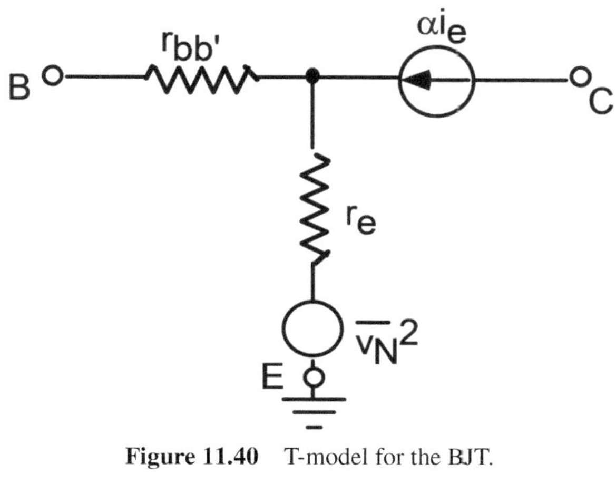

Figure 11.40 T-model for the BJT.

This noise current flows through the noiseless resistor r_e in Figure 11.40 to produce a noise voltage at the BJT input V_{BE}.

$$v_N = i_N r_e = V_T \left(2q/I_C\right)^{1/2}. \tag{11.45}$$

Applying this to the diff pair, we have two such noise sources in series in the input loop. Random noise sources add in an RMS sum, so the two in series produce $\sqrt{2}$ times more noise. Also, each transistor has a collector current I_C = $I_{EE}/2$. Thus, the input noise = $2V_T(2q/I_{EE})^{1/2}$ or

$$0.925\text{nV}/\sqrt{\text{Hz}}/\sqrt{I_{EE}} \quad (I \text{ is in mA})$$

roughly 1 nV/$\sqrt{}$Hz/$\sqrt{}$mA. Here we see that the *noise can be reduced when the emitter current is increased.*

There is a point of diminishing returns, however, when the base resistor r_{bb} dominates the noise. Also, there will be noisy base current flowing through the input source resistor that increases at higher currents. So, the above estimate from shot noise only gives the minimum for the noise of the diff pair.

If we take the base current into account, and have a finite generator resistance, R_S, there will be another series noise voltage due to $I_B(R_S + r_{bb})$. The total noise gets worse as the frequency approaches f_T because

$$\beta(f) = \beta_{DC} f_T / f \tag{11.46}$$

Thus, higher f_T will help reduce noise (unless it was obtained at the expense of r_{bb}).

$$\bar{v}^2_{n,total} = \left(\frac{0.925\ nV/\sqrt{Hz}}{\sqrt{I_{\text{EE}}}}\right)^2 + \left[\frac{(2r_{bb} + R_S)\sqrt{qI_{EE}}}{\beta(f)}\right]^2 + 4kT\left(r_{bb} + R_S\right) \tag{11.47}$$

We saw that the emitter degeneration made a big difference in the upper end of the dynamic range. But, it will also increase the noise at the input, raising the noise floor.

We can estimate how much the noise will be increased by the emitter resistors. Adding two resistors R_E in the base-emitter loop (Figure 11.41) will introduce a thermal (Johnson) noise spectral density of

$$\sqrt{2}\left(4kTR_E\right)^{1/2}. \tag{11.48}$$

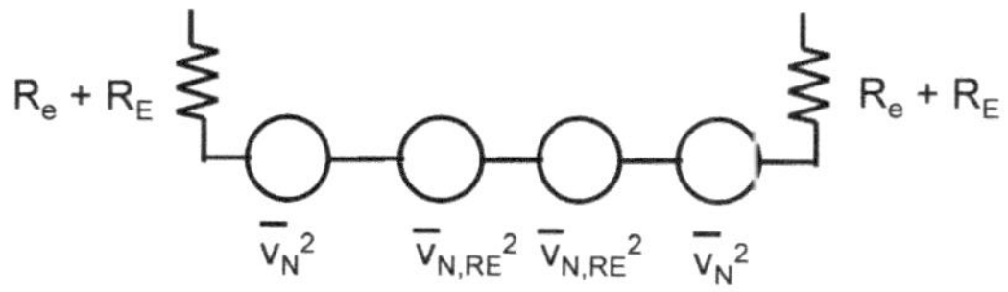

Figure 11.41 Additive RMS noise in the base-emitter loop caused by R_E.

This is equal to

$$v_{N,R_E} = (0.129nV/\sqrt{Hz})\sqrt{2R_E}. \tag{11.49}$$

Replacing R_E with σ, and adding the shot noise, we get the total noise approximately equal to

$$v_N = \sqrt{(1+2\sigma)}(0.925nV/\sqrt{HZ})/I_{EE}. \tag{11.50}$$

The dynamic range depends on the ratio

$$\frac{V_{1dB}}{V_{\text{MDS}}} \tag{11.51}$$

At the upper end of the dynamic range, the data for V_{1dB} in Table 11.1 can be estimated to increase by a factor of $(1 + 1.7\sigma)$. So, we can estimate that the dynamic range (DR) of the mixer will be increased by emitter degeneration according to:

$$DR \propto 20\log_{10}[(1+1.7\sigma)/\sqrt{(1+2\sigma)}]. \tag{11.52}$$

For example, if we choose σ = 10 and set I_{EE} = 1 mA, then the input referred noise will be increased from 0.925 to 4.25 nV/$\sqrt{}$Hz, a factor of 4.6. However, the peak input voltage will be increased by a factor of 18. So, there would be nearly a 12 dB improvement in the P_{1dB} dynamic range.

But is there another way to improve dynamic range without adding noise due to R_E?

$$Z_{LN} = j\omega L + \frac{1}{j\omega C_{gS}} + \frac{g_m}{C_{gS}}L \tag{11.53}$$

The source inductor in Figure 11.42 has the same effect on gain as R_E, but without adding more noise. It also has the benefit of adding a positive real part to the input impedance helping to avoid right-hand s-plane instabilities.

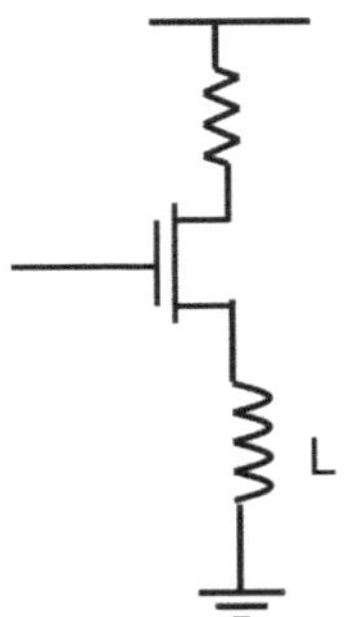

Figure 11.42 Source degeneration due to an inductor rather than a resistor.

11.12 Summary

In this chapter, a detailed description of the design and operation of a mixer was described. The topics below were addressed.

1. Understand operating principles of the mixer
2. What makes a good mixer?
3. Choices: nonlinear/switching mode; single/double balance; active/passive.
4. Specify performance: Gain, NF, P_{1dB}, TOI, SFDR, isolation.
5. Mixer examples – a few common ones were presented but there are numerous other possibilities
6. ADS was shown to be suitable for mixer analysis using harmonic balance. Refer to the mixer design guide for multiple applications.

References

[1] McWhorter, M., Scherer, D., and Swain, H. (1995) *EE344 High Frequency Laboratory* Stanford University, Chapter 6.

[2] Gilbert, B. (1995) Design considerations for BJT active mixers. *Analog Devices.*

[3] Hayward, W. (1994) *Introduction to Radio Frequency Design*American Radio Relay League, Chapter 6.

[4] Lee, T. H. (2004) *The Design of CMOS Radio Frequency Integrated Circuits, 2nd Edn.* Cambridge University Press, Chapter 13.

[5] Gray, P., Hurst, P., Lewis, S., and Meyer, R. (2009) *Analysis and Design of Analog Integrated Circuits, 5th Edn.* J. Wiley, Chapter 11.

[6] Minicircuits RF/IF Designers Handbook, www.minicircuits.com

[7] Maas, S. (1993) The diode ring mixer. *RF Design* 54–62.

Homework

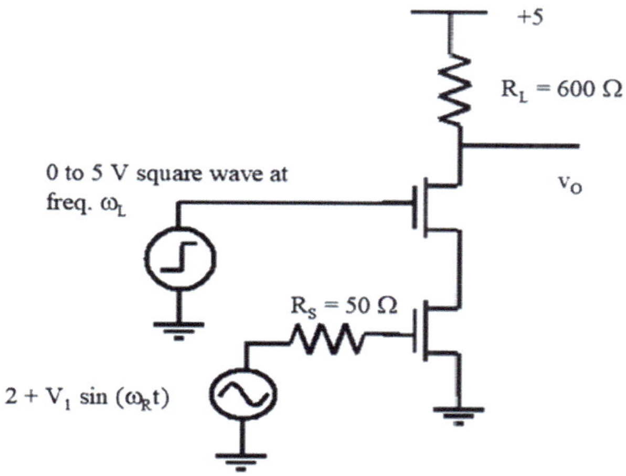

1a. The circuit above operates as a mixer. Lower MOSFET M1 is biased in saturation (active region) when the LO switch is on. You can assume that the RF input amplitude V_1 is small and that the LO input fully switches the upper MOSFET. Consider products up to the *second order*. Derive the conversion gain and DC, LO or RF feedthrough components, if any.

b. Using the device characteristics shown, and assuming $V_1 = 0.1$ V, calculate the IF output power.

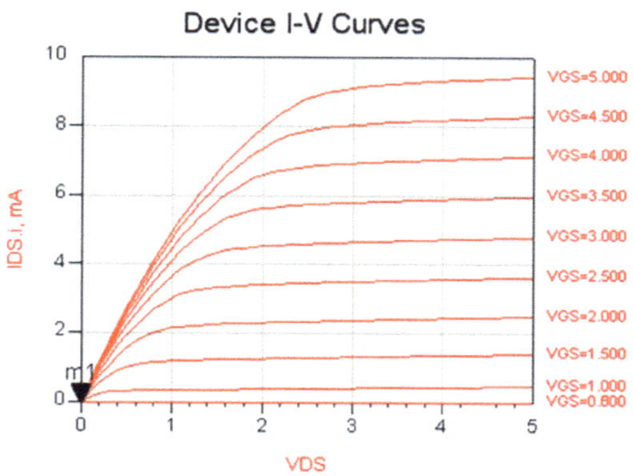

2. The circuit below is a mixer. You can assume that the RF input is relatively small and that V_{LO} is a square wave large enough to make the

transistor operate as a switch between cutoff and the ohmic region. V_T is set to the MOSFET threshold voltage so that the duty cycle of the LO is 50%.

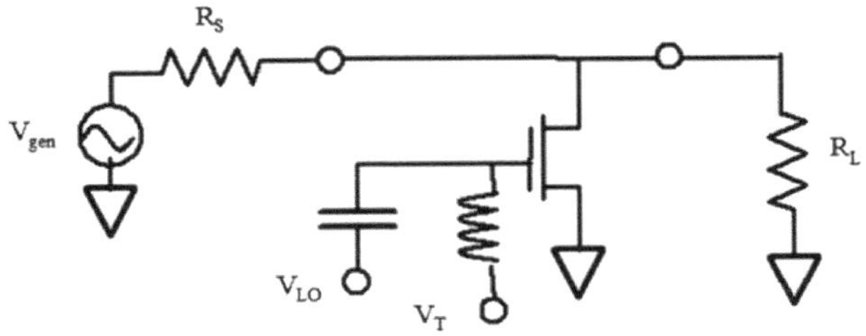

Consider all output frequency components up to the fourth order. Calculate the output power in dBm for each mixer output if the MOSFET were acting as a perfect switch. Assume R_S = R_L = 50 ohms and V_{gen} = $V_{RF}\cos(\omega_{RF}t)$. V_{RF} = 0.2 V.

Identify any LO or RF feedthrough components.

$$2\cos x\cos y = \cos(x-y) + \cos(x+y)$$
$$2\sin x\cos y = \sin(x-y) + \sin(x+y)$$

3. A MOSFET whose *I–V* characteristic is shown below is used as a mixer. Assume that V_{LO} is NOT large enough to cause the FET to behave as a switch. Thus, this is an old-fashioned nonlinear mixer.

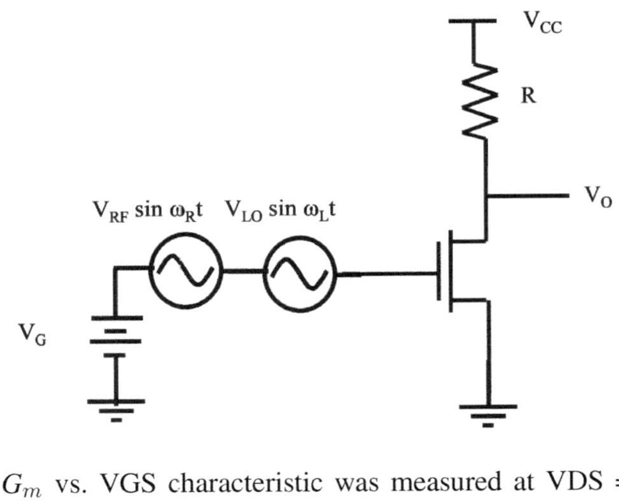

a. The G_m vs. VGS characteristic was measured at VDS = 1.25 V. In this type of mixer, the LO and RF signal must be multiplied through a nonlinearity. Thus, choose VGS so the G_m is signal dependent. Determine required values for V_G and R to bias the transistor properly. V_{CC} = 3.3 V

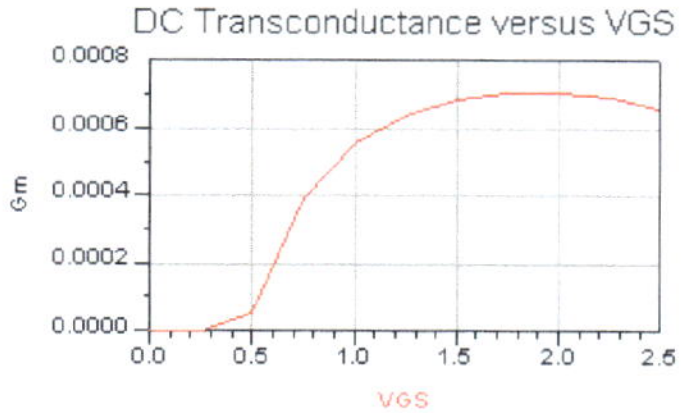

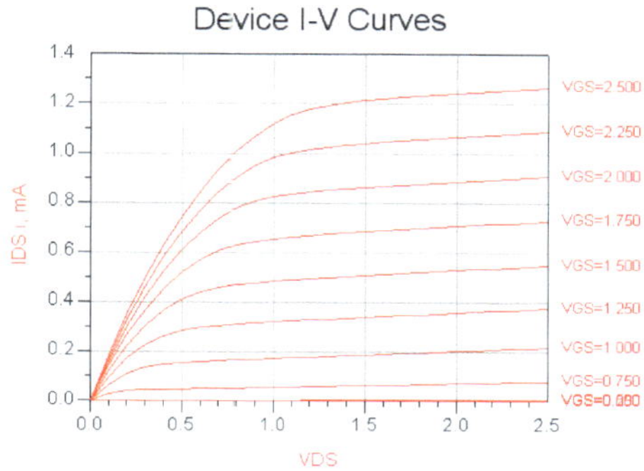

b. Fit an equation to the G_m region you have selected, then use that to predict the conversion gain of the mixer. Also identify any additional DC, first and second order mixing products.

4a. Derive the conversion power gain for this mixer. Assume that $R_S = R_L$ and that V_G is set at the FET threshold voltage.

b. Determine if there is any LO or RF feed through in this mixer. Why?

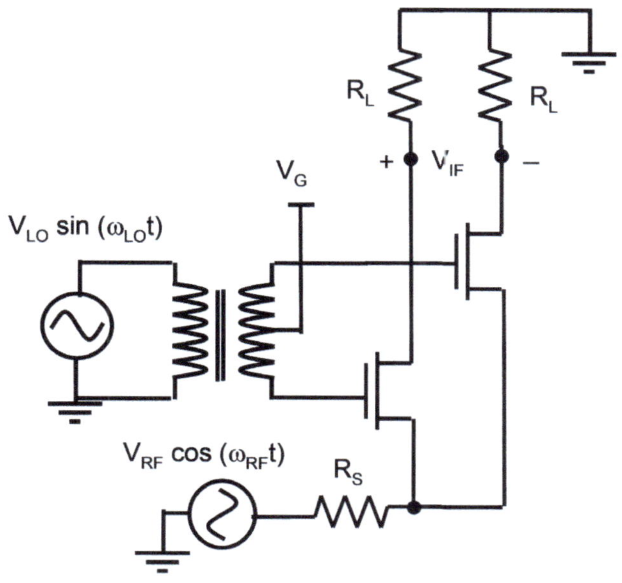

5. The circuit below can be used as a mixer.

a. Derive an equation for the conversion voltage gain V_{IF}/V_{IN} for each expected frequency component in the IF output (up to fourth order), the DC component of V_{IF}, and any RF or LO feedthrough components of V_{IF}. Assume V_{LO} is a square wave with sufficient amplitude to switch M2 fully on or fully off and neglect any high frequency elements in the transistor model. V_{GS} is the MOSFET DC gate bias voltage.

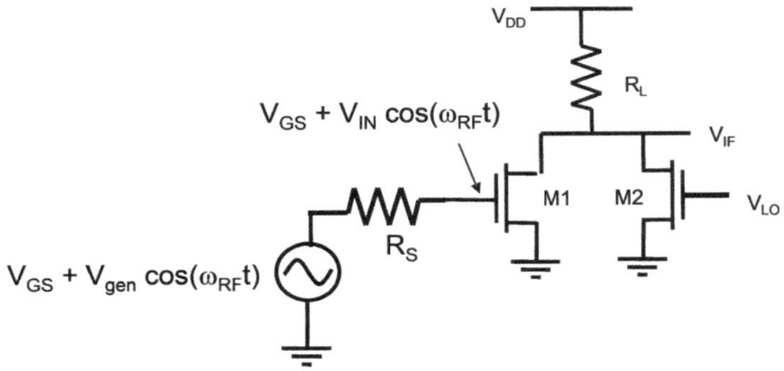

b. V_{DD} = 4 V; R_L = 1000 ohms; R_S = 50 ohms. Using the device characteristics from problem 1, choose a V_{GS} so that M1 is biased in saturation (active region) when the LO switch is off. Note that the maximum V_{DS} is about 2.5 V for these devices. Estimate G_m from the plot and evaluate the IF (second-order/up/downconversion) conversion power gain for this mixer using the results from part (a).

6. Use the ADS mixer design guide to simulate the mixer in problem 5.

a. Device. Use the MOSFET BSIM3 (0.25 um) model for the transistors as shown below. Use the NMOS I-V curve tracer to assist you with biasing. The RF input will require a DC offset. Note that the maximum V_{DS} is about 2.5 V for these devices. V_{DD} = 5 V.

b. Use the single-ended mixer templates to simulate conversion voltage gain V_{IF}/V_{RF} and transducer power gain. Use f_{LO} = 101 MHz and f_{RF} = 125 MHz. Modify the LO source to provide the DC bias and RF amplitude required to fully switch M2. Compare the result with your hand analysis in problem 1.

c. Determine P_{1dB} through simulation.

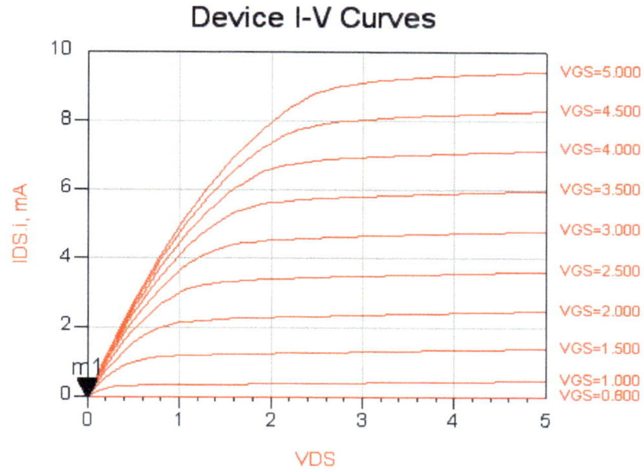

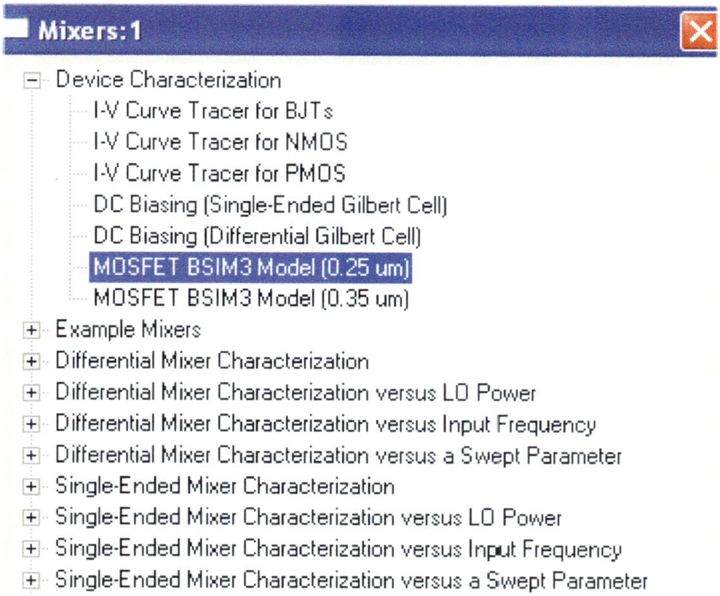

7. This passive circuit following the mixer is a diplexer. It provides wideband termination for the mixer and any reflected signals from the IF amplifier or filter. The desired IF signal passes through C_1L_1 with low loss

, but out-of-band images and spurs are absorbed in the resistors. C_2L_2 are parallel resonant at the design frequency.

Design this circuit for an IF frequency of 10.7 MHz, loaded $Q = 2$, source, load and $R = 50$ ohms. Simulate on ADS and plot S_{11} and S_{21} over the 1 to 200 MHz range.

To simplify the analysis, assume that the two resonators are isolated from each other. This is a reasonably good assumption because the series path presents a short circuit while the parallel resonant part an open circuit at the design frequency. Calculate Q values independently. Assume $R = Z_0$. Also, you must terminate both sides of the diplexer.

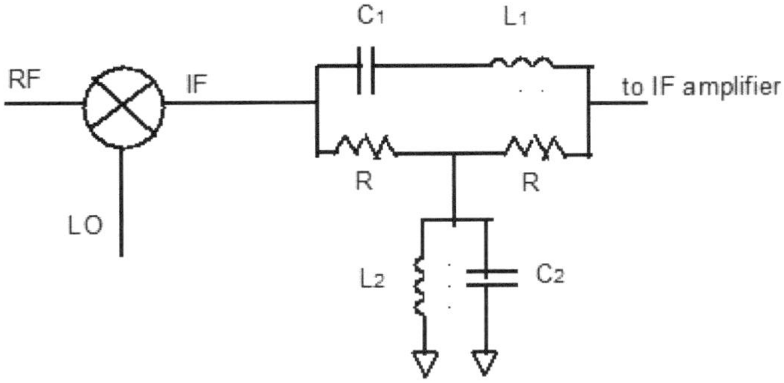

Appendix 11.1. Defining a Subnetwork in ADS

Subnetworks allow you to take advantage of the hierarchy possible in ADS. You can save circuits or portions of circuits that you plan to reuse as subnetworks which can then be accessed in the Component Library.[2]

Here are the steps:

Begin from your schematic of the circuit that you want to represent as a subnetwork. Note that you will need to include transistor models into the subnetwork if your circuit requires transistors.

1. Select the circuit or the part of the circuit that you want to turn into a subnetwork.

[2] The help file in ADS contains detailed instructions beyond these. Search for: Creating hierarchical designs

2. Edit > Component > Create hierarchy. Name the subnetwork. Ports will be added and a symbol will be created as seen in Figures 11.43 and 11.44.

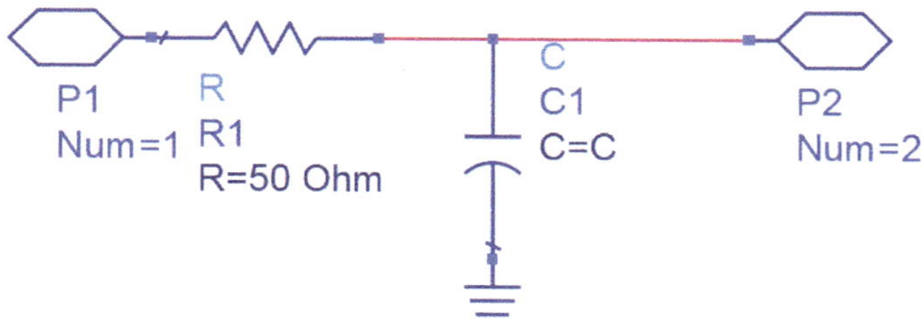

Figure 11.43 Simple RC lowpass filter with adjustable parameter for the capacitor.

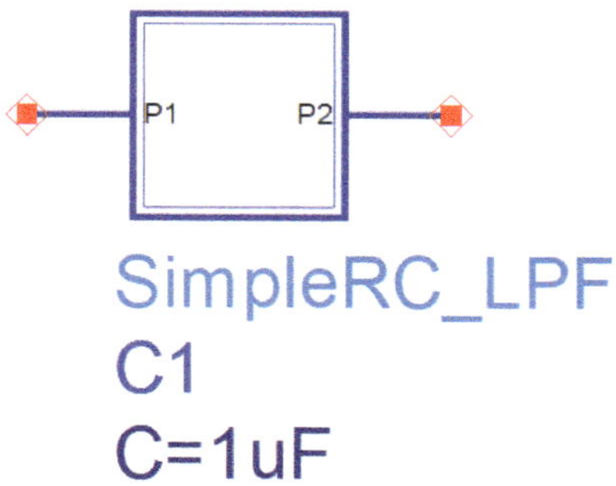

Figure 11.44 Default symbol for a two-port subnetwork.

3. To insert the subnetwork into another circuit:
   - Insert > Component > Component Library.
   - Select workspace libraries.
   - The named subnetwork should appear. Double click on the name and click in the new schematic in order to paste it in.
4. Sometimes you may wish to specify internal circuit parameters that can be set from outside the subnetwork. For example, you might want to sweep a variable.
   - Label the component inside the subnetwork with a variable, such as C in Figure 11.43.

- Starting from the subnetwork schematic: File > Design Parameters

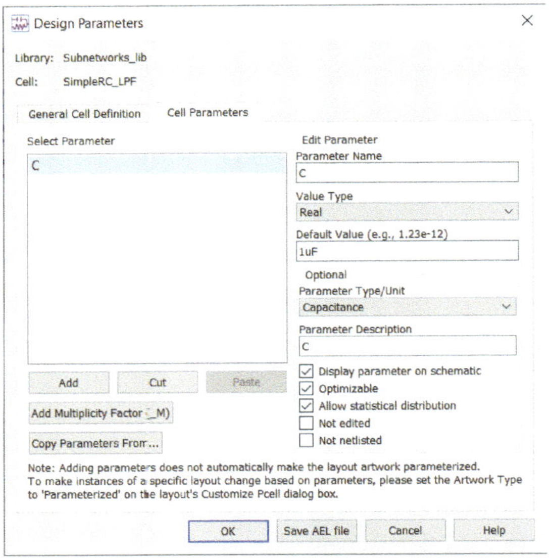

Figure 11.45 Design parameter panel.

- Figure 11.45 opens up.
- Name the parameter to be available outside the subnetwork (C in this example)
- Select type, default value, description.
- OK. Then the parameter should be visible outside the symbol as shown in Figure 11.44.

5. To view the internal schematic of a subnetwork, you can push into the symbol with the down arrow or pop back out with the up arrow shown in Figure 11.46.

Figure 11.46 Push into or out of the subnetwork with these arrows in the schematic.

12

Quadrature Signals and Image Reject Mixers

Quadrature signals are the basis for nearly all digital communications. In addition, the concept can be used to create mixers that can inherently reject image signals without additional filtering.

In this chapter, a basic introduction to the following topics will be presented:

1. Quadrature signal definitions and terminology
2. Applications for image rejection in receiver architectures
3. Generation of quadrature signals.

Much more extensive discussions can be found in some of the references.

12.1 Quadrature Signals

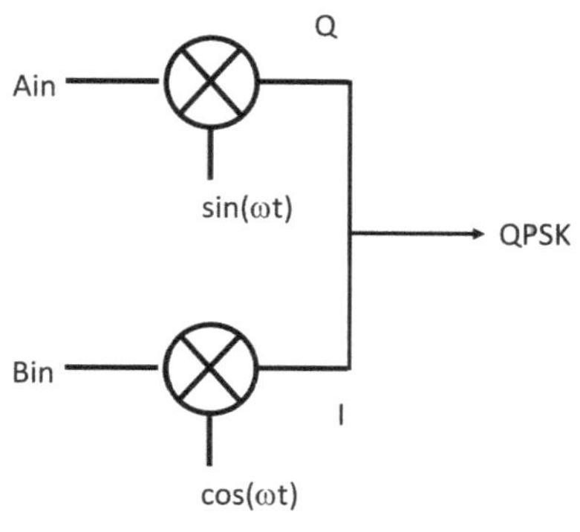

Figure 12.1 Generation of QPSK signals using two mixers and sin/cos functions.

In Figure 12.1 A_{in} and B_{in} could be serial digital signals. I stands for "in-phase" and Q for "quadrature". Q is 90° out of phase with I. QPSK is the abbreviation for quadrature phase shift keying. Note that modern communication systems may use many more phases and amplitudes than this, but for our purposes we will focus on applications with just the two phases and one amplitude. Either analog mixers or digital multipliers could be used depending on the application.

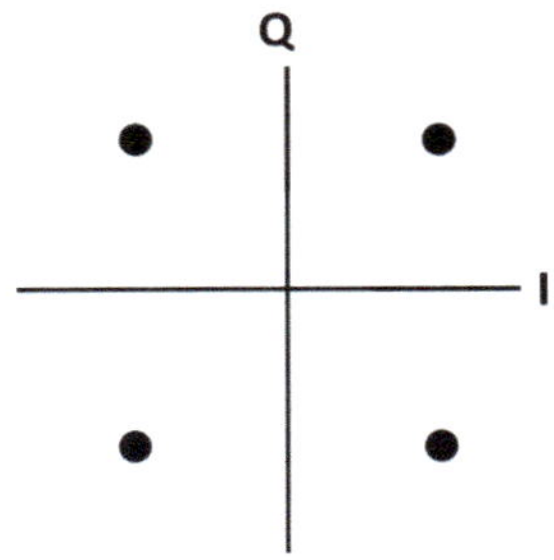

Figure 12.2 QPSK constellation.

When the I and Q signal outputs are summed, the result will be four possible amplitude and phases as represented by the dots in Figure 12.2.

To understand how this works, lets first define and draw the real signals, cosine and sine, in the I and Q plane [1]. I is real; Q imaginary in Figures 12.3 and 12.4. We see each is the sum of two complex signals. And, all real signals consist of positive and negative frequency components. The cosine function (Figure 12.3) has even symmetry and is real; the sine function (Figure 12.4) has odd symmetry and is imaginary. Equations (12.1) and (12.2) illustrate this mathematically.

$$\cos(\omega t) = \frac{e^{j\omega t} + e^{-j\omega t}}{2} \tag{12.1}$$

$$\sin(\omega t) = \frac{e^{+j\omega t} - e^{-j\omega t}}{2j} = j\frac{e^{-j\omega t} - e^{+j\omega t}}{2}. \tag{12.2}$$

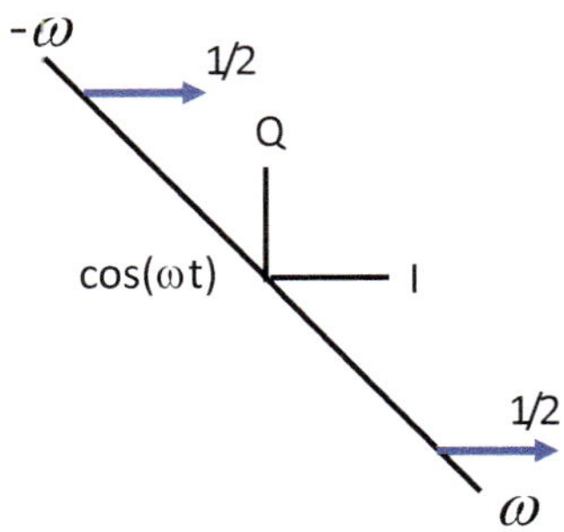

Figure 12.3 The cosine function is even symmetric and real.

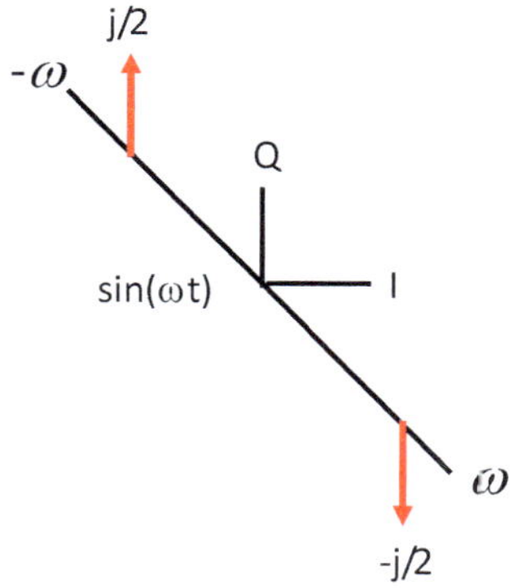

Figure 12.4 The sine function has odd symmetry and is imaginary.

What is meant by positive and negative frequencies? They differ by the direction of their rotation in the complex plane. Positive frequency rotates counter-clockwise; negative, clockwise as illustrated by Figures 12.5 and 12.6.

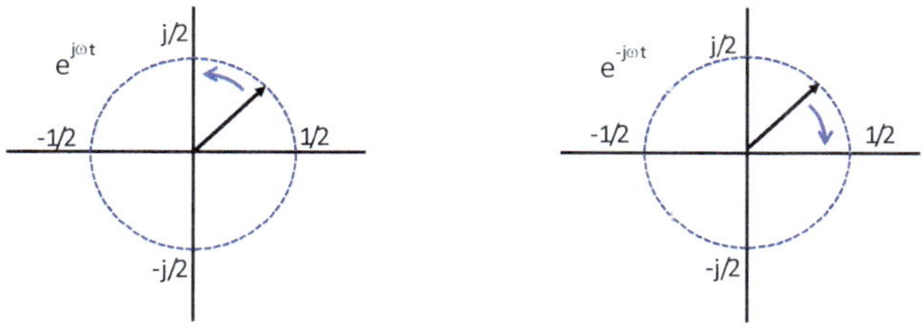

Figure 12.5 Description of positive and negative frequencies from cosine function.

Taking their vector sum of positive and negative frequency components, we find that the result is a real signal varying between −1 and +1.

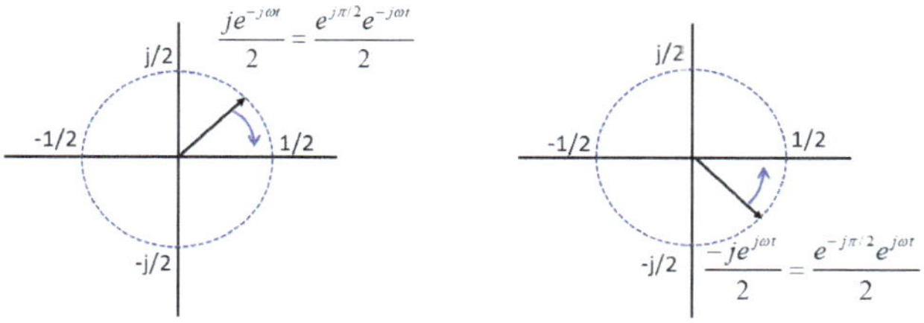

Figure 12.6 Positive and negative frequency sum for the sine function.

Again, the vector sum of these two signals produces a real signal varying from −1 to +1.

Next, what happens when we multiply by *j* or −*j*? Just rotate 90° CW or CCW accordingly, as shown in Figure 12.7.

$$j = e^{j\pi/2} \text{and} - j = e^{-j\pi/2}. \tag{12.3}$$

Example 12.1: Multiply sine function by *j*.

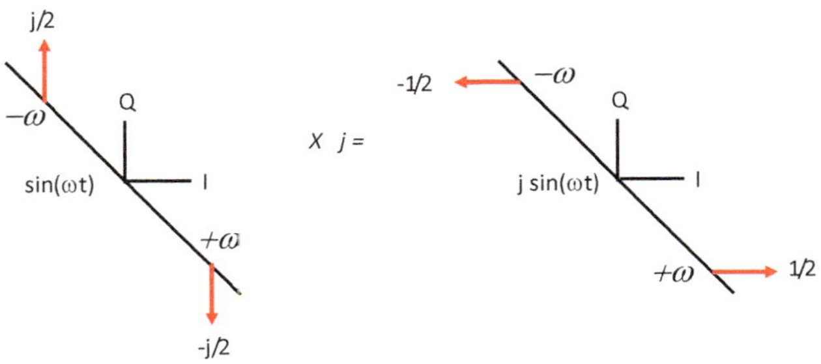

Figure 12.7 The entire spectrum is rotated 90° CCW.

It is important to realize that this is NOT the same as a 90° phase shift. Now, let's use this technique to demonstrate Euler's theorem.

$$e^{j\omega t} = \cos(\omega t) + j\sin(\omega t). \tag{12.4}$$

We observed in Figure 12.7 that the multiplication of sine by *j* produces two real components in the I–Q plane. Recall Figure 12.3 for cosine. When we add these two the negative frequency components cancel, and we are left with just a positive frequency $e^{j\omega t}$ as seen in Figure 12.8.

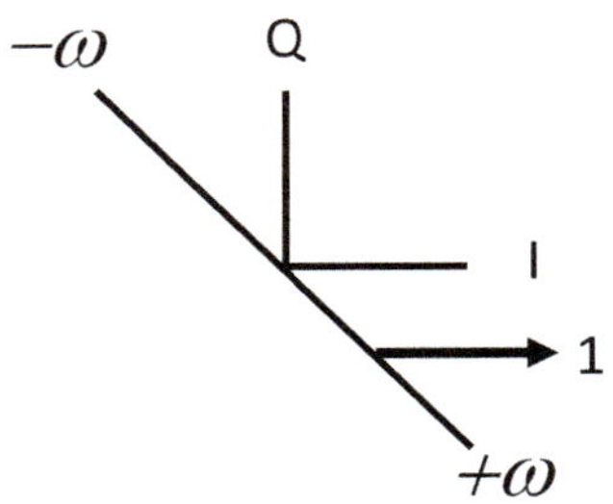

Figure 12.8 Vector sum results in a positive frequency signal $e^{j\omega t}$. This is referred to as a complex signal.

How does a -90° phase shift differ from multiplying by $-j$? Figure 12.9 illustrates Equation (12.5).

$$\cos(\omega t - 90) = \frac{e^{j\omega t}e^{-j\pi/2} + e^{-j\omega t}e^{+j\pi/2}}{2} = \frac{-je^{j\omega t} + je^{-j\omega t}}{2} = \sin(\omega t). \quad (12.5)$$

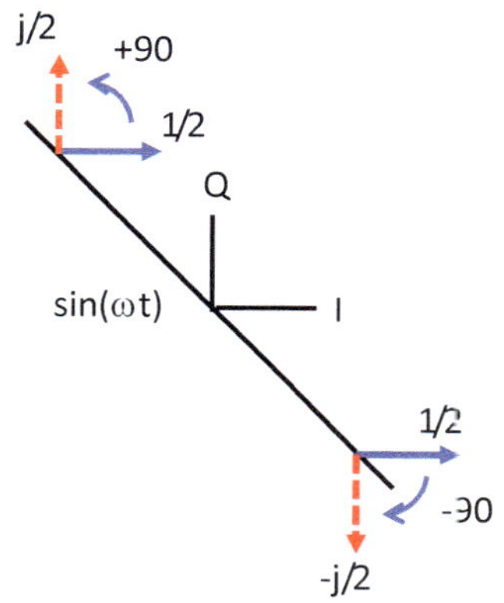

Figure 12.9 A -90° phase shift on the cosine function produces the sine function.

In a similar way, a -90° phase shift on the sine function yields $-$cosine, as seen in Figure 12.10 and Equation (12.6).

$$\sin(\omega t - 90) = j \times \frac{e^{-j\omega t}e^{j\pi/2} - e^{j\omega t}e^{-j\pi/2}}{2} = -\cos(\omega t). \quad (12.6)$$

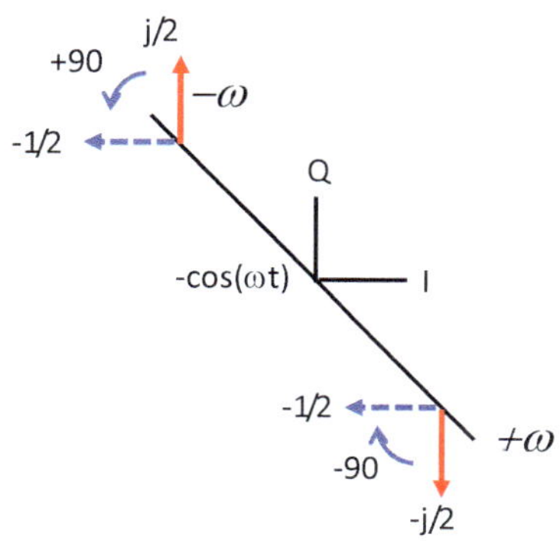

Figure 12.10 Result of -90° phase shift on the sine function.

12.2 Image Reject Mixers using I and Q signals

Recall that the image problem for downconverting mixers is not fully solved by the use of preselection filtering. Filters do not have adequate rejection and require extra space and power. Recall that an image signal often comes from an out-of-band source which may be another transmitter or might be due to a spurious signal generated in the receiver itself.

Now we have the basis to analyze the image rejection principle used in various mixers.

Mixers can be used to translate signals up or down in frequency. The downconverting mixers can either mix to a finite frequency (intermediate frequency or IF) or to baseband. The latter case is called direct conversion or zero IF. The discussion that follows applies to analog mixers with finite IF output frequency.

Now, let's apply the quadrature signal analysis to a downconverting mixer called the Hartley architecture (Figure 12.11) [2, 3].

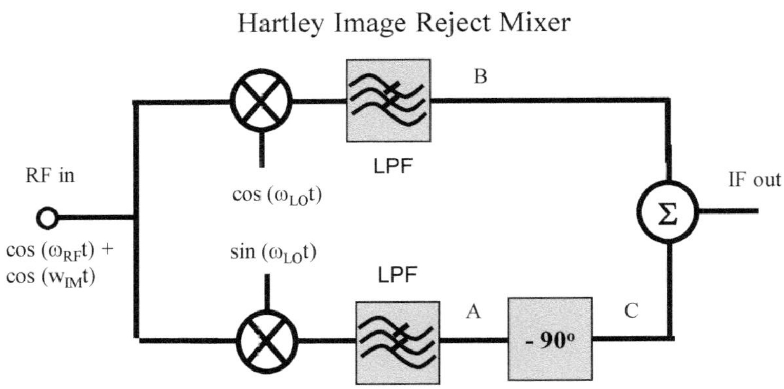

Figure 12.11 Hartley image reject mixer.

This mixer requires a finite IF such that f_{IF} is less than either the LO or the RF or image (IM) frequencies.

The branch with cos LO is the I or in-phase branch; the sin LO branch is the quadrature or Q branch. The low pass filter (LPF) is required in order to reject the upconverted output of the mixers.

Figures 12.12–12.14 illustrate that the image and desired RF signal are both downconverted to the same frequency and thus suffer from spectral overlap. The image signal will be cancelled by shifting the phase of the Q branch by 90° and adding to the I branch, as shown in Figure 12.15.

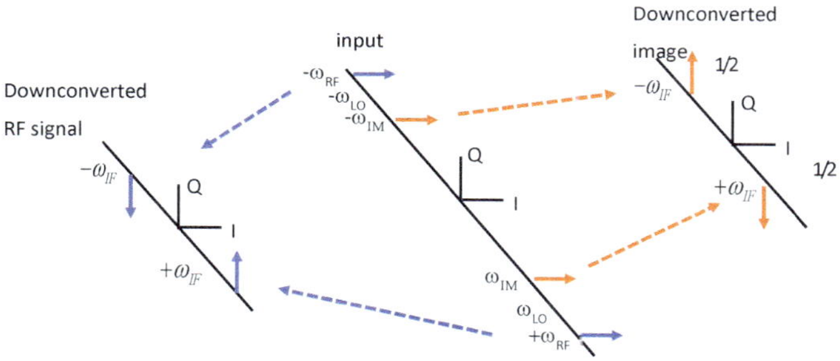

Figure 12.12 Signal at A. Quadrature path. Multiply $\cos(\omega_{RF}t)$ and $\cos(\omega_{IM}t)$ by $\sin(\omega_{LO}t)$ and the low-pass filter to remove the upconversion IF signal.

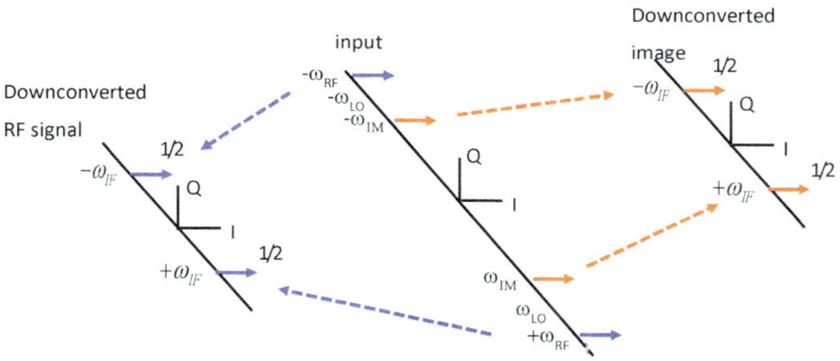

Figure 12.13 Signal at B. In-phase path. Multiply $\cos(\omega_{RF}t)$ and $\cos(\omega_{IM}t)$ by $\cos(\omega_{LO}t)$ and the low-pass filter to remove the upconversion IF signal.

Next apply a $-90°$ phase shift to the Q branch to produce the signal at C.

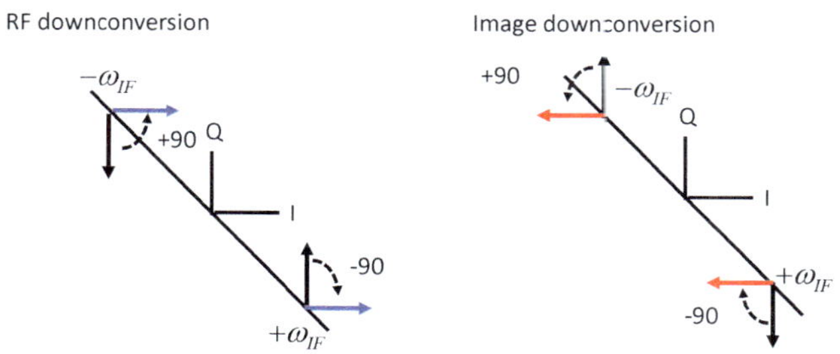

Figure 12.14 Apply $-90°$ phase shift in order to arrive at node C.

Finally, we add A and C producing I + Q – 90.

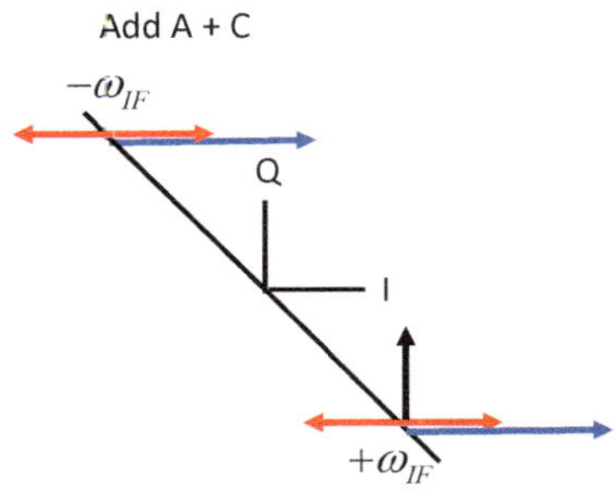

Figure 12.15 Images (red) are out-of-phase and cancel while the RF (blue) add-in phase to produce the desired IF output.

12.3 Transmit Upconverting Image Reject Mixers

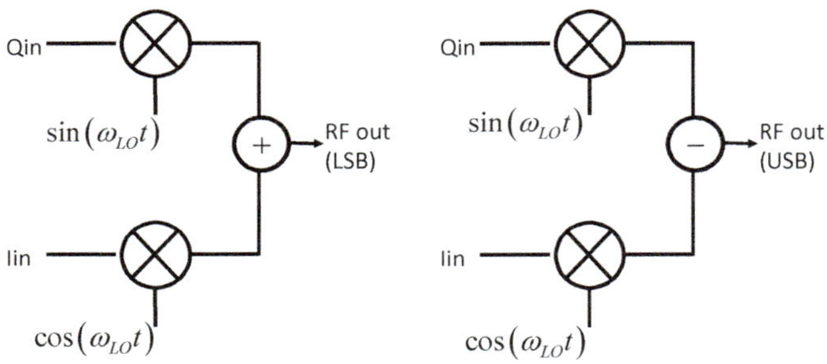

Figure 12.16 Digital data I_{in} and Q_{in} at frequencies $\omega_{IN}$are mixed to RF with a single sideband.

Here are two implementations for upconverting digital I and Q data to QPSK. One or the other sideband is suppressed.

$$\cos(\omega_{LO}t)\cos(\omega_{IN}t) + \sin(\omega_{LO}t)\sin(\omega_{IN}t) = \cos(\omega_{LO} - \omega_{IN})\,t \tag{12.7}$$

$$\cos(\omega_{LO}t)\cos(\omega_{IN}t) - \sin(\omega_{LO}t)\sin(\omega_{IN}t) = \cos(\omega_{LO} + \omega_{IN})\,t. \tag{12.8}$$

The input data would typically be generated by separating a serial data path into two channels with a serial-parallel converter. Often some signal conditioning will also be needed to reduce intersymbol interference [3].

12.4 Phase and Amplitude Requirements

The IR mixer technique requires accurate LO phase and amplitude matching between the I and Q LO channels to achieve a high degree of image rejection. To illustrate, the image rejection ratio is given by Equation (12.9) [3]. In this equation, ΔAcorresponds to the relative LO amplitude mismatch between I and Q. θ is the phase error in radians.

$$IRR = \frac{(\Delta A/A)^2 + \Delta\theta^2}{4}. \tag{12.9}$$

To reflect on how sensitive the IRR can be to these errors, look at the amplitude error in dB versus IRR (Table 12.1).

$$20\log_{10}(\Delta A/A). \tag{12.10}$$

Table 12.1 LO amplitude error associated with reduction in image reject ratio.

Amplitude error	IRR
0.5 dB	−31 dB
1 dB	−24 dB

Assuming a 0.5 dB amplitude mismatch, the additional error caused by phase mismatch:

Table 12.2 LO phase mismatch reduces IRR.

Phase mismatch (degrees)	IRR
1	−30 dB
5	−19.5 dB

12.5 Phase Shift Networks

How do we generate quadrature phases with sufficient accuracy for IR mixer applications?

1. A simple RC + CR low-pass + high-pass combination is adequate if precise amplitude matching is not needed over a wide frequency range. It requires source resistance (R_1 or R_g in Figure 12.17) $<<$ R. In this example, the network was designed for 100 MHz.

Low pass:

$$\left|\frac{V_1(j\omega)}{V_g}\right| = \frac{1}{\sqrt{1+\omega^2C^2(R_g+R)^2}} = \frac{1}{\sqrt{2}}. \tag{12.11}$$

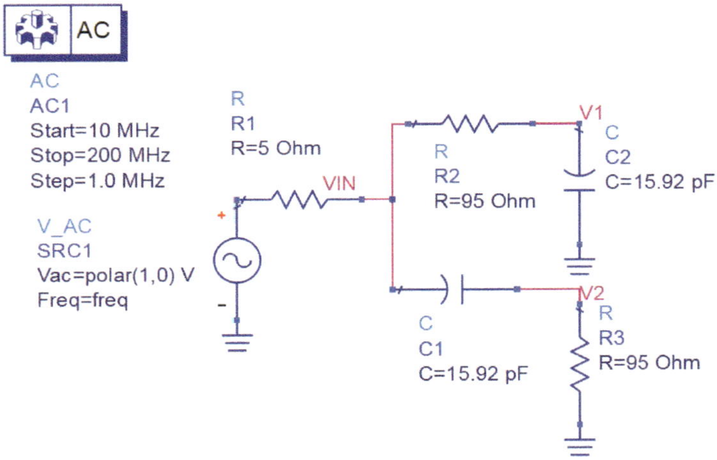

Figure 12.17 RC low-pass, CR high-pass network.

When

$$\omega_{3dB} = \frac{1}{C\,(R_g + R)}, \tag{12.12}$$

$$\theta(V1) = -\arctan\left[\omega C\,(R_g + R)\right] = -45^{\circ}@\omega_{3dB}. \tag{12.13}$$

High pass:

$$\left|\frac{V_2(j\omega)}{V_g}\right| = \frac{\omega RC}{\sqrt{1+\omega^2 C^2 (R_g + R)^2}} = \frac{R/(R_g + R)}{\sqrt{2}}@\omega_{3dB} \tag{12.14}$$

and

$$\theta(V2) = \frac{\pi}{2} - \arctan\left[\omega C(R_g + R)\right] = +45^{\circ}@\omega_{3dB}. \tag{12.15}$$

We can see in Figure 12.18 that

1. Phase difference is $\theta(V2) - \theta(V1) = \frac{\pi}{2}$ at all frequencies.
2. Amplitudes are equal only at ω_{3dB} and if $R_g << R$. Figure 12.18 shows how quickly the amplitudes become unequal as frequency changes.

This type of simple network may be adequate for LO phase generation if the mixer is a switching mode mixer. In that case, the mixer output amplitude is not extremely sensitive to LO amplitude, thus reasonably good

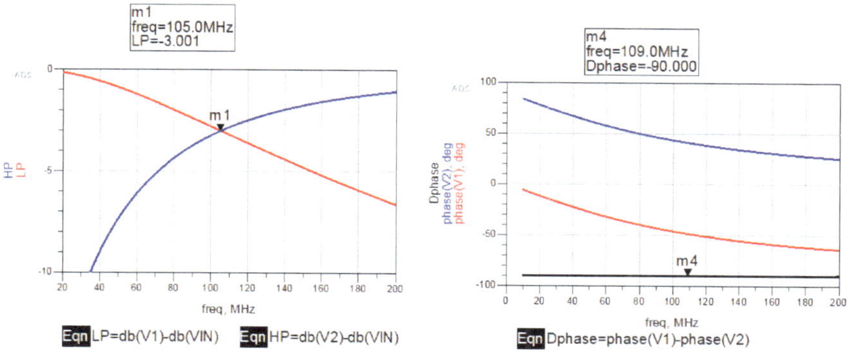

Figure 12.18 Amplitude and phase of RC/CR network versus frequency.

IF amplitude balance can be obtained over some bandwidth. And, the phase difference between the two paths is always 90°.

It would not be adequate, however, for RF or IF phase generation if any significant bandwidth is required. The IRR suffers from amplitude imbalance as shown above. Another IR mixer architecture, attributed to Weaver [4], requires accurate amplitude and phase matching for the IF signal path (Figure 12.20) [3].

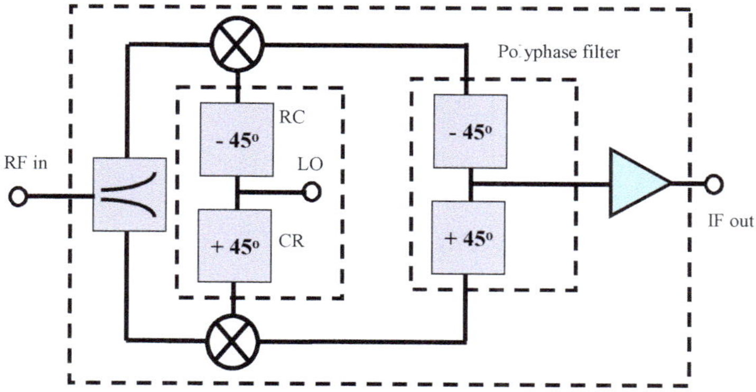

Figure 12.20 Weaver image reject mixer.

When better amplitude matching is needed, then *polyphase filters* can be a good choice. They can be cascaded (at some cost in amplitude) if wider frequency response is needed. See [5] and Figure 12.21.

Polyphase filters can be used for quadrature signal generation and also for image rejection. They discriminate between positive and negative frequency signals. There are a number of design considerations that are well described in the reference.

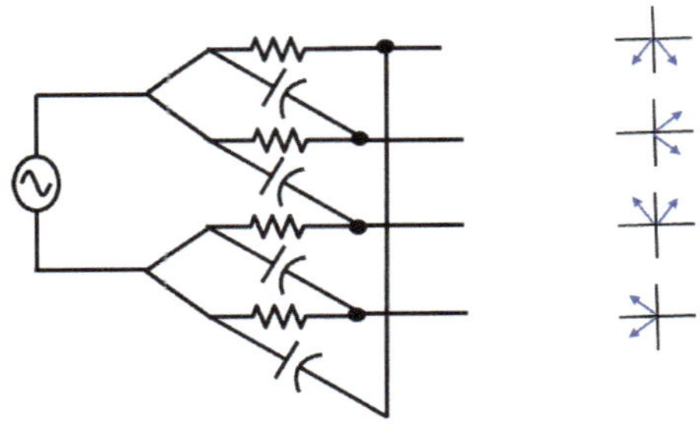

Figure 12.21 A polyphase filter can be used to generate differential quadrature phases with balanced amplitudes.

12.5.1 Digital phase generation

If quadrature LO signals for a switching mixer are needed, these can be generated with flip-flops in a divide-by-four configuration (Figure 12.22). This also has the benefit of swift transitions between on and off and a 50% duty cycle if the delays in the flip-flops are much less than the period. The disadvantage might be that a clock signal at four times the LO frequency must be generated with some expense in extra power.

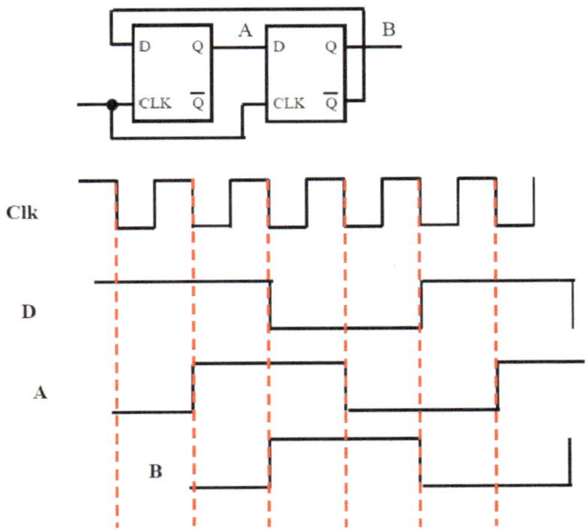

Figure 12.22 Quadrature phase generation using a divide-by-four shift register with feedback.

Outputs A and B yield the 90° and 180° phases. The $\bar{Q}$ outputs provide the inverted A and B for 270° and 0° respectively. This approach gives phase accuracy that is very independent of clock frequency.

12.5.2 Phase-locked loop

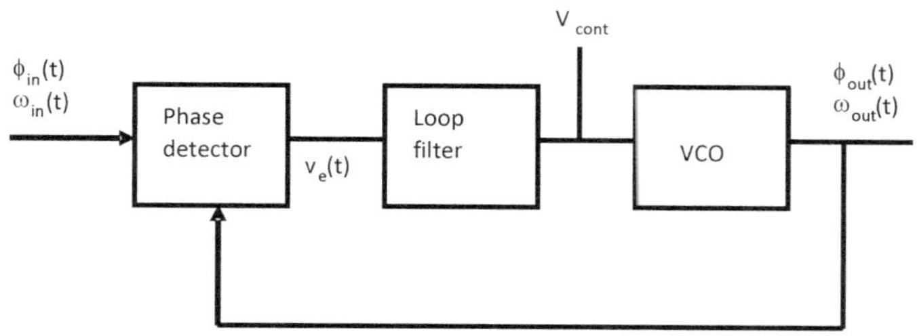

Figure 12.23 A phase locked loop can be used to provide a 90° phase shift.

The PLL phase detector adjusts the VCO control voltage until there is a 90° phase difference between the input and output. So, if the input is $\sin(\omega t)$ then the output will be $\cos(\omega t)$. This approach is also very broadband. However, the loop filter often must be located off-chip.

12.6 Summary

In this chapter we have included descriptions of:

1. Quadrature signal definitions and terminology:
   a. Defined positive and negative frequency
   b. Used complex exponentials and spectral drawings to illustrate single-sideband mixers.
2. Applications for image rejection in receiver and transmitter architectures:
   a. Up- and down-conversion.
3. Generation of quadrature signals using:
   a. Phase shift networks (RC/CR low-pass/high-pass), polyphase filters
   b. Divide-by-four shift register
   c. Phase locked loop (refer to references [3] and [6] for a detailed explanation of the PLL and design of stable loop filters.

References

[1] Lyons, R. G. (2004) *Understanding Digital Signal Processing* Prentice-Hall, ch. 8.
[2] Hartley, R. (1928) Modulation System *US Patent* 1,666,206.
[3] Razavi, B. (1998) *RF Microelectronics* Prentice-Hall, ch. 5.
[4] Weaver, D. K. (1956) A third method of generation and detection of single-sideband signals *Proc. IRE* 44, 1703–1705.
[5] Behbahani F. et al. (2001) CMOS mixers and polyphase filters for large image rejection *IEEE J. Solid State Cir.* 36(6) 873–886.
[6] Vaucher, C. S. (2000) An adaptive PLL tuning system architecture combining high spectral purity and fast settling time. *IEEE J. Solid State Cir.* 35(4), 490–502.

Homework

1. For upconversion, it is necessary to phase shift the baseband or IF input to the mixer as well as the LO, then sum the mixer outputs. The equation representing this is:

$$\cos(2\pi f_C t)\cos(2\pi f_O t) + \sin(2\pi f_L t)\sin(2\pi f_O t) \;=\; \cos[2\pi(f_O - f_C)t].$$

Figure 12.24 shows one architecture that will implement this function. It produces a lower sideband output ($f_O - f_C$), where f_O is the LO frequency and f_C is the IF input.

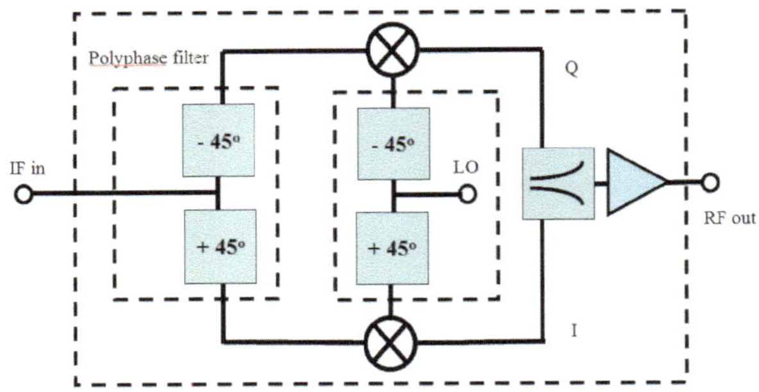

Figure 12.24 Upconverting image reject mixer.

a. Use the exponential form of the equation above to show that this architecture will produce a LSB output at positive and negative frequency.

b. Show how you can modify this to make it upper sideband and again demonstrate mathematically that it will work.
c. Model part (b) on ADS using harmonic balance, the ideal mixer library and single frequency sources with appropriate phases. Let f_O = 1 GHz and f_C = 100 MHz.

2. The block diagram in the figure below corresponds to an image reject quadrature mixer. If $f_{IN1} > f_L$ and $f_{IN2} < f_L$, demonstrate (prove) which input frequency will be cancelled at the output.

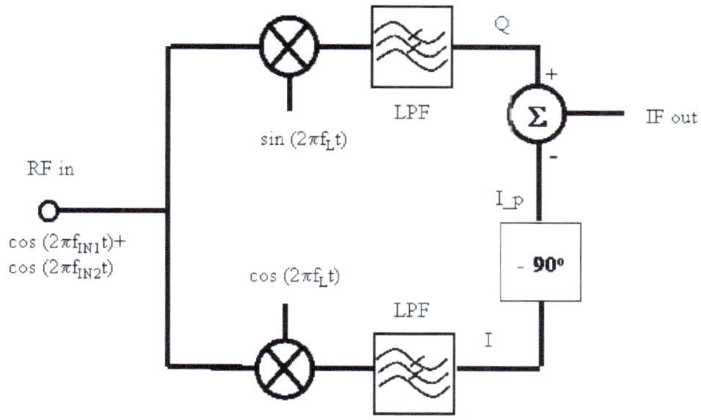

3.a. The block diagram in the figure below illustrates an upconverting mixer architecture. Using the three-dimensional graphical method, determine whether this mixer produces a upper or lower sideband output. Assume $f_{IF} << f_{LO}$.

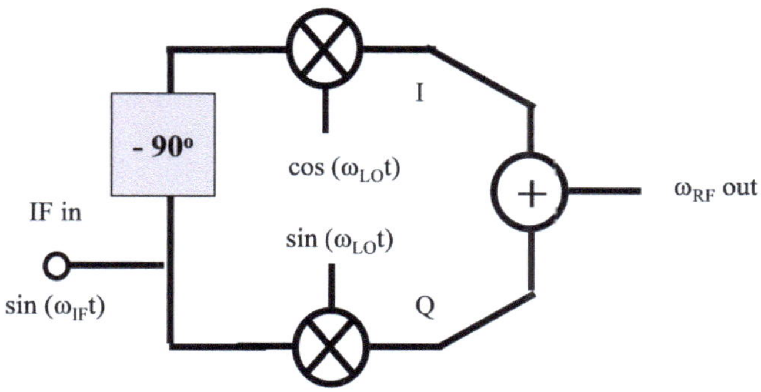

b. Show what would happen if you moved the -90° phase shifter from the input of the I mixer to the input of the Q mixer.

4. Use the ADS harmonic balance and the behavioral model library components (i.e. ideal mixer, phase shifter, 90° hybrid, etc.) along with single frequency voltage sources to simulate the mixers in parts (a) and (b) in problem 3. Use f_{IF} = 100 MHz and f_{LO} = 1 GHz.

5.a. The mixer as shown can be used for up- or downconversion. Show why one of the IF output frequencies provides better image rejection than the other.

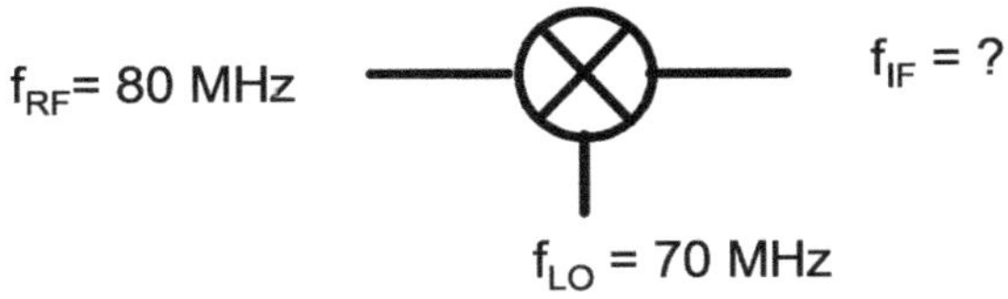

b. An LPF is added at the input. If the LPF preselector has f_{3dB} = 100 MHz and a roll-off rate of −40 dB/decade, calculate the image rejection that would be obtained with the best case of part (a).

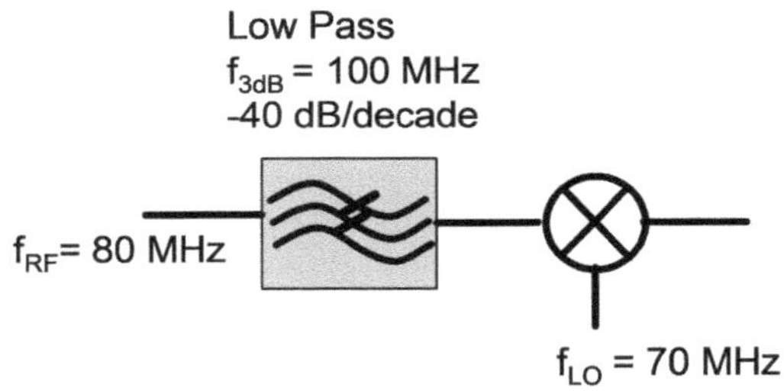

6. The lumped element pi section can be used as a phase shifting network, show through circuit analysis that the circuit below gives a 90° phase shift between V_{out} and V_g when $X_C = X_L = Z_0$. Simulate with ADS and determine the phase error obtained over a 20% bandwidth around ω_o = 100 MHz.

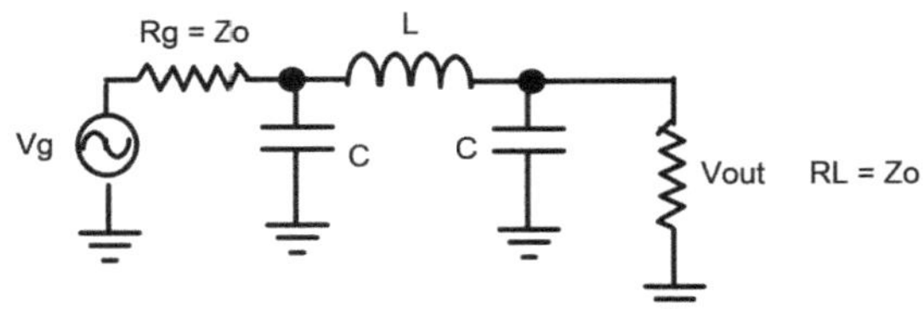

13

Resonators

There are many different circuit approaches for oscillators. In this chapter, we review and expand on the topic of resonant circuits. Resonant circuits are widely used when bandpass characteristics are required for interstage matching or filtering. Also, they form the core of many oscillator designs, establishing the frequency of the oscillator. Oscillator applications will be presented in Chapter 14.

Topics to be presented include:

1. Q and bandwidth of resonant networks
2. Insertion loss
3. Impedance transformation
4. Applications for bandpass filter
5. Temperature compensation.

In Chapter 1, definitions of Q were described as applied to individual lumped-element components and to simple LC resonant networks. Following a quick overview, the loaded Q of a bandpass type of resonator will be related to bandwidth and insertion losses.

13.1 Quality Factor, Q

In Chapter 1 reactive components such as capacitors and inductors were described with a figure of merit called Q. If we consider an example of a series resonant circuit (Figure 13.1). The resonant frequency is

$$\omega_0 = \frac{1}{\sqrt{LC}}. \tag{13.1}$$

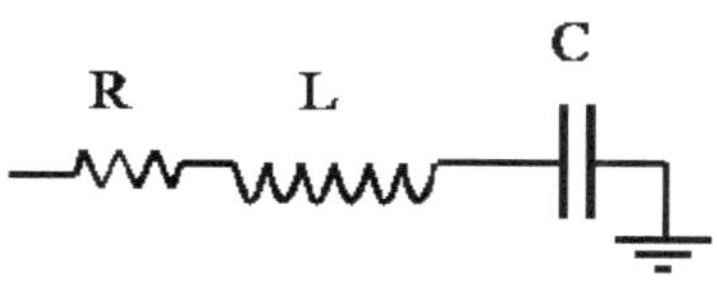

Figure 13.1 Series RLC resonant circuit

At resonance, the reactances cancel out leaving just a peak voltage, V_{pk}, across the loss resistance, R. Thus, $I_{pk} = V_{pk}/R$ is the maximum current which passes through all elements. Then,

$$Q = \omega_0 \frac{L I_{pk}^2/2}{I_{pk}^2 R/2} = \frac{\omega_0 L}{R} = \frac{1}{\omega_0 RC} \tag{13.2}$$

In terms of the series equivalent network for a capacitor shown above, its Q is given by:

$$Q = \frac{1}{\omega RC} = \frac{X}{R} \tag{13.3}$$

where we pretend that the capacitor is resonated with an ideal lossless inductor at frequency ω_0. X is the capacitive reactance, and R is the series resistance. Since this Q refers only to the capacitor itself, in isolation from the rest of the circuit, it is called unloaded Q or Q_U. The higher the unloaded Q, the lower the loss. Notice that Q decreases with frequency.

The unloaded Q of an inductor is given by,

$$Q_U = \frac{\omega_0 L}{R} \tag{13.4}$$

again X/R where R is a series resistance as described above. Note that Q is proportional to frequency for an inductor. The Q of an inductor will depend upon its implementation (core material if any, wire losses, coupling to a substrate). When applied to the resonator, we are more concerned with loaded Q which includes resistive losses of source and load as well at the components themselves.

It is easy to show that for a parallel resonant circuit, the Q is given by susceptance/conductance:

$$Q = \frac{B}{G} \tag{13.5}$$

where B is the susceptance of the capacitor or inductor and G is the shunt conductance.

When a resonant circuit is connected to the outside world, its total losses (let's call them R_P or G_P) are combined with the source and load resistances, R_S and R_L. Then, at resonance, when equal and opposite reactances X_L and X_C cancel, a voltage divider remains which reduces the output voltage, as shown in Figure 13.2, for example,

Figure 13.2 shows a parallel resonant circuit (C, L and R_P) connected to the outside source and load resistances; $G_P = 1/R_P$. The total Q of this

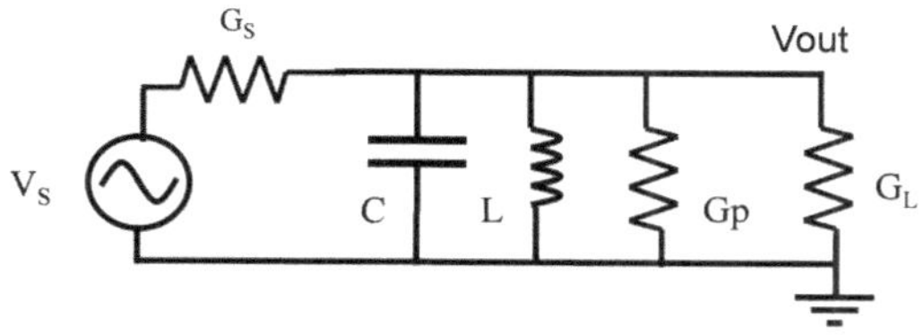

Figure 13.2 Parallel resonant RLC circuit with source and load.

circuit is called the loaded Q or Q_L and is given by

$$Q_L = \omega_0 C \left(R_P \| R_S \| \, R_L \right) \tag{13.6}$$

or

$$Q_L = \frac{\omega_0 C}{G_S + G_L + G_p} = \frac{\omega_0 C}{G_{total}} = \frac{\text{susceptance}}{\text{conductance}}. \tag{13.7}$$

The significance of this is that Q_L can be used to predict the bandwidth of a resonant circuit. We can see that higher Q_L leads to narrower bandwidth.

$$BW = \frac{\omega_0}{Q_L} \tag{13.8}$$

where

$$\omega_0 = \frac{1}{\sqrt{LC}} \tag{13.9}$$

1. So, large C will increase the loaded Q at a given resonant frequency and reduce the bandwidth.
2. Or, we could vary G_{total}. How?

A. Unloaded Q (Q_U) is an attribute of the passive L and C components and affects G_P. This will vary if we change L and C. And, $G_P = G_C + G_{ind}$,

$$\text{Capacitor: } Q_U \text{ (capacitor)} = \frac{\omega_0 C}{G_C}. \tag{13.10}$$

$$\text{Inductor: } Q_U \text{ (inductor)} = \frac{1}{\omega_0 L G_{ind}}. \tag{13.11}$$

Using this to set the bandwidth of a resonator is not a good idea. If Q_U becomes comparable to Q_L, then the loss in the resonator becomes very high, as we shall see.

13.2 Insertion Loss and Bandwidth

The resonator in Figure 13.2 can be used as a bandpass filter or matching network. In these applications, insertion loss can be important.

Consider that resonant LCR circuit. When $\omega = \omega_0$,

$$\frac{V_{out}\,(j\omega_0)}{V_S} = \frac{G_S}{G_S + G_L + G_P} < 1. \tag{13.12}$$

Typically, $G_S = G_L = G$. We need to find S_{21} to determine insertion loss. Recall that $|S_{21}|^2$ = transducer gain = power delivered to the load/available power from the source when the source and load are both Z_0.

First note that

$$\frac{Q_L}{Q_U} = \frac{G_P}{G_P + 2G} \tag{13.13}$$

and,

$$S_{21} = \frac{2V_{out}}{V_S} = 1 - \frac{Q_L}{Q_U} = \frac{2G}{G_P + 2G}. \tag{13.14}$$

So, insertion loss (dB) :

$$IL = 20 \log \left(1 - \frac{Q_L}{Q_U}\right) \tag{13.15}$$

Thus, obtaining narrow bandwidth where Q_L and Q_U are similar can be very lossy.

Consider an example: a circuit, parallel LCR, which includes a 500 ohm resistor (R_1) in parallel with the LC network, shown in Figure 13.3. This resistance represents the parallel equivalent loss due to both L and C. So, now we have a finite unloaded Q. L and C were selected for a resonant frequency of 1 GHz.

Note that insertion loss increases as loaded Q_L approaches Q_U (Figure 13.4),

$$S_{21}(\,\mathrm{dB}) = 20 \log \left(2V_{OUT}/V_{IN}\right) \tag{13.16}$$

Sweeping R_{LS}, we see that, at resonance, the reactances cancel, and we are left with a resistive divider.

$$V_{OUT} = V_{IN} \left[\frac{R_1}{(2R_1 + R_{LS})}\right] \tag{13.17}$$

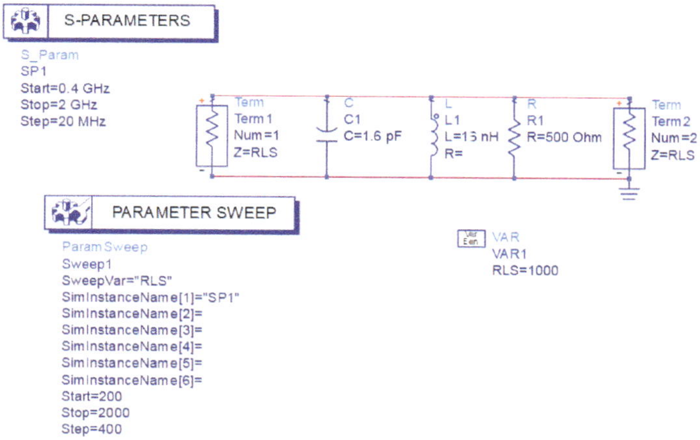

Figure 13.3 RLC resonant circuit with varying source and load resistances. R_1 is the parallel equivalent loss due to C_1 and L_1.

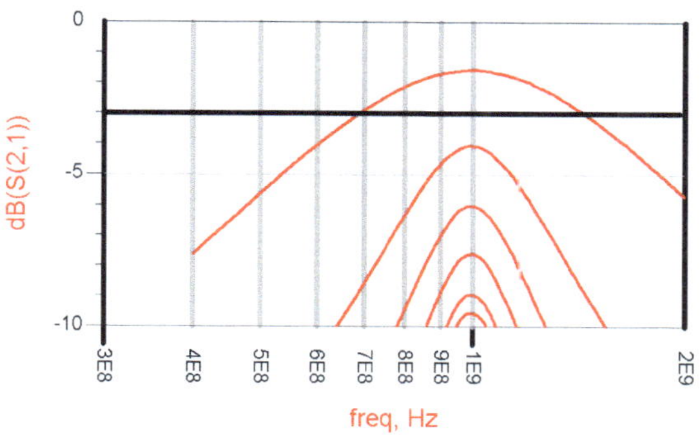

Figure 13.4 Insertion loss increases as loaded Q_L approaches Q_U.

So, we can set the Q_L and the bandwidth by adjusting the loading conductances or resistances,

$$BW = \frac{\omega_0}{Q_L} \tag{13.18}$$

However, we must make sure that $Q_U >> Q_L$ to avoid excessive losses.

So, if a really narrow bandwidth is required, the solution generally requires multiple resonators or more complicated bandpass filter approaches or mechanical resonators such as quartz crystal or surface acoustic wave filters.

How can you set Gs and G_L? Aren't these dictated by the generator and load?

Not necessarily! We can use tapped C matching circuits to transform source and load impedances to whatever we desire (assuming we don't increase loss too much by approaching Q_U).

But, first, we will introduce a convenient approximation which can be used to transform a parallel equivalent circuit intro a series equivalent circuit and vice-versa *at a single frequency.*

13.2.1 A useful series-parallel transformation

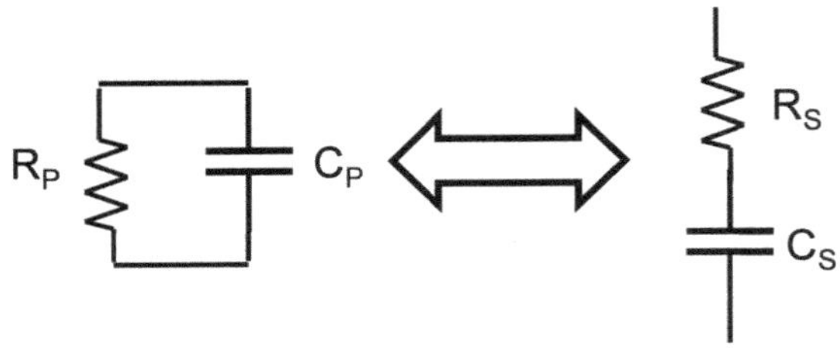

Figure 13.5 Series-parallel single frequency transformation.

A useful series-parallel transformation [1] is shown in Figure 13.5. Comparing the admittance Y_P of the parallel LC with the impedance Z_S of the series LC networks,

$$Y_P = \frac{1}{Z_S}. \tag{13.19}$$

Then, if each has the same unloaded Q,

$$\frac{R_P}{X_P} = Q = \frac{X_S}{R_S}. \tag{13.20}$$

It can be shown that,

$$\mathrm{Re}\,(1/Y_P) = \mathrm{Re}\,(Z_S) = \frac{R_P}{1+Q^2} \tag{13.21}$$

$$\mathrm{Im}\,(1/Y_P) = \mathrm{Im}\,(Z_S) = X_P = \frac{1}{\omega C_P}. \tag{13.22}$$

This leads to:

$$R_S = R_P\left(\frac{1}{1+Q^2}\right) \tag{13.23}$$

$$X_S = X_P\left(\frac{Q^2}{1+Q^2}\right). \tag{13.24}$$

If $Q > 10$:

$$\boxed{\begin{aligned} X_S &\cong X_P = X && (13.25)\\ R_P &\cong Q^2 R_S && (13.26)\\ R_P R_S &= X^2. && (13.27)\end{aligned}}$$

13.2.2 Tapped Capacitor Network

Now we return to the question of how do you control the loaded Q in order to obtain a certain bandwidth. The tapped capacitor network can be used for this purpose (Figure 13.6). It can transform an input resistance to a much higher impedance at the output which then can set the load and source resistances presented to the resonator. It also provides a good example of how the series-parallel transformation described above is used for circuit simplification.

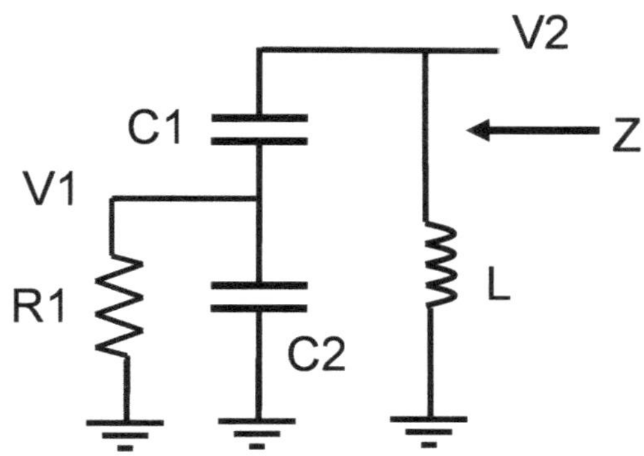

Figure 13.6 Tapped capacitor impedance transforming network.

Step 1: Convert the R_1, C_2 parallel network into a single series network as shown in Figure 13.7 using $R_P R_S = X^2$, Equation (13.27), or $R_P \cong Q^2 R_S$, Equation (13.23), valid if $Q > 10$. Here, $R_1 = R_P$ in the parallel side of Figure 13.5 . If we define $X_1 = 1/\omega C_2$ then,

$$R_{S1} = \frac{X_1^2}{R_1} = \frac{1}{R_1}\left(\frac{1}{\omega C_2}\right)^2 \tag{13.28}$$

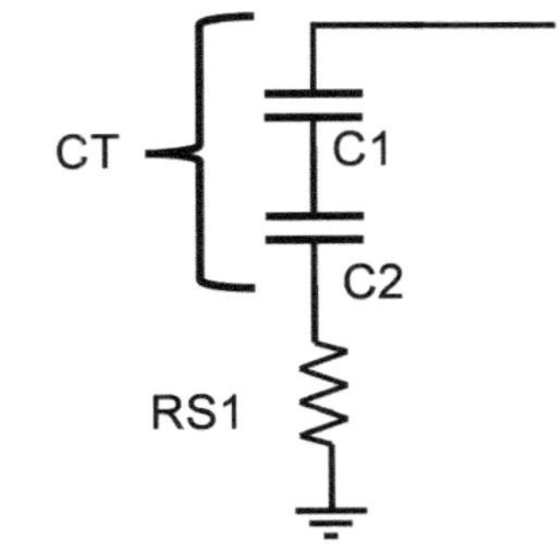

Figure 13.7 Convert to series equivalent.

Step 2: Combine the series combination of C_1 and C_2 into a single equivalent capacitance, C_T,

$$C_T = \frac{C_1 C_2}{C_1 + C_2}. \tag{13.29}$$

Step 3: Convert the series $\mathrm{R}_{S1} - \mathrm{C}_\mathrm{T}$ into the parallel equivalent in Figure 13.8 using $R_P R_{S1} = X_2$ in the opposite direction. Defining $X_2 = 1/\omega C_T$, we arrive at,

$$R_P = \frac{X_2^2}{R_{S1}} = \frac{1}{R_{S1}}\left(\frac{1}{\omega C_T}\right)^2. \tag{13.30}$$

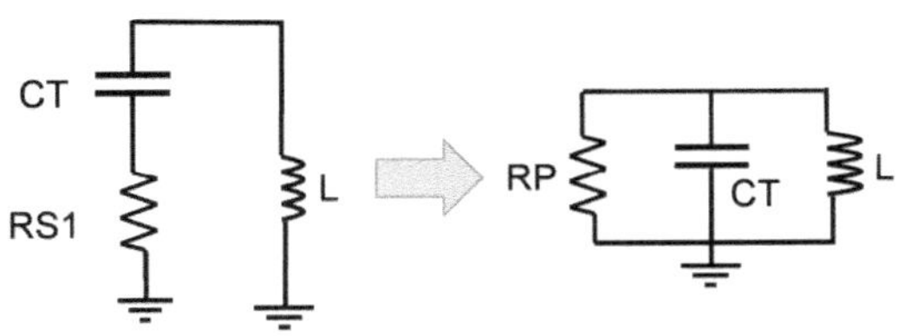

Figure 13.8 Final conversion back to parallel RLC.

After some further manipulation, it can be shown that

$$R_P = R_1\left(\frac{C_1 + C_2}{C_1}\right)^2 \tag{13.31}$$

and now,

$$\omega_0 = \frac{1}{\sqrt{C_T L}} \tag{13.32}$$

Or, if we define a parameter n as

$$n = \frac{C_1 + C_2}{C_1} \tag{13.33}$$

we get the transforming relationship,

$$\boxed{R_P = R_1 n^2} \tag{13.34}$$

similar to the turns ratio on a transformer. The input source resistance is multiplied by a factor n^2. However, it is much easier to implement because no core material or windings are needed. Similarly, the voltage is stepped up by a factor n.

$$\boxed{V_2 = nV_1.} \tag{13.35}$$

Example 13.1: Design a bandpass matching network, illustrated by the bandpass filter box in Figure 13.9, to transform 50 ohms to the input of an SA602A mixer. The SA602A is a bipolar double-balanced mixer with input amplifier, on-chip oscillator, and voltage regulator. While it was designed and first produced in the 1990s it is still available in surface mount packaging.

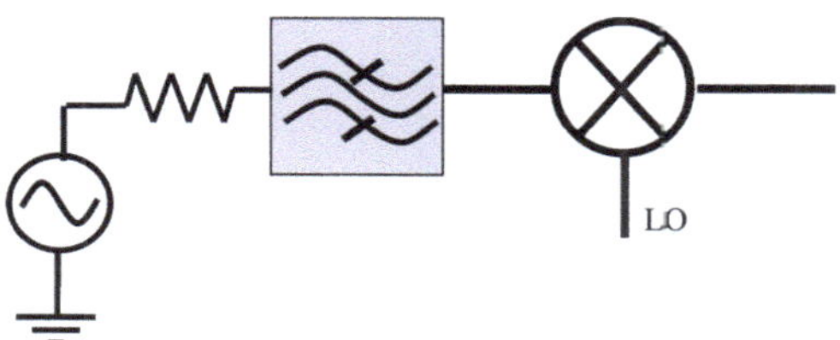

Figure 13.9 A bandpass filter at the input of a mixer is configured to also transform the 50 ohm source to the input impedance of the mixer.

The mixer input impedance is given on the data sheet as

$$Z_{IN} = 1.5\text{k} \parallel 3\text{pF for the SA602A IC.}$$

Choose a convenient frequency for this exercise: 159 MHz = 1×10^9 rad/s.

Use the tapped C procedure shown in Figure 13.10 to transform 50 ohms to 1500 ohms. Include the 3 pF into the equivalent series capacitance of C_1

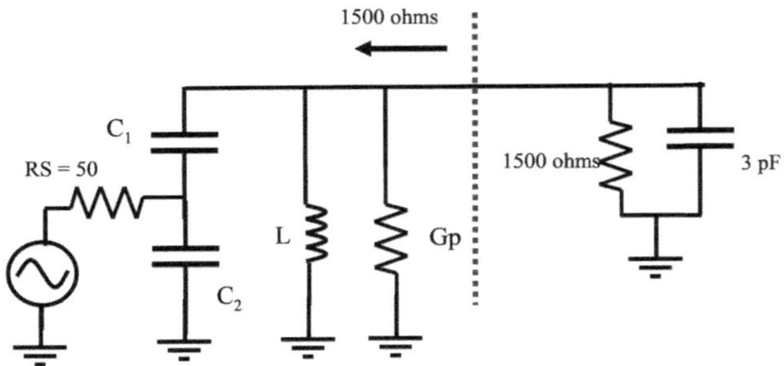

Figure 13.10 Tapped *C* input network.

and C_2. The equivalent parallel resistance, R_P, due to the unloaded Q of the components, should also be taken into account in calculating the loaded Q and the transformation ratio needed to match the source and load.

We need to specify a bandwidth to begin the process. Let's choose 10 MHz. From Equation (18),

$$Q_L = \frac{159}{10} = 16 \tag{13.36}$$

Q_U is generally limited by the inductor. The manufacturer's data generally specifies an unloaded Q at a certain frequency. For a T-30-12 powdered iron core material, Q_U of 120 is typical at this frequency. The external capacitors normally have a higher unloaded Q which can be neglected.

Our first design equations:

$$Q_U = \frac{1}{\omega_0 L G_P} = 120 \tag{13.37}$$

$$Q_L = \frac{B}{G_{total}} = \frac{1/\omega_0 L}{G_S + G_L + G_p} = 16. \tag{13.38}$$

Assume that $G_S = G_L = 1/1.5\text{k} = 6.7 \times 10^{-4}$. That is, design the tapped C network to provide a transformed impedance of 1.5k. This will get close to the solution needed.

So, now we have two equations with two unknowns. We can solve for L and for G_P.

$G_P = 2.04 \times 10^4$S or $R_P = 4900$ ohms.

$B = Q_U\ G_P = 2.44 \times 10^2$ S

$$L = \frac{1}{\omega_n B} = 41 \text{ nH}. \tag{13.39}$$

This is a rather small value and will require some care in layout to implement accurately on a PC board. Or it could be implemented without the core material and present a much higher Q_U, but for the purposes of our exercise, include Q_U = 120.

Check insertion loss:

$$IL = 20\log\left(1 - \frac{Q_L}{Q_U}\right) = -1.2 \text{ dB} \tag{13.40}$$

If we required less loss, then a wider BW or higher Q_U is necessary.

Next, combine R_P and R_L and design the tapped C network in Figure 13.11 to give 1.5k$||R_P$ = 1150 ohms for maximum power transfer.

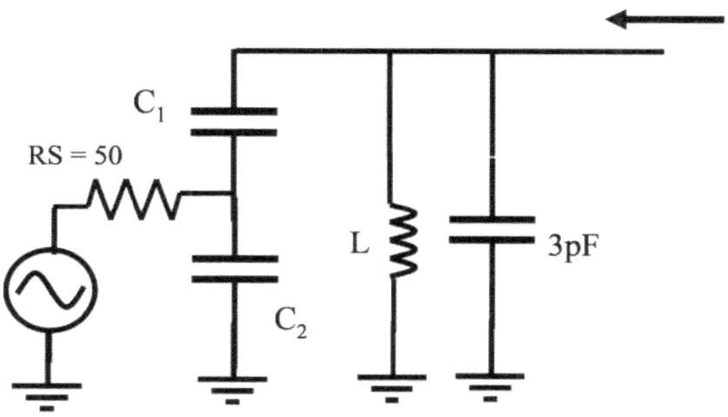

Figure 13.11 Input tapped C network which has absorbed the input capacitor and resistive load of the mixer.

Because L is known, we can calculate the required C_{total} to resonate at 159 MHz.

$$C_{total} = \frac{1}{\omega_0^2 L} = 24 \text{ pF}. \tag{13.41}$$

Now, deduct the 3 pF from C_{total} so that it is absorbed into the resonator. Design C_1 and C_2 for 21, not 24 pF.

The transformation ratio relates C_1 and C_2. It should transform 50 ohm source into the load impedance, 1500$||R_P$ = 1150 ohms.

$$n = \frac{C_1 + C_2}{C_1} = \sqrt{\frac{1150}{50}} = 4.8. \tag{13.42}$$

The series combination of C_1 and C_2

$$\frac{C_1 C_2}{C_1 + C_2} = 24 - 3 = 21 \text{ pF} \tag{13.43}$$

Two equations; two unknowns. Solve for C_1 and C_2,

$$C_1 = 27 \text{ pF} : C_2 = 103 \text{ pF}$$

Check the result with ADS, see Figure 13.12.

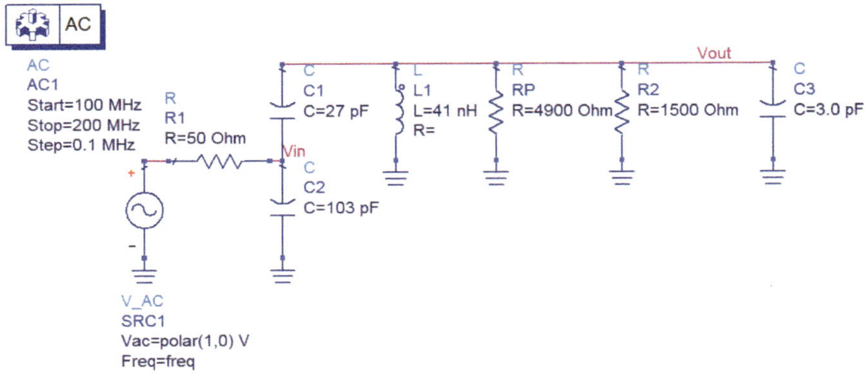

Figure 13.12 Schematic for tapped C matching network and input of mixer.

A small signal AC simulation is performed which includes the R_P and excess C of the load. If the load is matched correctly to the source, we should see half of the source voltage at node V_{in}, 0.5 V (available power). V_{out} should be 4.8 × 0.5 = 2.4 V.

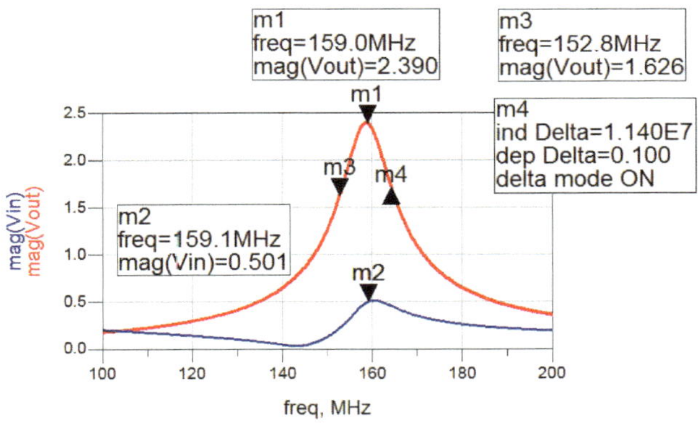

Figure 13.13 Plot of V_{in} and V_{out} of the tapped C example.

Everything looks as expected in Figure 13.13 except for the bandwidth, 11.4 MHz rather than 10 MHz. This is due to the effect of R_P. If it mattered, we could go back through again and include this effect and get a 10 MHz BW, but most applications like this one are not so critical. The important thing is to get the correct impedance transformation.

Of course, an L network, PI or T network could also have been used here with somewhat less flexibility in choosing loaded Q, as described in Chapter 5.

13.3 Tapped Capacitor Resonator

Figure 13.14 shows an example of a resonator that is configured as bandpass filter: 50 ohms to 50 ohms, 100 MHz. The tapped C transforms the 50 ohms up to 1000 ohms for this example. You can work through the design – similar to the earlier example. The result of the simulation using the S-parameter analysis in ADS is shown in Figure 13.15.

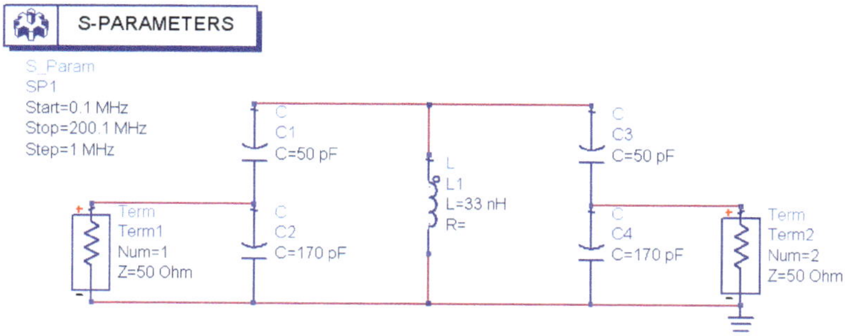

Figure 13.14 Tapped C bandpass filter.

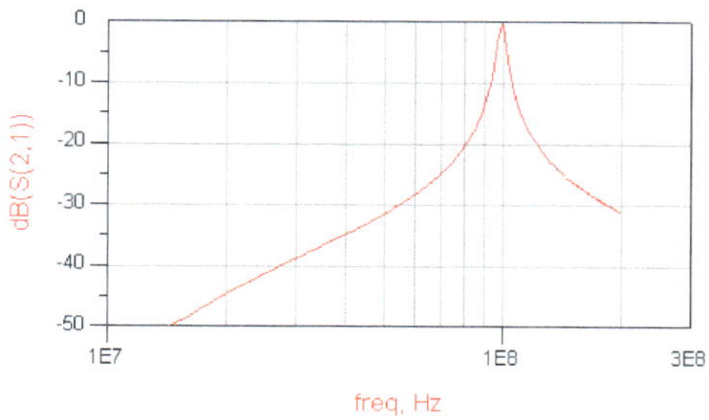

Figure 13.15 S_{21} of the bandpass filter.

13.4 Coupled Resonator [2]

To get better out-of-band attenuation, two identical resonators can be coupled with a series capacitor, as shown in Figure 13.16. Taking Q_L of one resonator and its series equivalent capacitance, C_T,

$$C_5 = \frac{C_T}{Q_L}$$

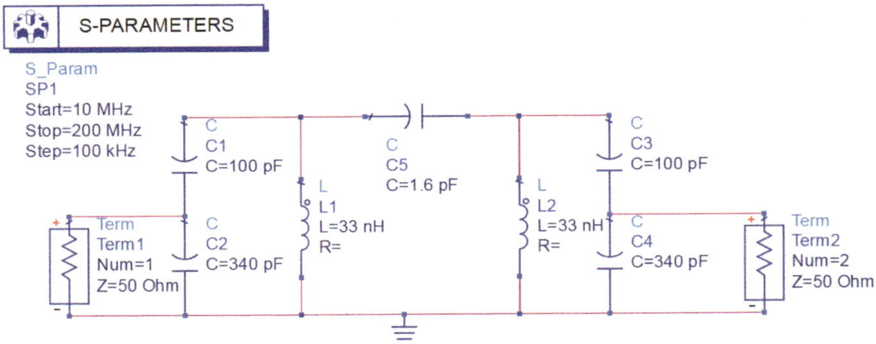

Figure 13.16 S_{21} is calculated using ADS S-parameter analysis.

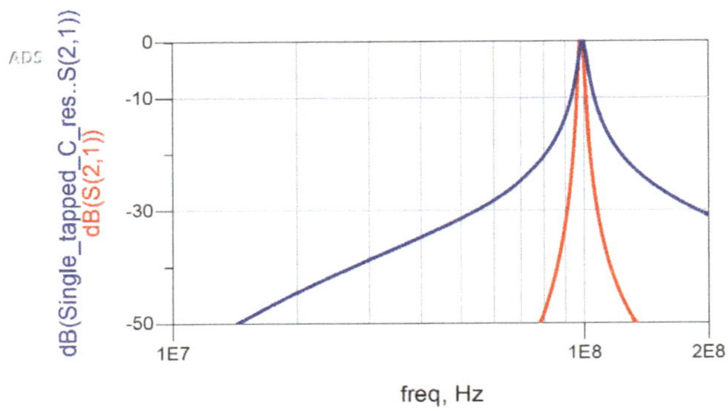

Figure 13.17 Single resonator compared with coupled resonator.

Note the improvement in the stopband rejection in Figure 13.17 provided by the coupled resonator compared with the single resonant tapped *C* circuit.

13.5 Temperature Compensation of Resonant Circuits

Oscillators are frequently used to set the transmit or receive frequency in a communication system. While many applications use a phase locked loop technique to correct for frequency drift, it is good practice to build oscillators with some attempt to minimize such drift by selecting appropriate components. Here we will discuss the properties of discrete capacitors and inductors. On-chip components will need characteristics that vary with the process type.

Inductors and capacitors often drift in value with temperature. Permeability of core materials or thermal expansion of wire causes inductance drift. Variations in dielectric constant with temperature in capacitors is the main source of drift for these components.

Temperature drift is expressed as a temperature coefficient in ppm/°C. The temperature range represents the recommended limits for the use of these capacitors.

13.5.1 Capacitors

The three most common types of dielectrics for RF capacitors are shown in Table 13.1.

Table 13.1 Temperature sensitivity of three common dielectrics.

Dielectric type	Temp coefficient (TC)	Temp range (°C)
C0G (or NP0)	±30 ppm/°C	−55 to +125
X7R (BX)	−1667 (+15% to −15%)	−55 to +125
Z5U	-10^4 (+22% to −56%)	10 to 85

Clearly, the Z5U is not much good for a tuned circuit and should be used for bypass and AC coupling (DC block) applications where the value is not extremely critical. At lower radio frequencies, polystyrene capacitors can be used. These have a −150 ppm/°C TC.

The C0G and X7R can be used in tuned circuits if their values are selected to compensate for the inductor drift.

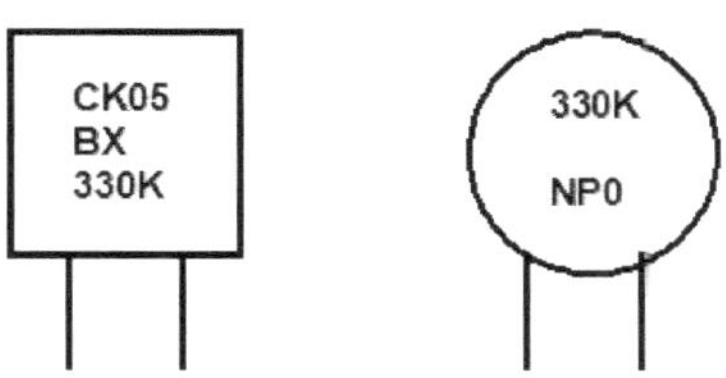

Figure 13.18 Two types of leaded capacitor profiles and labels.

The two leaded capacitors in Figure 13.18 illustrate the labels found on typical capacitors of the X7R and NPO types. The value is given by the numerals: 330. In this example, this is 33 pF. It goes first significant digit (3), second significant digit (3), and multiplier (10^0). The letter K is the tolerance, which is $\pm 10\%$.

As always, the parasitic inductance and self-resonance of any capacitor must be considered for RF applications.

13.5.2 Inductors

There are many types of inductor core materials which are intended for different frequency ranges, permeability, and TC. Powdered iron and ferrites are the two categories of these materials.

For example, one powdered iron, Type 12 (green/white) is useful from 50 to 200 MHz and gives Q_U in the 100–150 range. μ/μ_0 = 4. Manufacturer's data sheets can be found on the web that specify TCs for the many powdered iron and ferrite materials. This one has a weird TC vs. temperature behavior, but we are mainly interested in the 25–50 °C range for this example (Table 13.2).

Table 13.2 Temperature coefficients of a Type 12 powdered iron core.

Temperature range (°C)	TC (ppm/°C)
25–50	+50
50–75	−50
75–125	+150

So, how can you compensate for component drift?

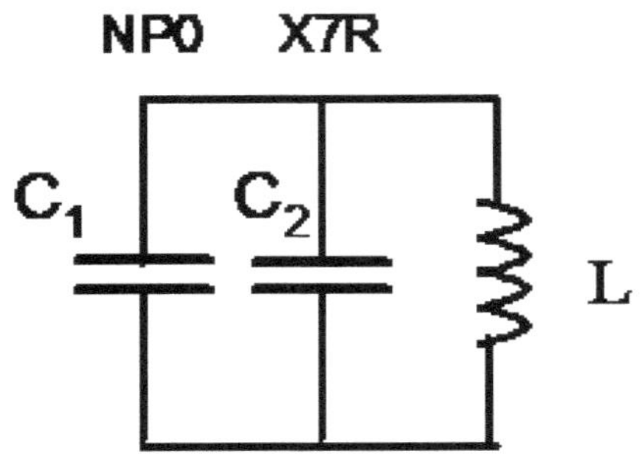

Figure 13.19 Compensation of component temperature drift.

The equation below shows how the TCs of individual components combine [3]. Suppose that the inductor was resonated with a drift free capacitor (NP0). The frequency drift will be –25 ppm/°C. If the design frequency is 100 MHz, this corresponds to a drift of 2.5 kHz/°C. However, the equation shows that you can set the total frequency TC (TCF) of a circuit to zero by combining capacitors with different TCs.

$$TCF = \frac{\Delta f}{f_0} = -\frac{1}{2}\left(TC_L + TC_{C1}\frac{C1}{C_{TOTAL}} + TC_{C2}\frac{C2}{C_{\text{TOTAL}}}\right).$$

Thus, if the inductor has a positive TC, you can correct for temperature drift with the right combination of non-drift and drifty capacitors. In this case, we want the total capacitance of C_1 and C_2 to have a net TC of –50 ppm/°C. The best oscillators will be designed with components with low intrinsic TCs so that you do not have to compensate them with different components having large and possibly unreliable TCs.

Summary

1. Distinctions between loaded and unloaded Q of RLC resonant circuits were presented.
2. Bandwidth and insertion loss of resonant RLC circuits were derived.
3. An easy transformation between series RLC and parallel RLC circuits was used to design tapped capacitor impedance transforming networks.
4. Simple bandpass filters using single and coupled resonators were presented.
5. A method for compensation of component temperature drift was described.

References

[1] McWhorter, M., Scherer, D., and Swain, H. (1995) *EE344 High Frequency Laboratory* Stanford University, ch. 4.
[2] Bowick, C. (2008) *RF Circuit Design*, 2nd Edn. Newnes, ch. 2.
[3] Hayward, W., Campbell, R., and Larkin, B. (2003) *Experimental Methods and RF Design* ARRL Press.

Homework

1. The resonator below could be used for a bandpass filter. If R_S = 50 Ω , L = 50 nH, and Q_U = 100, determine C_1 and C_2 such that the center frequency = 160 MHz and bandwidth = 10 MHz.

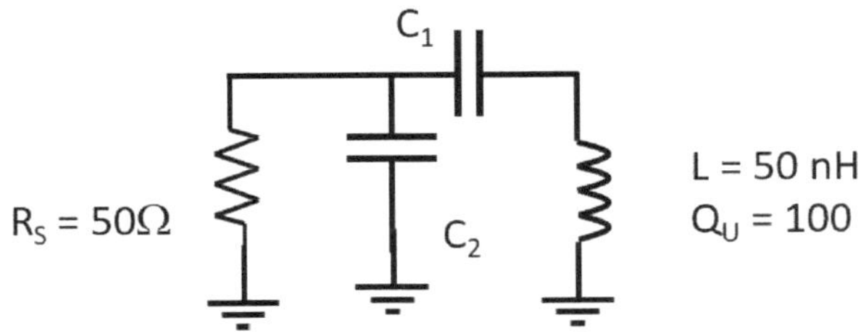

2. Redesign the bandpass filter such that the input and output *Rs* and R_L are 50 ohms.

3.a. A tapped *C* resonator is to be used as a filter. Design this for a center frequency of 80 MHz and 3dB bandwidth of 10 MHz. The inductor has an unloaded *Q* of 30 at this frequency. Transform the 50 ohm source and load resistance up to 450 ohms for this filter. Calculate the insertion loss.

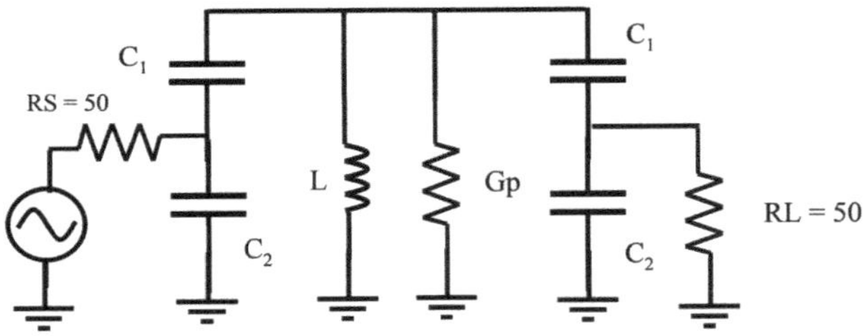

b. Use the equivalent circuit model for the closest 0603CS Coilcraft inductor value and simulate the frequency response of this circuit. The inductor models can be found at https://www.coilcraft.com/en-us/products/rf/ceramic-core-chip-inductors/0603-(1608)/0603cs/

c. Assume the inductor has a temperature coefficient of +100 ppm/°C. Find an appropriate combination of capacitors with C0G (NP0) and X7R (BX) dielectrics to reduce the frequency drift with temperature to zero.

14

Oscillators

This chapter presents an introduction to the basic fundamentals of oscillators: how they work and how to design and simulate them. Oscillators are contained within nearly any communication system or test equipment. They are inherently nonlinear. While hand design approximations are useful for the initial attempt, nonlinear simulation methods are essential for determining startup conditions, harmonic generation, and amplitude. Accuracy, however, requires well characterized nonlinear device models.

Goals:

1. Understand what makes an oscillator different from an amplifier
2. Present the design of LC Resonator-based oscillators and simulation methods
3. Amplitude limits of LC oscillators
4. Design of voltage-controlled oscillator (VCO)
5. How a crystal oscillator works
6. Ring oscillators .

Phase noise, another performance limitation, will be discussed in Chapter 15. A very brief introduction to phase-locked loops is found in Appendix 14.A.1.

14.1 Introduction: Oscillator Basics

An oscillator is intended to provide a sinusoidal output without need for an input. Positive feedback will be required as well as sufficient loop gain to ensure start-up. The noise inherent in the passive and active components is amplified as it feeds back on itself and eventually produces a sustained sinusoidal signal. The frequency will be determined by the configuration of a resonant circuit and its component values [1, 2].

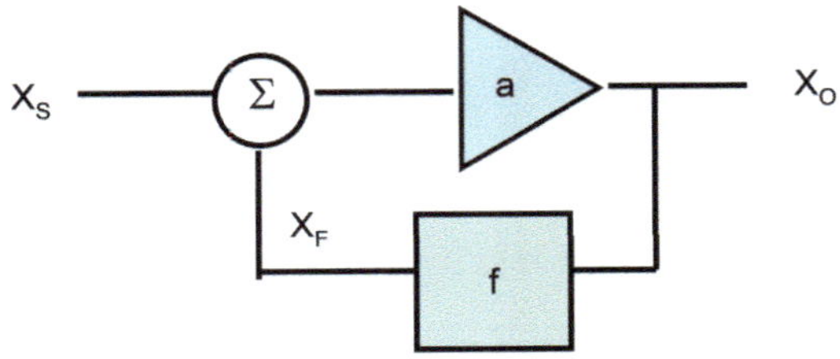

Figure 14.1 Block diagram of an amplifier with positive feedback.

A feedback amplifier is shown in Figure 14.1 with positive feedback. Gain with feedback is

$$a_F = \frac{a}{1 - af} = \frac{x_0}{x_S}. \tag{14.1}$$

What conditions are needed for oscillation?

$$|af| = 1 \ \angle af = 0. \tag{14.2}$$

This is called the Barkhausen criterion.

Also, if the oscillation is to be sustained with $X_s = 0$, no input, then

$$\begin{aligned} X_0 &= aX_F = afX_0 \\ X_F &= fX_0. \end{aligned} \tag{14.3}$$

Amplitude is controlled by the magnitude $|af|$. Frequency is controlled by the phase of af. Typically, the feedback block is frequency dependent - a resonator or filter or phase shift network. Considering a resonant circuit like a parallel RLC, the phase slope

$$\frac{\mathrm{d}\phi}{\mathrm{d}\omega} \tag{14.4}$$

will set the frequency (Figure 14.2).

You can see that a change in phase $\Delta\phi$ will result in a change in frequency $\Delta\omega$. A large phase slope produces less frequency variation for a given $\Delta\phi$. This could be important for stability of frequency absent external phase detection controls.

What would cause $\Delta\phi$?

A second way to study the operation of oscillators is to evaluate the characteristic equation: the roots of which are the circuit poles,

$$1 - a(s)f(s) = 0 \tag{14.5}$$

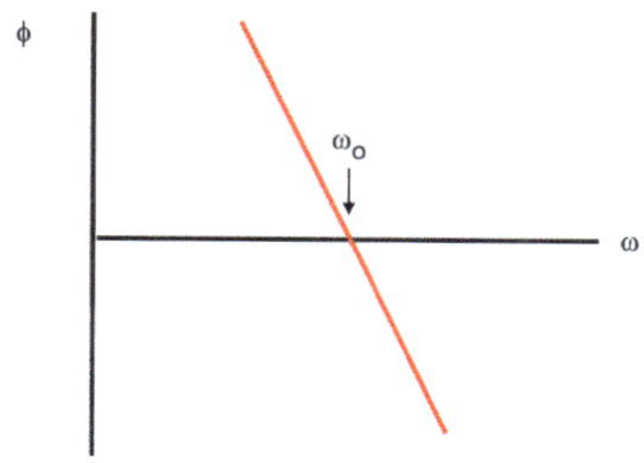

Figure 14.2 Phase vs. frequency.

For sustained oscillations at ω_0, we need roots on the $j\omega$ axis at $s = \pm j\omega_0$, as seen in Figure 14.3. This would be the case with a factor $\left(s^2 + \omega_0^2\right)$.

The inverse Laplace transform gives the solution; $\sin\left(\omega_0 t\right)$ or $\cos\left(\omega_0 t\right)$. In other words, an undamped sinusoid.

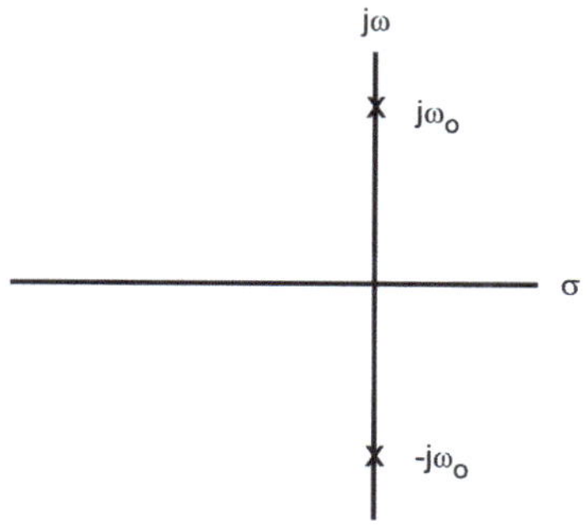

Figure 14.3 S plane showing the roots of the characteristic equation leading to sustained oscillation.

14.2 LC Resonator-based Oscillators

Let's examine how this might work with a simple parallel LCR resonator. Refer back to Figure 14.1. Let $a = g_m$ and $f = f(s)$. Insert the LCR circuit shown in Figure 14.4 in the $f(s)$ block [1].

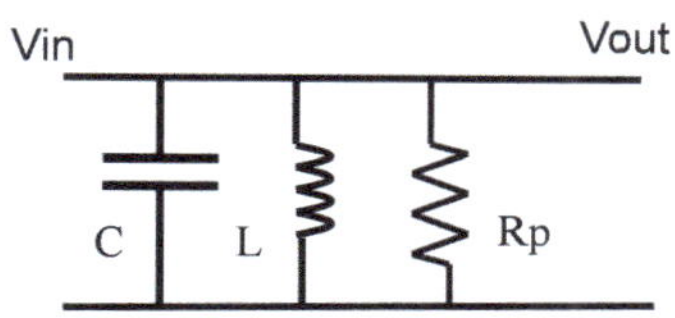

Figure 14.4 Parallel LCR resonator. Let $Y(s)$ be the admittance of this circuit.

We want

$$\frac{V_{out}}{V_{in}} = 1 = \frac{g_m}{Y(s)} \tag{14.6}$$

$$Y(s) = \frac{1}{sL} + sC + \frac{1}{R_P} = \frac{s^2LC + \frac{sL}{R_P} + 1}{sL} \tag{14.7}$$

$$s^2LCR_P + sL(1 - g_mR_P) + R_P = 0. \tag{14.8}$$

Then,

$$s = -\frac{(1 - g_mR_P)}{2CR_P} \pm \frac{1}{2LCR_P}\sqrt{L^2(1 - g_mR_P)^2 - 4LCR_P^2} \tag{14.9}$$

If $g_mR_P = 1$, then,

$$s = \pm\frac{1}{2LCR_P}\sqrt{-4LCR_P^2} = \pm j\frac{1}{\sqrt{LC}} = \pm j\omega_0 \tag{14.10}$$

Refer to Figure 14.5. When you utilize an LC tank circuit as a resonator to control frequency, you want:

- High Q resonator that provides good stability and low phase noise.
- The frequency can be adjusted by voltage if desired, by using varactor diodes in the resonator.
- A buffer amplifier is preferred to avoid the load impedance pulling the frequency.

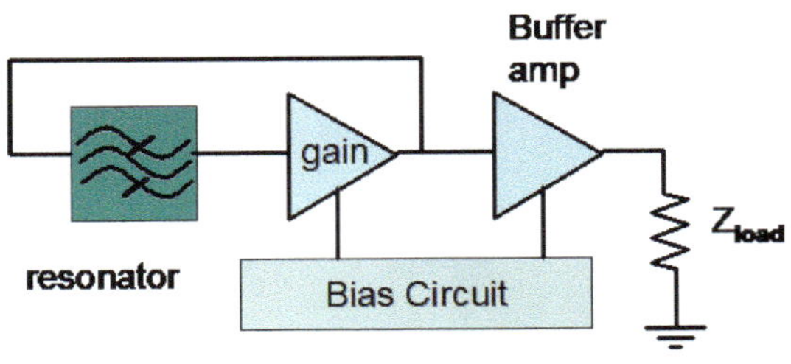

Figure 14.5 Block diagram of an LC oscillator with bias and buffer amplifier.

14.2.1 Circuit #1

Consider the tuned amplifier shown in Figure 14.6.

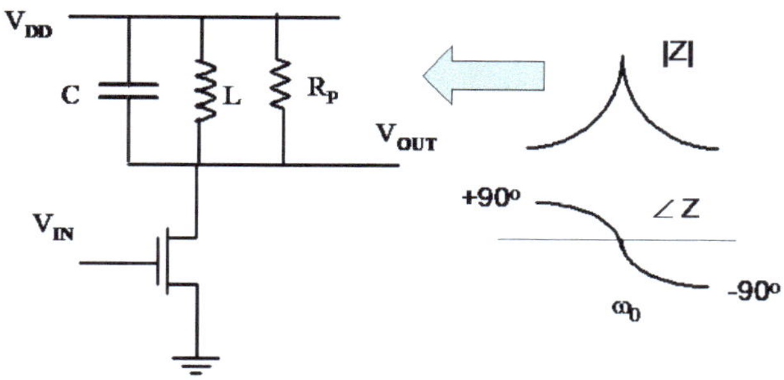

Figure 14.6 Tuned amplifier example.

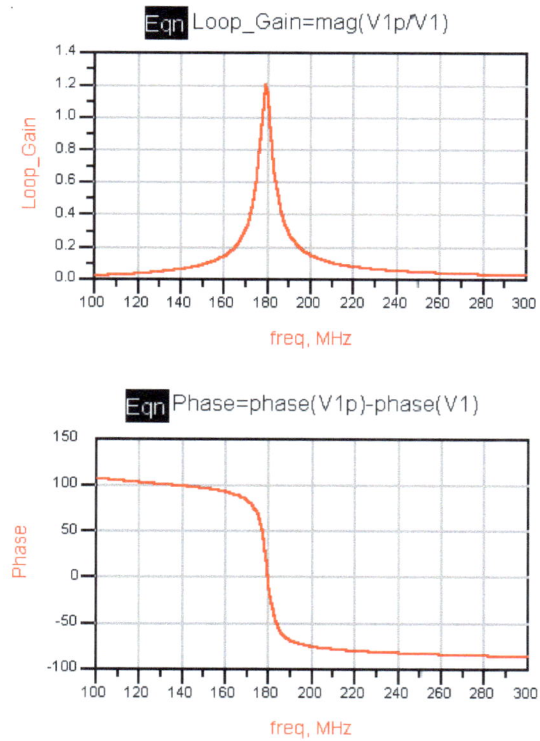

Figure 14.7 AC simulation of the tuned amplifier.

The impedance of the resonator peaks at R_p and the phase is $0°$ at ω_0. The susceptance of L and C cancel at resonance. Figure 14.7 shows the simulated loop gain and phase for the tuned amplifier.

We represent the MOSFET with its simplest small signal model (Figure 14.8).

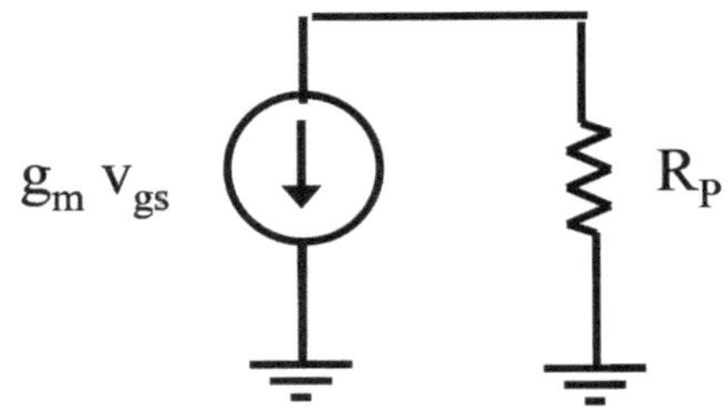

Figure 14.8 MOSFET small signal model with load RP.

The small signal gain is given by:

$$V_{out} = -g_m R_P v_{gs} \tag{14.11}$$

Notice the inversion between input and output. This produces a 180° phase shift for the stage, an inverter.

The large signal output will be a sine wave with a DC component equal to V_{DD}. Since there is very little DC voltage drop across the inductor, the average value of the output over one period must be equal to V_{DD}. The signal is out of phase with the input.

The maximum AC output voltage amplitude will be limited by either clipping (*voltage limiting*)

$$V_{out} = V_{DD} - V_{Dsat} \tag{14.12}$$

or by *current limiting*

$$V_{out} = IP_{D\max} \tag{14.13}$$

The two mechanisms have very different behavior. With *voltage limiting*, the output voltage begins to resemble a square wave. The odd-order harmonic distortion will increase. If the circuit is intended to provide good linear amplification or good spectral purity, this scenario is to be avoided.

With *current limiting*, the signal amplitude can be adjusted so that it never reaches clipping. It swings above and below V_{DD} without distorting. Always build the oscillator so that it current limits.

14.2.2 Circuit #2

The tuned amplifier can form the core of an oscillator. We need to add feedback and one more inversion.

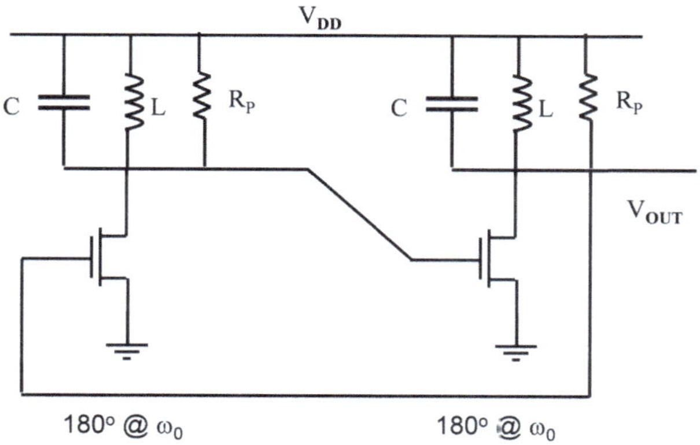

Figure 14.9 Tuned amplifier with two stages, each inverting.

If $(g_m R_P)^2 \geq 1$, this circuit will oscillate (Figure 14.9). It can only oscillate at ω_0, because only at that frequency will we have a total phase shift of $0°$.

The oscillations will begin when the noise inherent in the transistors is amplified around the loop. The strength of the oscillations will build exponentially with time. The small signal analysis doesn't provide a limit to this growth. Obviously, this is wrong. The amplitude will reach a limit either by voltage or current. The example in Figure 14.10 is current limiting,

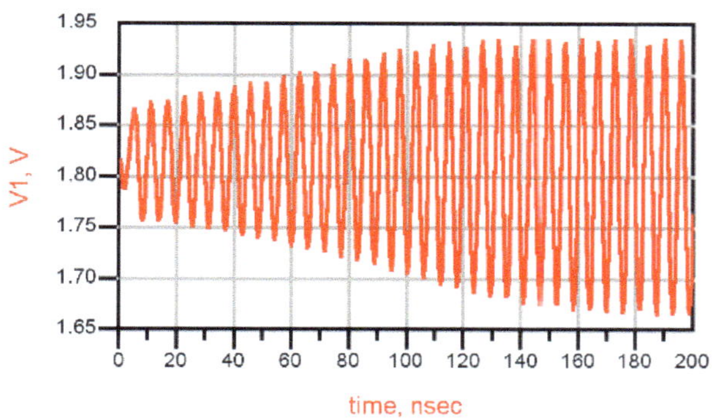

Figure 14.10 Transient simulation of the startup of oscillations and limiting amplitude due to current limiting.

$$V_{OUT} = I_D R_P. \tag{14.14}$$

This circuit is also known as the "cross-coupled oscillator". We can redraw it to look like that in Figure 14.11:

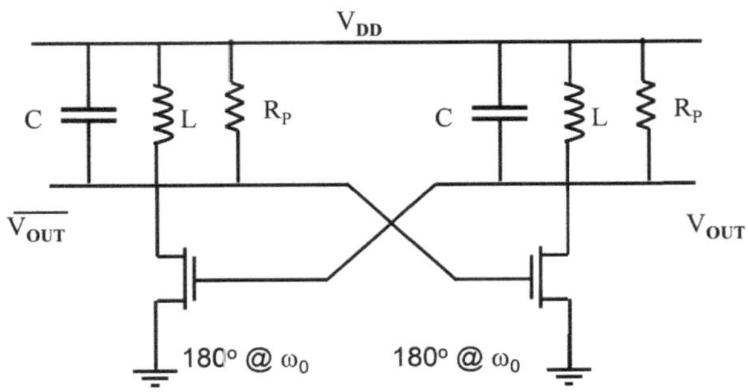

Figure 14.11 Redraw as a cross-coupled oscillator.

This representation emphasizes the differential topology. The two outputs are 180° out of phase. This can be very useful for many applications - driving a Gilbert cell mixer, for example. It has one major shortcoming, however.

Problem: amplitude control:

$$V_{out} = I_D R_P$$

V_{out} can be controlled by adjusting the widths of the active devices. This sets the maximum current that the device is capable of providing at a given V_{GS} since

$I_D \alpha \, (V_{GS} - V_T)$ for a deep submicron MOSFET.

However, V_{out} also will depend on V_{DD}, because the average $V_{GS} = V_{DD}$ in this circuit. That means that it may not always be possible to avoid voltage limiting. If the device width is reduced too far, there may not be sufficient gain for a reliable startup.

14.2.3 Circuit #3

One popular solution to achieve better amplitude control is to break the ground connection, connect the sources, and bias the cross-coupled pair with a current source (Figure 14.12).

Now, the amplitude is controlled by I_0. All of this current is steered between either the left or right side of the diff pair. Thus, the amplitude of

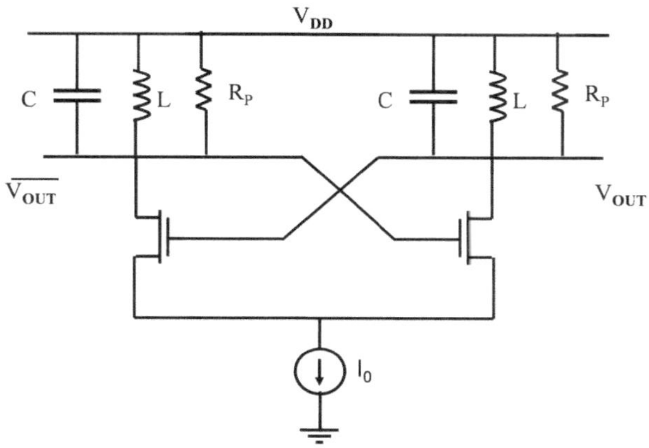

Figure 14.12 Control the amplitude with a current source.

the output will be I_0R_P. This is not perfect, because no transistor current source is ideal. The current will vary slightly with changes in V_{DD}, but it is much more stable than circuit #2. Also, the drain-substrate capacitances of the MOSFET vary with V_{DD} causing some frequency shift. Yet, it is much better for most applications than circuit #2.

The only drawback for this design comes from the current source noise. The channel noise of the device adds to the total noise of the amplifier, so the phase noise of this current-biased design is somewhat worse than that of circuit #2.

14.2.4 Circuit #4

Colpitts oscillator. Common-base configuration (Figure 14.13).

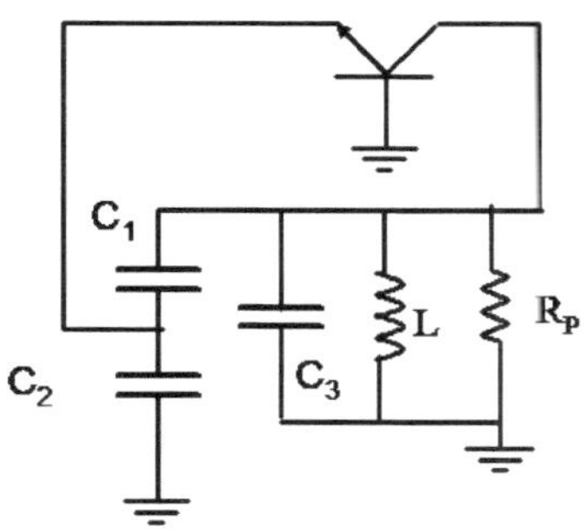

Figure 14.13 Common-base Colpitts schematic.

The active device is in a common base configuration. The open loop gain will be set by the g_m of the device, by Rp and the capacitive divider. The frequency of oscillation is determined by the resonator as in the previous examples. To analyze the open loop gain, let's open the loop at the emitter port shown in Figure 14.14. V1 is the input; V1p is the output from the loop.

This oscillator uses an LC resonator to set the oscillation frequency and a capacitive divider to establish the loop gain. The collector voltage is in phase with the input for a common base configuration. The goal is to determine the conditions where the open loop gain $= 1$ considering the voltage gain of the amplifier A, and the feedback factor $1/N$ The loading effects of the device and the unloaded Q of the resonator must be considered in the analysis.

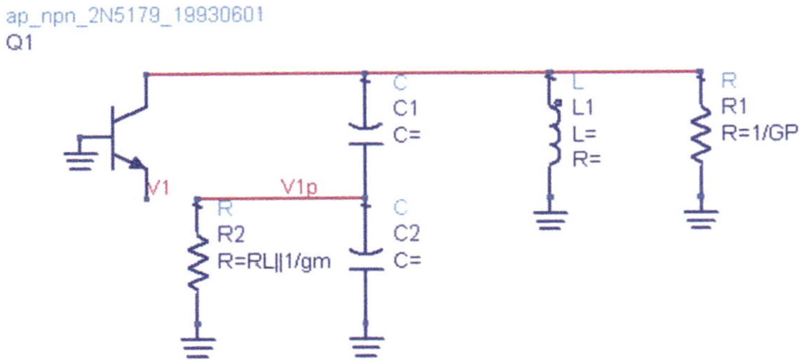

Figure 14.14 Open loop schematic of Colpitts including loading.

Here is a partial schematic - without biasing details. The C1C2 divider sets the feedback ratio. The loop is broken between $V1$ and $V1$p. R_L is the load resistance assuming that the output is taken from the emitter.

Assume $1/g_m \ll R_L$. Then, the loading effect of the device input impedance, $1/g_m$, is transformed up by the tapped C network to N^2/g_m represented by the symbol β. This is combined with G_P due to the unloaded Q of the resonator. Therefore,

$$\beta = \frac{c_1}{c_1 + c_2} = \frac{1}{N} \tag{14.15}$$

$$A = \frac{g_m}{G_P + g_m/N^2}. \tag{14.16}$$

Then the open loop gain is given by:

$$A_{OL} = \frac{V_1'}{V_1} = \left(\frac{g_m}{G_P + g_m/N^2}\right)\frac{1}{N} \tag{14.17}$$

Solving for N, we get:

$$N = \frac{g_m}{2G_P A_{OL}} \pm \frac{1}{2}\sqrt{\left(\frac{g_m}{G_P A_{OL}}\right)^2 - \frac{4g_m}{G_P}}. \tag{14.18}$$

This is bias dependent because $g_m = I_C/V_T$.

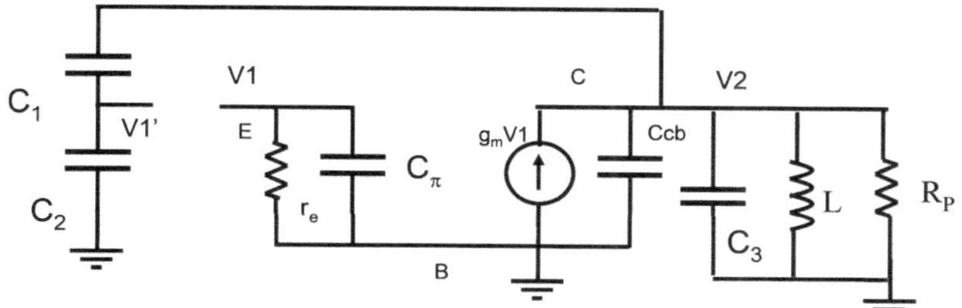

Figure 14.15 Small signal model of the device in the oscillator circuit.

We also need to consider the effects of the device capacitors on the frequency (Figure 14.15)

- Note that C_π is parallel to C_2.
- Also, C_{cb} is parallel to C_3 and C_T where

$$C_T = \frac{C_1 C_2}{C_1 + C_2} \tag{14.19}$$

and capacitor C_3 might represent the varactor diode or tuning capacitance.

How do you find the device capacitances? The data sheet will give you C_{cb} and f_T. From f_T you will find C_π

$$f_T = \frac{g_m}{2\pi\left(c_\pi + c_{cb}\right)} \tag{14.20}$$

Thus, subtract C_π from C_2 and C_{cb} from C_T.

For startup, we should choose an open loop gain greater than 1 . Choose 3 for example. Select I_{bias} and determine g_m. Evaluating the equation for N at an open loop voltage gain of 3, we obtain two answers. The larger N value limits loop gain by the tapped C ratio. The smaller N limits loop gain by the loading of the tank with N^2/g_m.

One further consideration is the phase shift caused by the parallel combination of C_2 and $r_e = 1/g_m$. There will be a slight phase shift that will move the total phase loop off the peak of the resonator.

- Not desirable: resonator has maximum phase slope at ω_0.
- The effect is minimized by making C_2 large and r_e large: $\frac{1}{\omega_0 c_2} << r_e$.
- Or, you can add resistance in series (Figure 14.16).

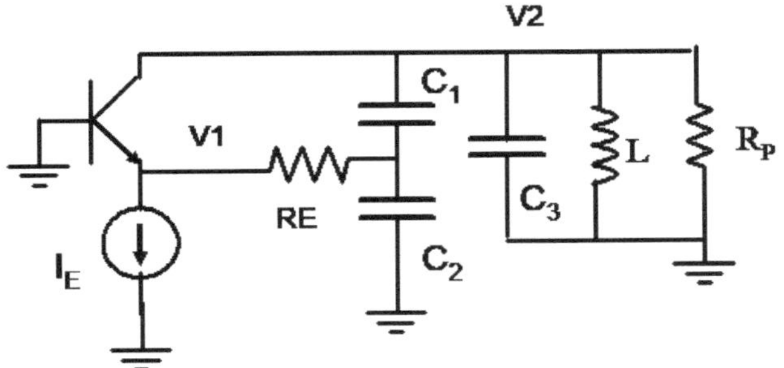

Figure 14.16 R_E can help reduce the phase shift by making $r_e + R_E \gg \omega C_2$.

However, this added resistance will also reduce the loop gain. The startup gain must be large enough so that the oscillator will not stop oscillating due to nonlinear g_m compression. That would cause a start-stop condition that would modulate the signal. This has been called "squeeging" in some literature.

14.3 Biasing and Bypassing

In order to simulate the open loop gain and the oscillator closed loop response, we first need to bias the transistor. The choice of collector current is arbitrary but the amplitude will grow in proportion to I_C. So, you need to choose V_{CC} and the base bias voltage to avoid voltage limiting due to forward biasing of the base-collector junction. In the example shown in Figure 14.17, $I_C = 0.5$ mA. The base voltage source $= V_{CC}/2$. Since the base current in the device is quite low, about 10μA, there is no need to dump a lot of current into the base bias resistive divider. R_2 is the emitter bias resistor. It also provides DC negative feedback to stabilize the collector current. L_2 is an RF choke, used to avoid AC gain degeneration due to R_2. It may not be feasible for implantation of integrated oscillators.

RF bypass capacitors should be appropriately chosen for the frequency in use. In this case, 165 MHz, A 1nF capacitor C_5 has roughly 1 ohm capacitive reactance at this frequency. Choosing much larger capacitors runs the risk of encouraging low frequency parasitic oscillations. Also, the inductive reactance above self-resonance can induce a negative resistance.

A high frequency bypass capacitor (C_7) is typically used where the DC source is connected. A small resistance would be added in series between the DC source and the L_1, R_3, C_7 node to provide a low-pass filter. This reduces oscillator output entering the power distribution on the board and also filters any high-frequency noise on that line from modulating the oscillator.

Where V_{CC} enters the board, a larger capacitor, typically tantalum or electrolytic, is appropriate to provide a charge reservoir on the PC board and to provide low frequency isolation from the power source.

14.4 Simulation Methods

14.4.1 AC simulation of open loop

AC is a linearized simulation, so the result is independent of the choice of V_{ac}. The schematic in Figure 14.17 and the results in Figure 14.18 are an example of the linear simulation. If you look at the ratio of V_{ep} to V_e, you find a gain of approximately 1.6 at the gain peak 161 MHz. Perhaps the frequency is off slightly due to C_{cb} being larger than expected. Also, open loop gain is lower than predicted by our simple hand analysis.

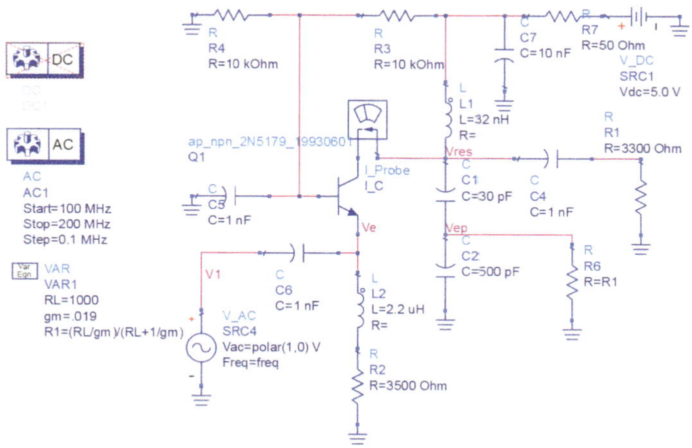

Figure 14.17 Open loop AC simulation of CB Colpitts.

Note that the peak OL gain simulated is less than the analysis predicted; 1.6 vs. 3. Replacing the device with a voltage controlled current source

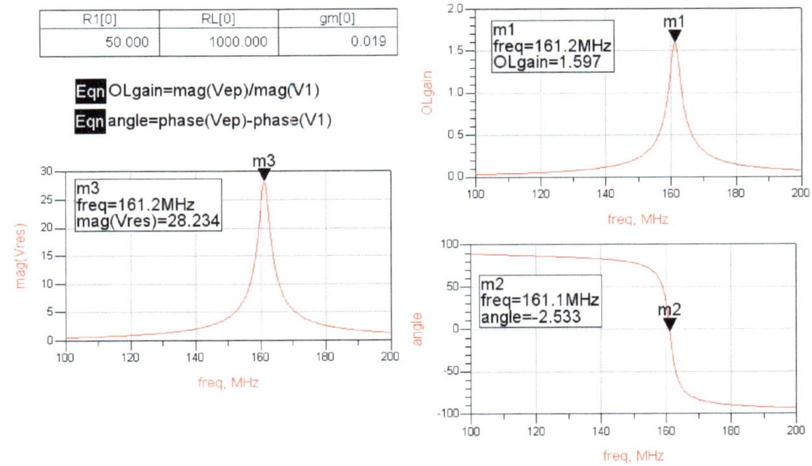

Figure 14.18 Output of open loop AC simulation of CB Colpitts.

(VCCS), the simulations and analysis agreed exactly. Evidently, our simplified small signal model of the BJT, which does not include internal base resistance, R_{bb}, is not good enough to get exact agreement.

Exercise: Use the design equations to verify the component values selected in Figure 14.17 for an oscillator frequency of 160 MHz. Assume $f_T = 1500$ MHz and $C_{cb} = 1$ pF.

14.4.2 Transient analysis of oscillator

Transient analysis can be used for a closed loop analysis of the oscillator (Figure 14.19). It uses the nonlinear model of the device. Starting the simulation requires that an impulse be delivered to the resonator. Otherwise, the oscillator will not start in simulation. A pulsed current source is ideal in that it defaults to an open circuit when inactive. Make sure the period of the pulse is longer than the simulation time. Otherwise, you will get multiple impulses making for a confusing output signal!

In Figure 14.20 we see that the oscillator has reached a limiting amplitude of 1.1 V. Note that the oscillator is current limiting, not voltage limiting because $V_{CB} = V_{res} - V_e$ never drops below 2 V. The resonator amplitude seems to be predicted by $V_{res} = IC^* \left(R_P || N^2/g_m\right)$, off by a factor of 2 from

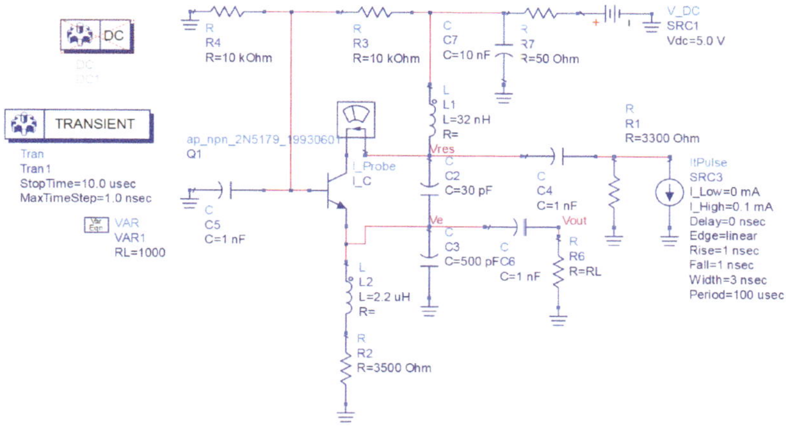

Figure 14.19 Schematic for a closed loop transient simulation of CB Colpitts.

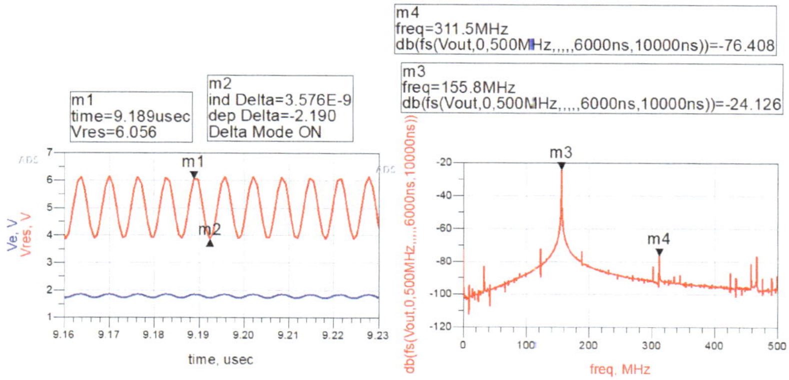

Figure 14.20 Output from CB Colpitts transient simulation.

Lee's predictions [3]. This may be explained by the rather low bias current which is not driving the device strongly into cutoff.

The frequency spectrum was made using the fs function in ADS. The syntax is:

$y = fs(\{x,$ fstart, fstop, numfreqs, dim, windowType, windowConst, tstart, tstop, interporder, transformMethod$\})$

To view the spectral content from a transient simulation, the fs function is used. fs(Vout, $0, 500$ MHz, , , , , 6 us, 10us) The parameters that are not of interest can be set to a default by replacing with commas. The 6us,

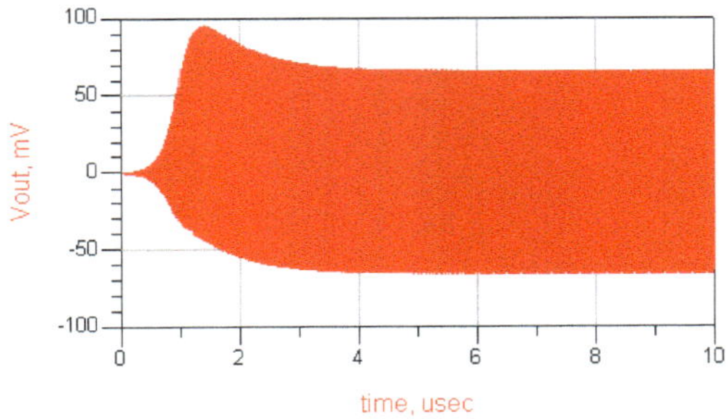

Figure 14.21 Transient analysis time domain plot. It takes many periods to reach a steady state amplitude.

10us parameters windows the fft to the end of the transient simulation data, thus avoiding the crazy startup stuff seen in Figure 14.21 which would give misleading spectral information. We see that the spectrum shows very low second harmonic content in the output.

14.4.3 Harmonic balance (HB) simulation

Harmonic balance (HB) simulation can also be used for closed-loop oscillator analysis (Figure 14.22). HB has the advantage that it calculates the steady-state solution directly and is relatively fast compared with transient analysis (when it converges). To simulate an oscillator, the ADS OscPort must be inserted into the signal path. The arrow must point in the output to input direction.

It is important to give HB a starting frequency that is close to what you expect to be the frequency of oscillation. Specify sufficient harmonics so that harmonic distortion can be calculated at the desired integer multiples of the fundamental frequency.

An emitter follower has been added to isolate the output loading from the resonator. The N ratio was modified to take into account the loading of the emitter follower.

The harmonic balance controller must be activated for oscillator analysis in order to recognize the OscPort, as shown in Figure 14.23.

A few specifics about setting up the OscPort.

1. **NumOctaves:** This specifies the total number of octaves that will be searched. Half are above; half below.

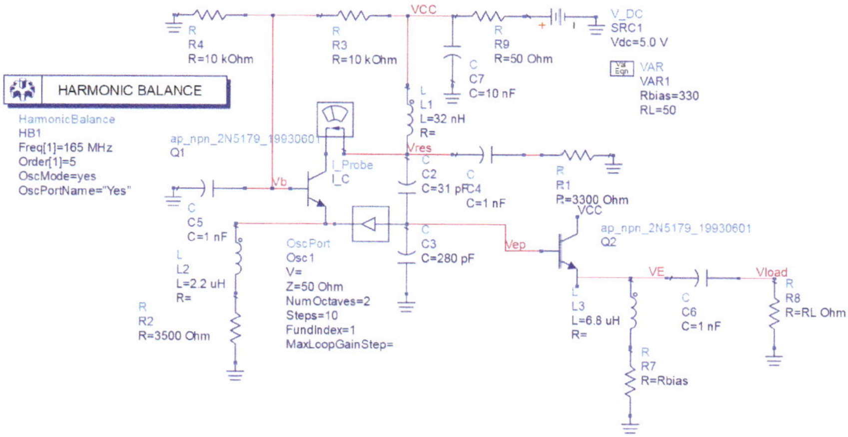

Figure 14.22 Harmonic balance simulation schematic. OscPort must be inserted in the feedback path.

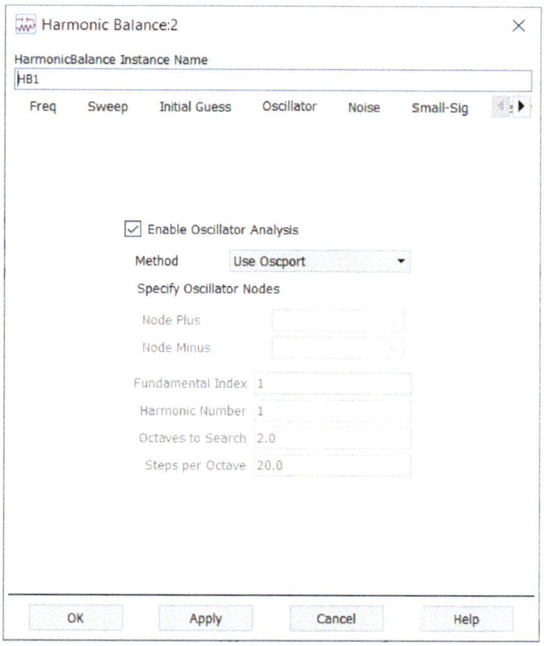

Figure 14.23 HB setup for oscillator simulation.

2. **Steps:** Number of frequency points per octave. If Q is high, many steps should be used to find the right frequency.
3. V **:** Amplitude. Don't specify. The program will figure that out.
4. Place the probe in the signal path pointing in the direction of positive gain around the loop. Figure 14.22 shows the placement that works well for the Colpitts.

The spectral output from the HB simulation is illustrated in Figure 14.24. Order = 5; five spectral components, fundamental plus four harmonics.

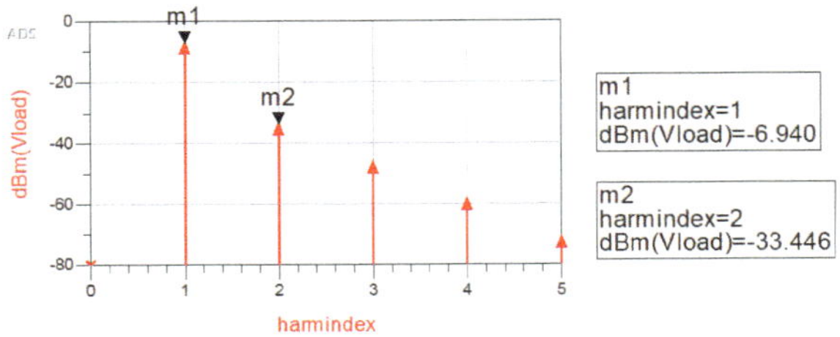

harmindex	freq
0	0.0000 Hz
1	164.0 MHz
2	328.1 MHz
3	492.1 MHz
4	656.2 MHz
5	820.2 MHz

Figure 14.24 The spectral output from the harmonic balance simulation of the example oscillator. Harmindex represents the harmonic frequency.

The output from a HB oscillator simulation can be either a frequency spectrum or a time domain representation of any voltage or current in the circuit. You can see that the spectral representation is much cleaner than the FFT generated from transient simulation. This is because HB calculates only at the harmonic frequencies.

The time domain is generated using the ts function. This is an inverse Fourier transform or something similar that is used to convert the harmonic frequency amplitudes and phases into a time domain sequence. Looking at the resonator voltage, and comparing with the base voltage, we can see in Figure 14.25 that the oscillator is current limiting. V_{cb} or V_{dg} must remain positive to avoid clipping.

In the case of the bipolar transistor, if $V_{cb} = 0$ or is negative, the base collector junction is forward biased - a low impedance at the collector will be experienced. For the commonbase Colpitts,

$$V_{cb} = V_{res} - V_b \tag{14.21}$$

As was previously mentioned, the opposite of current limiting is called voltage limiting. In that case, clipping limits the amplitude of oscillation.

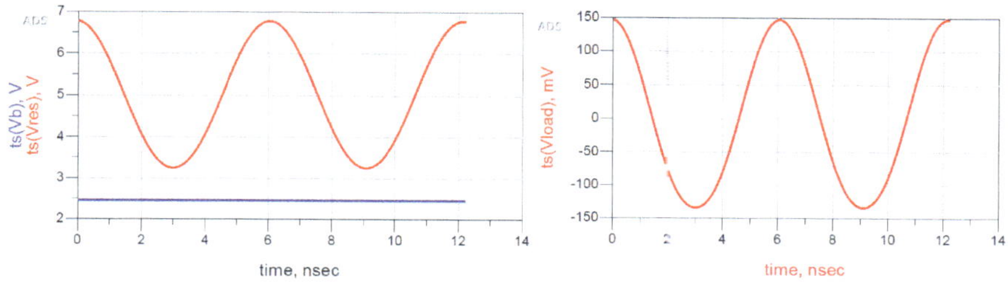

Figure 14.25 Time domain output from HB simulation uses the ts function.

This is undesirable because it generates a lot of harmonic distortion, plus the low impedance state when the device is driven into saturation or the ohmic region kills the tank Q for part of the cycle and is bad for FM or phase noise.

14.5 Amplitude Prediction

For some oscillator topologies, it is easy to predict the amplitude of the oscillator output. For example, the differential oscillator described in Figure 14.12 is biased by a current source, so the maximum current into the tank is known. If the R_P is also well controlled (not necessarily easy to do), then $V_{tank} = I_C R_P$. Other topologies like the Colpitts are not so easy. In Lee's book [3], he presents a describing function method for predicting oscillator amplitude. It works for the common base/gate Colpitts pretty well, but seems to fail completely when trying to predict amplitude of even the common collector/drain version.

In general, simulation provides your best chance of getting an estimate of the limiting behavior and thereby the amplitude of your oscillator under consideration.

What we can glean from Lee's work, however, is that a well-designed LC oscillator like the Colpitts must be current limited: that is, we allow the active device to go into cutoff for part of the period of oscillation. This reduces the effective g_m of the device, averaged over the period, and thus gives the needed gain compression. It also benefits phase noise as we will later see. The simulated drain current for a common collector Colpitts oscillator in Figure 14.26 illustrates that the device can be in cutoff for much of the period.

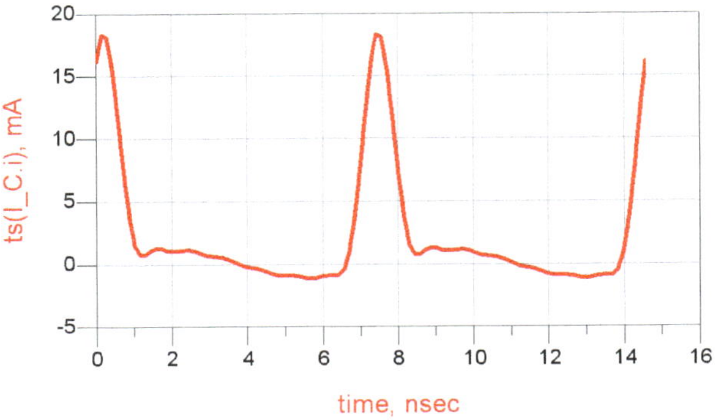

Figure 14.26 Collector current of the common collector Colpitts.

14.6 Buffer Amplifier Considerations

The output must be taken somewhere from the oscillator. Taking the output directly from the resonator will reduce the loaded Q and therefore may increase phase noise depending on whether the current can be increased to raise the oscillator power without breakdown of the device. Chapter 15 will delve into the phase noise problem in much detail.

For maximum power transfer, as in Equation (14.22), Q_L is reduced by a factor of 2:

$$G_L = G_P + \frac{g_m}{N^2} \tag{14.22}$$

To obtain lighter output coupling, a transformer could be used. Or take the output from the emitter as seen in Figure 14.22 making use of the tapped C as an impedance transformer.

The choice of amplifier circuit will depend on the application but the following considerations dictate the use of some sort of buffer amplifier between the oscillator and the load:

1. Isolation. Prevent the output load impedance from pulling the oscillator frequency, reducing resonator Q, or decreasing amplitude. This is especially important for an oscillator designed for an unknown load impedance that varies with application.
2. Power. The oscillator itself may not be capable of providing sufficient power to the load. For example, passive diode double balanced mixers require at least 7dBm($5\ \mathrm{mW}$. $0.71\ \mathrm{V}$ into 50ohm load).
3. High input impedance. Minimize loading on the oscillator.

14.6.1 Bandwidth

If the oscillator is used for a narrow bandwidth application (typically 10% or less), then a tuned amplifier can be designed as a buffer. For wider bandwidths than this, then untuned amplifiers are required.

14.6.2 Options

1. **Narrowband:** A tapped C tank circuit can be used for impedance transformation at the output of the amplifier. This can greatly reduce power dissipation if the supply voltage is sufficient.
2. **Wideband:**
    a. Inductive load on the drain or collector can be used as will be shown in the design example.
    b. Emitter or source follower.
    c. Wideband transformer coupling at output can reduce power required for the buffer.

14.6.3 Common emitter buffer amplifier

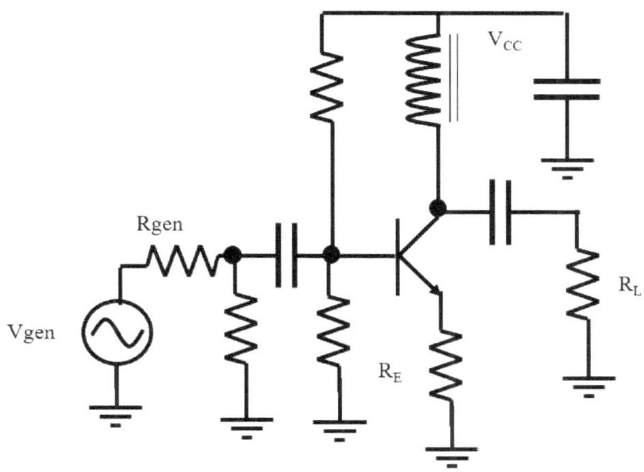

Figure 14.27 Common emitter class A buffer amplifier.

A common emitter class A buffer amplifier is shown in Figure 14.27. Note: The buffer amplifier is untuned. Thus, we need to avoid reduced conduction angle modes such as Class AB or Class B. The device should not be driven into cutoff. There is no low impedance termination for even-order harmonic frequencies. Harmonic termination is needed to form the half sine wave current waveform for Class B. A push-pull design with appropriate transformer could be used if Class B is desired.

The inductor will try to keep the current constant. Suppose the device current were to go to zero. Then the inductor will try to keep $I_{DC} = I_{CQ}$. So

$$V_{L,\max} = I_{CQ} R_L. \tag{14.23}$$

Then, to make the sine wave symmetric, the device must sink $2I_{CQ}$ so that the output voltage drops to zero. The maximum output power from the amplifier will be

$$P_O = \frac{V_{L\max}^2}{2R_L}. \tag{14.24}$$

To avoid clipping, the saturation of the device must be avoided. This presents a minimum requirement for the supply voltage:

$$V_{CC} > V_{CE,sat} + I_{CQ}\left(R_L + 2R_E\right) \tag{14.25}$$

Voltage gain from the circuit will be approximately

$$A_V \cong \frac{-R_L}{r_e + R_E}. \tag{14.26}$$

R_E can be used to adjust the gain. Or the transducer gain, G_T

$$G_T = \frac{\frac{V_L^2}{2R_L}}{\frac{V_{gen}^2}{2R_{gen}}}. \tag{14.27}$$

We must work within the limitations of the device chosen for the buffer.

- Maximum collector or drain current: I_{Cmax} or I_{Dmax}
- Maximum collector - emitter breakdown voltage, BV_{CEO}
- Maximum safe device operating temperature, T_{max}. This includes the self-heating and the temperature of the heat sink.

These can typically be found on the data sheet. A safe design does not run all the way to the limits of the device, so a reasonable margin needs to be provided. It may not be wise to exceed 75% of the recommended maximum values.

For the class A buffer amplifier, the peak values of voltage and current are:

$$V_{CE,max} = V_{CC} + I_{CQ} R_L \tag{14.28}$$

$$I_{C,max} = 2I_{CQ}. \qquad (14.29)$$

An emitter follower or source follower could also be used as a buffer amplifier. It can be directly DC coupled and is a simple design. However, it provides less isolation than the common emitter or common source buffer because of the gate-drain or base-collector capacitance.

Example 14.1: Let's design a buffer

Specs: 7 dBm output power (5 mW; 0.71 V in 50 ohms)
Source impedance = 100 ohms
Voltage output from oscillator = 0.5 V.
Suppose that the device has: $BV_{CEO} = 12$ V; $I_{C,max} = 100$ mA.

1. **Determine** I_{CQ} : We start with $R_L = 50$ ohms,

$$P_o = I_{CQ}^2 R_L/2 = 0.005 \text{ W} \quad I_{CQ} = 14.2 \text{ mA}$$

This gives us a maximum current of 28.4 mA, well within the limits of the device.

2. **Determine R_E:** This will be determined by the gain required,

$$A_v = 0.71/0.5 = 1.42 = -\tilde{g}_m R_L = \frac{-R_L}{r_e + R_E},$$

where $g_m = 0.55$; $r_e = 1.8$ ohms. So, $R_E = (R_L/Av) - r_e = 33$ ohms. Check the input resistance: $R_B \| R_E(\beta + 1) >> R_S$. Make sure the bias resistors (R_B) don't load down the oscillator or reduce the input amplitude significantly.

3. **Determine $V_{CC,min}$, $V_{CC,max}$:** This is the range of acceptable power supply voltages which avoid clipping and breakdown:

$$V_{CC,min} = V_{CE\,\text{sat}} + I_{CQ}(R_L + 2R_E) = 2.1 \text{ V}$$
$$V_{CC,max} = 0.75BV_{CEQ} - I_{CQ}R_L = 8.3 \text{ V}.$$

4. **Check for temperature rise:** Worst-case power dissipation $= P_D = V_{CC}I_{CQ}$ for class A.
Choose $V_{CC} = 5$ V.$P_D = 71$ mW (this will drop to approximately 66 mW at full output, but you must design for the worst case where the oscillator may fail to produce an output):

$$T_{max} = R_{TH}P_D + T_{ambient}$$

where R_{TH} is the thermal resistance (°C/ Watt).
Maximum temperature is usually 150°C. If you exceed this, then you must reduce V_{CC} or I_{CQ}.

5. **Stability:** Finally, you must check for stability over a wide range of frequencies. Use the S-parameter simulation mode in ADS and plot k and mag delta vs. frequency. If there are potential instabilities, you must deal with them using techniques that were detailed in Chapter 6.

How do you reduce power dissipation in a LO buffer amplifier? The bias current of the amplifier is directly related to the output voltage swing and R_L. If the supply voltage is fixed, then you can reduce current to save power.

1. **Increase R_L** : This may be practical in situations where the frequency is relatively low and interconnection lengths short. This is quite practical for on-chip implementations, but, if the oscillator and mixer are separated on different boards or distant from one another, then a transmission line must be used to interconnect them. This requires Z_0 typically in the $50 - 75$ ohm range.
2. **Use a transformer** to step up impedance at the drain or collector node. This is feasible if there is sufficient supply voltage to handle the extra voltage swing at the drain or collector and if the device is not close to breakdown.

Figure 14.28 shows an example of a wideband transformer that is used from HF through VHF frequencies.

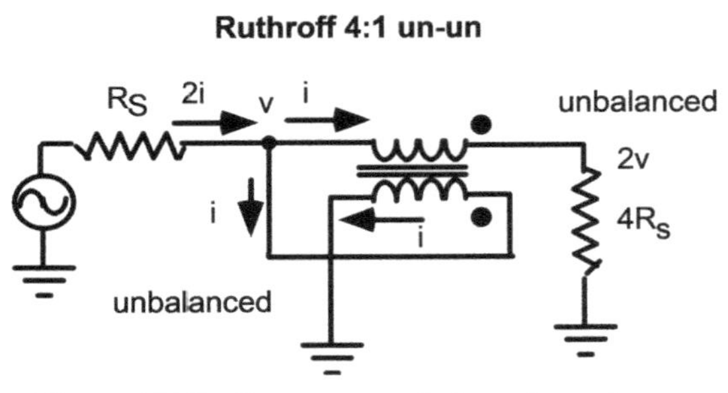

Figure 14.28 Four to one unbalanced transformer.

The circuit above provides a $4:1$ transformation between two unbalanced impedances. It works by bootstrapping the voltage from v at the left to $2v$ at the right. The transformer windings are connected in series. The current at the input is $2i$, split two ways. So the output current is just i. So, we get twice the voltage and half the current at the output. Of course, it can be used the other way to transform a lower R_L into $4R_L$ at the collector. The voltage at the high impedance end is twice that at the low impedance end, so if used at the drain of an amplifier, VCC and the breakdown voltage must be adequate.

The transformer is generally implemented using transmission lines, either coax or twisted wires. These are wrapped around a ferrite core to increase the common mode inductance and extend the bandwidth to lower frequencies.

3. **Tapped capacitor tuned amplifier:** If a narrowband buffer is adequate, then the tapped capacitor impedance transformer can be applied to the amplifier output (Figure 14.29). The loaded Q of the resonator must be small enough to provide the required bandwidth for the application.

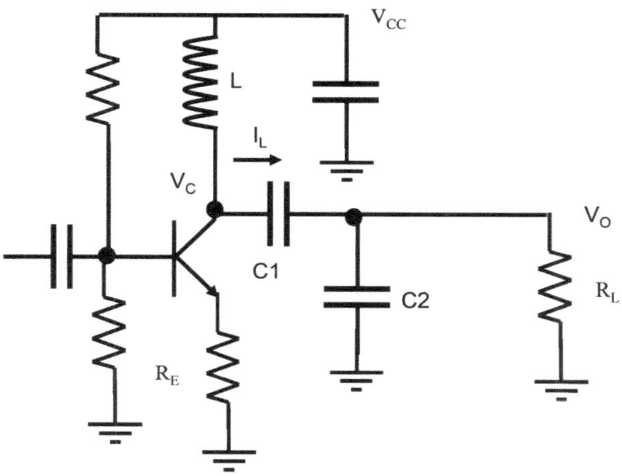

Figure 14.29 Tuned buffer amplifier.

The ratio $n = \frac{C_1+C_2}{C_1}$ applies to voltage and current and impedance as follows:

$$V_C = nV_O$$
$$I_L = \frac{V_O}{nR_L} \tag{14.30}$$
$$Z_C = \frac{V_C}{I_L} = n^2 R_L.$$

The collector voltage, V_C, will be scaled up by a factor of n, so one must have sufficient V_{CC} and breakdown voltage to scale by a large factor.

You can see though that bias current can be reduced by the factor $1/n$ when this is applied. If V_{CC} is fixed in the design, a net power savings is obtained.

14.7 Voltage Controlled Oscillator (VCO)

Many applications of oscillators require some frequency control. In older local oscillator applications, say for a radio, this might consist of an air-variable multiplate capacitor that tunes the resonator. It is often combined on the same rotating shaft with another for control of an input bandpass filter. More modern applications for resonator-based oscillators have used a varactor diode whose capacitance is adjusted by a reverse bias according to the equation:

$$C_j\left(V_R\right) = \frac{C_j(0)}{\left[1 + \frac{V_R}{\phi_B}\right]^m} \tag{14.31}$$

where $C_j(0)$ is the zero-bias junction capacitance, V_R is the tuning voltage (reverse bias), ϕ_B is the built-in voltage (0.7 V for silicon), and m is a grading parameter.

The parameter m varies with the junction doping profile (Table 14.1).

Table 14.1 Grading parameter vs. doping profile of a varactor diode.

$m = 0.5$	Abrupt junction
$m = 0.33$	Graded junction
$m > 0.5$	Hyperabrupt junction

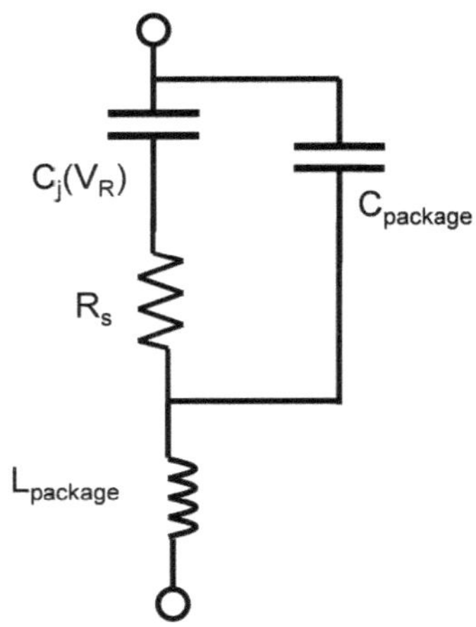

Figure 14.30 Equivalent circuit model of a varactor diode.

The varactor Q is determined by the series RC network shown in Figure 14.30. Thus, it is frequency and voltage dependent. On a data sheet for a discrete component, it is specified typically at $V_R = -4$ V and at a frequency of 50 MHz :

$$Q = \frac{1}{\omega RC_j\left(V_R\right)} \tag{14.32}$$

Varactor diodes are often specified by a ratio of capacitances at two voltages. Let's take an example - the MV104 diode. From this we can determine the exponent m,

$$\frac{c_j(-3)}{c_j(-30)} \cong 2.65 \tag{14.33}$$

If we assume $\phi_B = 0.7$ V for a silicon diode, then

$$\frac{c_f(-3)}{c_f(-30)} = \frac{[1 + 30/0.7]^m}{[1 + 3/0.7]^m} \tag{14.34}$$

Taking the log of both sides, we can solve for m : $m = 0.46$. Pretty close to the 0.5 that we would expect for an abrupt junction. Now using Equation (14.31), $Cj(0)$ can be determined to be 86pF. From this equation then, Cj at any bias voltage can be found.

14.8 Biasing the Varactor Diodes

Figure 14.31 shows an example from a CB Colpitts oscillator.

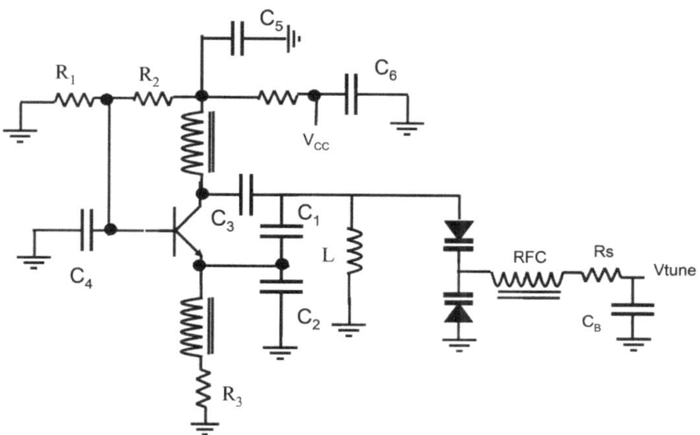

Figure 14.31 Schematic diagram of a varactor tuned Colpitts oscillator.

- You must isolate the tuning port from the resonator. The RF choke is intended to present a high impedance to the node between the two back-to-back varactor diodes so that the oscillator signal does not leak out the tuning port.
- Rs is needed to de- Q the RFC in case it resonates with a diode.
- Two varactor diodes, back-to-back, are used to make the $C - V$ characteristic symmetric and to avoid the possibility of forward bias.
- C_B is a bypass cap needed for isolation from outside. If the oscillator is used in a PLL, one must take care that the extra pole that C_B produced does not affect the loop phase margin.
- The tank inductor L keeps both ends of the varactor diode pair at ground DC potential so that each diode receives the same tuning voltage.

ADS can be used to demonstrate that this configuration provides equal and opposite voltages to the back-to-back diodes (Figure 14.32). Because of the back-to-back configuration, we get half of the capacitance of an individual diode.

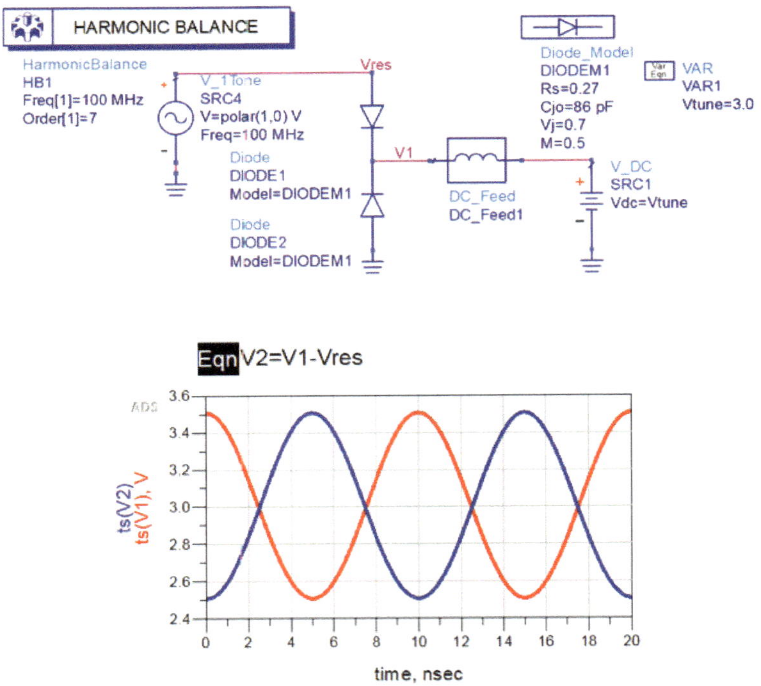

Figure 14.32 Back-to-back varactor diode voltages.

14.8.1 Tuning range of oscillator

The minimum (C_{min}) and maximum (C_{max}) diode capacitance, L, and the series combination of C_1 and C_2 will determine the tuning frequency range of the oscillator. Call the series capacitance C_T.

$$C_T = \frac{C_1 C_2}{C_1 + C_2} \tag{14.35}$$

Then

$$\begin{aligned} \omega_{\min} &= \frac{1}{\sqrt{L\,(C_T + C_{\max})}} \\ \omega_{\max} &= \frac{1}{\sqrt{L\,(C_T + C_{\min})}} \end{aligned} \tag{14.36}$$

Two equations determine L and C_T for a given frequency range.

14.8.2 Common drain Colpitts

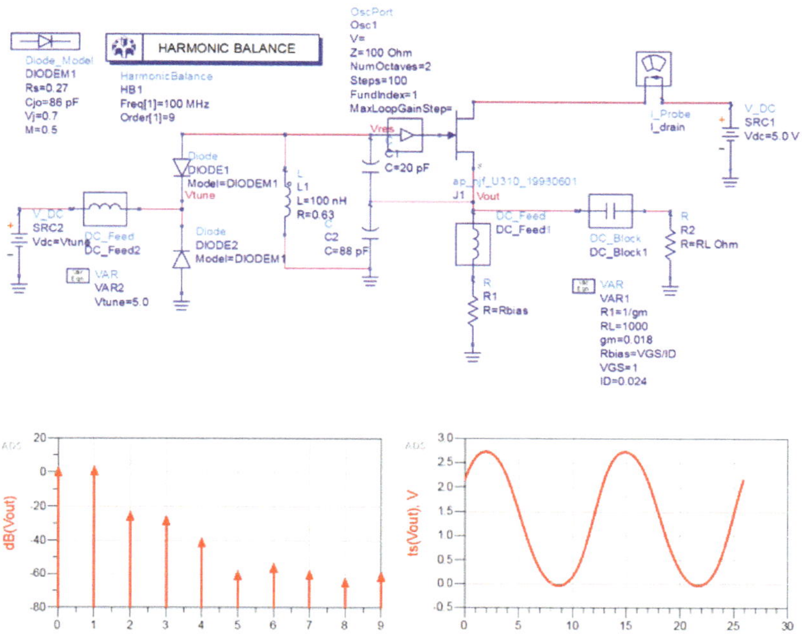

Figure 14.33 Common drain Colpitts example with varactor tuning diodes. JFET implementation.

Figure 14.33 presents a schematic for harmonic balance simulation of a common-drain Colpitts VCO at 100 MHz. The simulation output spectra and voltage are also shown. RF chokes are represented by ideal DC_Feed components and coupling capacitors by ideal DC_Block components. A real implementation would of course require replacing these with appropriately characterized real components. Note that the simulation setup for HB includes:

- A harmonic balance controller with order $=$ 9. Oscillator simulation must be activated in the controller.
- The OscPort is placed in the forward path between the top of the tapped capacitor resonator and the gate of the JFET.
- The varactor diode is modelled by the three parameters described in equation (14.31)
- A buffer amplifier was not included in this example.

One can see from the simulation output that the oscillator is producing a sine wave output with low harmonic content.

14.9 Crystal Oscillator

Crystal resonators can provide high stability with a reduced temperature coefficient. It acts as a mechanical resonator whose frequency is dependent on the thickness of the quartz. It yields a very high Q series resonance. Q on the order of 100,000 is typical. Crystal resonators can also oscillate at overtone frequencies, sometimes desirable when a higher frequency reference oscillator is needed.

The Pierce circuit below is a common one used when a quartz crystal oscillator is needed for an IC such as a PLL or microprocessor. A simple CMOS inverter, an inverting single transistor CE or CS amplifier can be used as the gain element needed for oscillation.

The gain element, perhaps on-chip, is above the dashed line in Figure 14.34. Rf is a feedback biasing resistor, often on-chip. R1 is used to limit the current through the crystal to prevent heating and consequent frequency vs temperature drift. This is not always required, but should be considered because excess current in the crystal could cause frequency drift due to heating.

C_1 and C_2 are external capacitors needed to obtain the correct oscillation frequency. All parallel resonant fundamental mode crystals are specified for a certain load capacitance C_L.

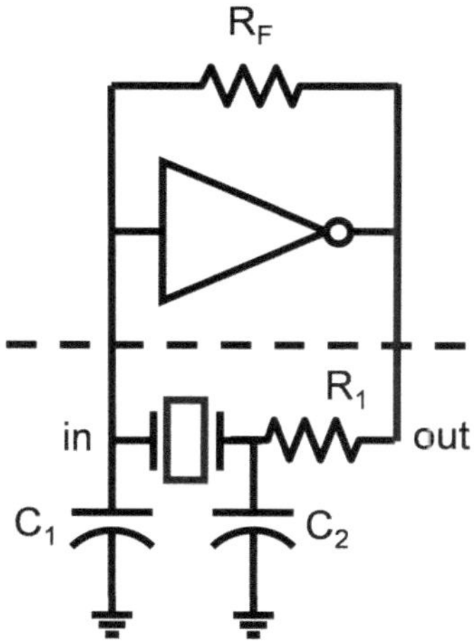

Figure 14.34 Pierce crystal oscillator.

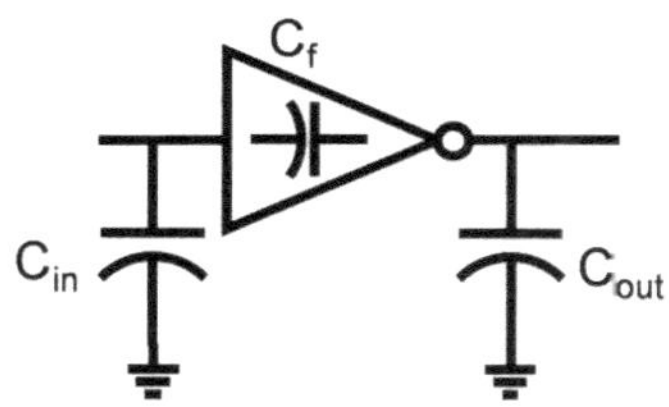

Figure 14.35 Parasitic capacitances of the oscillator must also be considered.

C_{in}, C_{out}, C_f are parasitic capacitances of the amplifier shown in Figure 14.35 (usually given on data sheet or can be calculated for a known discrete transistor amplifier). C_{in} and C_{out} may also include PC board capacitances.

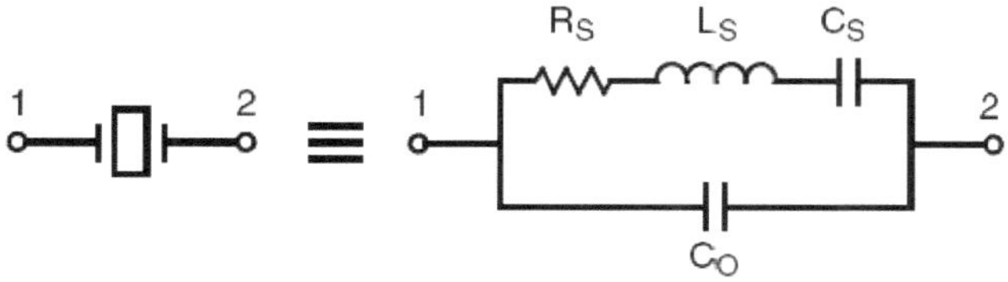

Figure 14.36 Equivalent circuit of the quartz crystal.

Figure 14.36 represents an equivalent circuit model for a quartz crystal. CS and LS are the motional capacitance and inductance respectively and R_S

is the internal series resistance. R_S is usually specified and can be used to calculate the power dissipation, $P = I^2 R_S/2$.

C_o is the crystal holder capacitance - usually given by the vendor data sheet. Typically, it is in the range $2 - 10$ pF.

To obtain the correct frequency:

$$C_L = \frac{C_1 C_2}{C_1 + C_2} + C_a + C_o + \frac{C_{in} C_{out}}{C_{in} + C_{out}} \tag{14.37}$$

C_1 can be a variable capacitor so that the frequency can be adjusted if necessary.

14.10 Ring Oscillators

Another type of oscillator can be made by connecting an odd number of inverting gain stages in a ring. This is more easily integrated on-chip than the resonator-based types. In order to start, the total loop gain must be greater than 1. However, in itself, it will not provide a stable frequency because the delays are temperature and voltage dependent. Therefore, it often must be accompanied by a phase-locked loop or delay-locked loop. A brief introduction to phase locked loops is provided in Appendix 14.A.1. These will stabilize the frequency through an external reference, often a crystal oscillator. The voltages at nodes A, B and C are drawn in Figure 14.37.

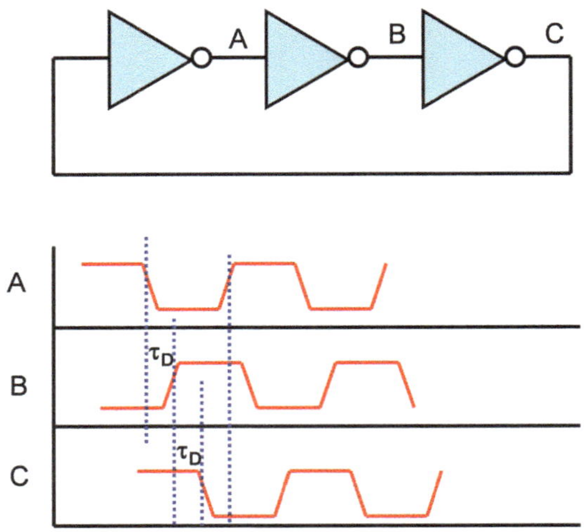

Figure 14.37 Simple inverter example of a ring oscillator. τ_D is the delay per stage.

Period of oscillation and frequency are related to the delay by:

$$T = 2n\tau_D, \tag{14.38}$$

$$f = \frac{1}{T}, \tag{14.39}$$

where n = number of stages.

14.10.1 Implementation of ring oscillators

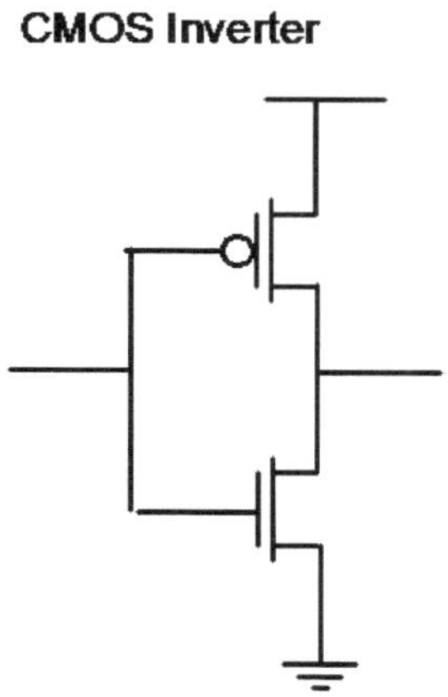

Figure 14.38 A CMOS inverter.

Figure 14.38 shows a CMOS inverter, which is a simple way of implementing a ring oscillator. We must have an odd number of stages, otherwise it forms a latch that hangs at either high or low.

Because ring oscillators are mainly used in mixed signal ICs where baseband digital and analog RF share the same die, a differential implementation is much more frequently utilized shown in Figure 14.39 The differential implementation provides:

- Common mode rejection of substrate coupled noise
- Easy to control the delay
- Can use an odd or even number of stages.

A current source is often used to control the time delay through the differential cell. Noise sources associated with the devices add to the timing jitter produced by oscillator phase noise. The jitter occurs because of

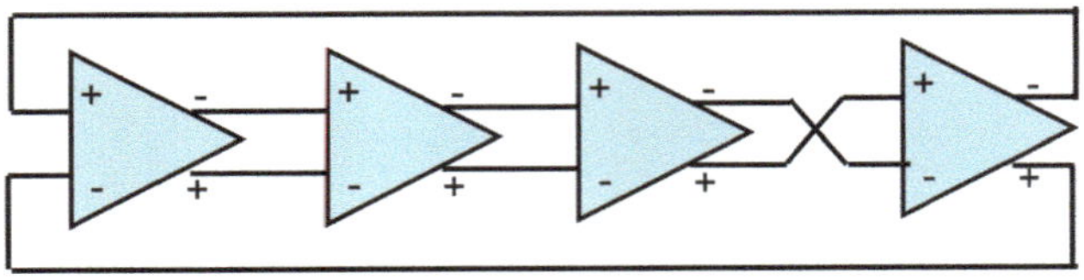

Figure 14.39 Twisted ring differential ring oscillator.

random phase variations that are converted to time delays at the crossover point.

When located in a phase locked loop, the oscillator frequency and phase are controlled by a filtered output from the phase detector. This output is proportional to the difference between the reference phase and the VCO phase. This voltage can then be used to set the delay of the ring oscillator loop [4] (see Appendix 14.A.1).

References

[1] Hayward, W. (1994) *Introduction to Radio Frequency Design* American Radio Relay League, ch. 7.

[2] McWhorter, M., Scherer, D., and H. Swain, H. (1995) *EE344 High Frequency Laboratory* Stanford University, ch 4.

[3] Lee, T. (2004) *The Design of CMOS Radio Frequency Integrated Circuits*, 2 nd Edn Cambridge University Press, ch 17.

[4] Weigandt, T. et al. (1994) Analysis of timing jitter in CMOS ring oscillators *IEEE Int. Symp. On Circuits and Systems*, Paper 4.27.

Homework

1. Design a *common-drain*_Colpitts oscillator at 100MHz using the U310 silicon JFET. This device has a pn junction gate and is normally on (depletion mode). The data sheet for the U310 JFET from InterFET can be found on the mouser.com webpage. There is also an ADS model for this device in the Analog Transistor Component Library.

a. Plot the DC characteristics of the U310 using the FET curve tracer in the ADS Mixer DesignGuide. The model for the U310 is in the Analog Parts library. Determine the threshold voltage and plot g_m vs. V_{GS}. Choose a bias condition for the device that keeps the device in saturation (the useful region for gain).

b. Perform a hand analysis of the open loop oscillator to determine loop gain as a function of bias current and $(C_1 + C_2)/C_1 = N$. Assume inductor $Q_U = 100$, $L = 100\text{nH}$ and a load resistance of $1\ \text{k}\,\Omega$. Account for the device C_{gs} and C_{gd} in your design.
c. Do a small-signal open loop AC analysis in ADS and compare the results with your hand analysis.
d. Use transient analysis or harmonic balance to perform a large signal simulation of the oscillator. Verify that your design produces a current limiting mode of operation.

2. Design a *common-collector* Colpitts voltage controlled oscillator using the 2 N5179 BJT and the MV2101 varactor diode. There is an ADS model in the Analog Transistor Library (pb_mot_2N5179_19921211). Look up the manufacturer's data sheet on-line. Use the model parameters for the varactor diode shown below. Assume the unloaded inductor Q is 100.

Supply voltage	$+5$ V
Tuning range	$134 - 137\text{MHz}$
Output power	-7dBm in 50ohm load
Second and third harmonic	-25dBc or better

a. Perform a hand analysis of the open loop oscillator to determine loop gain as a function of bias current and tapped capacitor ratio $(C_1 + C_2)/C_1 = N$. Do a small-signal open loop AC analysis in ADS and compare the results with your hand analysis.
b. Use harmonic balance to perform a large signal simulation of the oscillator and compare with the hand calculation. Show that the design is current limited, not voltage limited.
c. Take the output of the oscillator from the emitter, and design a buffer amplifier using the 2 N5179BJT that will provide -7 dBm into a 50 ohm load. Simulate the combination and determine whether the oscillator-amplifier meets the harmonic distortion spec.

Model parameters for the MV2101 varactor diode: ADS uses the SPICE diode model. The parameters below provide a good fit to the C – V behavior of the MV2101 in reverse bias. This model uses the equation below to describe the capacitance of a reverse-biased pn junction diode:

$$C(V) = CJO\left(1 - \frac{V}{VJ}\right)^{-M},$$

where CJO = diode capacitance at $V = 0, VJ$ = built-in voltage of the diode (default value 0.8 V$)$, M = grading coefficient (typically 0.5 for abrupt junction).

The diode equivalent circuit also includes a series resistance R_s due to the resistance of the semiconductor material and contacts. The varactor data sheet specifies this through the unloaded Q defined as

$$Q = \frac{\text{reactance}}{\text{resistance}} = \frac{1}{\omega C R_S}$$

at frequency ω and at a specific reverse bias voltage.

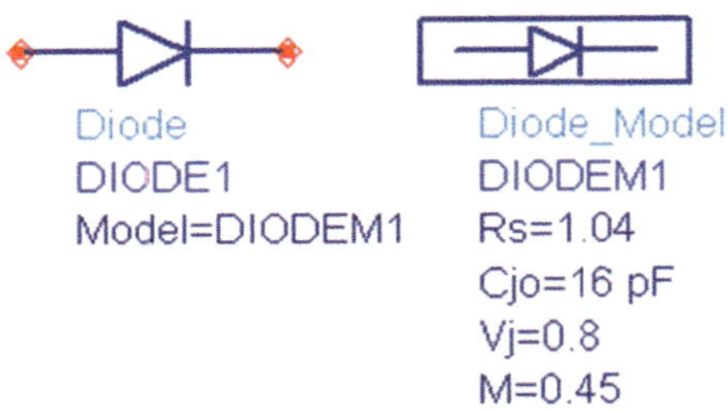

3. $L = 50$ nH, $QU = 20, g_m = 0.01$ S

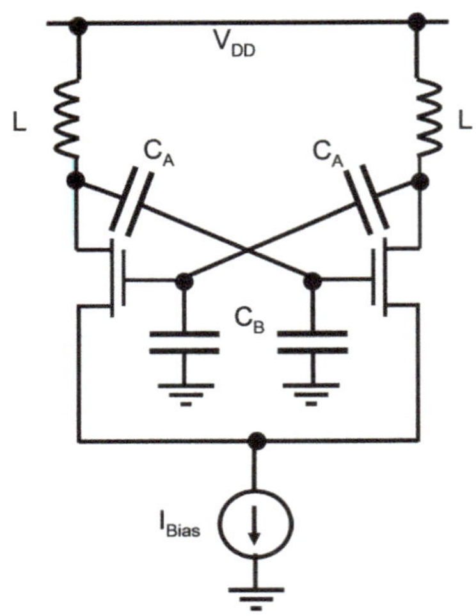

a. Determine C_A and C_B such that the open loop gain $=$ 4 and the oscillation frequency is 159MHz.
b. If $I_{\text{Bias}} = 0.5$ mA, estimate the amplitude of the voltage at one of the drain nodes.

4. Design a fixed frequency (165MHz) *common-base* Colpitts oscillator using the 2 N5179BJT. There is an ADS model in the Analog Transistor Library (pb_mot_2N5179_19921211). Assume the unloaded inductor Q is 100.

Supply voltage	+5 V
Output power	−7 dBm in 50 ohm load
Second and third harmonic	−25 dBc minimum

a. Perform a hand analysis of the open loop oscillator to determine loop gain as a function of bias current and tapped capacitor ratio $(C_1 + C_2)/C_1 = N$. Do a small-signal open loop AC analysis in ADS and compare the results with your hand analysis.
b. Use transient analysis or harmonic balance to perform a large signal simulation of the oscillator and compare with the hand calculation. Show that the design is current limited, not voltage limited.
c. Take the output of the oscillator from the emitter, and design a buffer amplifier using the 2 N5179BJT that will provide −7dBm into a 50ohm load. Simulate the combination and determine whether the oscillator-amplifier meets the harmonic distortion spec.

Appendix 14.1 A Very Brief Description of Phase Locked Loops

1. Definition. A PLL is a feedback system that includes a VCO, phase detector, and low pass filter within its loop. Its purpose is to force the VCO to replicate and track the frequency and phase at the input when in lock. The PLL is a control system allowing one oscillator to track with another. It is possible to have a phase offset between input and output, but when locked, the frequencies must exactly track.

$$\begin{aligned} \phi_{out}(t) &= \phi_{in}(t) + const \\ \omega_{out}(t) &= \omega_{in}(t) \end{aligned} \tag{A.1.1}$$

We inferred in our discussion of oscillators in this chapter that free-running oscillators are usually unsuitable for local oscillator or clock generation. A stable reference is needed to avoid drift and bit errors so that frequency and phase are locked against the reference phase.

The block diagram in Figure 14.A.1 shows how the three critical elements are connected. The PLL output can be taken from either V_{cont}, the filtered (almost DC) VCO control voltage, or from the output of the VCO depending on the application. The former provides a baseband output that tracks the phase variation at the input. The VCO output can be used as a local oscillator or to generate a clock signal for a digital system. Either phase or frequency can be used as the input or output variable.

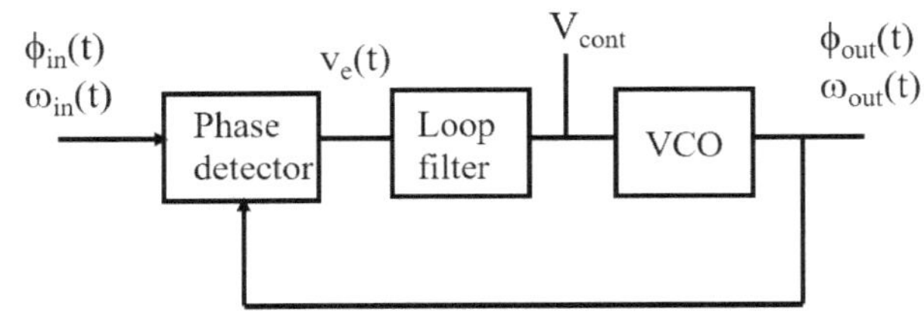

14.A.1. Block diagram of a basic PLL.

Of course, phase and frequency are interrelated by:

$$\omega(t) = \frac{d\phi}{dt}$$
$$\phi(t) = \phi(0) + \int_0^t \omega(t')\, dt' \quad \text{(A.1.2)}$$

Now consider the function of each of the elements in the loop.

2. Phase detector: compares the phase at each input and generates an error signal, $v_e(t)$, proportional to the phase difference between the two inputs. K_D is the gain of the phase detector $(\mathrm{V/rad})$.

$$v_e(t) = K_D\left[\phi_{out}(t) - \phi_{in}(t)\right] \quad \text{(A.1.3)}$$

As one familiar circuit example, an analog multiplier or mixer can be used as a phase detector. Recall that the mixer takes the product of two inputs. $v_e(t) = A(t)B(t)$. If,

$$A(t) = A\cos\left(\omega_0 t + \phi_A\right) \quad \text{(A.1.4)}$$

$$B(t) = B\cos(\omega_0 t + \phi_B) \tag{A.1.5}$$

Then,

$$A(t)B(t) = (AB/2)\left[\cos(2\omega_0 t + \phi_A + \phi_B) + \cos(\phi_A - \phi_B)\right] \tag{A1.6}$$

Since the two inputs are at the same frequency when the loop is locked, we have one output at twice the input frequency and an output proportional to the cosine of the phase difference. The doubled frequency component must be removed by the lowpass loop filter. Any phase difference then shows up as the control voltage to the VCO, a DC or slowly varying A C signal after filtering.

The averaged transfer characteristic of such a phase detector is shown below. Note that in many implementations, the characteristic may be shifted up in voltage (single supply/single ended). Also, there are many other implementations of phase detectors, a study in itself [1].

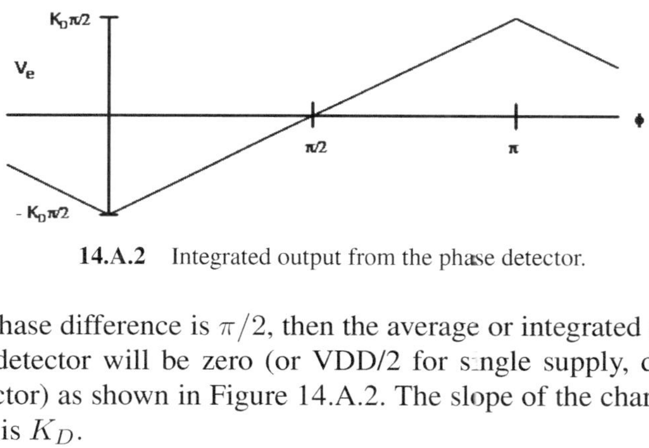

14.A.2 Integrated output from the phase detector.

If the phase difference is $\pi/2$, then the average or integrated output from the phase detector will be zero (or VDD/2 for single supply, digital XOR phase detector) as shown in Figure 14.A.2. The slope of the characteristic in either case is K_D.

3. VCO: In PLL applications, the VCO is treated as a linear, time-invariant system. Excess phase of the VCO is the system output.

$$\phi_{out} = K_O \int_{-\infty}^{t} V_{cont} dt'.$$

The VCO oscillates at an angular frequency, $\omega_{out.}$ Its frequency is set to a nominal ω_0 when the control voltage is zero. Frequency is assumed to be linearly proportional to the control voltage with a gain coefficient K_O or $K_{VCO}(\text{rad/s/v})$.

$$\omega_{out} = \omega_0 + K_O V_{cont}$$

Thus, to obtain an arbitrary output frequency (within the VCO tuning range, of course), a finite V_{cont} is required. Let's define $\phi_{out} - \phi_{in} = \phi_0$.

A control voltage V_{cont} will be necessary to produce this output frequency. The phase detector can produce this V_{cont} only by maintaining a phase offset ϕ_0 at its input. In order to minimize the required phase offset or error, the PLL loop gain, $K_D K_O$, should be maximized, since

$$\phi_0 = \frac{V_1}{K_D} = \frac{\omega_1 - \omega_0}{K_D K_O}.$$

Thus, a high loop gain is beneficial for reducing phase errors.

4. Loop filter: While the lowpass loop filter is needed to remove higher frequency components from a mixer type of phase detector, it also must be designed carefully for loop stability and dynamic response. A system level analysis is needed to see this, but is beyond the scope of this brief introduction [1][2]. Often, in such analysis, the phase margin in a Bode plot can be used to design such a filter. An interesting design of a third-order filter (potentially unstable loop phase) was presented in [3].

5. Frequency synthesis: The simple loop shown in Figure 14.A.1 is limited in its application. It can only track the input phase at the same frequency as the input. In general, if the PLL is to be used for a local oscillator, some means of changing the output frequency as a fixed ratio of input frequency is needed.

For the frequency synthesis application, we want to have ideally perfect phase tracking for phase and frequency steps. An output frequency that is a multiple of the reference frequency is obtained when digital frequency dividers are included in the reference and VCO feedback path. The phase detector will keep the phase and frequency equal at its inputs. The block diagram below is often referred to as an Integer-N Frequency Synthesizer.

Figure 14.A.3 illustrates how this might be accomplished. We will add a digital divider $1/N$ to the feedback path and a digital divider $1/M$ to the reference path.

The reference divider, modulus M, is used to reduce the frequency of the reference crystal oscillator, often a much higher frequency, to the frequency steps $\Delta\omega$ required for the application. The feedback divider, modulus N, is used to select the specific output frequency in steps N/M as shown in Equation A.1.7.

$$f_{out} = f_{ref}\frac{N}{M} \tag{A.1.7}$$

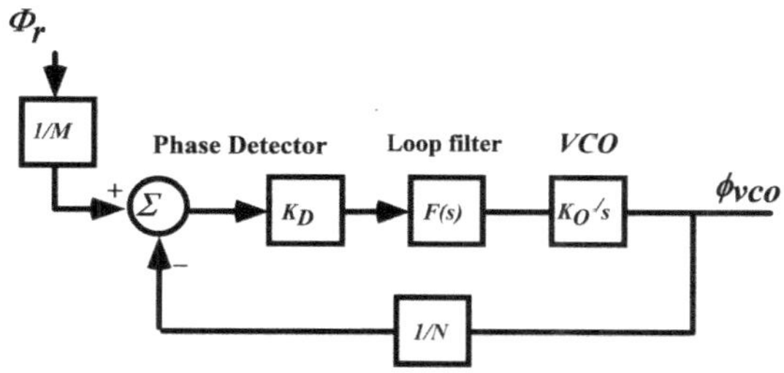

14.A.3. PLL with adjustable output frequency and phase.

This brief description is just the tip of the iceberg, but is sufficient to see in general how a PLL might be used to provide a stable adjustable frequency source.

References

[1] Lee, T. H. (2004) *The Design of CMOS Radio-Frequency Integrated Circuits, 2nd edn.* Cambridge University Press, Chapter 16.

[2] Razavi, B. (2001) *Design of Analog CMOS Integrated Circuits* McGraw-Hill, Chapter 15.

[3] Vaucher, C. S. (2000) An adaptive PLL tuning system architecture combining high spectral purity and fast settling time. *IEEE J. Solid State Cir*, 35(4), 490-502.

15

Low Phase Noise Oscillators

Noise is a limitation in amplifiers as we discussed in Chapter 9 on LNA design. It turns out that noise also is a serious problem with oscillators. Phase noise is a limiting factor in many RF systems. It is created by the time jitter of zero crossings of the generated sine wave. It is the source of reciprocal mixing. The local oscillator phase noise spectrum mixes adjacent channel interfering signals at some offset frequency into the IF passband. The interfering signal then shows up as a noisy replica, reducing the signal-to-noise ratio. Frequency synthesizers must be carefully designed to avoid adding to this problem. Reciprocal mixing is often a major limitation to the useable dynamic range of a receiver. Also, even without adjacent channel mixing, oscillator phase noise causes a reduction in signal-to-noise ratio in the desired passband.

In this chapter, our goal is to provide general insight into:

1. First order phase noise sources
2. Phase noise models and spectrum
3. Phase noise metric dBc/Hz
4. Oscillator designs that reduce phase noise
5. Phase noise simulation using ADS.

More detailed mathematically accurate modeling has been the source of many journal articles but will not be repeated in this chapter.

The voltage variation of oscillator voltage with time is described by Equation (15.1).

$$v(t) = V_s \cos[\omega t + \Delta\phi(t)]. \quad (15.1)$$

15.1 Sources of Noise

Noise is generated both by passive lossy circuit elements and by the active devices themselves [1].

Thermal noise:

$$\overline{v_n^2} = 4kTRB \tag{15.2}$$

BJT shot noise:

$$\overline{i_n^2} = 2qI_C B \tag{15.3}$$

MOSFET channel noise:

$$\overline{i_D^2} = 4kT\left(\frac{2}{3}g_m\right)B \tag{15.4}$$

B(or Δf) represents the measurement bandwidth.

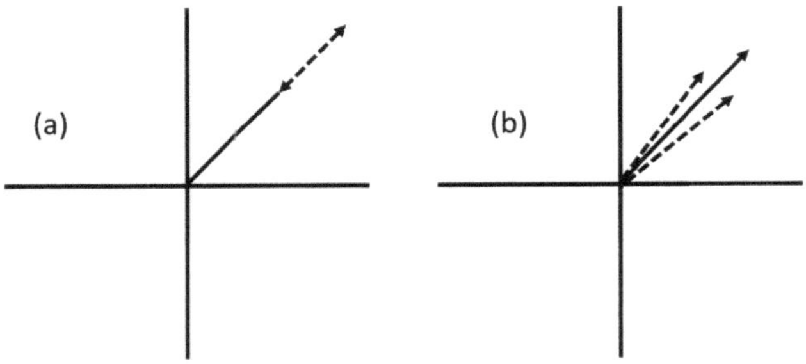

Figure 15.1 Complex plane. The cosine function rotates the vectors in a counterclockwise direction. (a) amplitude noise; (b) phase noise.

Due to equipartition, half of the noise is amplitude and half is phase (refer to Figure 15.1). However, oscillators all have some limiting action that suppresses amplitude noise. Phase noise has no restoring mechanism so must be dealt with by design or feedback. In this regard, an LC tank resonator attenuates harmonics and thermal noise away from the carrier. R_p accounts for the losses in the resonator (Figure 15.2).

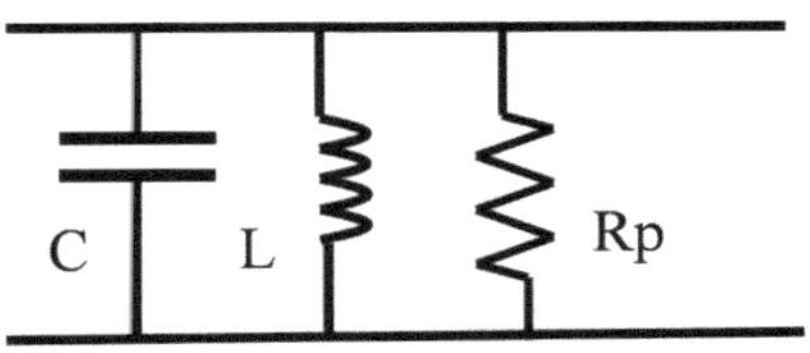

Figure 15.2 Parallel LC resonator showing loss R_p.

15.2 Phase Noise Spectrum

If we were to take the output of our oscillator under test into a phase detector, providing the other input of the detector with a very stable reference signal (crystal oscillator for example), then the output at baseband would look somewhat like Figure 15.3.

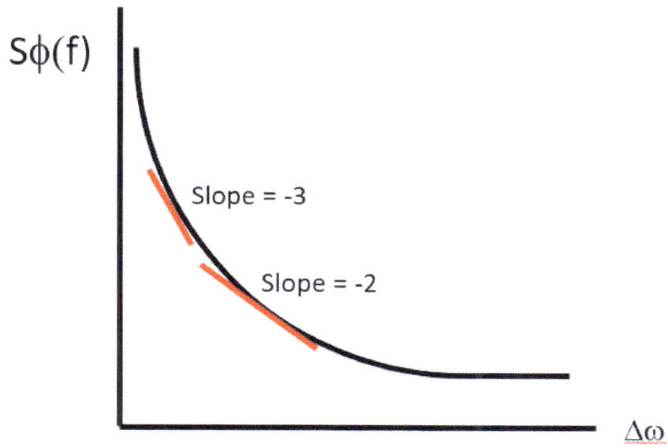

Figure 15.3 Power spectral density of phase fluctuations at offset frequency $\Delta\omega$ from the carrier [1, 2, 3].

The slope $= -2$ region comes from the thermal noise source (Equation (15.2)) being shaped by the LC resonator. Using the frequency response of the LC resonator, the noise power spectral density due to this source can be shown [2, 3] to be

$$\frac{\tilde{v}_n^2}{\Delta f} = 4kTR_P \left(\frac{\omega_0}{2Q\Delta\omega} \right)^2, \tag{15.5}$$

where R_P is the loss resistance of a parallel LC resonator. Note that while the loss is compensated by the g_m of the active circuit in order for oscillation to occur, the noise contributed by R_P continues to be a major contributor to phase noise. Comparing with Figure 15.3, we see that this simple model only describes the square law part of the observed spectrum. As the offset frequency, $\Delta\omega$, approaches zero, it predicts infinite noise voltage. For large $\Delta\omega$, it predicts zero noise. No noise floor is predicted.

Before going on to revise this model, it is common practice to express oscillator noise as a single-sideband noise-to-carrier ratio, $L(\Delta\omega)$, shown in Figure 15.4, normalized to a 1 Hz bandwidth, Δf. The carrier power (at $\Delta\omega = 0$) is P_{sig}

$$L(\Delta\omega) = 10\log\left[\frac{\bar{v}_n^2/\Delta f}{\bar{v}_{sig}^2}\right]. \tag{15.6}$$

In terms of signal power,

$$P_{sig} = \frac{\bar{v}_{sig}^2}{R_P} = \frac{V_{pk}^2}{2R_P}. \tag{15.7}$$

Then,

$$L(\Delta\omega) = 10\log\left[\frac{2kT}{P_{sig}}\left(\frac{\omega_0}{2Q\Delta\omega}\right)^2\right] \tag{15.8}$$

becomes

$$L(\Delta\omega) = 10\log\left[\frac{4kTR_P}{V_{pk}^2}\right]\left(\frac{\omega_0}{2Q\Delta\omega}\right)^2. \tag{15.9}$$

This noise-to-carrier ratio has a peculiar unit designation: dBc/Hz at an offset frequency from the carrier, $\Delta\omega$. It is a negative number since the noise power is much less than the signal power. Figure 15.4 shows how this is defined.

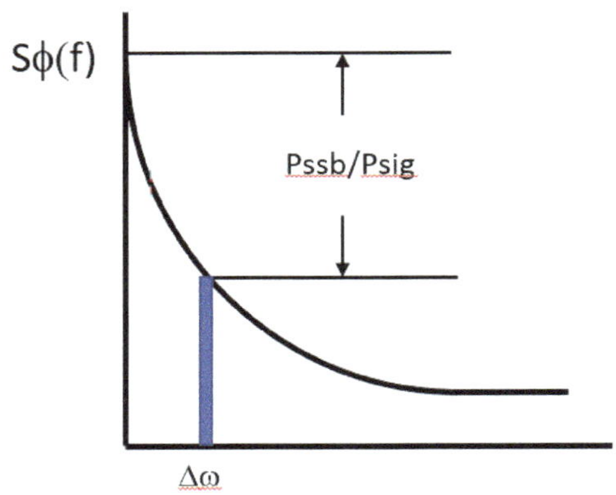

Figure 15.4 Noise-to-carrier or signal ratio defined in a 1 Hz noise bandwidth at offset from the carrier, $\Delta\omega$[1].

From Equation (15.9) it would appear that the key to improving $L(\Delta\omega)$ would be to increase Q and V_{pk}. If V_{pk} were unlimited, this might be true.

However, the physical limits of the active devices by saturation and breakdown mandate that we must fix the voltage swing as seen in Equation (15.10). The best results will occur when this is at its highest voltage or current while maintaining the oscillator in the current limited regime.

$$V_{pk} = I_{bias} R_P \tag{15.10}$$

If $4kT/V_{pk}^2$ is a constant, then only Q and R_P remain to be adjusted in order to improve phase noise. In terms of the series loss R_S,

$$R_S = \frac{R_P}{Q^2 + 1} \tag{15.11}$$

Then,

$$L(\Delta\omega) \propto \frac{R_P}{Q^2} \propto \left(\frac{Q^2 + 1}{Q^2}\right) R_S \propto R_S \tag{15.12}$$

Considering on-chip spiral inductors, $Q > 4$ is easily achieved with sensible layout. Therefore, in light of Equation (15.12), high Q with high Rs will not help. High Q alone will raise voltage swing, so the oscillator current I_{bias} must be reduced in order to prevent breakdown, therefore keeping the output power constant. R_S is the only parameter that matters.

For example, if R_S or R_P is reduced by a factor of 2 , phase noise is reduced by 3 dB when V_{pk} remains fixed by increasing I_{bias} by 2 times [4, 5]. This insight can be used to evaluate oscillator designs that can improve phase noise performance by their topological construction.

15.3 Oscillator Figure of Merit

A commonly accepted quantity, figure of merit (FOM), can be used to compare various oscillators. This FOM is defined as

$$FOM = L(\Delta\omega) - 10\log\left(\frac{\omega_0}{\Delta\omega}\right)^2 + 10\log(P) \tag{15.13}$$

where P is the power dissipation of the oscillator in milliwatts (not the RF output power), ω_0 is the oscillation frequency, $\Delta\omega$ the frequency offset, and $L(\Delta\omega)$ the phase noise measured at the offset frequency. It is used to measure how effectively the power consumption of the oscillator is utilized. As shown in the previous example, for a fixed Q, if R_S is doubled, R_P is doubled (Equation (15.11)) and I_{bias} is reduced by half in order to maintain the same voltage swing. So, FOM remains the same.

As an alternate illustration with increased Q, if R_S were to be doubled, in order for Q to be doubled the inductance must increase by 4 times. Then R_P will increase by 8 times. I_{bias} must be reduced by 8 times as well to keep the same voltage swing. Because $P = V_{DD}I_{\text{bias}}$ the power dissipation will go down by roughly 9 dB. However, R_S will double and thus $L(\Delta\omega)$ will increase by 3 dB (Equation (15.12)). So, the net improvement in FOM is 6 dB. Thus, there is some benefit for higher Q provided reducing power is more important than phase noise [4, 5].

15.4 Leeson's Equation

But what is missing from Equation (15.9)? Figure 15.3 shows that there is a noise floor and a f^{-3} region that must be considered. The active device and any post-amplifiers also contribute to a noise floor. A fudge factor F can be included to account for observed noise floor power. A noise floor corner frequency is also needed. Also, the $1/f^{-3}$ region, due to upconverted $1/f$ noise, must be added. The $1/f$ noise occurs in most active devices due to surface states or interface states with slow time constants. Equation (15.9) was modified by Leeson [6] to become the classic empirical description of noise-to-carrier ratio:

$$L(\Delta\omega) = 10\log\left\{\frac{2FkT}{P_{sig}}\left[1+\left(\frac{\omega_0}{2Q\Delta\omega}\right)^2\right]\left(1+\frac{\Delta\omega_{1/f^3}}{|\Delta\omega|}\right)\right\} \tag{15.14}$$

The $1/\Delta\omega^2$ corner frequency occurs at $\Delta\omega = \omega_0/2Q$. $\Delta\omega_{1/f^3}$ is an empirical fitting parameter to model the $1/f^{-3}$ corner frequency. F, an oscillator noise factor of a current biased differential CMOS LC oscillator can be given as

$$F = 1 + \frac{2\gamma I_{\text{bias}} R_P}{\pi V_{pk}} + \gamma\frac{g_{m\_\text{bias}} R_P}{4} \tag{15.15}$$

Here, the first term is the noise contribution of R_P, previously discussed, the second term is the phase noise contributed by the differential pair. γ is the FET noise factor. The third term is due to the bias current source where g_{m_bias} is the transconductance of that device [7]. Additional noise would also be introduced with a post-amplifier.

As reasonable as this approach seems to appear, it doesn't give good agreement with measurement or provide much insight beyond the benefit of increased P_{sig}. With the constraint of fixed voltage swing, this brings us back to the role of R_P or R_S.

One problem is that the model assumes linearity (reasonable because the noise power is much lower than the carrier) but also assumes time invariance. The latter is not a good approximation because, as we saw in the previous chapter, the device in a well-designed oscillator will be in cutoff for a significant part of the cycle. Hajimiri and Lee [2, 3] noted that oscillator phase noise generation is a time-varying process. Noise sensitivity depends on the specific time at which the noise pulse occurs. They proposed a modified noise model which is linear and time-varying. A noise impulse response was defined, although they admit it is difficult to calculate in most cases.

15.5 Oscillator Designs that Reduce Phase Noise

Let's define some characteristics of a good oscillator design:

1. Resonator designs that can maintain reasonable Q while reducing R_P or R_S.
2. Large signal power. If the signal can be increased without adding more noise or causing device saturation or breakdown, then the noise to carrier ratio will improve.
3. Increase in resonator voltage swing through implementation and/or power combining techniques.
4. In cutoff for most of the cycle. Less sensitive to impulse noise. Colpitts-based oscillators running current limited are good in this regard.

A few more words on the above:

Let's assume that the inductor limits Q. This is typically true at frequencies below 10 GHz or so. How does the inductor affect Q_U and R_P at a given frequency?

For a coil inductor (whether on a magnetic core or in air) with N turns, we can assume that the series resistance increases in proportion with length of the wire, therefore with N. The inductance will increase in proportion to N^2. Thus, Q_U increases in proportion to N.

What about R_P ?

$$R_p = \frac{X^2}{R_s} = \frac{\omega^2 L^2}{R_s} \tag{15.16}$$

It should increase as N^3. Based on our observation in Sections 15.2 and 15.3 , increasing R_P will also require a reduction in bias current. A careful tradeoff must be calculated to determine the benefit of larger inductance if any.

This is an ideal version and, in reality, it may not scale as favorably. The skin effect will cause R_S to increase with frequency for example. Also, if we consider a spiral inductor on an RFIC, the inner turns have less area than the outer ones. They contribute less inductance per turn than a solenoidal coil, so the scaling is not obvious and requires E/M simulation to determine Q_U and R_P.

Another consideration is what Q definition should be used? There are several:

$$Q = \omega L / R_s \tag{15.17}$$

$$Q = \frac{\omega \times \text{ Energy Stored}}{\text{Power Dissipated}} \tag{15.18}$$

$$Q = -\,\omega_0 \frac{d\theta}{d\omega}\bigg|_{\omega=\omega_0} \tag{15.19}$$

Analysis of oscillators has shown that the last definition, the phase slope with frequency, correlates best with phase noise performance. While these definitions may predict the same Q for simple resonators (series or parallel RLC), they do not agree very well with more complicated topologies or with transmission line resonators. Thus, phase slope is less dependent on the implementation details.

15.5.1 How can we improve P_{sig}?

Rather than elaborating on the more detailed and accurate mathematical descriptions in the literature, let's consider a few oscillator designs that adhere to the above general criteria.

We saw in Chapter 14 that $V_{res} = 2I_{BIAS}R_P$ is an approximation of the voltage across the resonator for the Colpitts oscillator. Therefore, we can increase signal power by increasing the resonator voltage or current.

$$P_{sig} \alpha V_{res}^2 \text{ or } I_{BIAS}^2 \text{ or } R_p^2 \tag{15.20}$$

Any increase in I_{BIAS} or V_{res} will help, but our device may begin to clip or break down. There are limits to how much voltage and current a device can handle safely. Simply increasing R_P will increase V_{res}, but may reach these limits in the device. So, a better approach is to modify the topology so that more power can be put into the resonator without exceeding the limits of the transistor. There are many such attempts in the literature of oscillators to do this, but three such approaches will be briefly reviewed here:

1. Tapped inductor [4].
2. Transformer power combining [8, 9]
3. Clapp oscillator [10].

15.5.2 A baseline: LC differential CMOS oscillator

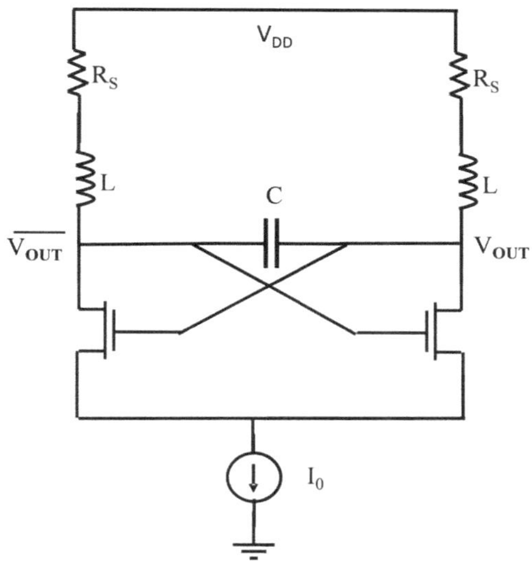

Figure 15.5 LC differential oscillator. Details of the tail current source not shown.

The oscillator shown in Figure 15.5 fits well with the previous discussion. V_{OUT} will be controlled by I_o and R_s.

15.5.3 Tapped inductor oscillator

The resonator on the left in Figure 15.6 produces a parallel equivalent load resistance of R_P. This was used in our baseline differential LC oscillator. But now, on the right, a tapped inductor is introduced, essentially an autotransformer with turns ratio N.

- This has the effect of reducing the collector voltage (or drain voltage), $V_C = V_{res}/(N+1)$.
- It also reduces the load resistance at the collector, $R_L = R_p/(N+1)^2$.

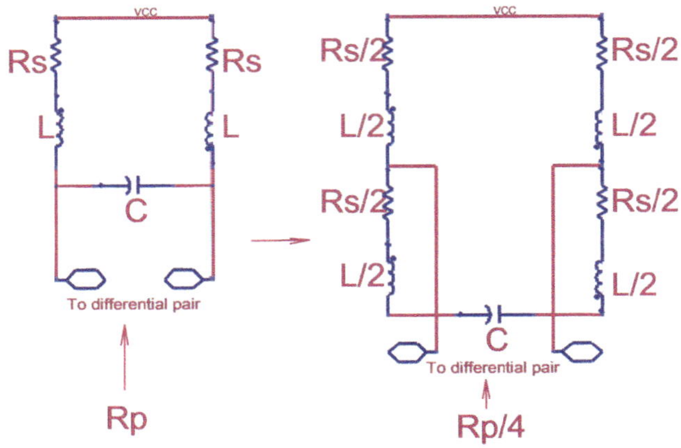

Figure 15.6 The tapped inductor oscillator tank [4, 5]. Copyright 2003 IEEE. Reprinted with permission [4].

If the tapping ratio was $1:1$, as seen in Figure 15.6, then $V_C = V_{res}/2$ and $R_L = R_p/4$. Thus, we can now increase I_{BIAS} by a factor of 4 , increasing P_{sig} by 16 times, without increasing the voltage swing at the collector. Of course, the noise contribution of the transistor, shown by the factor F in Equation (15.15), also increases with I_{BIAS}, so the net result is a 4 times improvement in phase noise to carrier ratio (6 dB). Scaling to higher N values may be beneficial, but it depends on how R_p scales with N. This can only be determined accurately by measurement or E/M simulation of the inductor.

With on-chip spiral inductor implementations, the inductor area consumes a lot of space. Thus, it is of benefit to construct the inductor so that only one is needed. Figure 15.7 shows one such layout of a tapped inductor.

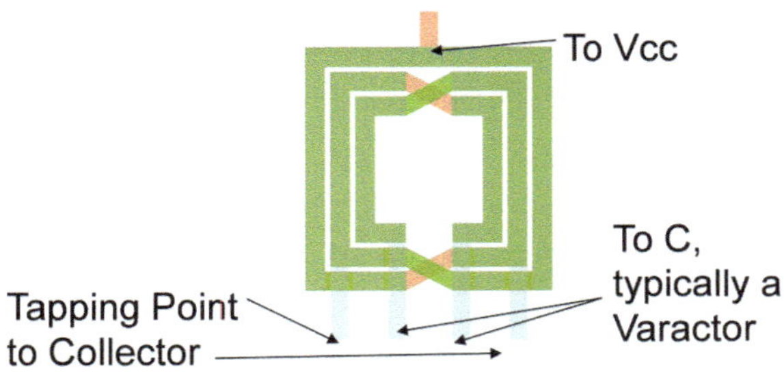

Figure 15.7 The tapping points were designed for the 1 : 1 ratio using the momentum EM simulator. Copyright [2003] IEEE. Reprinted with permission [4].

This implementation used a SiGe bipolar transistor process from Maxim Integrated Products [4]. A phase noise of -128dBc/Hz at 1MHz offset was observed at 2.4GHz. $FOM = 179.5$

Also, from Equation (15.15), the tail current source has an impact on phase noise [11]. According to this reference, the major noise frequency at the current source is at the second harmonic of the oscillator fundamental. They suggest adding a capacitor in parallel with the current source as a filter for the even harmonics.

To show relative improvement, another LC VCO was fabricated in an InGaP-GaAs HBT process from Triquint Semiconductor. A 1GHz standard LC differential oscillator and a tapped LC oscillator were fabricated on the same chip to measure the improvement due to the tapped L. At an offset of 1MHz, phase noise of -134 dBc/Hz and -138 dBc/Hz were measured, an improvement of 4 dB [5].

15.5.4 Transformer power combining oscillator

In order to reduce the phase noise of the oscillator, a large voltage swing is needed. However, this avenue is limited by supply voltage and breakdown or saturation (BJT) or triode region (MOS) operation which would kill the Q of the resonator. An integrated transformer can be used to couple multiple (N) identical oscillators, as seen in Figure 15.8 [8, 9]. The same voltage swing can be retained because inductance and R_p is reduced by the same factor N. In the example of four coupled oscillators, each oscillator can run at four times the bias current while retaining the same voltage swing. Therefore, according to

$$L(\Delta\omega) \propto \frac{1}{I_{bias}^2 R_P Q^2}, \tag{15.21}$$

each oscillator will have four times lower phase noise (6 dB) when the voltage swing proportional to $I_{\text{bias}}\, R_P$ is constant. In addition, now four oscillators are coupled together, so the net phase noise will be reduced by another 6 dB for a total 12 dB improvement. The cost, however, is that the power consumed by the oscillator will increase by 16 times.

15.5.5 Clapp oscillator [10]

To simplify the analysis [5], assume that the inductor L and R_s have the same values as the simple LC Colpitts. Therefore, Q will also remain the same. The supply voltage must also be assumed to be the same,

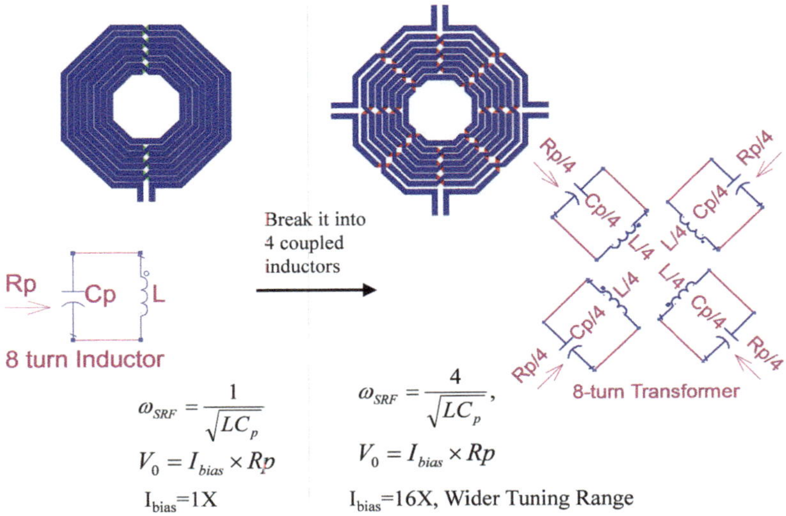

Figure 15.8 Transformer coupling concept to reduce phase noise [5][8][9].

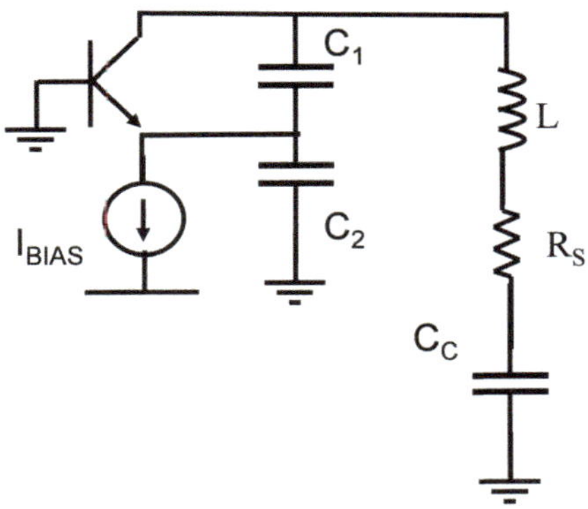

Figure 15.9 The Clapp oscillator is a variation of the Colpitts, which, similar to the tapped inductor oscillator, will reduce R_p without degrading Q.

$$Q = \frac{\omega_0 L}{R_P} \tag{15.22}$$

For the simple LC,

$$R_P = \left(Q^2 + 1\right) R_S. \tag{15.23}$$

Next consider the modified resonator that the Clapp oscillator provides (Figure 15.9),

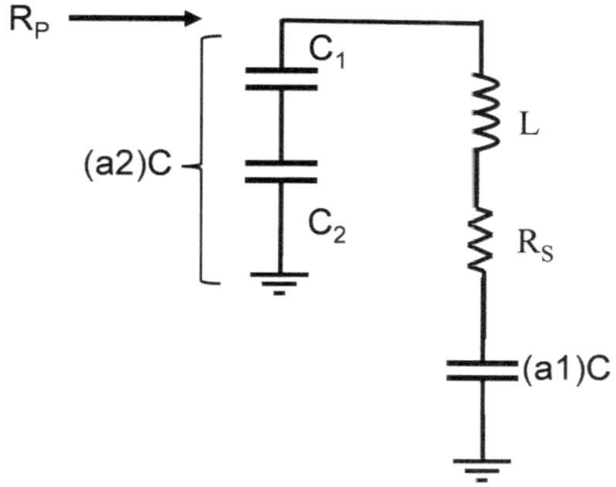

Figure 15.10 Clapp resonator. C and L have the same values as in the simple LC Colpitts.

Figure 15.10 represents a schematic diagram of how the modified resonator can be modeled.

$$(a_2)\,C = \frac{C_1 C_2}{C_1 + C_2}. \tag{15.24}$$

The resonant frequency of this network can be shown to be

$$\omega_{0\_Clapp} = \frac{1}{\sqrt{\left(\frac{a_1 a_2}{a_1 + a_2}\right) LC}}. \tag{15.25}$$

To have the same frequency as the simple LC resonator $a_1 + a_2 = a_1 a_2$. Combining L and $(a_1)\,C$ into an equivalent reactance X_{eq} in series with R_s,

$$X_{eq} = \sqrt{\frac{L}{C}}\left(1 - \frac{1}{a_1}\right). \tag{15.26}$$

Using the series to parallel transformation at the oscillation frequency, R_P for the Clapp resonator is

$$R_{P\_Clapp} = \left\{\left(\frac{X_{eq}}{R_s}\right)^2 + 1\right\} R_s \cong Q^2 R_s \left(1 - \frac{1}{a_1}\right)^2. \tag{15.27}$$

Or, in terms of the original R_P of the simple LC Colpitts,

$$R_{P\_Clapp} = R_P \left(1 - \frac{1}{a_1}\right)^2. \tag{15.28}$$

So, a smaller a_1 will reduce R_{P_clapp} and thus improve phase noise by allowing I_{bias} to increase. The scaling factor a_2 will need to be increased to keep $a_1 a_2$ constant and retain the same resonant frequency. Reasonable values of capacitance can be obtained in this range, and Q will not be degraded (Table 15.1).

Table 15.1 Scaling of R_P for three capacitance selections. Note that the second entry produces similar scaling as the 1 : 1 tapped inductor.

a_1	a_2	Scale factor
1.46	3.16	$0.1R_P$
2	2	$0.25R_P$
3.41	1.41	$0.5R_P$

Other variations of this design also exist while scaling the inductor value [12]. Keeping L constant allows for easier comparison or R_P with the LC Colpitts design.

15.6 Phase Noise Simulation with ADS

Harmonic balance simulation in ADS can be used for oscillator phase noise analysis. Chapter 14 described how to use HB for oscillator analysis using the OscPort tool. That provided a spectral analysis up to the maximum order selected. One can also enable nonlinear noise simulation in order to mix small-signal noise sources with the oscillator fundamental and harmonics. Then a phase noise output in $\mathrm{dBc/Hz}$ can be generated. Some guidelines:

In the HB controller:

- Enable nonlinear noise
- Oversampling should be greater than 4
- Enable small-signal and "use all small-signal frequencies"
- Noise(1) settings (Figure 15.11).

Figure 15.12 shows Noise(2) settings. Add the labeled node for phase noise analysis.

Figure 15.13 shows our common-base Colpitts oscillator from Chapter 14.

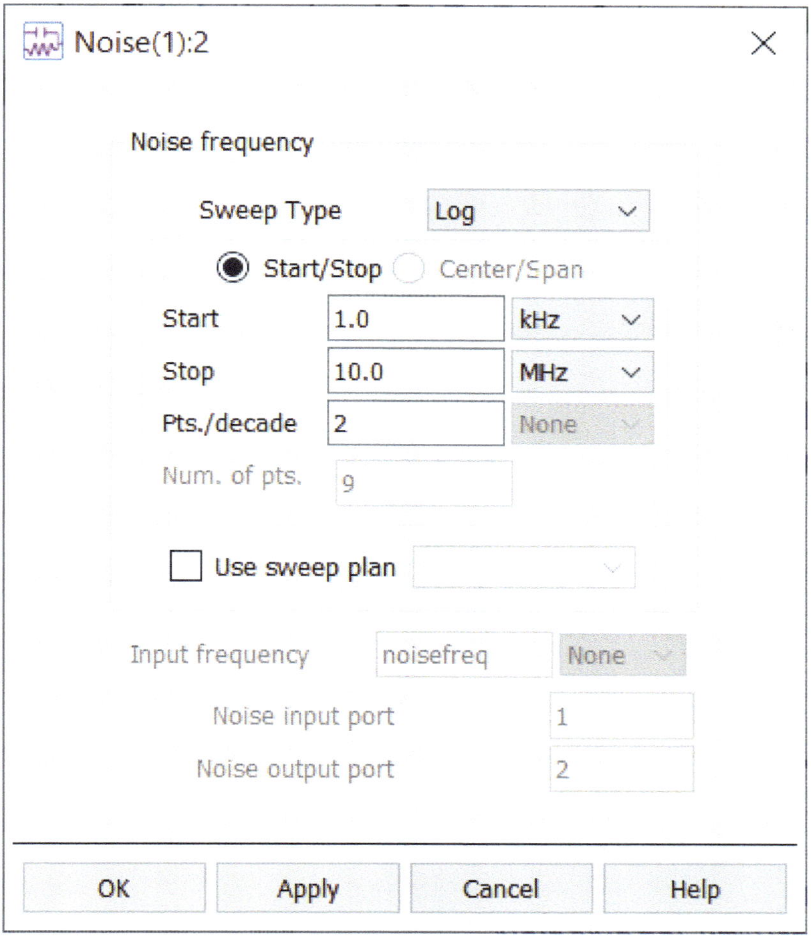

Figure 15.11 Noise(1) settings.

Then simulate and plot vout.pnmx on a log plot (Figure 15.14).

Note that the slope = 2. The $1/f$ region is not included in this, nor is the white noise floor.

"Most of the nonlinear devices in the semiconductor libraries do not have stable $1/f$ noise parameters. Instead, noise sources must be added to model the $1/f$ noise." "To model added white noise, select a V_Noise (Noise Voltage Source) component if you know the rms noise voltage, and an

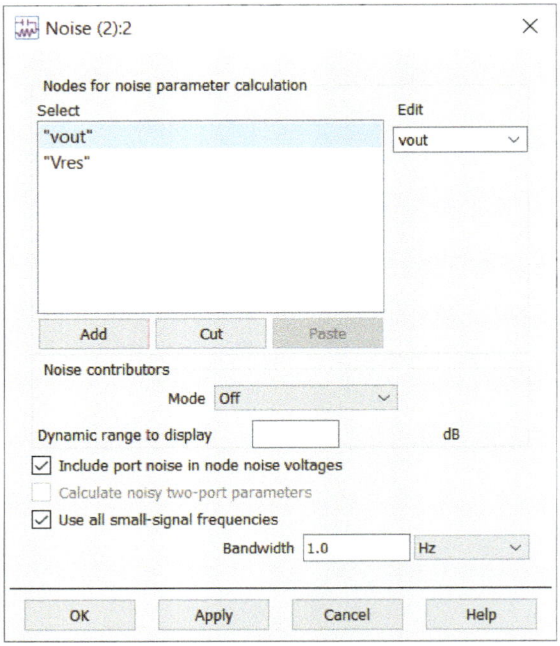

Figure 15.12 Noise(2) settings.

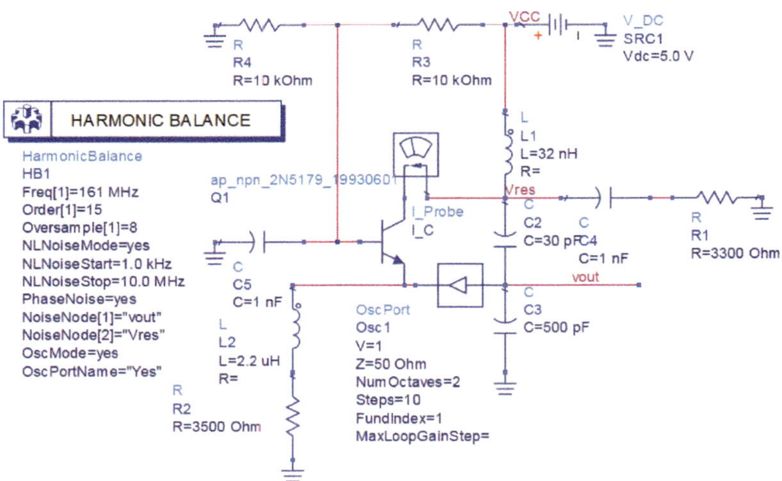

Figure 15.13 Colpitts oscillator phase noise simulation.

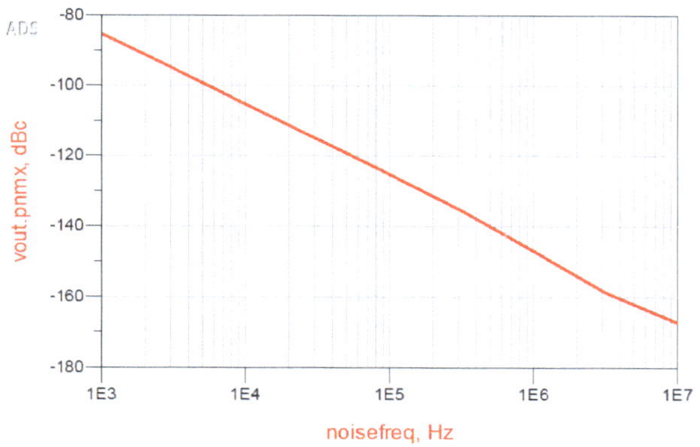

Figure 15.14 Plot of phase noise (dBc/Hz) due to noise mixing vs. offset frequency $\Delta\omega$ for the oscillator in Figure 15.13.

I_Noise (Noise Current Source) component if you know the rms noise current (these are found in the Noise and Controlled Sources palette)." [13]

References

[1] McWhorter, M. (1995) EE344 *High Frequency Laboratory* Stanford University, ch. 5.

[2] Lee, T.H. and Hajimiri, A. (2000) Oscillator phase noise: A tutorial *IEEE J. Solid State Circ.*, 35(3), 326-336.

[3] Lee, T.H. (2004) *The Design of CMOS Radio Frequency Integrated Circuits*, 2nd Edn., Cambridge University Press, ch. 18.

[4] Lai, P.W., Dobos, L., and Long, S. (2003) A 2.4GHz low phase-noise VCO using on chip taped inductor *IEEE ESSCIRC* 505-508.

[5] Lai, P.W. (2006) "Low voltage, low phase noise and wide tuning range integrated oscillator design *PhD Dissertation* University of California, Santa Barbara.

[6] Leeson, D.B. (1966) A simple model of feedback oscillator noise spectrum", *Proc. IEEE* 54, 329-330.

[7] Hegazi, E., Rael, J., and Abidi, A. (2004) *The Designer's Guide to HighPurity Oscillators* Kluwer Academic Publishers.

[8] Lai, P.W. and Long, S. (2005) A low phase noise InGaP-GaAs HBT transformer power combining VCO *IEEE CSIC Symposium* pp. 81-84.

[9] Lai, P.W. and Long, S. (2006) A 5GHz CMOS low phase noise transformer power combining VCO *IEEE RFIC Symp.* pp.133-136.

[10] Clapp, J. (1948) An inductive-capacitive oscillator of unusual frequency stability *Proc IRE* 36, 356-358 and 1261.
[11] Rael J. J. and Abidi, A.A. (2000) Physical processes of phase noise in differential LC oscillators *Proc. IEEE CICC* pp. 569-572.
[12] Hayward, W. (1994) *Introduction to Radio Frequency Design* American Radio Relay League, ch. 7.
[13] Pathwave ADS help menu. Search for: "oscillator phase noise analysis" Phase Noise Using OscPort.

Homework

1. Choose one of the oscillator design homework problems in Chapter 14. Perform a phase noise analysis on it in ADS.

16

Power Amplifiers

In this chapter, we will present the basics of power amplifiers and how they differ from the small signal designs that were previously discussed in Chapters 6, 7, and 9. These were often based on measured S-parameters. Matching of the input and output ports was dictated by stability and gain considerations and tradeoffs.

While stability and gain are still important for power amplifiers, other conditions must also be met. These include:

- Device limitations due to breakdown and saturation
- Large signal output match
- Design for efficiency, power-added efficiency
- Thermal resistance and temperature limitations
- Class A operation
- Operation at reduced duty factor:
    - Harmonic requirements
    - Class B, C, F, inverse-F.

16.1 Device Limitations

The operating bias point is shown as V_{DQ} and I_{DQ} in Figure 16.1. The blue line, a small-signal load line, represents a possible small signal conjugate match. This leads to limitations on voltage swing, ΔV and current swing ΔI.

$$\begin{aligned} \Delta V &<< V_{DQ} \\ \Delta I &<< I_{DQ} \end{aligned} \tag{16.1}$$

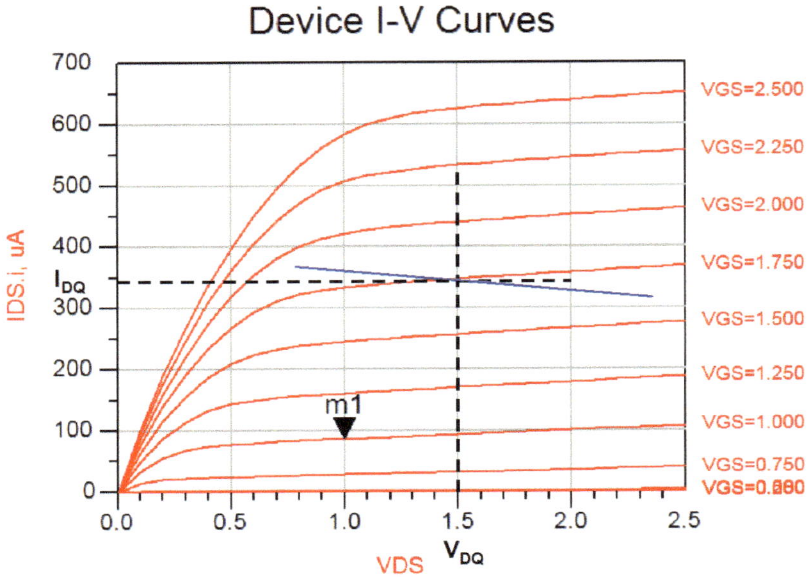

Figure 16.1 Device *I–V* characteristics of a typical small area discrete MOSFET.

While not important for SS amps, it is crucial for power amps that these be maximized. ΔV and ΔI should be significant when compared with V_{DQ} and I_{DQ}.

Power amp objective: Get the largest ΔV and ΔI without:

1. Clipping occurring when the voltage swing extends into the saturation region or into breakdown. This produces a large gain compression which produces distortion.
2. Damaging the device:
    - Avoid breakdown
    - I_{MAX} must be within device specs
    - Dissipated power must not overheat the device.

16.1.1 Breakdown

Every device has maximum voltage and current limits. Breakdown voltage:

- Electric field large enough to generate electron-hole pair $qV > E_{gap}$ of the semiconductor.

- Electrons injected into the channel from source or into collector from emitter are accelerated in a high field (Figure 16.2):

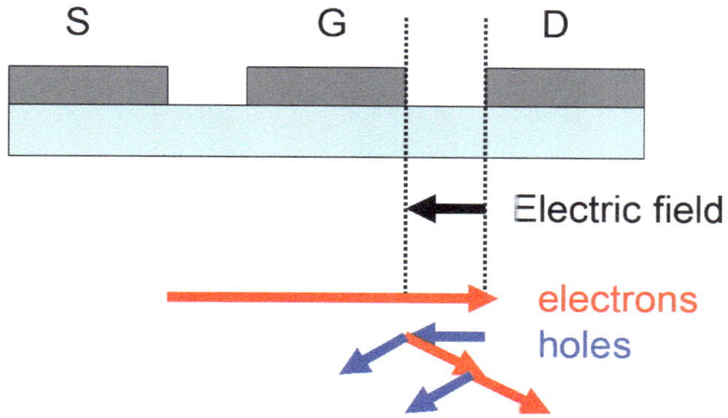

Figure 16.2 Cross section of an FET showing region of high electric field.

  - Collisions transfer energy to Si atoms. Electron is released, accelerates
  - creates hole.
  - Electron-hole pairs are then accelerated further generating more electrons, holes.
  - This leads to rapid increase in current for voltages beyond breakdown.

AVALANCHE BREAKDOWN!!!

16.1.2 Maximum current

GaAs FET:

$$I_{DSS} \text{ for a GaAs FET } = I_D @ V_{GS} = 0$$

$$I_D \propto qnv_{\text{sat}} \tag{16.2}$$

$$n \propto (V_{GS} - V_T)^m. \tag{16.3}$$

Also, you must avoid forward gate conduction on a MESFET or PHEMT.

Si MOSFET:

- I_{max} specified by foundry or manufacturer.
- Also, you must avoid gate oxide breakdown.

BJT:
I_{max} is limited by a collector electric field profile when the mobile charge is approximately the same as the fixed charge,

$$\mathrm{E} = \frac{q\,(N_D - n)\,x}{\varepsilon} \tag{16.4}$$

$$\text{electric field} \rightarrow 0$$

$$\text{transit delay} \uparrow$$

$$f_T \downarrow$$

We will discuss thermal limitations later. Heat generation will limit the operation of all device types.

16.2 Conjugate Match Revisited

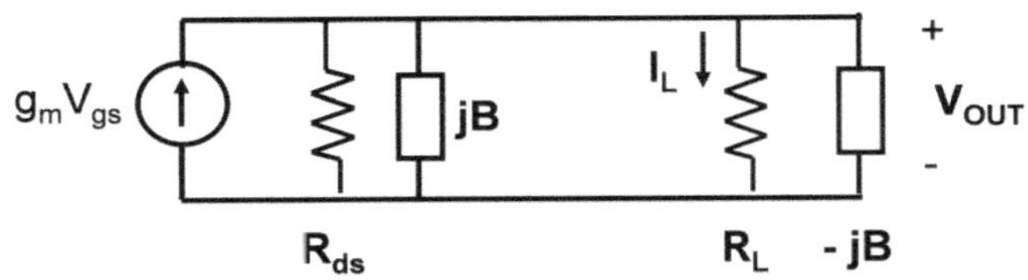

Figure 16.3 Small signal model of the FET output network.

Refer to the equivalent circuit in Figure 16.3. We have learned that maximum gain would occur when $jB = -jB$ and

$$R_L = R_{dS}.$$

While this does not account for stability considerations, if we are thinking in terms of load lines, this is a possible small signal match condition. But what happens when we have limitations on voltage and current?

Example 16.1: Suppose that for this device,

$$V_{max} = 10\ \mathrm{V}, \quad I_{max} = 1\ \mathrm{A}, R_{dS} = 100\Omega.$$

If we consider a conjugate match to require

$$R_L \| R_{dS} = 50\Omega = R'_L$$

Then,

$$V_{OUT} = I_{max} R'_L = 50 \text{ V!}$$

clearly, I_{max} can't be reached because

$$V_{max} = 10 \text{ V}$$

Then, the maximum output power would be limited to:

$$P_{OUT} = \frac{(V_{max}/2)^2}{2R'_L} = \frac{25}{100} = 1/4W.$$

16.2.1 Large-signal load line

Instead, consider a large-signal load line (Figure 16.4). It would be constructed between the maximum current, I_{max}, reached at V_{Dsat}, and the maximum allowable voltage, V_{BR} when $I_{DS} = 0$.

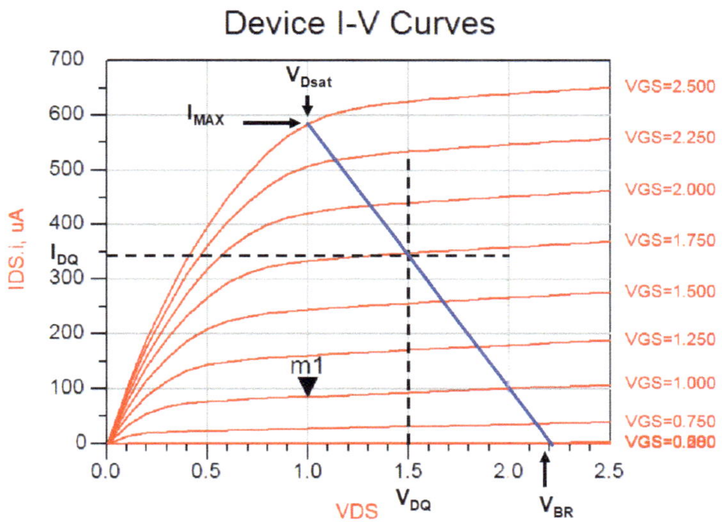

Figure 16.4 This device $I - V$ demonstrates how a large-signal load line could be determined to maximize power to the load.

Back to our previous example,

$$R_L = \frac{V_{max}}{I_{max}} = 10\ \Omega$$

$$P_{out} = \frac{25}{20} = 1.25 \text{ W}.$$

You can see that the large signal load line will have a slope of $1/R_L =$ 10 ohms. This uses the maximum capability of the device more realistically while providing five times more output power.

We have now shown that different criteria are used for output matching a power amp than a small signal amp.

You may have noticed that the large signal load line doesn't extend to $V_{DS} = 0$. To avoid excessive distortion, we must also take into account the "knee" voltage (V_{Dsat} or V_{CEsat}). Clipping will occur if the drain voltage swing extends into the ohmic region of the device characteristic.

Thus, our definition of the large signal load line resistance must take this into account:

$$R_L = \frac{V_{BR} - V_{knee}}{I_{max}} \tag{16.5}$$

for maximum voltage swing.

16.3 Efficiency

Efficiency, as defined in Equation (16.6), is also important. The input and output variables are shown in Figure 16.5. Low output power compared with DC input power means that there will be a lot of wasted power that must be removed as heat. Also, the DC power requirements will be increased, meaning less operating time from a battery powered source. This dissipated power, P_D, in Equation (16.7), can be a thermal limitation on the device that could limit its potential output power to well below the maximum.

$$\eta = \frac{P_{OUT}}{P_{DC}} \times 100\% \tag{16.6}$$

$$P_D = P_{DC} - P_{OUT}. \tag{16.7}$$

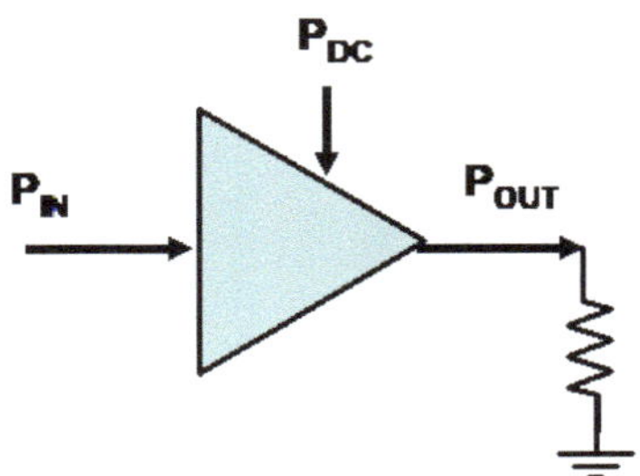

Figure 16.5 Definitions of input and output power of a generic power amplifier.

The typical maximum temperature of a device is constrained by reliability – it will affect the lifetime between failure. A safe T_{MAX} for most devices would be 150°C.

16.3.1 Power-added efficiency (PAE)

A more restrictive definition of efficiency includes how much input power, P_{IN} is required to obtain the required output power, $P_{OUT}.P_{IN}$ would come from the driver amplifier, also a factor in the overall power budget.

$$PAE = \frac{P_{OUT} - P_{IN}}{P_{DC}}. \tag{16.8}$$

Power gain should be at least 10 dB to avoid significant reduction in PAE.

Example 16.2: Suppose an output power, $P_{OUT} = 10$ kW, is required for an FM broadcast transmitter application.

Table 16.1 Effect of efficiency on the DC power and dissipated power.

$\eta(\%)$	P_D(kW)	P_{DC}(kW)
90	1.1	11.1
50	10	20
25	30	40
10	90	100

We see from Table 16.1 that low efficiency leads to extreme amounts of power that must be removed from the amplifier. Not a practical condition. So, how do we improve efficiency? This will be an ongoing discussion in this chapter and will be addressed for each type of amplifier considered.

16.4 Thermal Resistance and Temperature Limitations

We had previously asserted that a safe maximum junction temperature would be of the order of:

$$T_J \leq 150°\text{C}$$

Why?

16.4.1 Reliability

Device failure mechanisms are strongly activated by temperature. MTTF is the median time to failure and E_a is the activation energy. The activation energy might be determined by the metallization system used for the contacts to the transistor. You can see that the lifetime has an exponential relationship with temperature:

$$MTTF \propto e^{-Ea/KT} \tag{16.9}$$

16.4.2 Thermal resistance model

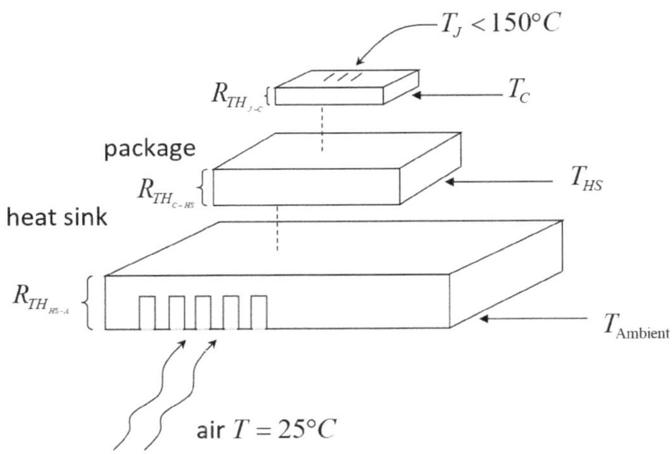

Figure 16.6 A thermal model including the device chip, package and a heat sink.

A thermal resistance model is shown in Figure 16.6. Thermal resistance models the relationship between the power dissipation and temperature rise of a device. It is a sort of Ohm's law for heat, as shown in Figure 16.7. T_J is the junction temperature, $R_{TH,J-C}$ is the thermal resistance of the chip itself, $R_{TH,C-HS}$, chip to heat sink, and $R_{TH,HS-A}$, heat sink to ambient. The ambient temperature is T_A.

$$T_J = (R_{TH,J-C} + R_{TH,C-HS} + R_{TH,HS-A})\, P_D + T_A. \tag{16.10}$$

R_{TH} has units of °C/ Watt.

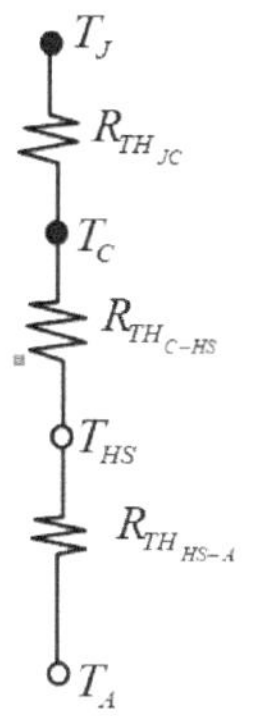

Figure 16.7 Resistor model for thermal resistance between a junction and the ambient temperature.

16.5 First PA: Class A

The Class A amplifier is most similar to a small-signal amplifier. V_{DQ} and I_{DQ} are set so that the amplifier is always on when there is no signal at the input. We can define a conduction angle as the ratio of conduction time or "on-time" within each cycle to the period T.

$$\text{Conduction angle} = \frac{\text{"on" time}}{T} \cdot 2\pi = 2\pi. \tag{16.11}$$

Case 1: Resistive load: A resistive load is a bad idea for a power amplifier because it acts as a voltage divider between the desired output, absorbing some of the useful output power. In Figure 16.8, R_L is seen to be in parallel

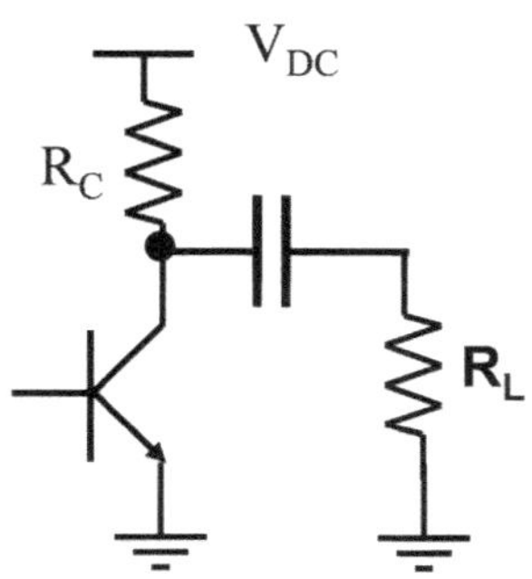

Figure 16.8 Class A amplifier with a resistive load.

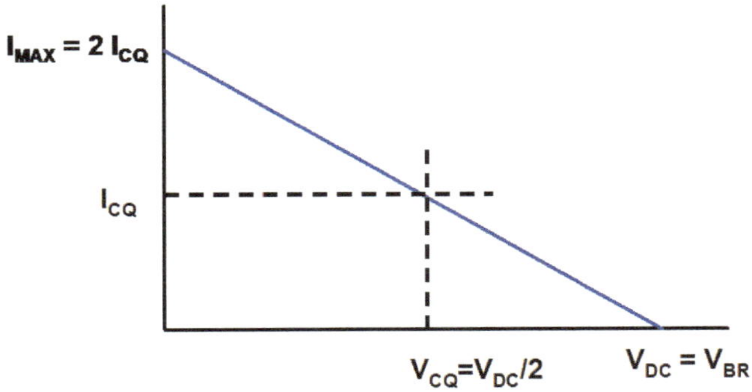

Figure 16.9 Load line for a Class A amplifier. Slope is $1/\left(R_L \| R_C\right) = 1/R_P$ for the case with resistive biasing and an AC coupled load.

with R_C. The load line is drawn in Figure 16.9. It shows the range of currents and voltages on the collector of the device for the given bias conditions, V_{CQ} and I_{CQ} and the resistive loading due to R_C in parallel with R_L.

Neglecting V_{knee}, we find

$$R_P = \frac{V_{BR}}{I_{max}} = \frac{V_{DC}}{2I_{CQ}} \tag{16.12}$$

$$I_C(\theta) = I_{CQ} + I_m \sin\theta \quad (\theta = \omega t). \tag{16.13}$$

The maximum value of $I_m = I_{CQ}$ where the voltage just begins to clip.

We can see that the average DC component is I_{CQ} and the fundamental component of current is I_m or, more generally, we can use Fourier integrals to determine P_{DC} and P_{OUT}. The DC term (a_0) can be used to calculate power dissipation:

$$P_{DC} = \frac{V_{DC}}{2\pi} \int_0^{2\pi} I_c(\theta) d\theta \tag{16.14}$$

$$= \frac{V_{DC}}{2\pi} \int_0^{2\pi} \left[I_{CQ} + I_m \sin\theta\right] d\theta = V_{DC} I_{CQ}. \tag{16.15}$$

For 100% duty factor, we obtain constant DC power and constant input current. You can also use Fourier integrals to find coefficients for fundamental and harmonics, $T = 2\pi$:

$$a_0 = \frac{1}{T}\int_{-T/2}^{T/2} f(x)dx \tag{16.16}$$

$$a_n = \frac{2}{T}\int_{-T/2}^{T/2} f(x)\sin(nx)dx \tag{16.17}$$

$$b_n = \frac{2}{T}\int_{-T/2}^{T/2} f(x)\cos(nx)dx \tag{16.18}$$

Calculate the current amplitude at the fundamental frequency $\omega : n = 1$

$$a_1 I_m = i_{OUT}(\omega) = \frac{1}{\pi}\int_{-\pi}^{\pi} I_m \sin^2\theta d\theta = I_m \tag{16.19}$$

because

$$\left[\int_{-\pi}^{\pi} \sin^2\theta d\theta = \pi\right] \tag{16.20}$$

Then the output power P_{OUT} is

$$P_{OUT} = \frac{1}{2}\operatorname{Re}\{V_m I_m^*\} = \frac{1}{2}V_m I_m \tag{16.21}$$

where

$$V_m = \frac{V_{DC}}{2}; I_m = \frac{V_m}{R_L}. \tag{16.22}$$

For purposes of our discussion, let's assume $R_L = R_C$. In this hypothetical case, there is no external load resistor AC coupled in Figure 16.8.

Thus:

$$P_{OUT} = \frac{1}{2}\frac{V_{DC}}{2}\frac{I_{\max}}{2} = \frac{V_{DC}I_{\max}}{8} \tag{16.23}$$

$$P_{OUI} = \frac{I_m^2 R_L}{2} = \frac{I_{CQ}^2 R_L}{2} = \frac{V_{DC}^2}{8R_L} \tag{16.24}$$

because

$$I_{max} = 2I_{CQ}, V_{BR} = V_{DC}, \text{ and } R_L = \frac{V_{BR}}{I_{max}}. \tag{16.25}$$

In terms of device limitations, the maximum output power is

$$P_{OUT} = \frac{1}{2}\frac{I_{\max}^2}{4} \cdot \frac{V_{BR}}{I_{\max}} = \frac{V_{BR}I_{\max}}{8}. \tag{16.26}$$

Next, what about harmonics? $n > 1$?

$$I_m a_n = \frac{1}{\pi}\int_{-\pi}^{\pi} I_m \sin\theta \sin n\theta \, d\theta = 0 \tag{16.27}$$

There is no nonlinearity in $I_c(\theta)$, therefore no harmonic currents. What about efficiency? If $I_{CQ} = \frac{V_{DC}}{2R_L}$, then,

$$P_{DC} = V_{DC}I_{CQ} = \frac{V_{DC}^2}{2R_L}. \tag{16.28}$$

Therefore,

$$\boxed{\eta = \frac{P_{OUT}}{P_{DC}} = \frac{1}{4} \quad (25\%) \text{ max!}}$$

If we AC couple the load resistor as shown in Figure 16.8, we have a current divider. The efficiency can be much worse. So, the Class A with resistive load is a bad idea for any power application that requires significant efficiency.

For RF applications we can do much better.

Case 2: Inductive load:

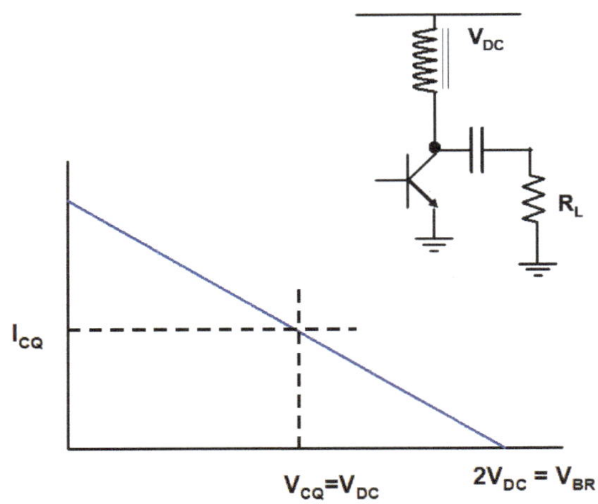

Figure 16.10 Current-voltage characteristic with inductive load.

The inductor, as shown as the load for the Class A amplifier in Figure 16.10, must maintain an average voltage over one complete cycle of V_{DC}. Then, I_{CQ} must be selected so that the peak voltage does not exceed V_{BR} and the peak current does not exceed I_{max} when the amplifier is driven to full output.

The large-signal load line will have the same slope as for the resistive load Class A; however, with the inductive load, the DC supply voltage is half that of the resistive case. The voltage swing is V_{DC}. The output power is the same as Case 1:

$$R_L = \frac{V_{BR}}{I_{\max}} = \frac{2V_{DC}}{I_{CQ}} \tag{16.29}$$

$$P_{OUT} = \frac{V_{DC}^2}{2R_L} = \frac{V_{DC}I_{\max}}{4} = \frac{V_{BR}I_{\max}}{8}. \tag{16.30}$$

However, as shown above, V_{DC} is half that in Case 1. Therefore, the DC input power required to obtain this output power is reduced by half. The maximum efficiency is therefore doubled:

$$\eta = \frac{1}{2}(50\%)$$

$$P_{OUT}\text{ (Case 2)} = P_{OUT}\text{ (Case 1)}$$

but

$$P_{DC}(\text{Case 2}) = \frac{1}{2}P_{DC}(\text{Case 1})$$

because:

$$V_{DC}(\text{Case 2}) = 1/2V_{DC}(\text{Case 1})$$

Thus, the inductive feed allows the amplifier to produce the same output power with half the supply voltage. This also applies to tuned amplifiers.

Alternatively, we can get two times more P_{OUT} for the same V_{DC} if the device has sufficient breakdown voltage (twice the voltage amplitude; twice the R_L).

However, what about a more typical situation where we have a large range of signal powers to be amplified? How much power gets dissipated as heat in the device? Repeating Equation (16.7), Figure 16.11 shows the relationship between P_{DC}, P_{OUT} and I_M from zero output to the maximum (Equation (16.30)):

$$P_D = P_{DC} - P_{OUT}.$$

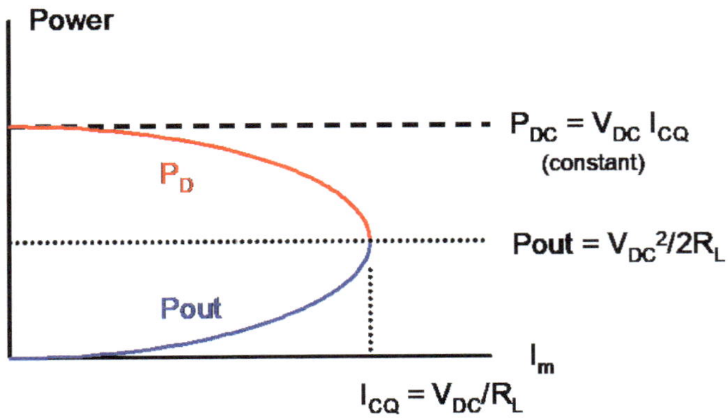

Figure 16.11 The relationship between P_{DC}, P_{OUT} and I_M from zero output to the maximum output.

We have doubled the efficiency (now 50%), but still have maximum power dissipated in the device at zero input because of the 100% duty factor ($2pi$ conduction angle). This condition is clearly undesirable for power amplifiers where high powers may be required. However, for a driver stage which requires at least 10 dB less output power and high linearity, Class A may be suitable.

16.5.1 Class A power amplifier summary

1. Device limitations (V_{BR} and I_{MAX} and T_{MAX}) constrain the design for a PA:

$$\boxed{R_{opt} = \frac{V_{BR} - V_{knee}}{I_{MAX}}} \quad \text{not } \Gamma^*_{OUT}!$$

   Large signal load line match.
2. $T_{max} = 150°\text{C}$ for reliable operation.
3. Waste power, $P_D = P_{DC} - P_{OUT}$ is converted to heat. It must be removed.
4. Efficiency, $\frac{P_{OUT}}{P_{DC}}$, or PAE, $\frac{P_{OUT}-P_{IN}}{P_{DC}}$ are critical for PAs.

For Class A with same output power, same R_L (Table 16.3).

Table 16.2 Comparison of Class A power amplifiers with resistive DC feed and inductive DC feed.

	Resistive DC feed (1)	Inductive feed (2)
V_{DC}	V_{DC1}	$V_{DC2} = V_{DC1}/2$
η	25% max	50% max
P_{OUT}	$\frac{V_{DC1}^2}{8R_L}$	$\frac{V_{DC2}^2}{2R_L} = \frac{V_{DC1}^2}{8R_L}$
V_{DC} max	up to V_{BR}	up to $V_{BR}/2$
$R_L = R_{opt}$	$\frac{V_{BR}}{I_{MAX}} = \frac{V_{DC1}}{2I_{CQ}}$	$\frac{V_{BR}}{I_{MAX}} = \frac{V_{DC2}}{I_{CQ}} = \frac{V_{DC1}}{2I_{CQ}}$

16.6 Higher Efficiency Power Amplifiers: Reduced Conduction Angle

Efficiency limitations for Class A: At best 50%; at worst 0%.

For really high power or battery applications, Class A efficiency is not good enough.

How do we improve this? Reduce the conduction angle. Allow the device to be in cutoff for part of the cycle by changing the bias conditions.

First example: Class B: Device is biased right at cutoff. Duty factor is reduced to 50% (conduction angle p). (Linearity considerations can be more favorable with a Class AB bias point in between A and B. It depends on the device.)

Consequences:

1. P_{DC} is reduced for same P_{OUT}
2. Harmonic termination is now required
3. RF drive signal drives device into conduction
4. More drive is required for Class B than Class A. Typically, 6 dB of gain is lost since the input voltage and current swing must be doubled.

Preliminary observations:

1. P_{OUT} roughly the same between Class A and B
2. Class B reduces P_{DC} by factor of $\frac{\pi}{2}$
3. Class C provides rapidly increasing efficiency, but rapidly decreasing P_{OUT}.

16.6.1 Class B

Suppose the device, in this case a BJT, is biased at cutoff.

Case 1: Single-ended Class B:

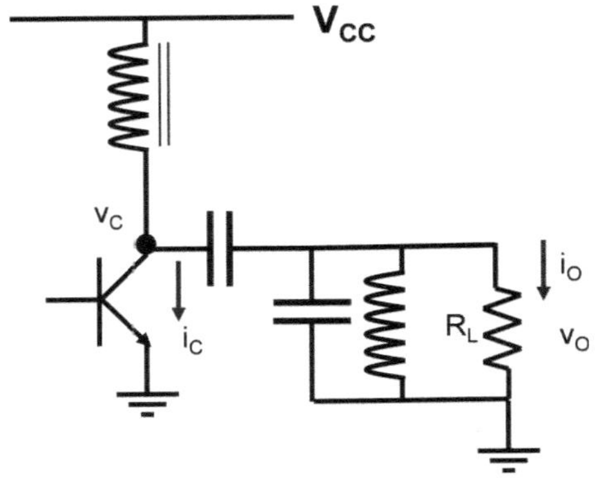

Figure 16.12 Single-ended Class B amplifier schematic.

A schematic of a single-ended Class B amplifier is shown in Figure 16.12. i_O = amplitude of the fundamental component of i_C. The LC resonator shorts harmonics. Only the fundamental component of the current can appear at the load resistor. The collector current is a half cosine because the device conducts only 50% of the time. The average value of $v_C = V_{CC}$. The total supply current is

$$i = i_C + i_O. \tag{16.31}$$

Base bias level is adjusted so that the conduction angle $= \pi$. The collector current is a half cosine function (Figure 16.13).

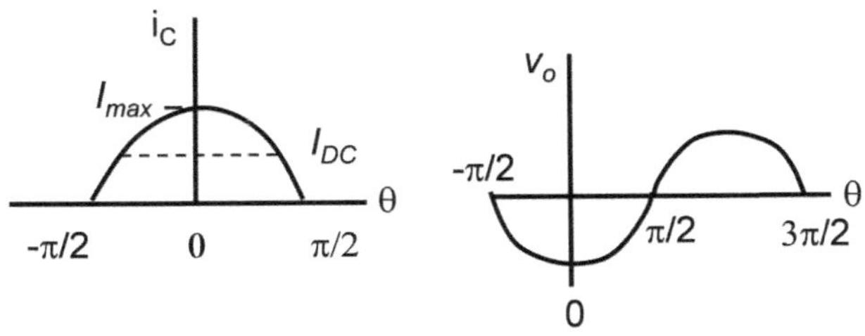

Figure 16.13 The collector current is a half cosine function. The output voltage is a full inverted cosine because the LC network at the collector shorts all harmonics and the inductor restores the following half cycle.

$$i_C = I_{\max}\cos(\theta) \text{ for } -\pi/2 \le 0 \le \pi/2 \tag{16.32}$$

$$i_C = 0 \text{ for } \begin{array}{l} -3\pi/2 \le \theta \le -\pi/2 \\ \pi/2 \le \theta \le 3\pi/2 \end{array}. \tag{16.33}$$

Since current is not a complete sine wave, we must calculate DC power and AC fundamental and harmonics from Fourier components:

$$a_0 = \frac{1}{2\pi}\int_{-\pi/2}^{\pi/2} I_{max}\cos\theta d\theta = \frac{I_{max}}{2\pi}\underbrace{\left[\sin\frac{\pi}{2} - \sin\left(-\frac{\pi}{2}\right)\right]}_{2} \tag{16.34}$$

$$\boxed{= \frac{I_{max}}{\pi} = I_{dc}} \tag{16.35}$$

$$a_1 = \frac{1}{\pi}\int_{-\pi/2}^{\pi/2} I_{max}\cos^2\theta d\theta = \frac{I_{max}}{\pi}\left[\frac{\theta}{2}\right]_{-\pi/2}^{\pi/2} = \frac{I_{max}}{2} = I_1. \tag{16.36}$$

I_1 is the fundamental current component,

$$V_1 = I_1 R_L \tag{16.37}$$

$$\begin{aligned} a_2 &= \frac{I_{max}}{\pi}\int_{-\pi/2}^{\pi/2}\cos\theta\cos 2\theta d\theta = \frac{I_{max}}{2\pi}\int_{-\pi/2}^{\pi/2}[\cos 3\theta + \cos\theta]d\theta \\ &= \frac{I_{max}}{2\pi}\sin\theta\Big|_{-\pi/2}^{\pi/2} + \frac{I_{max}}{6\pi}\sin 3\theta\Big|_{-\pi/2}^{\pi/2} = \frac{I_{max}}{\pi} - \frac{I_{max}}{3\pi} \\ &= \frac{2I_{max}}{3\pi} = I_2 \end{aligned} \tag{16.38}$$

I_2 is the second harmonic current. Higher harmonics can be calculated in the same way:

$$\begin{aligned} a_3 &= \frac{I_{max}}{\pi}\int_{-\pi/2}^{\pi/2}\cos\theta\cos 3\theta d\theta = \frac{I_{max}}{2\pi}\int[\cos 4\theta + \cos 2\theta]d\theta \\ &= \frac{I_{max}}{8\pi}\sin 4\theta\Big|_{-\pi/2}^{\pi/2} + \frac{I_{max}}{4\pi}\int\sin 2\theta\Big|_{-\pi/2}^{\pi/2} = 0 \end{aligned} \tag{16.39}$$

ω odd terms $= 0$; ω even terms finite, but shorted by LC resonator.

Thus, the total current,

$$I_{max} \cong I_{dc} + I_1 + I_2 + I_4 + \cdots \tag{16.40}$$

This must be equal to the peak current that the device is capable of providing. Choose the device accordingly.

The current that is actually delivered to the load is the fundamental frequency component $I_1 = \frac{I_{\max}}{2}$. This is the current that must be used to determine the optimum load resistance.

Figure 16.14 shows the amplitudes of the fundamental and harmonic currents (relative to $I_{\max} = 1$) as a function of the conduction angle.

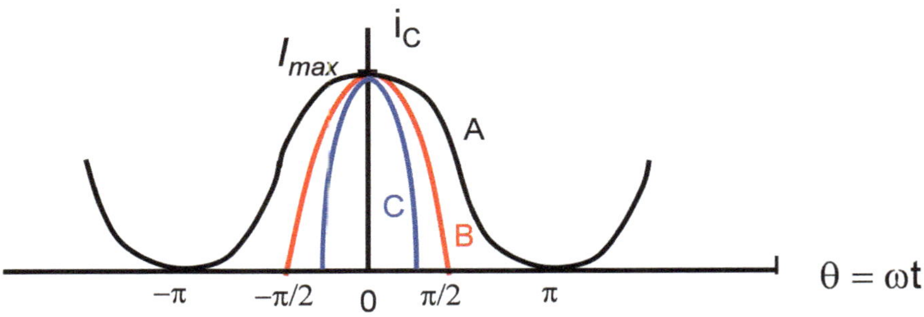

Figure 16.14 Device collector current for Class A, B, and C duty factors.

Class C reduces the conduction angle even more; P_{DC} and P_{OUT} are both reduced. Efficiency approaches 100% as P_{OUT} approaches 0! Large harmonic currents are required that put more strain on the device. Not a very effective means of increasing efficiency, plus it is quite nonlinear.

Choosing load resistance: The best load resistance is the large signal load resistance that allows full voltage and current swing at the device collector or drain. To get the full voltage swing across R_L, $V_O = V_{DC}$, (neglecting knee voltage V_K), we must choose

$$R_L = \frac{V_{DC}}{I_1} = \frac{2V_{DC}}{I_{max}} \tag{16.41}$$

Or, when V_K is considered,

$$R_L = \frac{V_{DC} - V_K}{I_1} = \frac{2\left(V_{DC} - V_K\right)}{I_{max}} \tag{16.42}$$

I_2 (and higher harmonics of the current) are assumed to be shorted by the LC resonator or bias network. Note that this load resistance is the same as was optimum for Class A.

Now, the output power, DC power and efficiency can be calculated for Class B. Recall that $I_l = I_{\max}/2$.

$$P_0 = \frac{1}{2}I_1^2 R_L = \frac{I_{max}^2 R_L}{8} = \frac{1}{2}\frac{V_{DC}^2}{R_L} \tag{16.43}$$

$$P_{DC} = \frac{I_{max}V_{DC}}{\pi} = \frac{2V_{DC}^2}{\pi R_L} \tag{16.44}$$

$$\eta = \frac{P_0}{P_{DC}} = \frac{V_{DC}^2/2R_L}{2V_{DC}^2/\pi R_L} = \frac{\pi}{4} \quad (79\%) \tag{16.45}$$

The efficiency is much greater than Class A: 79% vs, 50% !

Case 2: Push-pull:

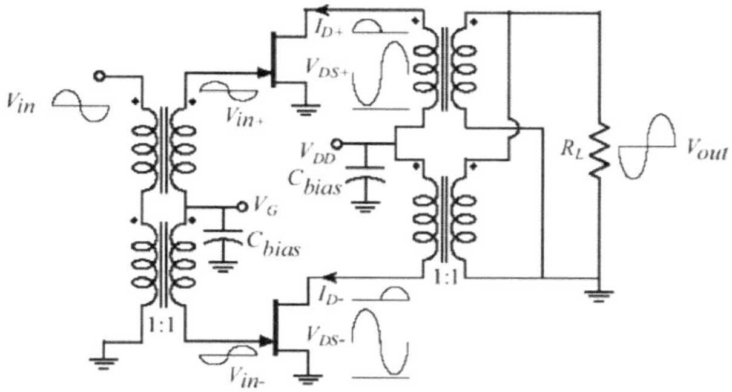

Figure 16.15 Push-pull Class B schematic.

A schematic of a Class B push-pull system is shown in Figure 16.15. The devices conduct on alternate half cycles to provide fully sinusoidal v_O as shown in Figure 16.16. A transformer or balun is needed at the input and output. The transformer or balun on the output side also has the benefit of producing an even-harmonic short.

Push-pull is all about balun design. Transmission line baluns and transformers with ferrite loading can be made to work well at 500MHz and below. PC board balun structures are physically large. Unless a balanced amplifier is really necessary (for bandwidths of an octave or greater, for example), there is no need to incur this added complication. The single ended Class B works fine.

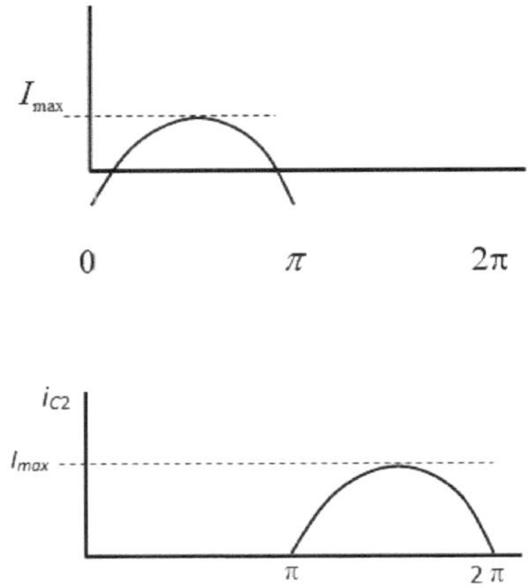

Figure 16.16 Collector currents of each device in push-pull Class B.

16.6.2 Harmonic termination

We have seen that I_C has a strong second harmonic current amplitude,

$$I_2 = \frac{2I_{\max}}{3\pi} = 0.21 I_{\max} \tag{16.46}$$

all I_n for $n =$ even are present; all $I_{n+1} = 0$ for odd harmonics.

How do we provide the even harmonic short? To get half sinusoidal current, at least I_2 must be shorted. In addition to the push-pull balanced design, there are at least two more ways to provide a low impedance at even harmonics.

Case 1: RFC DC feed

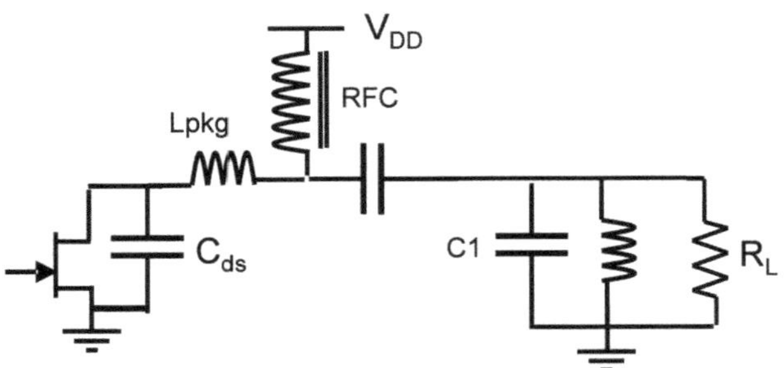

Figure 16.17 Output network of a reduced conduction angle power amplifier.

Referring to Figure 16.17, where an RF choke is used for the DC feed, we are depending on either or both the device output capacitance C_{ds} and the tank capacitance C_1 to short the harmonics. If C_{ds} is large, the harmonic short happens by default at device; if L_{pkg} is small, C_1 can provide the termination required by Class B at $2f_0, 4f_0$.

Case 2: Transmission line feed
At higher frequencies with low parasitic packages or no package, a quarter-wave line, $\frac{\lambda}{4}@f_0$ line works well. This also serves as the bias feed as seen in Figure 16.18. The bypass capacitor at V_{DC} produces a short circuit for fundamental and harmonic currents.

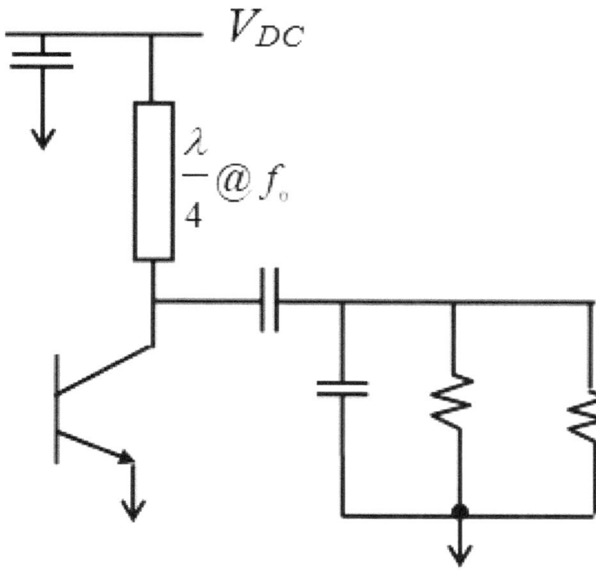

Figure 16.18 Reduced conduction angle amplifier with transmission line bias feed.

Recall that for a quarter-wave line, a short circuit is converted into an open circuit every quarter wave and a short circuit every half wave (Figure 16.19). This helps provide the harmonic currents required for reduced conduction angle operation. We will see later that the open-circuit at the odd frequencies also can be used beneficially.

Are there costs to using reduced conduction angle?

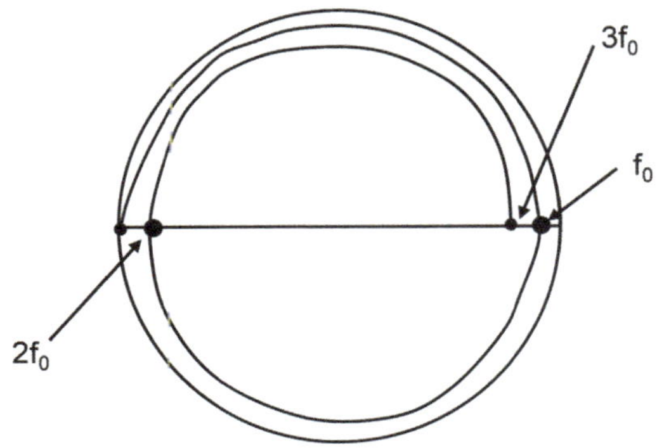

Figure 16.19 Trajectory of a quarter wave shorted line on the Smith chart.

16.6.3 Input drive

We must still achieve the I_{max} of the device to maintain peak current. For a MOSFET or HEMT in Class B, this requires that the gate voltage swing is double that of Class A because the negative swing must go below cutoff for half of the period, as seen in Figure 16.20. This requires a 6 dB increase in drive power - a problem for maintaining PAE unless the device has very high f_{max}.

This is not as severe with BJT or HBT because of the nonlinear base voltage - collector current relationship.

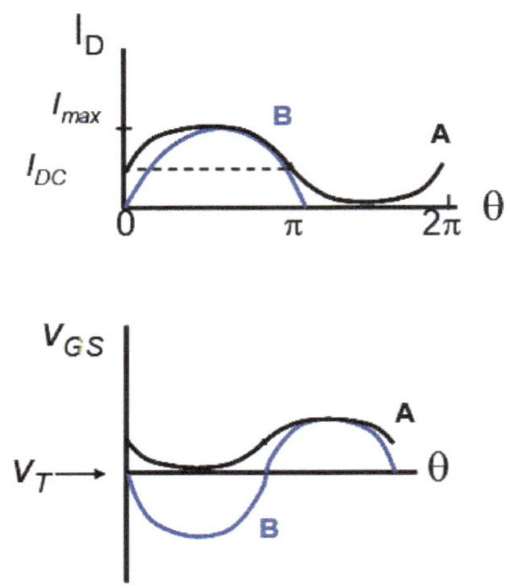

Figure 16.20 Gate-source voltage and drain current comparing Class A and B.

16.6.4 Nonlinear gate capacitance

The gate capacitance of all FETs can be quite nonlinear. Specifically, it drops very fast as the device is biased below threshold into cutoff. Then, if the drive current from the source remains roughly constant, the voltage swing on the gate in the negative direction can be much larger than in the positive direction. This is detrimental to breakdown because the maximum drain voltage occurs at the same time that the gate voltage is at minimum. Thus, V_{dg} is increased. It also produces a strong even harmonic voltage on the gate. Figure 16.21 shows the gate capacitance as a function of gate-source voltage for a GaN HEMT.

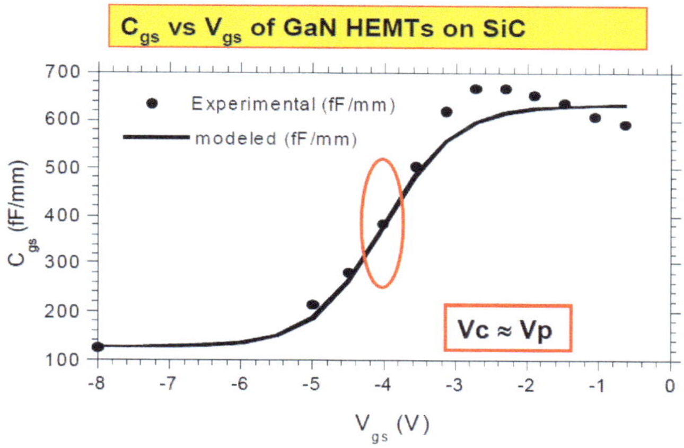

Figure 16.21 Nonlinear gate capacitance of a GaN HEMT on SiC.

An even harmonic trap has been found to be effective in reducing the impact of these factors. By shorting the even harmonic voltage, the gate voltage can remain either sinusoidal (Figure 16.22) or, if some odd harmonics are possible, can even become slightly square [1, 2]. The phase of the second harmonic short has been shown to be critical for achieving efficiency in GaN HEMT power amplifiers [3, 9].

16.7 Power Output Capability

Before going on, we need to introduce another metric for PA performance, the power output capability, or POC [4]. Recall that efficiency increases as conduction angle is reduced. This benefits the thermal problem of removing

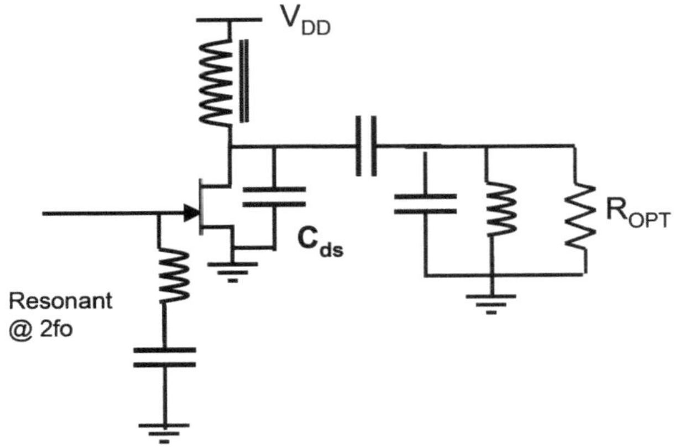

Figure 16.22 Even harmonic trap on the gate can make gate voltage more sinusoidal.

dissipated heat, but can also affect how much voltage and current stress the device.

Since devices are expensive, we want to get as much output power as we can from the maximum voltage and current that the device can provide. Hence, POC is defined to compare maximum output power with maximum voltage and current on the device.

$$\boxed{POC = \frac{P_{OUT}}{V_{D,\max} I_{D,\max}}.} \tag{16.47}$$

We find in Table 16.3 for the linear operating Classes that there is very little difference. Class AB is slightly better because output power can increase by up to 0.5 dB over Class A or B. However, we see Class C is a disaster - too little power output for the stress on the device.

Table 16.3 Comparison between reduced conduction angle, efficiency and power output capability.

Class	Conduction angle	Maximum drain efficiency	Power output capability
A	2π	$< \mathbf{50\%}$	$\mathbf{1/8}$
AB	$\pi < \theta < 2\pi$	$50 - 75\%$	$\sim \mathbf{1/8}$
B	π	$< \mathbf{79\%}$	$\mathbf{1/8}$
C (nonlinear)	$< \pi$	$< \mathbf{100\%}$	$<< \mathbf{1/8}$

Is there still a path to higher efficiency? Class B can give us up to 79%. Can we do any better? Not with a linear amplifier. Classes A, AB, B are fairly linear if properly designed. We saw that Class C does not use the device power output capability effectively, plus it is highly nonlinear.

To achieve better efficiency than Class B, we must intentionally overdrive the input to drive output power into saturation. This, with the correct harmonic tuning, can produce efficiencies well above 79% but at the cost of linearity.

16.8 Class F

What if we could produce a square wave voltage at the collector or drain?

$$V_1 = \frac{1}{\pi}\int 2V_{DC}\cos\theta d\theta = \frac{2V_{DC}}{\pi}[\sin(\pi/2) - \sin(-\pi/2)] = \frac{4}{\pi}V_{DC} \tag{16.48}$$

or 1.28 V_{DC} as opposed to V_{DC} for Class B. This gives us a large increase in fundamental output power for the same V_{DC} and I_{DC} :

$$\begin{aligned} I_1 &= \frac{I_{\max}}{2} \quad I_{DC} = \frac{I_{\max}}{\pi} \\ P_{out} &= \frac{1}{2}V_1 I_1 = \frac{1}{2}\left(\frac{4V_{DC}}{\pi}\right)\left(\frac{I_{\max}}{2}\right) = \frac{V_{DC}I_{\max}}{\pi} = V_{DC}I_{DC} \end{aligned} \tag{16.49}$$

So, $P_{out} = P_{DC}$! 100% efficiency.

Then all we need to do is produce an ideal square wave voltage. How? Unfortunately, it can't be done. We would need all odd harmonics in voltage and only even harmonics in current. But, this requires infinite bandwidth to produce all harmonics, and all devices have an upper frequency limitation, f_{max}. Also, we have neglected the knee voltage.

However, we can apply a finite number of harmonics. We can show that each harmonic, properly applied, increases the efficiency toward 100% [5]. Table 16.4 shows the improvement in efficiency with just a third harmonic enhancement. Figure 16.23 presents what the ideal voltage and current might look like for an ideal Class F with infinite bandwidth and all harmonics properly terminated.

Consider just the third harmonic:

$$V_D(\theta) = V_1\cos\theta - V_3\cos 3\theta \tag{16.50}$$

If $V_3/V_1 < 1/9$, the voltage waveform has a single peak. At $1/9$,

$$V_{Peak} = V_1 - V_3 = 8V_1/9. \tag{16.51}$$

The reduction in peak voltage allows us to increase V_1 by increasing R_L. In general, we can increase the fundamental voltage from

$$V_1 = V_{MAX}(ClassB) \tag{16.52}$$

to a higher value,

$$V_1 = V_{Peak}/\kappa \tag{16.53}$$

where

$$\kappa = \frac{V_{peak}}{V_1}. \tag{16.54}$$

Table 16.4 Comparison of Class B vs. Class F efficiency.

V_3/V_1	K	$P_{\text{out}}, F/P_{\text{out}}, B$	η
1/9	8/9	1.125	88.4%
1/6	0.866	1.154	90.7

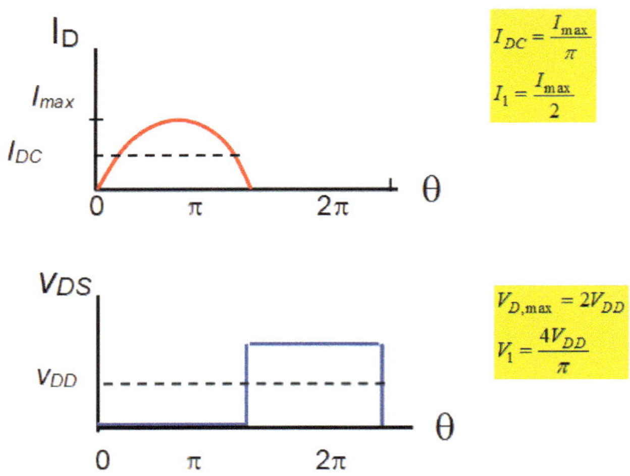

Figure 16.23 Voltage and current waveform for ideal Class F. Copyright [2007] IEEE. Reprinted with permission from [3].

Note that P_{DC} is not affected by squaring the voltage. We overdrive the amplifier using a higher R_L value to obtain the odd harmonics.

Adding higher harmonics will further increase efficiency. So, retain the even harmonic short as in Class B, but open circuit as many odd harmonics as possible. Overdrive the amplifier to clip the voltage so that we approach a square wave. Table 16.5 shows the potential benefit of adding even harmonics up to the fifth.

Table 16.5 Relationship between efficiency and number cf harmonics provided for a Class *F* amplifier.

Class F Harmonics	1 (class A)	2	3	4	5
Ideal drain efficiency %	50	71	82	87	91
Power output capability	0.125		0.144		0.151

16.8.1 Class F circuits

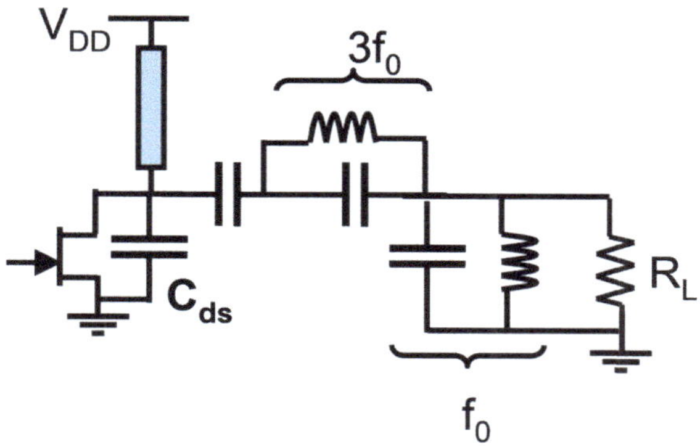

Figure 16.24 The Class *F* circuit gives low Z_L at even harmonics and high Z_L at $3f_0$.

The Class *F* circuit in Figure 16.24 gives low Z_L at even harmonics due to the quarter-wave drain bias feedline and high Z_L at $3f_0$. The parallel resonant circuit in series presents a high impedance at three times the fundamental frequency.

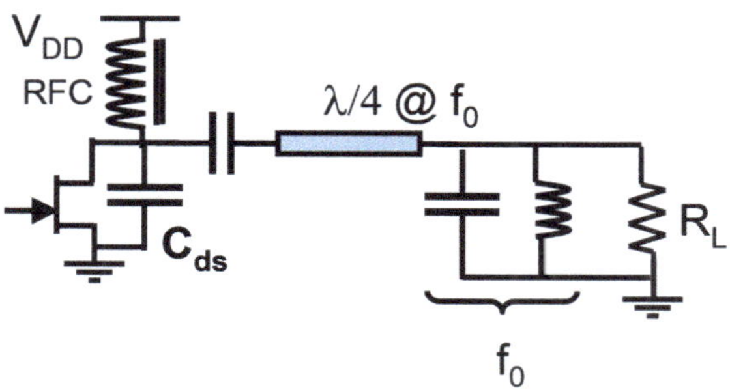

Figure 16.25 Class F circuit with quarter-wave transmission line.

The circuit shown in Figure 16.25 gives low Z_L at even harmonics and high Z_L at odd harmonics due to the impedance transforming nature of the quarter wave line (assuming the capacitive reactance of the tank C is small at harmonic frequencies).

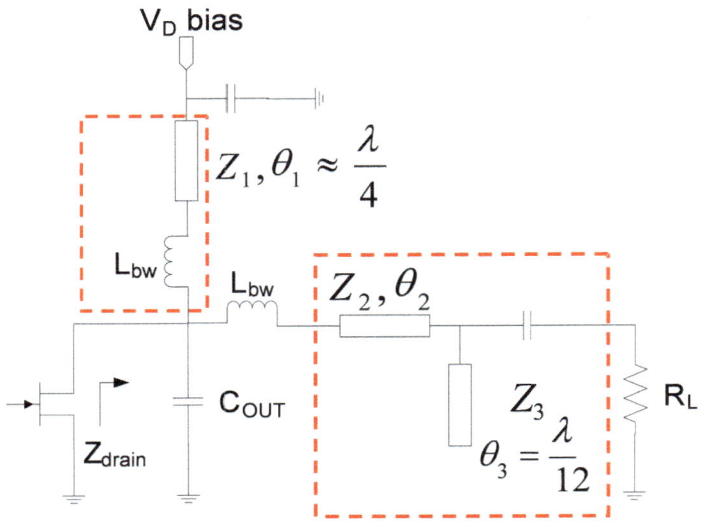

Figure 16.26 Class F design using transmission lines and including the effect of bondwires [3]. Copyright [2007] IEEE. Reprinted with permission from [3].

The output matching network in Figure 16.26 is designed to absorb C_{OUT} and interconnect inductance while providing the correct fundamental and harmonic resistances at the intrinsic drain of the transistor [9]. The network can be tuned for a variety of R_L values to accommodate a range of supply voltages. The bondwire inductance on the drain pad can be incorporated into the quarter-wave drain bias transmission line, giving low even-order harmonic termination. The Z_2, Θ_2 and Z_3 network matches R_L while absorbing the bondwire inductance and C_{OUT}. It also provides a high impedance at the third harmonic, as seen in Figure 16.25.

A plot of the impedance presented to the intrinsic drain of Figure 16.24 is shown in Figure 16.27. The markers indicate the simulated impedances at the harmonic frequencies.

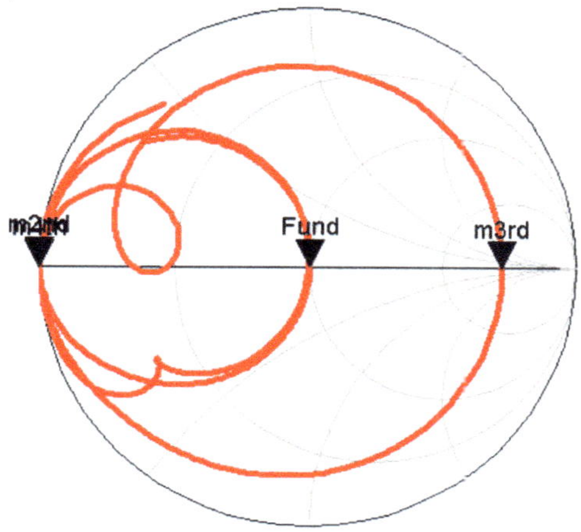

Figure 16.27 Plot of the impedance presented to the intrinsic drain of Figure 16.24.

16.8.2 Inverse F circuit

There is also an inverse-F $\left(\mathrm{F}^{-1}\right)$ mode in which the current is shaped into a square wave. It has the opposite harmonic requirements: high Z_L for even; low Z_L for odd harmonics [5].

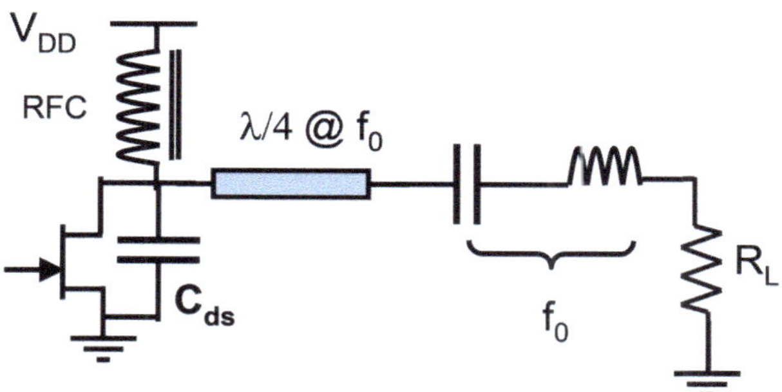

Figure 16.28 Circuit for an inverse-F amplifier.

The circuit of an inverse-F amplifier is shown in Figure 16.28.

The quarter-wave transmission line will present a high impedance at even harmonics. C_{ds}, if sufficiently large, will provide a low impedance at odd harmonics. The ideal drain voltage and current waveforms for inverse-F is shown in Figure 16.29.

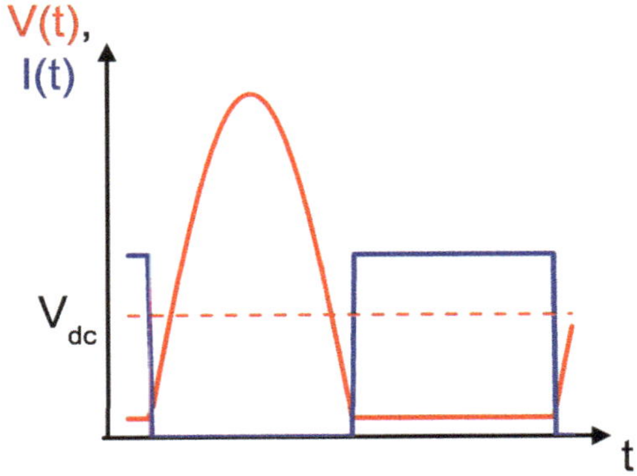

Figure 16.29 Voltage and current waveforms for an ideal inverse-F amplifier.

16.8.3 Class E PAs

Class E PAs are very popular in that many papers have been written on Class E [6, 7]. Class E can be very effective for most devices below 100MHz, however, there are serious power and frequency limitations above 1 GHz for LDMOS, HBT, and GaAs PHEMT devices. Class E delivers a lower power output capability than AB or F.

The GaN HEMT is clearly the best device for Class E. It provides low C_{DS}, high f_T, and high breakdown voltage. The peak voltage and currents can be shown to be:

$$V_{D,\,\max} = 3.56 V_{DD} \tag{16.55}$$

$$I_{D,\max} = 2.86 I_{DC} \tag{16.56}$$

yielding a power output capability of 0.098.

16.8.3.1 Design equations [7]

Design equations for jX and jB are shown in Figure 16.30. In addition,

$$R_L = \frac{8V_{DD}^2}{P_{OUT}\left(\pi^2 + 4\right)} \tag{16.57}$$

$$f_{E,\max} = \frac{P_{OUT}}{2\pi^2 C_{DS} V_{DD}^2}. \tag{16.58}$$

Higher V_{DD} gives higher load resistance which is desirable for efficient implementation of the output network.

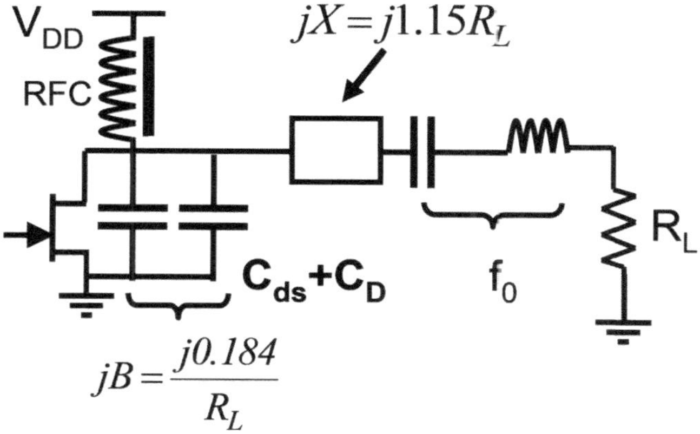

Figure 16.30 Typical Class E circuit implementation.

The maximum Class E frequency of operation is limited by C_{DS}, the drain capacitance. While it appears that higher power leads to higher frequency, this is not the case. Consider V_{DD} fixed. Then, P_{OUT} scales with I_{max}. The I_{max}/C_{DS} ratio is a constant, thus the upper frequency is really determined by the device. Devices such as the GaN HEMT with a large I_{max}/C_{DS} ratio perform at higher frequency than other devices such as LDMOS which have rather low I_{max}/C_{DS} [8].

It is also evident that Class E stresses the device more than Class F or inverse F. Both the peak voltage and current are higher, thus POC at best is 0.098 rather than 0.159 .

16.9 Summary

In this chapter, we presented an overview of the design of power amplifiers:

- Device limitations due to breakdown and saturation
- Large signal load line output match
- Design for efficiency, power-added efficiency
- Thermal resistance and temperature limitations
- Class A operation
- Operation at reduced duty factor to improve efficiency
    - Harmonic requirements
    - Class B, C, F, inverse-F.

References

[1] White, P. (1998) Effect of input harmonic terminations on high efficiency Class-B and Class-F operation of PHEMT devices, in *IEEE MTT-S Int. Symp. Dig.* 3, 1611-1614.
[2] Gao, S., Butterworth, P., Ooi, S., and Sambell, A. (2006) High-efficiency power amplifier design including input harmonic termination, *IEEE Microwave and Wireless Components Letters*, 16(2), 8183.
[3] Schmelzer, D. and Long, S. (2007) A GaN HEMT Class F amplifier at 2GHz with $>$ 80%PAE *IEEE J. Solid State Cir.* 42(10), 2130-2136.
[4] Raab, F. H. (2001) Maximum efficiency and output of Class-F power amplifiers, *IEEE Trans. Microwave Theory Tech.* 49, 1162-1166.
[5] Raab, F. (2001) Class E, C and F power amplifiers based upon a finite number of harmonics, *IEEE MTT* 49(8), 1462-1468.
[6] Sokal, N. and Sokal, A. (1975) Class E - A new Class of high-efficiency tuned single-ended switching power amplifiers, *IEEE J. Solid State Cir.* SC-10(3), 168-176.
[7] Raab, F. H. (1977) Idealized operation of the Class E tuned power amplifier, *IEEE Trans. Circuits and Systems* CAS-24(12) pp 725-735.
[8] Xu, H. et al. (2006) A high-efficiency Class E GaN HEMT power amplifier at 1.9GHz. *IEEE Microwave and Wireless Components Letters* 16(1), 22-24.
[9] Schmelzer, D. (2010) wide bandwidth, high efficiency switch-mode power amplifiers, PhD Dissertation, UCSB.

17

Power Amplifiers: Part 2

In this chapter, additional topics related to the design of power amplifiers will be presented.

- PA impedance matching – large signal
  - Load pull contours
  - Package parasitics
  - Large signal input match
- Simulation resources in ADS
- PA bias circuits: design for stability.

17.1 Power Amplifier Large-signal Impedance Matching

How does mismatch affect output power? We saw in the previous chapter that the large signal load line provided a guideline on choice of R_L. Beyond that, techniques referred to as load pull (Figure 17.1) and source pull can be used to better evaluate the effects of output and input mismatch respectively.

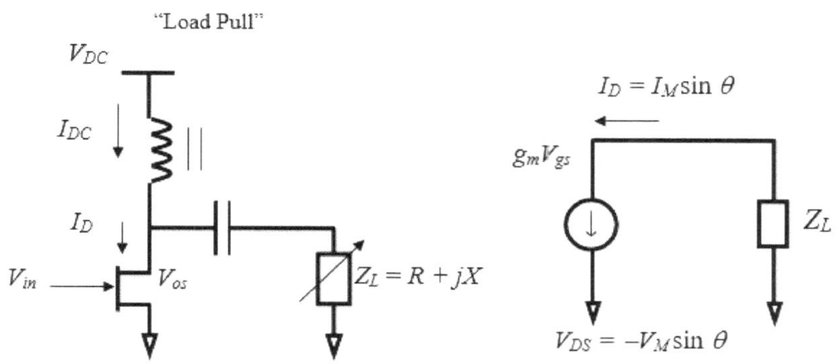

Figure 17.1 Load pull consists of using variation of Z_L to investigate power output.

$$\text{Let } Z_L = R_{opt} = \frac{V_{DC}}{I_{MAX}/2} = \frac{V_{DC}}{I_{DC}}. \tag{17.1}$$

This is the optimum power match (neglecting knee voltage) for classes A and B. R_{opt} is plotted in Figure 17.2 on a Smith chart. Figure 17.3 presents the maximum drain voltage and current for an inductive bias feed Class A amplifier. P_{opt} in Equation (17.2) is the corresponding output power delivered to a real load $R_{\text{opt.}}$.

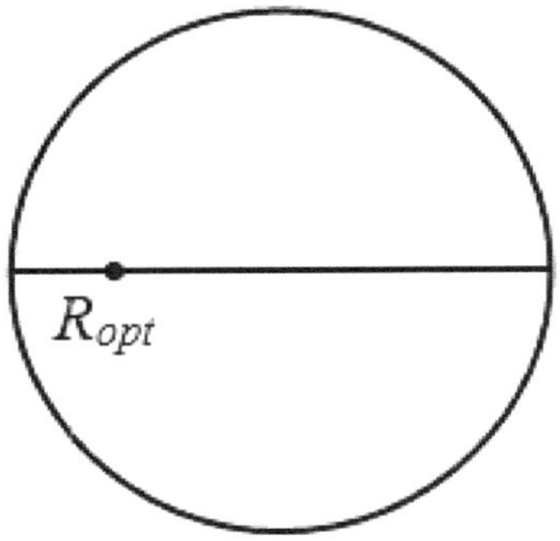

Figure 17.2 Example showing location of R_{opt} on a Smith chart.

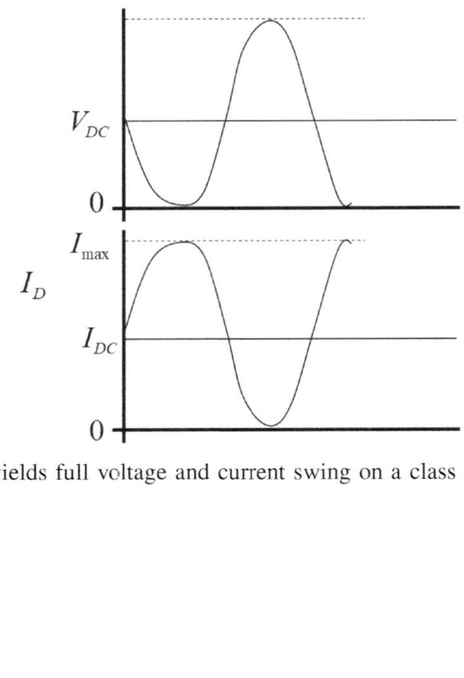

Figure 17.3 R_{opt} yields full voltage and current swing on a class A amplifier. V_{knee} has been neglected.

$$
\begin{aligned}
P_{opt} &= \frac{V_{DC}I_{DC}}{2} \\
&= \frac{V_{DC}^2}{2R_{opt}} = \frac{1}{2}I_{DC}^2 R_{opt}
\end{aligned}
\tag{17.2}
$$

To determine the influence of a mismatched load on the power amplifier output power and also efficiency, we must trace out a contour on the Smith chart that will give a power of P_{opt}/p. Here, p is the power reduction factor. R_{opt} is the initial load resistance [1].

There are two solutions, one with variation in R_{opt}; one with variation in jX or jB. The two extremes of R_{opt} are shown in Figure 17.4 where they intersect the center line of the Smith chart.

a. Resistive terminations:

$$
(1)\ pR_{OPT} \begin{cases} I_m = \dfrac{V_{DC}}{pR_{opt}} \\ V_m = V_{DC} \end{cases} \quad \mathrm{P_o} = \frac{1}{2}\frac{V_{DC}^2}{pR_{opt}} = \frac{P_{opt}}{p} \tag{17.3}
$$

$$
(2)\ R_{opt}/p \begin{cases} \mathrm{I}_m = I_{\max}/2 = I_{DC} \\ V_m = I_{DC}R_{opt}/p \end{cases} \quad \mathrm{P_0} = \frac{1}{2}\frac{I_{DC}^2 R_{opt}}{p} = \frac{P_{opt}}{p} \tag{17.4}
$$

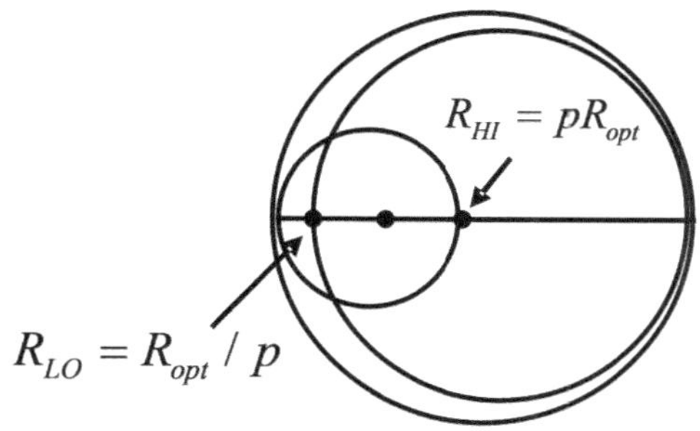

Figure 17.4 Contour of constant power, P_{opt}/p.

b. Series reactance:

When $R = \frac{R_{opt}}{p} = R_{LO}$, add jX(Figure 17.5).

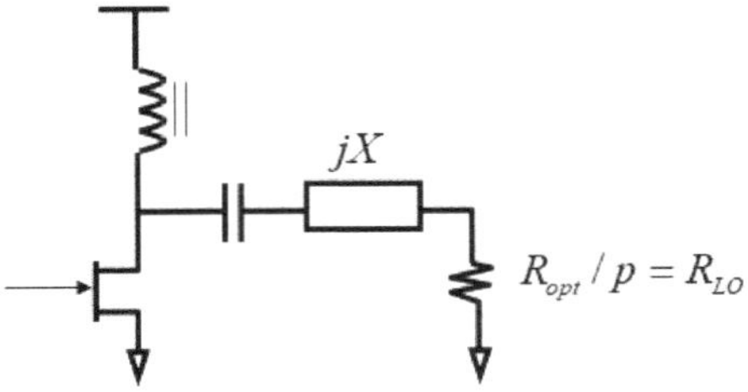

Figure 17.5 Variation of jX around R_{LO}.

$$|V_M| = I_{DC}\left|R_{LO} + jX\right|. \tag{17.5}$$

In this limit, when $R = R_{LO}$, we have power limited by I_{MAX}. Voltage $V_m < V_{DC}$. So, series reactance will increase the voltage swing but not affect current. The impedance will follow the constant resistance circle seen in Figure 17.4 until jX reaches a maximum, jX_m as shown in Equation (17.6) and Figure 17.7.
This is valid up to $\pm X_m$, where

$$|R_{LO} + jX_m| = R_{opt}. \tag{17.6}$$

c. Shunt susceptance:
When $R = pR_{\text{opt}} = R_{HI}$, add jB (Figure 17.6).

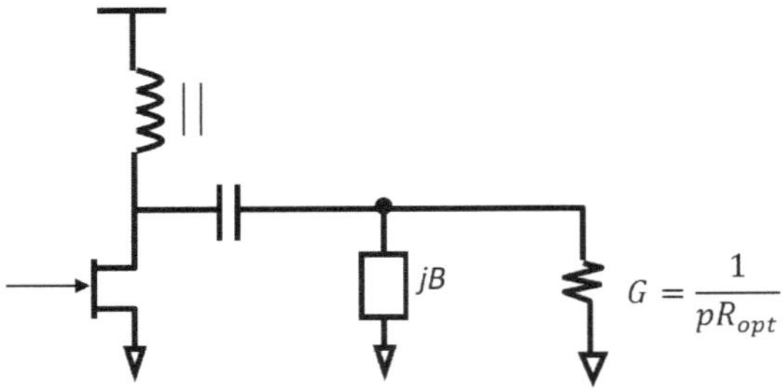

Figure 17.6 Variation of jB around R_{HI}.

At this limit, power is limited by the voltage swing $V_m = V_{DC}, I_D < I_{MAX}$. Shunt susceptance can be added. It will cause current to increase until

$$|G + jB_m| = \left|\frac{1}{R_{opt}}\right|. \tag{17.7}$$

The variation of jX and jB trace out paths on the Smith chart called load pull contours. They follow constant r or g circles. It can be shown that the intersection points of these circles represent X_m or B_m. The load pull contour takes the place of gain circles for power amplifier designs. Typically, these are displayed with varying output powers (Figure 17.7).

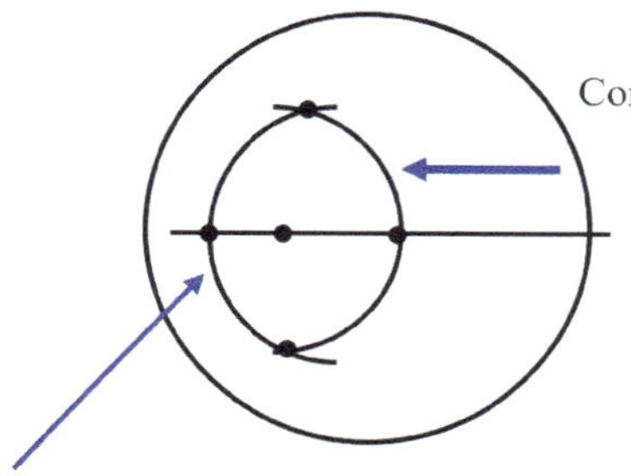

Figure 17.7 Load pull contours for variation of reactance and susceptance.

17.1.1 Package parasitics

At higher frequencies, device capacitances and package capacitance and inductance can be significant. In many cases, C_{OUT} can be incorporated into the matching network if L_{pkg} is not too large. Note that the load match must be achieved at plane A, not plane B in Figure 17.8.

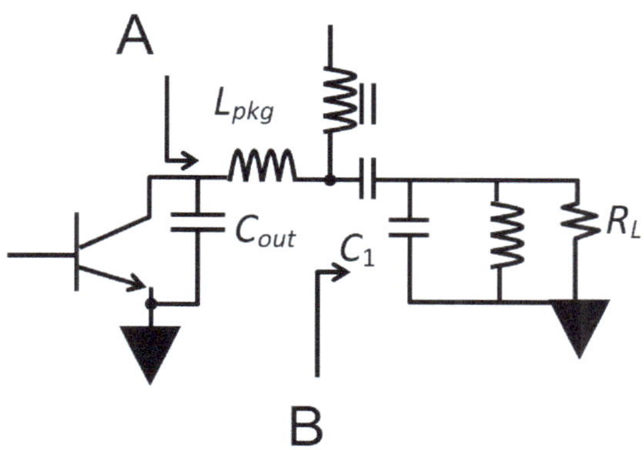

Figure 17.8 Package inductance becomes part of the output network.

What does the package do to Z_L? We require that $Z_A = R_{opt}$, so there is a problem – the output capacitance and the package will transform R_{opt} into something else, as seen in Figure 17.9. So, R_{opt} must be compensated to account for the package.

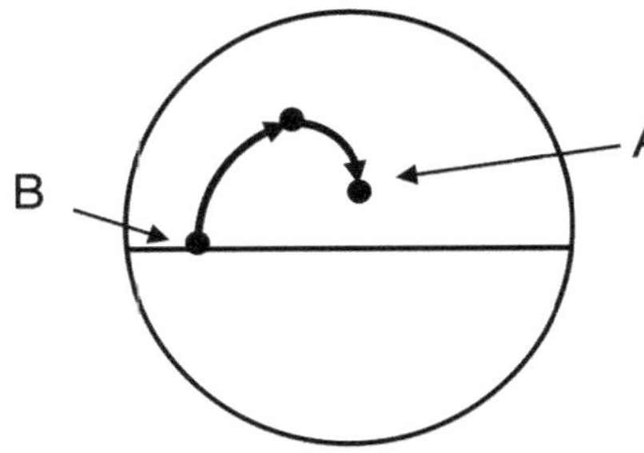

Figure 17.9 In this figure we see the result of a large jxL and small jbc.

17.1.2 Input matching

For PAs, we often have large voltage or current variations on the input, needed to drive the device to full output. This can lead to large input impedance variation with drive power, P_{IN}. Non-linear gate capacitance contributes to this. So, what procedure can be used to design input network?

First approximation: use small signal S-parameter calculation to find Γ_{IN} at quiescent bias. This can serve as a starting condition, a baseline. But, of course, the S-parameter model doesn't include any non-linearity.

When a suitable large-signal device model is available, then Z_S or Γ_{IN} could be found more accurately using a source pull method by harmonic balance simulation.

Figure 17.10 illustrates how the amplifier can also be evaluated with a manual or automated source pull setup. The amplifier can be driven to a large signal level representative of normal operation. The tuner can be adjusted to minimize the input reflected power as measured on the network analyzer.

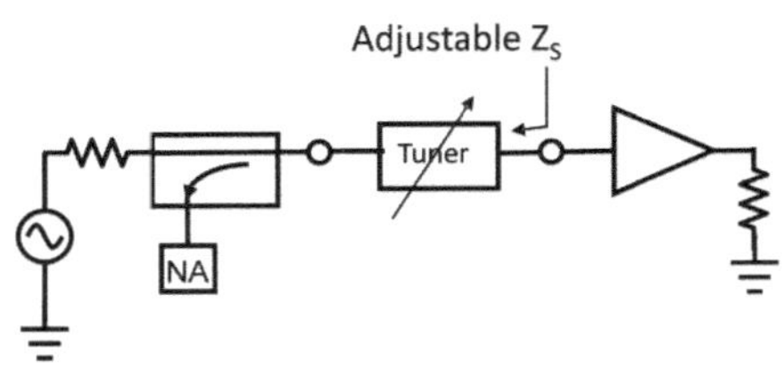

Figure 17.10 Experimental source pull measurement.

17.2 Simulation Methods for Power Amps

At small signal levels, most linear power amps will behave like a small signal amplifier. S-parameter simulation may be useful as a starting point for estimating input impedance.

However, power amps always become nonlinear at some input drive level. Clipping, compression, and distortion are inevitable. Thus, we need to use a nonlinear simulator to predict these effects. Pathwave Keysight ADS gives us two options:

- Harmonic balance, a frequency domain method
- Transient analysis, a time domain method.

Since we are designing the amplifier mainly in the frequency domain, harmonic balance makes the most sense and is usually more efficient in simulation time (when it converges properly). Harmonic balance was described in Chapter 8. More detail can be found in the help files on ADS, *harmonic balance simulation-basic setup* [2].

Using harmonic balance with parameter sweeps, you can determine key performance measures such as:

- Gain vs. P_{IN}
- Efficiency vs. P_{IN}
- Output power vs. P_{IN}
- Distortion.

17.2.1 Simulation approach

Use the amplifier design guide found in the amplifier design guide menu on ADS. There are extensive simulation panels that can be adapted to your design that will evaluate:

- Power vs. frequency
- Output spectrum
- Gain, efficiency vs. power
- Two-tone distortion measurement
- Source and load pull simulations.

Setting up such simulations along with the needed equations is a difficult and tedious task, so use of the design guide can make this much faster and easier (Figure 17.11).

These templates are very extensive and can save much time. You can replace the generic amplifier model used in these schematics with your own.

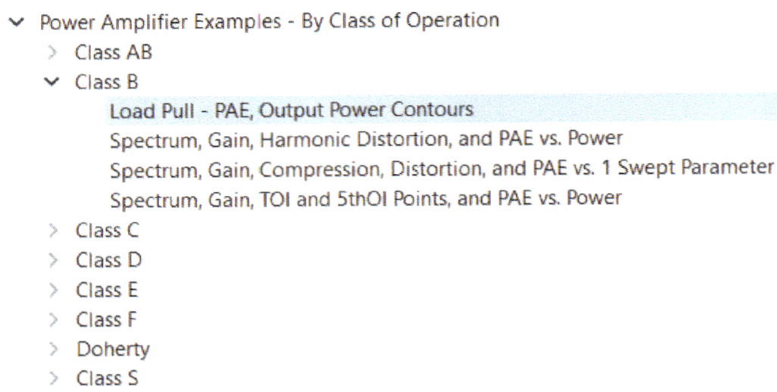

Figure 17.11 Power amplifier design guide options in Pathwave ADS for specific classes of amplifiers.

For example, Figure 17.12 shows one schematic set up for class B – PAE and output power sweep. The amplifier model includes a quarter-wave drain bias line and employs a generic GaAs FET with the Statz model.

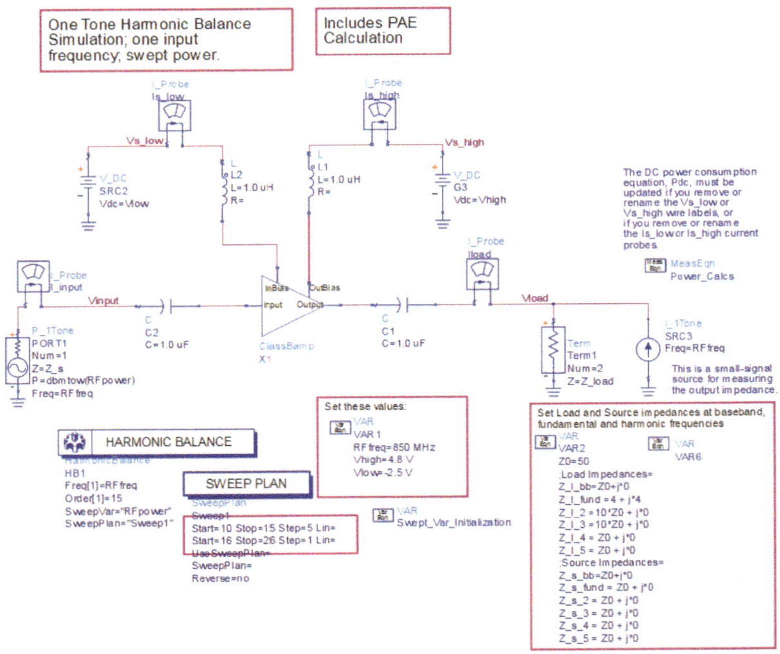

Figure 17.12 Harmonic balance simulation schematic. One of many.

Figure 17.13 shows how the power added efficiency is increasing with output power. This occurs when the amplifier is driven toward a reduced conduction angle. In Figure 17.14, you can see how as drive power is increased, the drain voltage becomes more square in shape and the current resembles more the half-cosine shape.

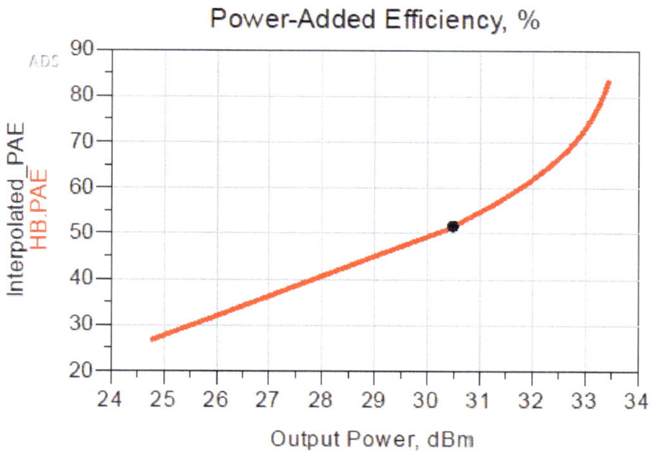

Figure 17.13 A plot of PAE vs. output power from the simulation shown in Figure 17.12.

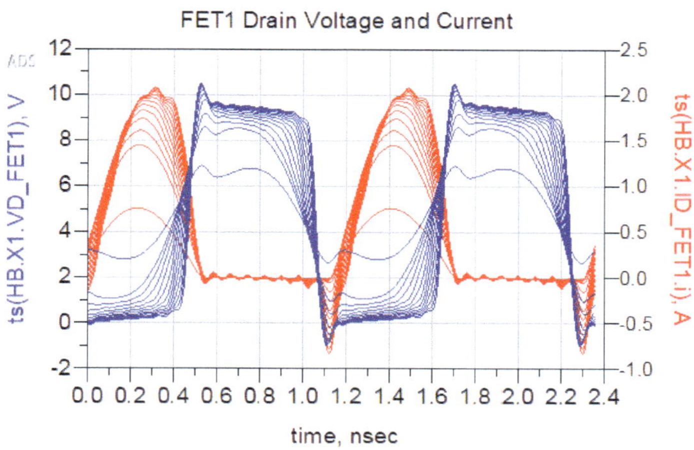

Figure 17.14 Simulation of FET drain voltage and current with stepped drive power.

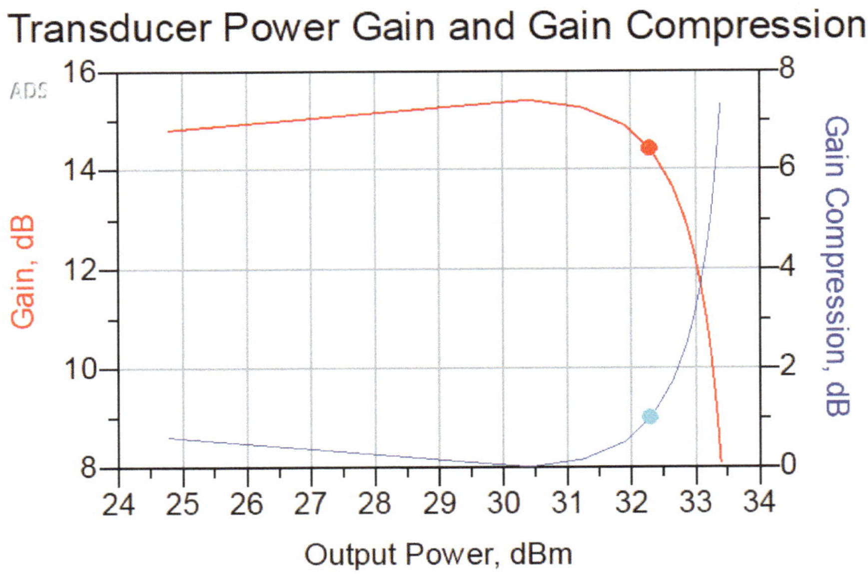

Figure 17.15 Plot of transducer power gain and gain compression vs. output power.

In Figure 17.15 it is clear that the higher efficiency that the Class B amplifier can provide is at the expense of linearity.

Anything above 1 dB gain compression would be unacceptable for a linear amplifier. Nevertheless, the efficiency is much superior to class A up to that point. Plus, the amplifier is in cutoff when there is no input power.

For applications requiring linear power amplifiers, techniques such as predistortion can be used to extend the useful operating power range and thus increase average efficiency.

17.2.2 Source pull

Input source pull simulation templates are also provided in the amplifier design guide.

Figure 17.16 shows an example again using a generic class B amplifier with a GaAs FET with the Statz model. This comes from the PA Design Guide under HB 1Tone_Source Pull.

Marker m3 can be moved to any of the points of source impedance and the PAE and Power Delivered is presented along with the corresponding

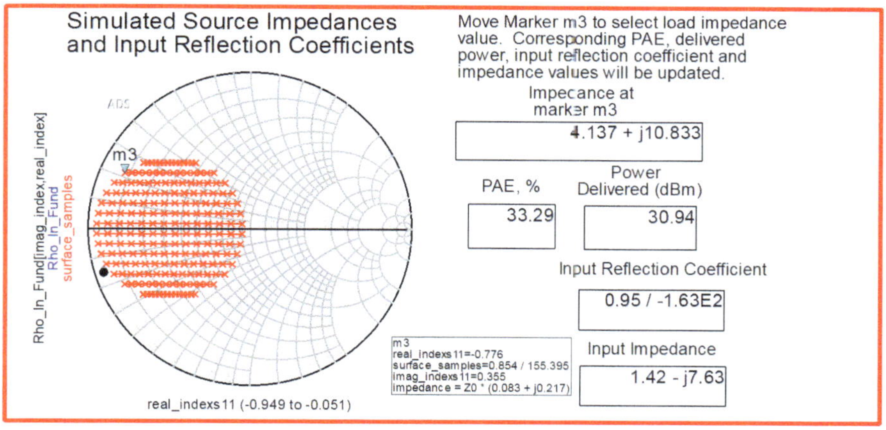

Figure 17.16 In this display template, a Smith chart with a cluster of source impedances is shown.

input impedance to the amplifier. The corresponding power gain can also be simulated for a given S_{11} by modifying the input impedance (with a Z_{11} block or specific S_{11}) on the PAE Power Sweep schematic.

17.3 Power Amp Bias Circuits

Power amps may often be harder to stabilize than small signal amps.

- Large voltage swing causes capacitance and g_m variation with time (Figure 17.17). Your SS stability predictors are not necessarily accurate.
- Common mode feedback is more serious (Figure 17.18). Power amps have higher ground, I_{DC} currents.

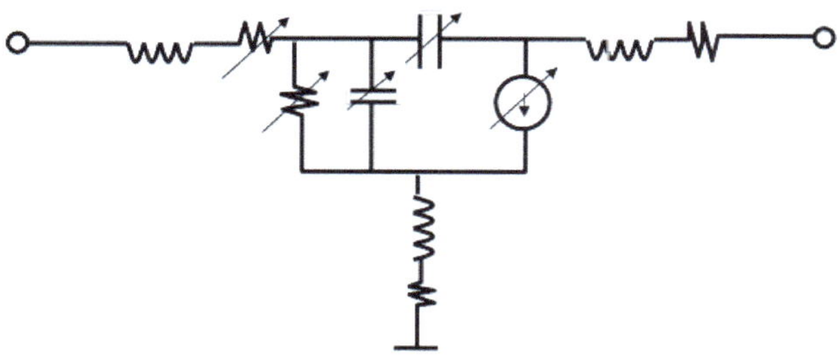

Figure 17.17 Nonlinear device model.

Figure 17.17 reveals a typical large signal model. All the elements with arrows are dependent on the input and output voltages and currents. Lots of nonlinearity.

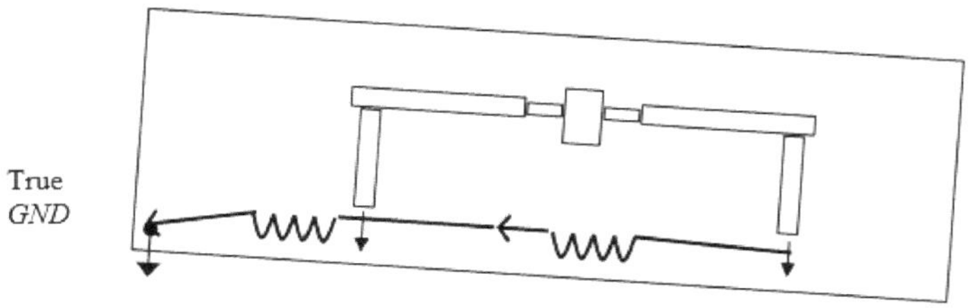

Figure 17.18 Layout on circuit board. Ground connection may add a lot of common mode inductance.

Impedance levels are often lower, much lower, in power amplifiers. Thus, current levels in normal operation are much higher. Better power supply bypassing is needed to provide adequate isolation between input and output. Otherwise, unintended feedback paths could possibly destabilize the amplifier.

High current transients are beyond what a DC regulated supply can provide. Inadequate bypassing produces "memory effects", long time constant transients in amplitude which will add to distortion. Therefore, you need to bypass for both LF and HF.

RF chokes are frequently used at lower frequencies to provide isolation between the DC power system and the amplifier ($< 1\text{Ghz}$). $\lambda/4$ lines are useful at higher frequencies for the same function.

$$\text{Rule: } |Z_{choke}| >> |Z_L|$$

Why? Don't want to lose power into your DC feed system.

How can we provide low inductance bias feed and still have good control of L at low frequencies? Typically, a more complicated decoupling system is required for power amplifiers to avoid resonances and oscillation (which could be done by introducing loss). The trick is to do this without losing power or efficiency.

Figure 17.19 shows one approach for the drain or collector bias feed. R_1 is used for damping low frequencies. A more extensive analysis can be found in the homework problem for this chapter.

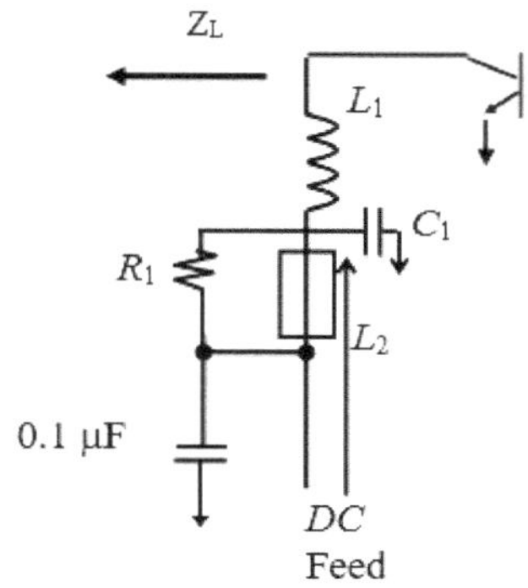

Z_{L_1}	$10 \bullet \lvert Z_L \rvert$
R_1	$\approx \lvert Z_L \rvert$
Z_{C_1}	$\frac{1}{10} \lvert Z_L \rvert$
L_2	Ferrite Bead

Figure 17.19 Multi-element bias feed.

17.3.1 Potential problems

1. Typically large Z_{choke} is in conflict with the requirements of amplitude modulation; the envelope of the RF signal varies greatly at modulation frequencies up to several MHz for wideband digital signals. This requires small Z_{choke}.
2. Stability.
   Oscillation: If no series R in the bias network, the device feedback capacitance and bias inductances can cause negative resistances. A pi-network (Figure 17.20) that is formed will have a $180°$ phase shift at some frequency. In that case, we then can have an oscillator.

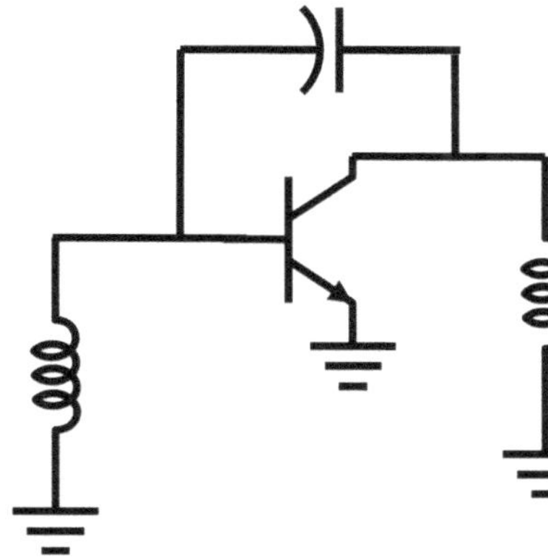

Figure 17.20 Inductive biasing with device feedback capacitance yields a pi network.

To avoid this, good damping at low frequency is needed to keep loop gain < 1. This must be done with care so that you do not throw away gain or efficiency unnecessarily.

The use of stability circles as described in Chapter 6 for analysis of power amplifier stability is questionable. If the amplifier does not pass that test, then for sure it will not be useable. However, the developers of Pathwave Keysight ADS have introduced another stability probe, the WS_Probe that is based on traditional and more recent feedback analysis. When a circuit includes a feedback path, one must be careful to ensure that the magnitude of the loop gain af is never equal to one [3].

$$A_{CL} = \frac{a(s)}{1 - a(s)f(s)} \tag{17.8}$$

Here, A_{CL} is the closed loop gain, a the open loop gain, and f the feedback factor. Such analysis dates all the way back to H. Bode with Bode or Nyquist plots. The WS_Probe enables this analysis under the following limitations:

1. There must be an identifiable feedback loop in which to insert the probe. This may be less than obvious in some circuits because it can also depend on the layout of the chip or PCB. EM simulations would be needed to extract an equivalent network corresponding to layout.
2. Unless the large signal device model is accessible in an equivalent circuit form including nonlinear device capacitances, then the device(s) internal feedback loop can't be probed. Using simple S-parameter or equation-based device models (Gummel-Poon, Curtice, etc.) does not provide this access. In that case, the WS_Probe analysis is not that much superior to using the k-factor because extra feedback paths are being ignored.
3. Keysight has provided videos [4] and application notes [5] that describe how to use the probe and what it will reveal regarding stability. Nyquist plots can show whether there are right-hand plane poles.

17.3.2 Experimental measurement for stability

Test #1:

Figure 17.21 describes a technique to probe the amplifier by checking the output for spurious frequencies on the spectrum analyzer. This is not exhaustive, but if the amplifier fails this simple test, you must revise the design.

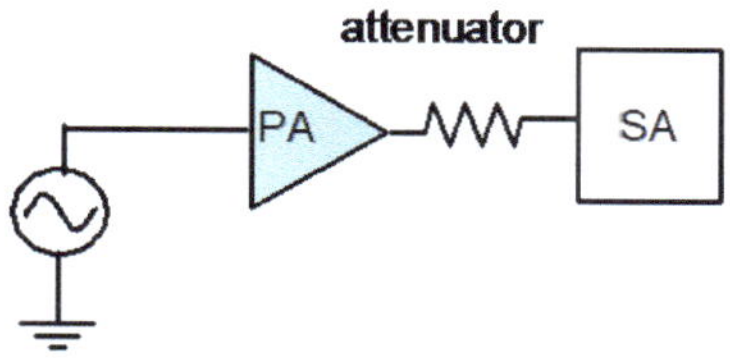

Figure 17.21 Simple experimental test for stability.

1. Set the generator at the design frequency, f_0.
2. Turn generator on.
3. Are there any outputs at other frequencies besides f_0 ?
4. Turn generator off. Is there any output at all with no input?

Test #2:

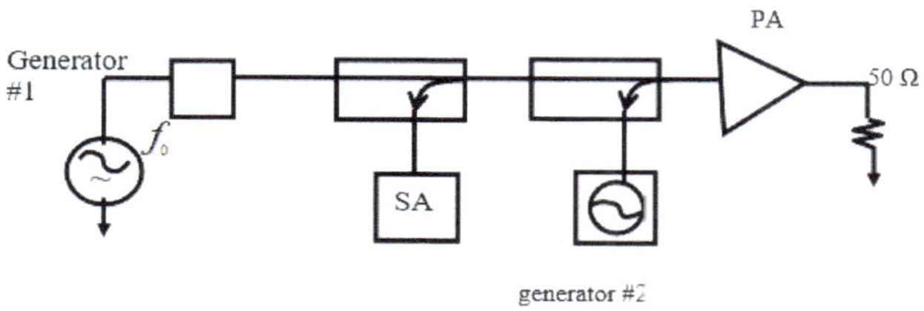

Figure 17.22 Experimental approach to evaluating hidden stability problems.

A more accurate test is shown in Figure 17.22. This can reveal if there are stability problems caused by low frequency bias networks. Zoom in with the spectrum analyzer (SA) around f_0. Vary frequency of generator 2 around f_0. Look for peaks in reflected power (Figure 17.23).

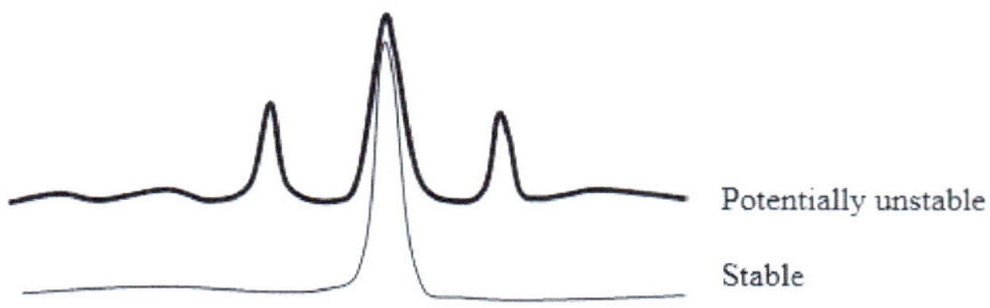

Figure 17.23 The spectrum will reveal sidebands if there are low frequency instabilities.

A wideband spectral display is also needed to reveal out-of-band low or high frequency instabilities.

17.4 Summary

PA impedance matching techniques:

- Load pull contours are used instead of gain circles to map power output and power gain as a function of load impedance.
- Package parasitics can transform load impedances. It's the impedance at the collector or drain that is critical to achieving optimum performance.
- Large signal input match will be different from small signal. Harmonic balance simulation method can be used to predict the proper input source impedance.

Design for stability. Bias circuits and associated components must be designed to limit low frequency gain. Harmonic balance methods must be used to predict large signal stability of the amplifier over a wide frequency range. Experimental methods also exist to probe for instabilities.

References

[1] S. Cripps, (2006) *RF Power Amplifiers for Wireless Communications*, 2nd Edn. Artech House.

[2] Pathwave ADS help files are included with the simulator help menu. Search under harmonic balance simulation - basic setup.

[3] Gray P.R. et al. (2009) *Analysis and Design of Analog Integrated Circuits*, 5th Edn. Wiley, Chapter 9.

[4] https://youtube.com/playlist?list=PLtq84kH8xZ9Gz-1_VvqWsWeGwzMnKjRwY

[5] (2021) *Designing for Stability in High Frequency Circuits* www.keysight.com.

Homework

BJT power amplifier design project: Goals:

Design and simulation of a 450MHz power amplifier using the NE85634 BJT. Note that this particular device may no longer be available, but the design procedure can be used for other devices and is a necessary exercise for any PA design.

Specifications

Power added efficiency	Better than 30% at $P_{1\ dB}$
Gain	Greater than 10 dB
Frequency	450MHz ± 30 MHz
Output power	At least 20 dBm (saturated)
Power supply	+8 volts

Design a power amplifier at 450 MHz based on the NE85634 BJT. Bias the device with $V_{CE} = 8$ V and $I_C = 20$ mA. Note that this device is bilateral and potentially unstable at this frequency.

Use the nonlinear device model from the NEC data sheet for all simulations. The model can be seen below. It can be constructed as a schematic that can then be included in an amplifier subnetwork. Use Pathwave/Agilent ADS S-parameter and harmonic balance simulation modes for this project. The harmonic balance simulator is needed for large signal simulations since it uses the full nonlinear model evaluating all of its voltage and current dependent model elements at all harmonic frequencies specified.

- To begin, you should bias the device as a Class A amplifier at $I_C = 20$ mA There is a BJT curve tracer template that can be accessed from the ADS Main Window: *File > New Design > Schematic Design Templates > BJT, curve tracer*. This DC simulation can determine the correct V_{BE} needed to obtain the required I_C for purposes of simulation. Bias the device through ideal DC feeds and DC blocks at this stage in the design.
- Run an S-parameter simulation on the transistor; plot stability factor k vs. frequency from 10MHz to 2GHz. Refer to the Stability and Gain (Chapter 6) You will see that the device is unstable at lower frequencies ($k < 1$). Add some stabilization element(s) so that the device is unconditionally stable, at least above 250MHz. You should consider adding any resistive stabilization elements on the input side, since in a power amplifier application any losses on the output side will hurt your efficiency. Additional lower frequency stabilization will employ the bias networks you will be designing later. It will now be possible to match the device for power and efficiency.
- Design an output matching network that will provide the correct large signal load line impedance at the collector of the device. An L network will provide lower Q than a Pi or Tee network and thus should be

less critical to fabricate. You should represent lumped components with their high frequency model. According to the C_{OB} vs. V_{CB} plot on the data sheet, at 8 V DC bias, the output capacitance is about 1.1pF. To obtain the correct match at the collector, you will need to include the effect of this device and package capacitance in your network design. The bondwire inductive reactance at this frequency is small and can be neglected.

- Simulate the amplifier with the output network using the PA Design Guide. Drive the base with a 50ohm generator (see the ADS HB tutorial in chapter 11). Verify that you can obtain the required output power and collector efficiency (P_{out}/P_{DC}) at 450 MHz. Be careful not to overdrive the amplifier.
- Next, design the input match with the output matched at the large signal load line. The small signal Γ_{IN} does not provide a good match at large signal. You must use a harmonic balance simulation and include a directional coupler on the input of the amplifier to determine a good match when the input is driven with roughly Pout/10 where Pout is your expected output power at the 1 dB compression point. Sweep the real and imaginary parts of the input impedance (implemented by a Z1P_Eqn from the Eqn-Based Linear menu as shown below. Find the combination that provides the least input reflected power. The real part is swept by changing variable r. Param Sweep is used to vary a parameter in the simulation. The parameter must be defined first in a Var block. The initial value there doesn't matter. The imaginary part can be swept the same way. Fix r, then sweep x. Find the minimum ratio of $\mathrm{Vref}[1]/\mathrm{Vfwd}\ [1]$ (the brackets select the index for the fundamental frequency).
- Replace the DC blocks. In the case of power amps, you do not want the capacitive coupling to be good from DC to light. There is increased possibility of low frequency oscillation if the coupling caps are too big. Also, the excess inductive reactance may detune your matching networks. Choose a value no larger than is required to get to the series resonance of the DC blocking cap at the operating frequency.
- Replace the DC feeds. Proper impedance termination at all frequencies is necessary for a stable amplifier. The matching network does not provide appropriate termination out of band, so you must rely on the DC biasing network to kill gain at low frequencies. The double decoupling network shown below is well suited for providing wideband isolation between the amplifier and power supply. While this approach can be used on both input and output sides of the amp, it is most critical on

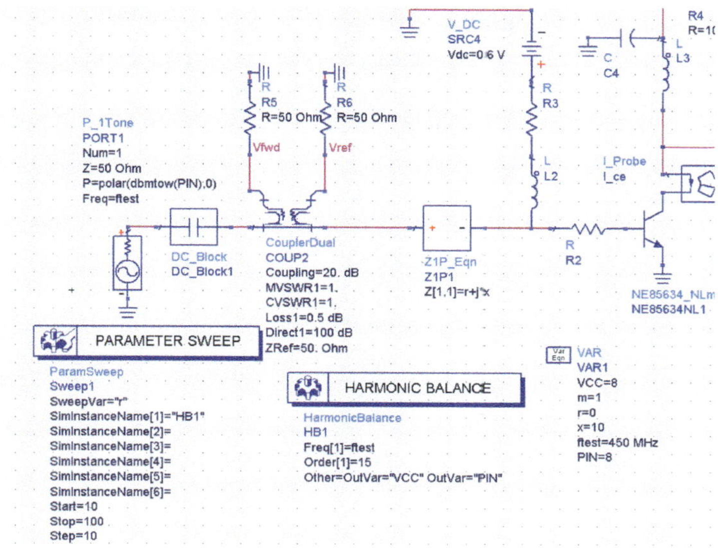

the output to avoid resistive losses in the bias network that will hurt efficiency.

Note: When laying out your board, provide separate connections to V_{CC} for the base and collector bias networks. This will allow you to measure I_C and calculate the efficiency of the amplifier without considering the base bias network (which is not very efficient).

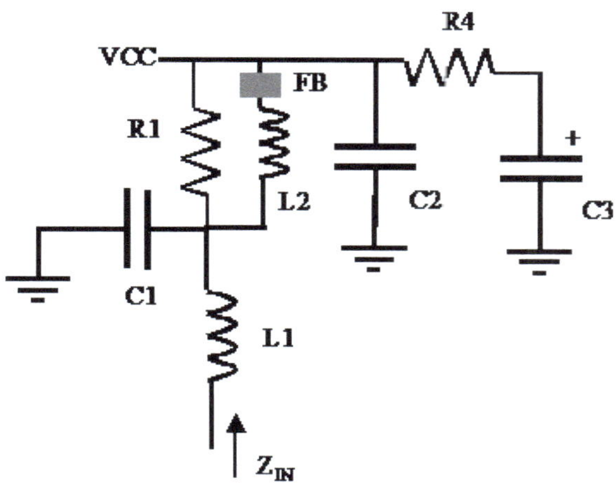

Double decoupling network: A purely resistive element in the collector bias network would kill the efficiency of the PA because of the voltage drop at high collector currents. So, shunt R1 with L2 to provide a DC feed. FB is a ferrite bead which helps to DeQ the inductor resonance with C1. L1 provides a high impedance at the design frequency. C1 should be small enough to allow the amplifier to see the R1, L2 network at low frequencies (under 100 MHz). C2 and C3 provide for a low impedance at low frequency, necessary to follow envelope variation when two or more signals are applied to the input. R4 is a small resistor ($\angle 5$ohms) needed to de- Q any unexpected resonances caused by the interaction of C2 and C3. C3 is a large (10 uF) tantalum capacitor needed to supply any excess currents at the modulation rate. Envelope variation at the modulation frequency puts extreme demands on a DC power supply which often can't be met without suitable bypassing. Recommended starting values for the components are shown in the table. You may need to modify these to obtain $k > 1$ at all frequencies.

R1	R1 $\cong$ Ropt (large signal load line resistance)
L2	| X $\cong$ Ropt at 10MHz or below
L1	| X $\cong$ 10Ropt
C1	| X $\cong$ Ropt10 at operating frequency
C2	Low frequency bypass cap(s). 0.1 uF or so. May need more than 1 in parallel if there is a low frequency oscillation problem.

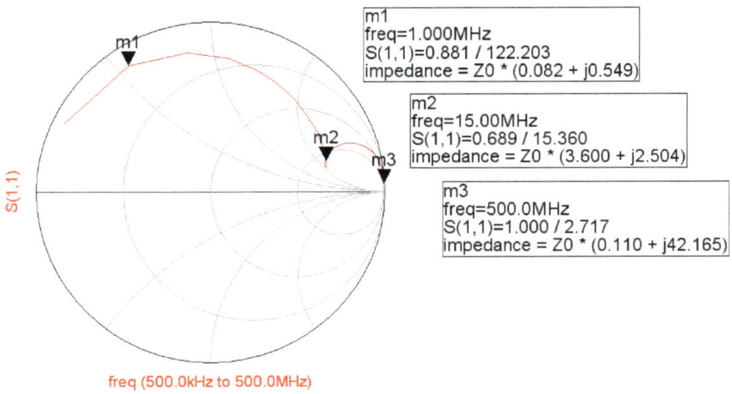

Plot of S_{11} representing typical Z_{IN} for the couble decoupling network. We see large inductive reactance at 500 MHz. In the 1 to 15 MHz range, we see a low Q inductor which should be effective in stabilizing the amplifier against oscillation. And, the network does not dissipate significant DC power, therefore efficiency will not be degraded.

- The base biasing source will use the current mirror shown below [1]. Since the average collector current is nearly constant for a Class A amplifier, the current mirror bias current I_{C2} can be set to provide the RF BJT collector current I_{C1}. Because the transistor areas are different, the two currents will not be the same, thus you must correct for this with a DC simulation. R4 is part of the base bias network and can be used to provide stability at low to mid frequencies. If you use R4, you may need to compensate for the voltage drop by choosing an $\mathrm{R3} = \mathrm{R4} * \mathrm{A}$ where A is the area ratio or current ratio I_{C1}/I_{C2}.

Clearly, there needs to be thermal feedback to avoid runaway. Thus, we must compensate for the temperature drift by attaching Q2 to the RF power transistor case. Thermal grease can be used for this purpose. Then V_{BE} will track with temperature. Include a large (10uF) tantalum capacitor between VCC and ground near where VCC enters the board. This will help provide a good low frequency AC short on the board.[1]

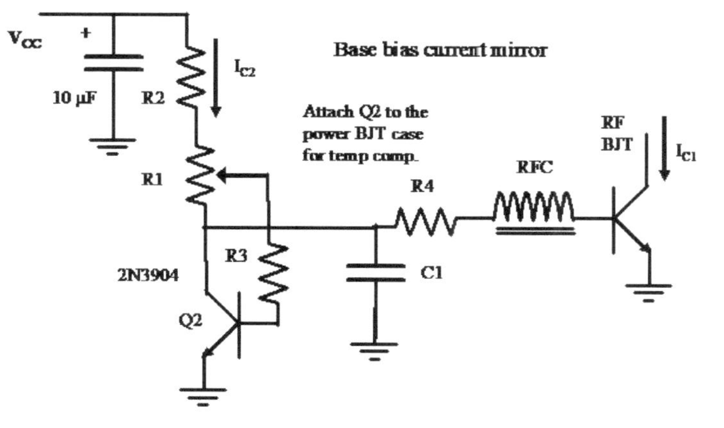

[11] Jarvinen, E., Kalaio. S., Matilainen, M. (2001) Bias circuits for GaAs HBT power amplifiers. IEEE Int. Microwave Symp, Digest Paper TUIF-28.

The RFC inductance should be about $10\times$ the real part of Γ_{IN}. Include R4 and the RFC in your wideband stability simulation. $C1$ is a RF bypass capacitor.

- Perform another S-parameter stability simulation. Verify that the amplifier will be stable at $450\ \mathrm{MHz}$ and over a wide frequency range (10 MHz to 2GHz), at least in small signal mode. You will need to include some additional stabilization element(s) that provide resistive loading at low frequencies but not at high frequencies if stability at low frequencies is a problem. Power amplifiers are even more susceptible to instability than small signal amplifiers because the large voltage and current variations cause the device capacitances and gm to vary with time. Simulate the k factor as usual at the amplifier input and output. $|\mathrm{S}11|$ and $|\mathrm{S}22|$ should remain less than 1. Verify that the gain, center frequency, power and efficiency still meet or exceed specifications after any changes you may have made.

Finally, carry out large signal HB simulations to verify that your amplifier meets the design specs.

Nonlinear model for NE85634. The nonlinear model can be constructed as a schematic using the parameters given.

```
* FILENAME:        NE85634.CIR
* NEC PART NUMBER: NE85634
* LAST MODIFIED:   11/97
* BIAS CONDITIONS: Vce=5V to 10V,  Ic=20mA to 20mA
* FREQ RANGE:      0.1GHz to 3.0GHz
*
*                            CCBpkg
*                    .-------||--------.
*                    |          CCB     |
*                    |      .--||---.   |       COLLECTOR
* BASE               |      |        | |
*            __Tb_   |      |   /----o-o-----o----o
*  o---------|_____|-o-LB---o--|Q1   |        |
*                    |      B   \   --       --
*                    |          |   -- CCE   -- CCEpkg
*                    --CBEpkg E o----'       |
*                    --         |            |
*                    |         LE            |
*                    `----------o------------'
*                               |
*                               -
*                              | |  Te
*                              | |
*                               -
*                               |
*                               o EMITTER
*
*       CCB    = 0.087pF        LB  = 1.2nH          Tb/Te:
*       CCE    = 0.16pF         LE  = 1.2nH          z=60 ohms
*       CCBpkg= 0.18pF                               l=50 mils
*       CCEpkg= 0.18pF                               a=0.0001
*       CBEpkg= 0.01pF                               f=0.9GHz
*                                                    k=5.0
*                  c b e
.SUBCKT NE85634/CEL 2 1 3
Q1 2 6 7 NE85600
CCB 6 2  0.087E-12
CCE 2 7  0.16E-12
LB  4 6  1.2E-9
LE  7 5  1.2E-9
TB 1 0 4 0 Z0=60 TD=9.63E-12
TE 5 0 3 0 Z0=60 TD=9.63E-12
CCBPKG 4 2 0.18E-12
CCEPKG 2 5 0.18E-12
CBEPKG 4 5 0.01E-12
.MODEL NE85600 NPN
+( IS=6e-16       BF=120          NF=0.978       VAF=10          IKF=0.08
+ ISE=32e-16      NE=1.93         BR=12          NR=0.991        VAR=3.9
+ IKR=0.17        ISC=0           NC=2           RE=0.38         RB=4.16
+ RBM=3.6         IRB=1.96e-4     RC=2           CJE=2.8e-12     VJE=1.3
+ MJE=0.5         CJC=1.1e-12     VJC=0.7        MJC=0.55        XCJC=0.3
+ CJS=0           VJS=0.75        MJS=0          FC=0.5          TF=10e-12
+ XTF=6           VTF=10          ITF=0.2        PTF=0           TR=1e-9
+ EG=1.11         XTB=0           XTI=3          KF=1.56e-18     AF=1.49 )
.ENDS
```

Index

About the Author

Stephen I. Long received his B.Sc. degree in engineering physics from the University of California, Berkeley, and his M.Sc. and Ph.D. degrees in electrical engineering from Cornell University, Ithaca, NY. From 1974 to 1977 he was a Senior Engineer at Varian Associates, Palo Alto, CA. From 1978 to 1981 he was a member of the technical staff at Rockwell International Science Center, Thousand Oaks, CA. He has been a member of the Electrical and Computer Engineering department of the University of California, Santa Barbara since 1981. He is currently Professor Emeritus and has served as Vice Chair, Graduate Advisor, and Undergraduate Program Director. His research interests included the design of very high-speed digital integrated circuits, high performance devices and fabrication technologies, and high frequency analog integrated circuits for wireless and fiber-optic communications.

Dr. Long received the IEEE Microwave Applications Award in 1978 for development of InP millimeter wave devices. In 1988 he was a research visitor at GEC Hirst Research Centre, UK. In 1994 he was a Fulbright research visitor at the Signal Processing Laboratory, Tampere University of Technology, Finland and a visiting professor at Electromagnetics Institute, Technical University of Denmark. In 1999, he was a visitor at Hewlett-Packard HP/Eesof Division, Santa Rosa, CA.